Hydrometallurgy

Hydrometallurgy

Principles and applications

Tomáš Havlík

Cambridge International Science Publishing Limited
in association with
Woodhead Publishing Limited

CRC Press

Boca Raton Boston New York Washington, DC

WOODHEAD PUBLISHING LIMITED
Cambridge, England

Published by Cambridge International Science Publishing Limited in association with Woodhead Publishing Limited

Cambridge International Science Publishing Limited, 7 Meadow Walk, Great Abington, Cambridge CB21 6AZ, England
www.cisp-publishing.com

Woodhead Publishing Limited, Abington Hall, Granta Park, Great Abington, Cambridge CB21 6AH, England
www.woodheadpublishing.com

Published in North America by CRC Press LLC, 6000 Broken Sound Parkway, NW, Suite 300, Boca Raton, FL 33487, USA

First published 2008, Cambridge International Science Publishing Ltd, Woodhead Publishing Limited and CRC Press LLC

British Library Cataloguing in Publication Data
A catalogue record for this book is available from the British Library.

Library of Congress Cataloging in Publication Data
A catalog record for this book is available from the Library of Congress.

Woodhead Publishing ISBN 978-1-84569-407-4 (book)
Woodhead Publishing ISBN 978-1-84569-461-6 (e-book)
CRC Press ISBN 978-1-4200-7044-6
CRC Press order number: WP7044

The publishers' policy is to use permanent paper from mills that operate a sustainable forestry policy, and which has been manufactured from pulp which is processed using acid-free and elementary chlorine-free practices. Furthermore, the publishers ensure that the text paper and cover board used have met acceptable environmental accreditation standards.

English translation by Cambridge International Science Publishing Ltd
Typeset by Thymus Solutions Ltd, Mumbai, India
Printed by TJ International Limited, Padstow, Cornwall, England

Contents

This book is devoted to the life and work of Univ. Prof. Dr.-Ing. Dr. h.c. mult. ***Roland Kammel*** *– a friend, to whom I am very grateful not only for his support in my professional field but especially for his human approach.*

Introduction

The procedures used for processing minerals containing metallic elements to metals of the required purity for specific applications are generally referred to as extraction metallurgy. A large number of procedures have been developed for this purpose and, at first sight, it is quite difficult to determine the optimum method of extraction of metals. A suitable guide may be the concept in the first sentence, i.e., the purity of produced metals often depends on the procedures used for processing primary and secondary raw materials. Some of the processes can be used only for producing relatively contaminated metals, whereas others are quite efficient in producing metals of almost 100% purity.

Each process can be used to produce metals with a wide range of purity. To understand how a specific process may lead to the extraction of metals with the required composition, it is necessary to take into account theoretical considerations regarding the principles controlling the rate and extent of chemical reactions taking place in the process. Thermodynamics defines the final, i.e., equilibrium state of these reactions and can be used to study how the final state can be changed by changing the given conditions, such as temperature, the pressure and composition of gaseous, liquid and solid components in the given system. On the other hand, the kinetics defines the rate at which the equilibrium state is established and this also indicates the reaction time required for the realisation and completion of the essential chemical reactions.

The investment and production costs are a very important factor and determine the selection of the optimum process. Metals are traded on the open market and, consequently, from the commercial viewpoint it is not possible to determine the specific method of production of metals of the given composition with respect to the actual price of a specific production process. In addition, the prices are also controlled by other aspects, such as the cost of environmental protection, processing and marketing of secondary products, recycling, etc. These factors develop dynamically and in many cases it is difficult to forecast their contribution to the total price of metal, even for the near future. In any case, only the efficiently mastered theoretical fundamentals and applications of the most advanced achievements of science to the process of production of metals and also all other operations and the logistics of these processes lead to more efficient production enabling the producer to be successful in the market. Therefore, special attention is given to the search for and development of new, often unconventional methods of producing and processing metals, or combinations of these metals. An alternative method to the existing pyrometallurgical processes is the hydrometallurgical extraction of nonferrous metals. Some metals, such as uranium, zinc, gold or aluminium oxide, etc, are extracted completely or mostly by the hydrometallurgical method, whereas in other methods the application of this method is more difficult because of objective reasons. This is so in the case of copper, which is one of the most important nonferrous metals. Therefore, this book is concerned especially with the hydrometallurgical method of obtaining copper from its sulphide minerals, in particular, with one of its most important stages – leaching.

The individual chapters deal in a logical manner with this method, in order to understand the entire range of the problems of leaching copper sulphide minerals. After initial introduction, subsequent chapters review the interesting sulphide minerals, present in the leaching process as the raw material, semifinished products, or the leaching product. This is followed by the description of the thermodynamics of leaching of copper sulphides from the general viewpoint and also with respect to practical application, using the potential–pH diagrams.

Attention is then given to the leaching kinetics, again from the general viewpoint and with respect to specific applications in sulphide leaching.

The applications and the current state of the problem of leaching copper sulphides are dealt with in subsequent sections of the book. Special attention is given to the behaviour of sulphur in the leaching process, as one of the most important and process-controlling factors. Final sections describe interesting technological procedures which were used or are being used on the pilot plant and production scale, and prospects for the future are also discussed.

The book is based mainly on the fundamental and cited literature. Although basic knowledge of inorganic and physical chemistry is essential, together with the knowledge of the theory of metallurgical processes, the book also presents the main concepts to such an extent that it can be used as a textbook for students of all stages of metallurgy and related disciplines.

The book is intended not only for students but also for a wide range of experts, working in the hydrometallurgy of nonferrous metals. It is constructed in such a manner as to ensure that the general conclusions may also be applied to similar processes in metallurgy or applied chemistry. The author will be delighted if this is the case.

Chapter 1

CURRENT SITUATION IN COPPER PRODUCTION

Copper has always played a significant role in the history of mankind and directed development so significantly that one entire era of the development of mankind is referred to as the Bronze Age. Figure 1.1 shows the history of application of copper and copper alloys BC.

At the present time, the amount of copper produced annually is approximately 12 000 kt and continuously increases. Figure 1.2 summarises the trend in the increase of production of refined copper on the worldwide scale. However, this trend does not take into account the production of copper in the countries of the former Eastern Bloc [1].

The current production of copper is concentrated mainly in the processing of sulphide (mostly chalcopyrite or mixed) concentrates by the pyrometallurgical method. The method consists of two operations: melting, including the production of raw copper, and refining, ensuring the production of refined metal with the purity of at least 99.9% Cu.

The pyrometallurgical production of copper (general flow chart is shown in Fig. 1.3) includes the following operations of production of pure copper from sulphide concentrates.

Roasting. The sulphide concentrate is roasted at a strictly controlled temperature with limited access of air in order to remove part of sulphur by roasting. Subsequently, the resultant roasted product is melted. At present, the roasting process is no longer used in pyrometallurgical production of copper because of the introduction of advanced autogenous processes.

Production of matte. In this operation, the roasted product is melted in a shaft furnace at a temperature of approximately 1200 °C, resulting in melting of the sulphides and the formation of the so-called copper matte. The molten tailings form a slag in the

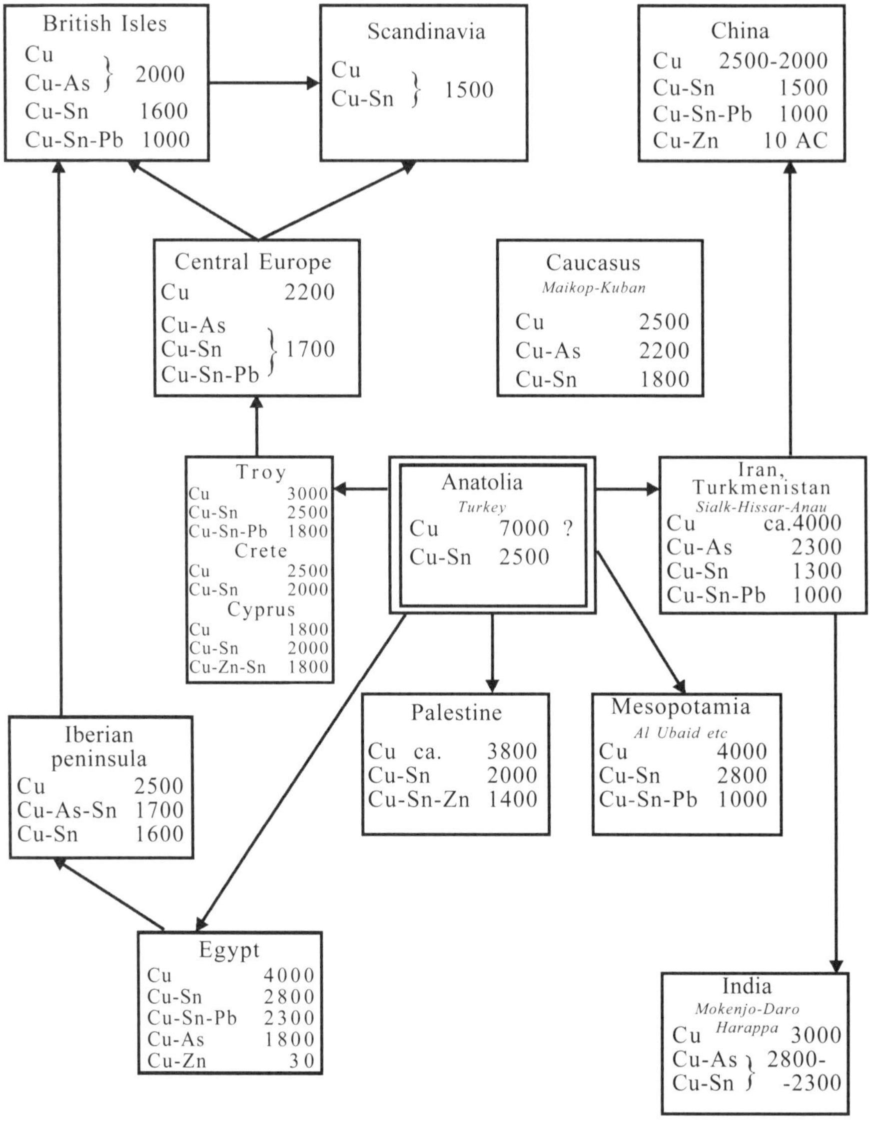

Fig. 1.1. Historic review of the application of copper and copper alloys.

presence of slag-forming additions, enabling separation of undesirable impurities.

Refining of the matte. The copper matte is in fact a melt of copper and iron sulphides. The iron is separated in processing in a converter by blowing air with oxygen, or pure oxygen, resulting in the oxidation of iron because of its higher affinity to oxygen and

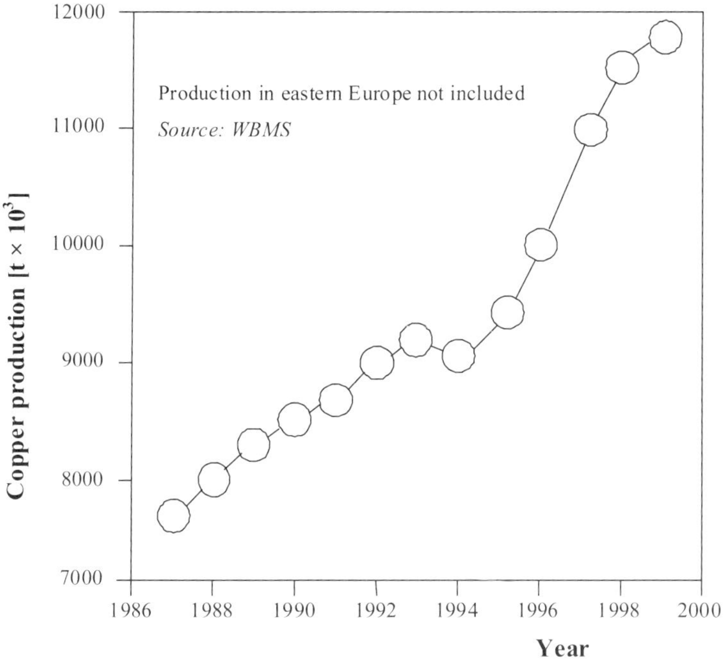

Fig. 1.2. Development of production of refined copper worldwide.

subsequent transfer of the iron oxide into the slag in the presence of slag-forming additions.

Convertering. In a converter, sulphide sulphur is removed by blowing air or oxygen into the melt, and sulphur reacts to form volatile oxides and copper is transferred into converter copper.

Fire refining. The aim of this operation is to remove the residual sulphur in a rafination furnace in two stages: in the first stage, sulphide is oxidised into volatile oxides by air or enriched air, and the second stage is characterised by the removal of the oxygen bonded with a metal in the first period, using birch logs, or with gaseous hydrocarbons. This operation is referred to as pooling.

Electrolytic refining. In this operation, all the residual impurities are removed from copper by electrolysis. Many of these impurities, trapped in the so-called anode sludge, are important components and, consequently, the anode sludge is further treated in order to recover them. Electrolytic refining results in the removal of impurities such as silver, gold, platinum metals, selenium, tellurium, nickel, arsenic, bismuth, lead, etc.

Although the general flow chart of pyrometallurgical production of copper, shown in Fig.1.3, is relatively simple and cheap and has been used for many years, it greatly differs from optimum requirements. The most important shortcomings include:

Unsuitable thermal balance. Some of the processes, such as roasting, matte refining and converter treatment, are exothermic, whereas matter formation is endothermic. The heat, generated by exothermic processes, is not utilised in endothermic processes resulting in an unsuitable thermal balance. However, in modern practices, oxidising smelting in suspension or bath smelting reactors, roasting, matte formation and separation take place in the same unit and in connection with each other and the whole process is highly exothermic.

Unsuitable design of the system. The flame furnaces are far less efficient than the shaft furnaces from the viewpoint of the transfer of heat and matter. They do not ensure efficient contact of the hot gases with the charge and produce excessive amounts of flue dust, carried away by the gases, because of the use of the dust charge.

Ineffective manipulation with materials. The liquid converter slag is usually recycled in the flame furnace in order to remove copper from the slag. Since the slag contains a large amount of magnetite, the gradual buildup of the latter requires, after some time, shutting down the furnace and the removal of magnetite.

Contamination of the environment. The copper melting plants produce large amounts of sulphur oxides emitted into the atmosphere. Whilst the sulphur dioxide, formed in roasting and converter treatment, is relatively concentrated and may be used for the production of sulphuric acid, the sulphur dioxide formed in the flame furnace is characterised by a low concentration (0.1–0.2%). The latter is processed either to produce sulphuric acid or is neutralised with limestone. The large volume of the gases, emitted into the atmosphere from this source, is a serious environmental problem.

Copper losses. The copper losses into the slag in the flame furnace are proportional to the richness of the matte. Therefore, in order to minimise the losses, copper-rich matte is used only seldom.

Processing and liquidation of waste. The metallurgical production of copper is characterised by the formation of very large amounts of slags with very large quantities of copper. However, the concentration of copper is low and this prevents efficient processing of the slags. In addition to the copper, the waste dumps

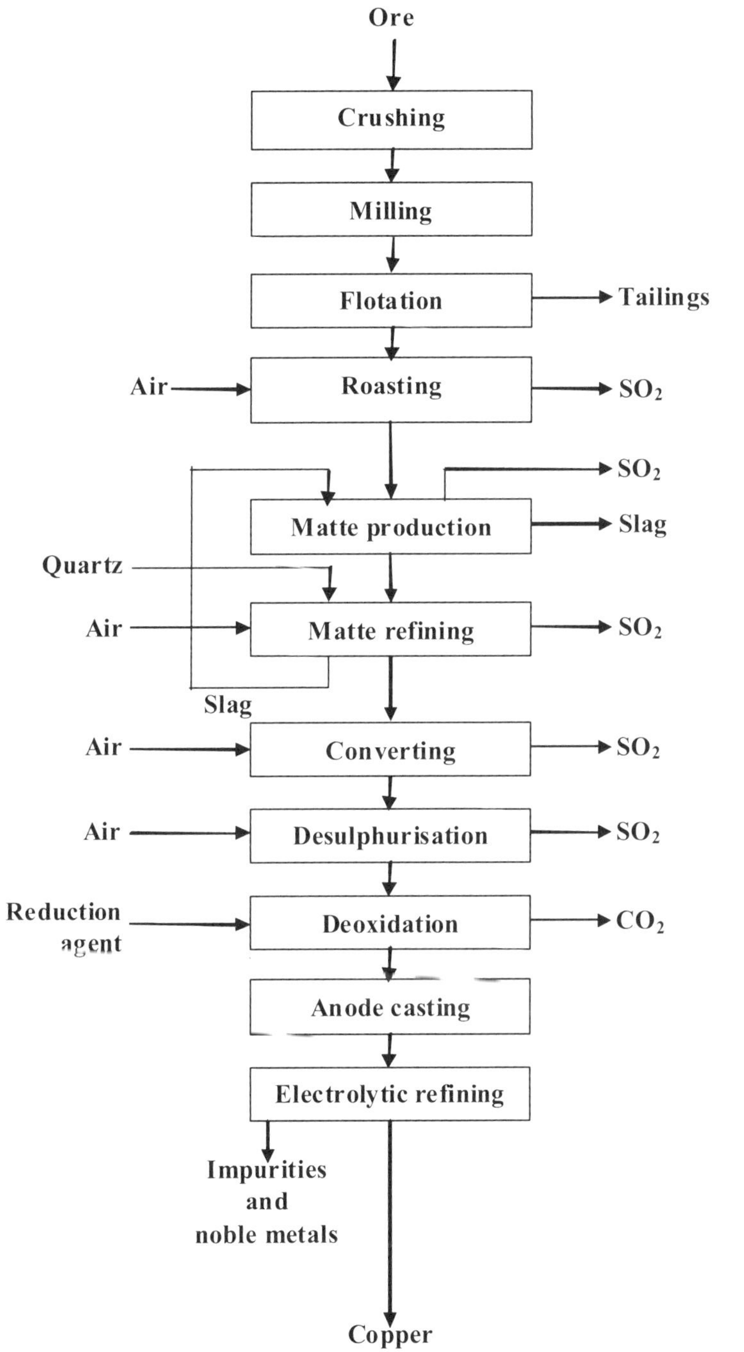

Fig. 1.3. General scheme of pyrometallurgical production of copper.

also contains a large amount of iron which, for economic reasons, cannot be processed into big iron. The electrolytic refining of copper is also accompanied by the formation of anode sludge and raw nickel sulphate requiring further processing and, consequently, higher production costs. The unavoidable formation of sulphuric acid by processing of gaseous sulphur oxides is a very important aspect.

High production costs. Although the electrolytic refining of copper is relatively cheap, it is a very slow operation resulting automatically in high energy consumption and production costs.

The attempts to improve the current situation have been made in various directions, mainly by the development of continuous smelting, improvement of furnace design, reduction of emissions, increase of the efficiency of extraction of copper, improvement of the efficiency of the thermal balance and of electrolytic refining, etc. The most important problem of the pyrometallurgical production of copper is the formation of volatile oxides of sulphur and the emission of these oxides into the atmosphere and, therefore, the need for further processing, in most cases into sulphuric acid, or also secondary products as, for example gypsum, ammonium sulphate, etc. The production of sulphuric acid requires further investment and the product is relatively dangerous from the ecological viewpoint and, consequently, must be efficiently stored and distributed. The unavoidable operation of conversion of sulphur oxides into sulphuric acid improves the production costs: for example, in 1989, the introduction of desulphurisation equipment in American copper plans increased of the cost of 1 kg of copper by 0.17 US dollars [2]. The ratio of the expenditure to profit is very unstable and, in fact, the marketing of sulphuric acid controls the amount of profit in pyrometallurgical plants because the amount of produced sulphur or sulphur compounds is not small. Chalcopyrite, $CuFeS_2$ contains 34.94 wt.% of sulphur which, in comparison with the mass of the metal fraction, forms 69.81 wt.% SO_2 and also 106.87% H_2SO_4. The consumption of copper from primary sources in Europe, i.e. sulphide concentrates of copper, in 1994 was 1968 kt and the processing of these concentrates resulted in the production of 1374 kt of SO_2 and 2103 kt of H_2SO_4 [4]. Since these values described the production in Europe, it may be assumed that the world production amount is at least twice as high. Of course, to these values it is necessary to add the amount of the sulphur oxides or acids, produced by the processing of sulphides of other metals, such as Pb, Zn, Sn, etc. Unfortunately, at the present time, sulphuric acid is almost impossible to sell and, the therefore, the copper melting plans are in a difficult situation.

Generally, it may be concluded that the production of nonferrous metals is accompanied by two fundamental problems – protection of environment and energy requirement. In recent years, marked by worldwide economic recession, it was also necessary to take into account the problem of complex utilisation of all products of the process with the minimum financial requirement. It is generally known that at present the conventional rich deposits of ores of nonferrous metals have been almost completely exhausted. The need for processing of lean ores is associated with another complication – the complex nature of these ores. It is therefore necessary to develop economical methods of processing lean and complex ores for which the conventional pyrometallurgical methods of processing are no longer effective.

In addition to the conventional natural sources, other, unconventional starting materials are becoming also increasingly important for the production of nonferrous metals. The most important starting materials include secondary resources, formed in industrial activity and include complex materials from pyrometallurgical processes, such as slags, dust, anode and galvanic sludge, sulphide matte, alloy scrap from processing, spent electrolytes and leaching agents, ashes, dross and other industrial waste. Another important material is the waste of elkectronic and electronic equipment because of the content of noble and also rare-earth metals. The extent of recycling of these materials is continuously increasing. All these facts, together with the increasing pressure on the protection of environment and water systems, will soon result in the situation in which one of the direct aims of the newly formed legislative, preventing contamination of the environment, will be the industry of extractive metallurgy.

Many of these problems have been solved successively by an alternative approach to pyrometallurgical production of nonferrous metals, including copper, i.e. the hydrometallurgical method of production of nonferrous metals.

Hydrometallurgy is based on two main steps: the transfer of metal or metals from an ore or concentrate into the solution; the process is referred to as leaching, and the selective extraction of the metal from the solution-the operation based on precipitation methods, or liquid extraction. The general diagram of hydrometallurgical processes is shown in Fig. 1.4.

The modern era of hydrometallurgy started at the end of the 19th century. The 1960s and the beginning of the 1970s of the 20th

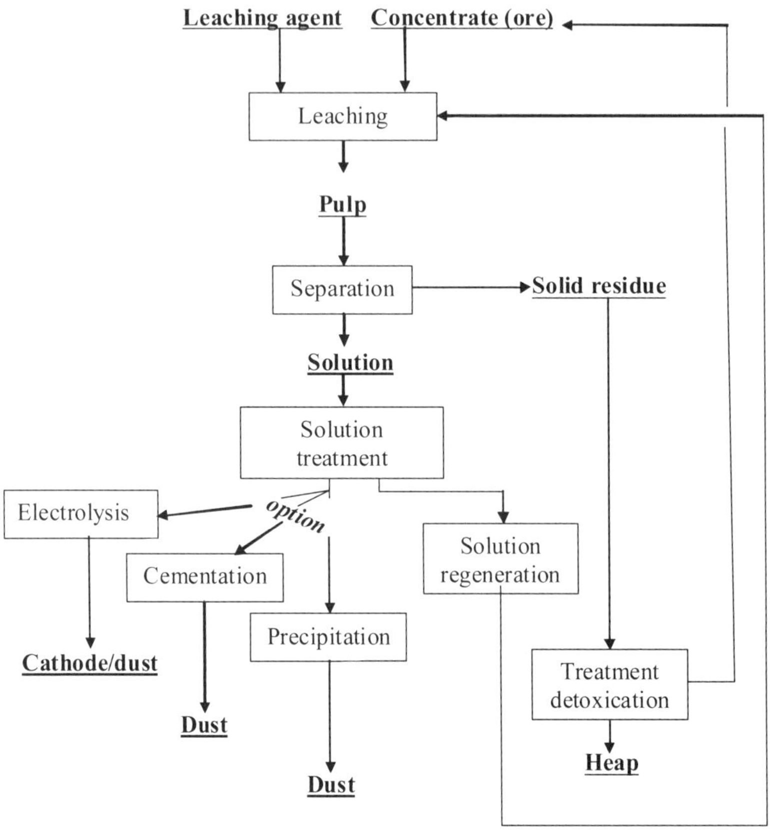

Fig. 1.4. General scheme of the hydrometallurgical process [4].

century are characterised by a large increase of the extent of research and development of hydrometallurgical processes in the area of production of nonferrous metals. One of the reasons for this situation was the extensive and flexible possibilities of producing metals from solutions into which they were extracted. This may be carried out using practically all methods of classical analytical chemistry which may also be adapted for application in the industrial process. In reality, it is the hydrometallurgists who have been utilising recently the large amount of information from chemistry, collected by inorganic chemists at the end of the 19th and the beginning of the 20th century.

Of course, hydrometallurgy does not offer a solution for extracting nonferrous metals by the only general method. The leaching methods are highly individual and depend not only on the type of process but also on the complexity of the process, starting materials and also on their physical, physical–chemical, chemical and mineralogical properties.

The chemical reagents, used in hydrometallurgy for the dissociation of minerals, have different abilities to attack the processed material. At the present time, there are many different hydrometallurgical methods of production of nonferrous metals from primary raw materials. However, because of the previously mentioned facts, only several hydrometallurgical plants are in operation. Despite this situation, the basic and applied research in the hydrometallurgy of nonferrous metals is still highly intensive and promising, especially because of the availability of lean and more complex primary sources, environmental aspects of pyrometallurgy and hydrometallurgy, and the legislative pressure on environmental protection.

Comparison of the pyrometallurgical and hydrometallurgical processes shows that pyrometallurgy was more successful in the processing of rich bulky sulphide ores in shaft furnaces because these systems are maximally economical from the viewpoint of heat exchangers: the cold charge descending from the top is heated by the rising hot gases from the shaft. However, shaft furnaces are past, of no essential importance anymore. The problems with dust removal are also minimised because in the charge the ore is in the form of pieces. Gradually, with the exhaustion of the deposits of these rich ores, attention has been given to the processing of leaner and complex nonferrous metal ores. Therefore, the hydrometallurgical processes should became more interesting.

The overall summary of some other characteristics from the viewpoint of the shortcomings and advantages of the pyrometallurgical and hydrometallurgical processes [5, 6] is presented in Table 1.1.

It appears that the detailed considerations of the overall position of hydrometallurgy and its inclusion in industrial application are not simple. Of course, the hydrometallurgical processes which have been investigated quite efficiently in the laboratory and pilot plant conditions, cannot compete commercially with the pyrometallurgical processes. This is caused by the less efficient economic parameters in comparison with the pyrometallurgical processes, although the differences are very small. Most importantly, the new hydrometallurgical plants require a large investment which is not feasible in the period of economic recession. Because of these facts, hydrometallurgy has been displaced into the sphere of

Table 1.1. Comparison of some characteristics of pyrometallurgical and hydrometallurgical processes.

Process	Pyrometallurgy	Hydrometallurgy
Processing of rich ores	more economical	less economical
Processing of lean ores	unsuitable from the energy viewpoint	suitable with selective leaching
Processing of polycomponents ores	difficult separation of components – unsuitable	suitable, flexible, production of secondary products
Process economics	suitable for largest plants	small capacity production; smaller investment
Processing of secondary sources	unsuitable in most cases	suitable
Separation of pure components	not possible	possible
Processing of sulphide ores	polution of environment with SO_2 which must be treated to H_2SO_4 with low concentration	SO_2 does not form, elemental S is produced
Reaction rate	high due to high temperature	low due to low temperature
Througput of material	very high unit throughput	low, for low-volume production
Handling of material	quite difficult with liquid metals, slag, matte	solutions and pulp can be easily transported in a pipeline
Contamination of environment	problems with exhausts, dust, noise	no atmospheric pollution, problems with effluents
Solid residues	large amount of residues, can be stored in a dump	mostly fine, may contaminate environment
Toxic gases	many processes generate toxic gases	gases are generated by only a small number of processes and can be cleaned
Charge	suitable for inhomogeneous charges	sensitive to change of charge; generallgy charge should be homogeneous
Production features	quite simple; procedure is not complicated	more sophisticated technology; control and regulation more complicated

processing secondary materials and the processing of lean complex ores using which the problem can be efficiently solved. However, hydrometallurgy cannot be a universal medicine for extractive metallurgy. Hydrometallurgy should play a complimentary (not competing) role in the dressing of raw materials for pyrometallurgy. In addition to this, hydrometallurgy is capable of closing the gap between geology, mineralogy, mining engineering, processing methods, metallurgical techniques, materials science and industrial design in such a manner as to use more efficiently hydrometallurgy in primary production and other branches of the industry. Only the complex understanding of the entire problem can ensure in the final analysis economically efficient production of selected nonferrous metals whilst fulfilling all the environmental requirements. However, it should be stressed that some of the metals are produced on a large extent or completely by hydrometallurgy, for example, zinc, gold, uranium or aluminium oxide as the main intermediate product for the production of aluminium.

1.1. Copper hydrometallurgy

'Hydro' means water, 'metallurgy' is the production of metal, i.e. hydrometallurgy is therefore the science and method of aqueous methods of extracting metals from their ores. In recent years, non-water solutions have also been used for this purpose, Therefore, hydrometallurgy includes extraction methods in which metals, salts of metals or other compounds of metals are obtained by chemical reactions from aqueous or non-aqueous solutions. In the normal conditions, the hydrometallurgical processes are realised in the temperature range 25–250 °C, and the overall pressure may vary from several kilopascals (vacuum) to more than 5000 kPa.

Although many experts treat hydrometallurgy as a new production method, this is not the case. Table 1.2 shows the chronology of copper hydrometallurgy [7]. Already at the beginning of the 20th century, Greenawalt published a monograph concerned with the hydrometallurgy of copper [8].

Sulphide raw materials are the most widely used and the processing of these materials in the form of lean ores requires milling and flotation. These result in the formation of a finely ground concentrate representing the starting material. Of course, these materials cannot be charged into the shaft furnace because the furnace would be immediately clogged up and would prevent the flow of reaction gases and products. This is the reason for the development of horizontal furnaces, heated with fossil fuel. However, the result was the extremely unfavourable situation in the area of contamination of the atmosphere, high consumption of energy and excessive formation of flue ash. This is another reason

Table 1.2. History of hydrometallurgy of copper

Copper source	Leaching system	Copper obtaining	Place	Year
Mining water	Leaching in mine	Fe cementation	Rio Tinto, Spain	1670
Poor ore	Leaching in mine	Fe cementation	Rio Tinto, Spain	1752
Mine water	Leaching in mine	Fe cementation	Strafford, USA	1820
Mine water	Leaching in mine	Fe cementation	Ducktown, US	1860
Ore	Calcine, heap	Fe cementation	Rio Tinto, Spain	1876
Mine water	Leaching in mine	Fe cementation	Butte, USA	1886
Oxide ore	Tank leaching	Fe cementation	Clifton, USA	1892
Flotation tailings	Pilot plant Roasting/leaching production	Fe cementation	Butte, USA	1912
Oxide ore	Tank leaching	Electrolysis	Ajo, USA	1915
Oxide ore	Tank leaching	Electrolysis	Chuquicamata, Chile	1915
Oxide flotation tailings	Vat leaching NH_3/CO_2	Precipitation of CuO by water steam	Kennecott, USA	1916
Poor ore	Vat leaching NH_3/CO_2	Precipitation of CuO by water steam	Clumet/Hecla, USA	1916
Oxide ore	Vat leaching H_2SO_4	Precipitation by SO_2	Anaconda, USA	1920
Mixed ore	Vat leaching $Fe_2(SO_4)_3$	Electrolysis	Inspiration, USA	1930
Oxide ore	Vat leaching H_2SO_4	Electrolysis	Panda, Zaire	1930
Mixed ore	Leaching+flotation	Fe cementation	Miami, USA	1934
Oxide ore	Dump leaching H_2SO_4	Solvent extraction, electrolysis	Ranch Bluebird, USA	1968
Oxide residues	Vat leaching H_2SO_4	Solvent extraction, electrolysis	Nchanga, Zambia	1974
Concentrate	Vat leaching NH_3 (Anaconda)	Solvent extraction, electrolysis	Anaconda, USA	1974
Concentrate	Vat leaching NH_3 (Anaconda)	Solvent extraction, electrolysis	Anaconda, USA	1974
Concentrate	Vat leaching NH_3 (Anaconda)	Precipitation by SO_2, pilot plant	Tucson, USA	1974
Concentrate	Vat leaching H_2SO_4+HCl (Duval®)	Electrolysis	Sierita, USA	1977
Concentrate	Leaching with $CuCl_2$+NaCl+NaBr (Intec®)	Electrolysis	Australia	1994
Concentrate	Leaching with $CuCl_2$+NaCl (Hydrocopper®)	Cu_2O precipitation	Pori, Finland	2001

for the examination of the possibilities of using hydrometallurgical methods of processing sulphide concentrates.

In the case of copper, special effort has also been made to find the optimum hydrometallurgical procedure but the results have not as yet been satisfactory. The most important obstacle is the fact that the starting material for the production of copper contains approximately 30% Cu which is only 50% of the zinc content in the starting zinc material produced extensively by hydrometallurgical procedure. The presence of other components in the raw materials for the production of copper (Fe, S, SiO_2 and other minor components) results in considerable problems in the hydrometallurgical production of copper.

At present, there is a competition in the production of copper between hydrometallurgy and pyrometallurgy in the following areas [9]:

Extraction of copper: using conventional melting–refining processes, 98–99% of copper is produced from the initial charge of the concentrate. This indicates that for the hydrometallurgical process of production of copper to be attractive for the industry, the yield must be very high. The copper melting plants are characterised by a higher degree of extraction by dumping of the slag with the iron/copper ratio of approximately 100. The leaching residue from the hydrometallurgical production of copper with this the ratio would be highly suitable. But unfortunately, even at almost 100% yield of copper from the concentrate it is very difficult to obtain such a ratio. The unwashed filter cake, containing 25% moisture and 60 g/l of copper, contains 1.5% of dissolved copper and also part of non-leached copper. This shows that it is imperative to innovate of the methods of production of dissolved copper from moist filter cakes, of course without the formation of other problems with the equilibrium of water in the system.

Problem with iron: the typical copper concentrate releases approximately 1 t of iron per every tonne of produced copper, which is transferred into the slag. If this slag from pyrometallurgical production of copper contains 40% of iron, it is transferred to the dump. However, the leaching residue (with 30% of ferrous jarosite [10–12]) is not acceptable for storage on dumps because the residue is considerably finer than the slag and, in addition to this, it also contains acid solutions leaching the heavy metals and having a detrimental effect on the environment.

Problems with sulphur: hydrometallurgy offers the optimum solution of the problem of sulphur because it prevents the formation of any sulphur dioxide. At present, there are many hydrometallurgical processes of copper production accompanied by the formation of elemental sulphur. In other processes, it is necessary to consider the formation of sulphates which may be a saleable product, for example, ammonium sulphate, or may be dumped, for example, gypsum or the basic ferrous sulphate salts.

Extraction of noble metals: the copper concentrate usually contains large quantities of silver and gold. In the conventional melting processes, both these metals transfer in the process of refining into the anode sludge with a wide concentration range. In older metallurgical processes, the noble metals remain in diluted residue from which they are difficult to extract.

Toxic waste: toxic metals often found in the process of production of copper include arsenic, antimony, bismuth, lead, zinc, mercury and others. In conventional pyrometallurgy, these elements are transferred into dust and outgoing gases which in older plants contaminated the atmosphere and environment. In more advanced plants, these metals are produced to a certain extent for commercial purposes and partially are eliminated in the slag. However, these elements are still a source of contamination of the environment. In the hydrometallurgical methods, the elements do not penetrate into the atmosphere, but there is still a problem (although perhaps not so acute) with the potential contamination of water. It is therefore necessary to develop cleaning technologies or advanced prevention methods.

At least these and also other problems must be efficiently solved by the proposed hydrometallurgical method of production of copper, if this method is to be capable of competition. However, it is obvious that the changing composition of the raw materials often requires the application of methods which were not economical in the past but it has been shown gradually that these raw materials cannot be processed by any other method. At present, the amount of copper produced by hydrometallurgy increases on the worldwide scale and represents approximately 20% [13]. Figure 1.5 shows the development of production of copper by hydrometallurgy and, for comparison, also gives the development of the total production of copper already presented in Fig. 1.2.

As indicated by the above considerations, the production of copper by any method is a complicated and demanding process. Of course, this also relates to the hydrometallurgical method of production of copper which has its specific features which depend on the processed raw material and also on the type of extraction

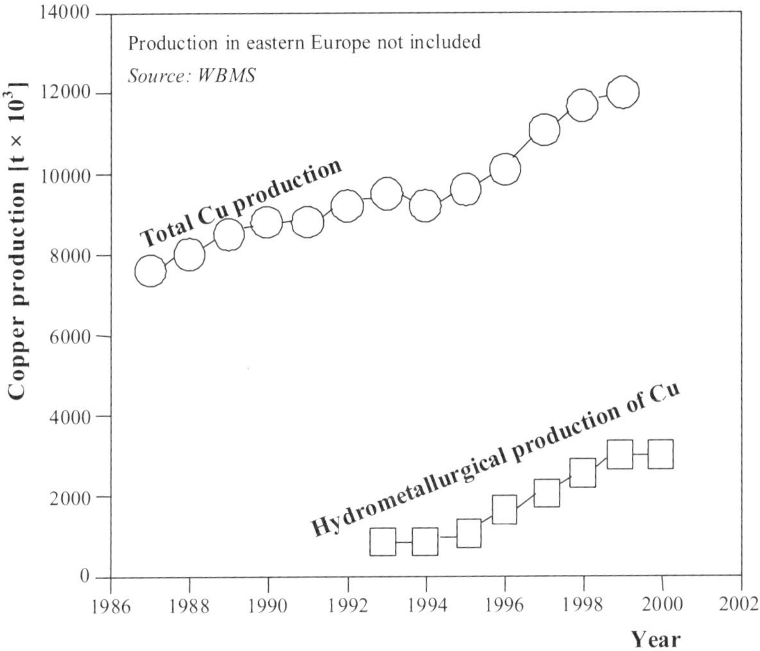

Fig. 1.5. Development of production of copper by hydrometallurgy [14].

agents used. In the individual stages of the hydrometallurgical production of copper, it is likely that the most important stage is the stage of the direct transfer of the metal into the solution, i.e. leaching, because it determines the rate and efficiency of transfer of metals into the solution and, consequently, a large part of the economic parameters of the entire process. The efficiency of the leaching process should be determined by examining the process from the thermodynamic aspect, i.e., to verify whether there are any agents which could interact together. Another important step is the determination of the kinetic conditions of the process, i.e. the duration and the type of conditions in which these reagents react, and also the reaction mechanism is important. On the basis of this information it will be then possible to propose and optimise the process of hydrometallurgical production.

References

1. MacMillan A., Norton K.: *Metall*, 52, 7-8, 1998, 429-437.
2. Rothfeld L.B., Towle S.W.: *Eng. Min. Journal*, 190, 10, 1989, 32-34.
3. Crowson P.: *Erzmetall*, 47, 10, 1994, 611-617.

4. Havlík T., Škrobian M.: In: Hydrometallurgy, Industrial Applications of X-ray Diffraction, Marcel Dekker, Inc., Chung F.H., Smith D.K. eds., chapter 11, 303–317, 2000.
5. Habashi F.: A Textbook of Hydrometallurgy, Met. Extr. Quebec, Enr., 1992
6. Gupta C.K., Mukherjee T.K.: Hydrometallurgy in Extraction Processes, vol. I, CRC Press, 1990.
7. Arbiter N., Fletcher A.W.: Copper Hydrometallurgy – Evolution and Mile-stones, Hydrometallurgy – Fundamentals, Technology and Innovations, Hiskey, Warren eds., Soc. Min. Met. Explor., Colorado 1993, 549–565.
8. Greenawalt W.E.: The Hydrometallurgy of Copper, McGraw-Hill Book. Comp., 1912.
9. Peters E.: *Hydrometallurgy*, 29, 1992, 431–459.
10. Swinkels G.M., Berezowski R.M.G.S.: The Sherritt-Cominco process - Part I: The process, CIM Bulletin, 71, 1978, 105–121
11. Kawulka P., Kirby C.R., Bolton G.L.: The Sherritt-Cominco process - Part. II: Pilot plant operation, CIM Bulletin, 71, 1978, 121–130.
12. Maschmeyer D.E.G., Milner E.F.G., Parekh B.M.: The Sherritt-Cominco process - Part. III: Commercial implications, CIM Bulletin, 71, 1978, 131–139.
13. McCoy M.: Sulfuric market rocked by copper, Chemical & Engineering News, October 4, 1999, 13–17
14. World Bureau of Metal Statistics, 1998.

CHAPTER 2

ORE MINERALS

The extraction of metals from ores in hydrometallurgical processes takes place at relatively low temperatures, up to approximately 200 °C. If the process takes place at temperatures up to 100 °C and the total pressure of 0.1 MPa, it is referred to as leaching in the normal conditions (i.e., temperature and pressure). If the process is realised at higher temperatures, it automatically requires a higher total pressure. The process then takes place in a pressure reactor, i.e. autoclave, and is referred to as pressure leaching.

The rate of the reaction may be low and usually depends on the type of the phases containing a metal. For example, in acid oxidation leaching, copper is produced by leaching from chalcosine Cu_2S at a considerably higher rate than from chalcopyrite $CuFeS_2$. This shows that the type and nature of minerals present in the concentrate is the important parameter of the hydrometallurgical processes. In pyrometallurgy, this is not so important.

The type of leaching used for transferring a metal into a solution depends on the chemical nature of the mineral in the concentrate. Nonferrous metals and, consequently, also copper are present in a large number of minerals but only several of them are of interest for hydrometallurgical processes. This interest is caused by two main factors – the occurrence of the given raw material in the industrially interesting amount, and the capacity for reacting in the given hydrometallurgical system.

The majority of the minerals solidify in a crystallographic system, i.e. the main construction particles, atoms or ions are distributed in the three-dimensional system whose idealised form is referred to as the spatial lattice. Since the real materials always contain some defects, the real crystal lattice is referred to as the crystal structure. The crystallographic lattice is formed by the repetition of the motive of the single cell which is then characteristic of some groups which can be used to define the individual minerals. These matters are studied by mineralogy.

The single cell is defined by the lattice (structural) parameters, which depend on crystal symmetry. The internal structure can be described completely by six constants, three length constants *a*, *b*, *c*, which are used to determine the Miller indexes, and by the angles between the individual axes α, β, γ.

In the unit cell, the atoms are arranged with the lowest possible energy which means that the closest atomic spacing is preferred. Since in many minerals the atoms are arranged more efficiently than the ions, these distances are referred to as the ion radius. In the case of elements which may have several valencies, each type of ion has a different ion radius. The outer areas of the ions are the volumes of electron charges and heavier elements are characterised by a tendency to be larger rather than light one, but in the case of the elements with similar atomic numbers the anions are larger than the cations.

In certain solids, for example, in oxides of a light metal, such as MgO, the size of O^{2-} is considerably greater than that of Mg^{2+} and, consequently, the ion spacing in the lattice is determined by the distance of the closest arrangement of the oxide ions. Because of their size, the magnesium atoms cannot fit between the spherical ions of O^{2-}. This is known as close packing and, in general, these lattices are formed by the layers of spherical anions arranged in such a manner that each ion occupies the smallest possible volume and they are closely packed. There are two possible methods of such arrangement, hexagonal close-packed and cubic close-packed lattice. The ratio of the radii of the anions and cations determines whether close packing is possible.

The mineralogical lattice may be regarded as the arrangement of the anions and cations occupying the holes between them. Some holes are in tetrahedral arrangement between four anions around the central point of the hole and are referred to as the tetrahedral holes. Other holes are located between six anions with octahedral arrangement around the central point, i.e. the octahedral holes. Generally, the most symmetric arrangement of the anions around the cation is three-dimensional with 3, 4, 6 or 8 anions arranged on the tip of the triangle, tetrahedron, octahedron or cube around the central ion.

The oxides of the MgO type are characterised by the ion bond between the cation and the anion but the physical properties indicate a different type of bond. The structure of the lattice is determined by electrostatic forces between the regions with opposite charges, and the MgO has the structure of NaCl in which every ion is surrounded by octahedrally arranged six ions of a different type. In many other minerals, the bonds between the cations and the anions are covalent and the arrangement of one type of atom around another type is determined by the orientation of covalent bonds in space. This is the case of many sulphides of which many have the form of semiconductors. Consequently, the properties of the minerals are determined by the energy level of the electrons rather than by atomic bonds.

The sulphides of alkali metals and alkali rare earth metals are of the ion type and the structure is similar to oxides. The sulphides of other metals have basically a covalent bond and many of them show the physical properties typical of alloys rather than salts. They are semiconductors, the surface has metallic shine and higher reflectivity and many of them have the composition which do not correspond to the normal valency. This is a typical example of the sulphides of copper. For example, covellite, CuS, has the structure in which part of sulphur is bonded in S_2^{2-} ions and the mineral is a diamagnetic metallic conductor in which copper is present in the form of Cu^+.

The sulphide can be described efficiently using a pure ion model, in which the ions are treated as charged spheres of a defined radius. The model can be used for predicting the structural type and substitution of the cations more efficiently than in the case of oxides. Of course, this does not provide information on the electronic structure or the properties which depend on the behaviour of electrons. Although information on the valency of the sulphides is useful, a significant contribution to understanding the behaviour of sulphides in processing is the application of the theory of crystal fields proposed in 1960. Later, the theory of molecular orbits and band theory were developed. The band theory can be used for multiatomic compositions but it is not very efficient when quantifying the crystals containing multiatomic unit cells. These aspects of the mineral chemistry of sulphides helped significantly in understanding the theory of hydrometallurgy of the sulphides and have been described in specialised literature, for example [1].

A large number of equilibrium diagrams of binary sulphides of metals are available. The copper–sulphur system contains several minerals which will be described later. Since the metal–sulphur bond in many sulphides is of the strongly covalent nature, including sharing of the electrons, the metal is capable of forming only a limited number of ones and these form only in specific directions in the appropriate space. The most common arrangement are the four bonds oriented in the tetrahedral direction and six bonds, with octahedral arrangement. The metallic atoms are present in tetrahedral or octahedral holes in a sulphur skeleton.

The sulphide phase may contain more than one metal. From this viewpoint, copper forms a number of polymetallic sulphides, and combinations of the same metals also form a large number of different sulphides. Copper forms with iron a very large number of sulphides of which chalcopyrite $CuFeS_2$ is best known. These sulphides will also be described later.

Chalcopyrite has a relatively simple structure of the ZnS type and also a diamond lattice. Carbon forms four tetrahedral covalent bonds and in the diamond crystal every atom is bonded with four neighbours with equal spacing, and the bond continues through the entire crystal. If the carbon atoms are substituted by zinc a sphalerite structure forms in which every zinc atom is surrounded in tetrahedral arrangement by four sulphur atoms and every sulphur atom by four zinc atoms. If the zinc atoms are replaced by copper and iron atoms, a chalcopyrite structure forms. However, if the entire cell of ZnS takes part in substitution, the chalcopyrite crystal does not form by repetition in all three dimensions. The cell, produced from the unit cell of ZnS, must be repeated in one direction in order to produce the unit chalcopyrite cell. The resultant chalcopyrite structure is based on the ZnS superstructure, sphalerite. There are also quaternary compounds with a tetragonal structure, for example, the structure of stanite, Cu_2FeSnS_4, formed by the substitution of half of the iron atoms into the unit cells of chalcopyrite by tin.

The metals requiring hexagonal arrangement cannot form spharelite or other tetragonal structure. They usually form a structure of the NiAs type. In the structure, every atom has six neighbours of a different type, but since the arsenic atoms are surrounded by six nickel atoms on the tips of the trigonal prism, the nearest neighbours of the nickel atom are six arsenic atoms arranged around in the tetrahedral structure. In addition to this, every nickel atom has two identical atoms with a relatively close packing. NiAs contains six nickel atoms with the Ni–As spacing of 0.243 nm, and two nickel atoms with the Ni–Ni spacing of 0.252 nm. The properties of the structure of nickel arsenide are such that it may form solid solutions, including transitional metals.

Pyrrhotite, $Fe_{1-x}S$, has the NiAs structure but does not have the stoichiometric composition FeS because there is a shortage of iron in the unaffected sulphur lattice. These sulphides will also be described later. The structures of pyrite and marcasite (both FeS_2) differ from the structure of pyrrhotine. They contain discrete groups S_2^{2-} in which the sulphur atoms have a covalent bond. The S–S spacing is 0.21 nm. The pyrite structure is derived from the lattice of sodium chloride (face-centred cubic lattice) in which the atoms of iron and the centres of the S_2 groups occupy the positions of sodium and chlorine.

The NiAs structure has been derived for a large number of compounds of the MeX type in which Me is a transition metal and X is the element of subgroup VIII B, Sn, As, Sb, Bi, S, Se, Te. The pyrite structure is characteristic of the compounds of type MeX_2 of the same metals. In the compounds such as arsenopyrite, FeAsS, the lattice is characterised by lower symmetry but is still derived from the pyrite structure. The marcasite structure is characterised by less symmetric arrangement of the same structural units, Fe and S_2, based on the lattice of sodium chloride but with a lower symmetry.

In hydrometallurgy, in particular the hydrometallurgy of sulphides, it is necessary to pay attention, in addition to sulphides, also to other types of minerals, especially oxides and silicates. This is caused by the fact that they may also be present as an impurity in the charge for leaching but in the case of oxides they are often present as an intermediate product of the product of leaching.

In hydrometallurgy, two types of oxides are important: binary oxides which contain only one metal, and complex oxides, containing several metals. Both types have mostly an ion structure and the metal atoms are characterised by high co-ordination numbers, often 6 or 8. The most important structures of the binary oxides, considered in hydrometallurgy, are presented in Table 2.1. The co-ordination numbers of the rutile structure indicate, for example, that each metal atom has a group of six oxygen atoms around it and every oxygen atom has three metal atoms.

In addition to this, some oxides have a silicate structure, layered structures (MoO_3, As_2O_3), chain structures (HgO, SeO_2, Sb_2O_3), and molecular structures (RuO_4, OsO_4). This is of considerable importance for practice, for example, OsO_4 because of its molecular structure evaporates at 130 °C and then separates from other metals of the platinum group. RuO_4 melts at 25.5 °C and dissociates at 108 °C. On the other hand, the melting and boiling points of SiO_2 are 1610 and 2230 °C, corundum, α-Al_2O_3, 2015 and 2980 °C and the melting point of tenorite, Cu_2O, is 1326 °C. These are typical values of substances with a strong ion bond.

Table 2.1. Structures of binary oxides.

Type	Structure	Co-ordination number		Examples
		Me	O	
MeO_3	ReO_3	6	2	WO_3
MeO_2	Rutile	6	3	TiO_2, MnO_2, SnO_2
	Fluorite	8	4	ZrO_2, ThO_2, UO_2
MeO	Sodium chloride	6	6	MgO, CaO, NiO
	Wurtzite	4	4	ZnO
Me_2O_3	Corundum	6	4	Al_2O_3, Fe_2O_3, Cr_2O_3
Me_2O	Cuprite	2	4	Cu_2O

In the case of binary oxides, presented in Table 2.1, all the metallic ions in the compounds are characterised by the same valency and environment. If a binary oxide contains a metal in two oxidation degrees, then there are two types of metallic ions, for example, Pb^{2+} and Pb^{4+} in Pb_3O_4 and these may have a different environment because in this case their co-ordination numbers are 3 and 6. The complex oxides may also have a structure which is regular as in a simple binary oxide. Substances with one metal in two valences should, however, be regarded as complex oxides.

Complex oxides contain more than one metal. In some complex oxide structures, the environment of individual metallic ions greatly differs so that the difference in their size or in the charges between the ions does not enable the formation of a stable structure for one metal in two valances. For this reason, there are a large number of structures of complex oxides which the binary oxides cannot have. One of the most important is the spinel structure. The mineral spinel $MgAl_2O_4$ solidifies in a cubic structure and has 8 entities in an elementary cell.

A large number of compounds solidifies in the spinel structure AB_2Z_4, where Z is the bivalent anion, A is the bivalent metal, and B is a trivalent metal. The anion Z may include O, S, Se, Te. The following metals may be found in complex oxides with a spinel structure:

A – Mg, Cr, Mn, Fe, Co, Ni, Cu, Zn, Cd, Sn
B – Al, Ga, In, Ti, V, Cr, Mn, Fe, Co, Rh

although only some combinations of these metals are known. The spinel structure is based on a system of layers of O^{2-} ions, each with a radius of 0.132 nm. These layers are arranged in such a manner that each layer is almost closely packed in the face-centred cubic system so that the oxygen ions in the layers 1, 3, 5, etc. are in the lines at the same distance from the original oxygen atoms. There are two types of holes between oxygen atoms in the spinel structure – octahedral – there are 32 of these sites in the unit cell, and tetrahedral – 64. In the spinel, all the Mg^{2+} ions are in the tetrahedral position (or A) and all Al^{3+} are in the octahedral position (or B). The equilibrium of the electrical charge continues and consequently there are more holes than metal ions in the structure. Only eight tetrahedral and sixteen octahedral positions are occupied. This is known as the 'normal' spinel structure. The positions, occupied by the metallic ions, are characterised by regular arrangement and form the added face-centred cubic sublattice 'interwoven' by an oxide lattice.

In the ferrites, Fe^{3+} substitutes Al^{3+} in the structure, and Me^{2+} may be substituted by a bivalent cation with an ion radius in the range 0.06–0.1 nm. The 'normal' spinel structure, i.e., all Me^{2+} ions are in positions A and all Fe^{3+} ions are in positions B, is typical of some ferrites, such as $ZnFe_2O_4$ and $CdFe_2O_4$, which are paramagnetic. Other ferrites have an 'inverse' spinel structure in which all Me^{2+} ions are in the positions B so that 8 Fe^{3+} ions are in position A and 8 in positions B. A suitable example of such an inverse spinel structure are Fe^{2+}, Co^{2+} and Ni^{2+} ferrites which are highly magnetic. The ions in the octahedral positions are probably statistically distributed between the total number of available positions.

The normal and inverse spinel structures should be regarded as extreme structures because in most cases they are present as transition structures. In $MnFe_2O_4$ 80% of Mn^{2+} ions are in the positions A and 20% in positions B, Fe^{3+} ions are distributed in an equilibrium manner between positions A and B. In $MgFe_2O_4$ only 10% of magnesium is in position A. The distribution of the bivalent metals in ferrite is not determined by their ion radius because the largest ions Cd and Mn are found mainly in lower tetrahedral positions. The formation of the normal or inverse structure depends exclusively on the preferential selection in the detailed environment, for example, zinc is characterised by tetrahedral co-ordination and nickel by octahedral co-ordination.

Iron is one of the most important metals in hydrometallurgy because it is present everywhere in ores not only as a mechanical addition but also forms part of the structure of the most frequently processed minerals, for example chalcopyrite $CuFeS_2$, and must be available in the process. If in the normal conditions of temperature and pressure the ion precipitates from the solution because of a change of pH, then it usually forms the amorphous compound $Fe(OH)_3$. After drying it slowly solidifies in the sludge as α-FeOOH (goethite) which, after dehydration, forms α-Fe_2O_3 (haematite). The oxide and hydroxides of trivalent iron and their structure are described in Table 2.2. Although Fe_3O_4 and γ-Fe_2O_3 have a spinel structure and may easily transform to a different iron compound, there are only a small number of common features between their structures and the structures of other iron oxides and hydroxides. Many transformations between these compounds are topotactic, i.e. the product of their transformation has the same structure as the initial compound. Possible transformations are shown in Fig.2.1.

The Fe_3O_4 and γ-Fe_2O_3 oxides are very similar because of their spinel structure. The cubic elementary cell contains 32 close-packed oxygen atoms with a face length of approximately 0.85 nm. FeO has the structure of sodium chloride, in which the cations and anions have the octahedral co-ordination with 32 Fe^{2+} ions in octahedral positions, as in the case of the spinel. Fe_3O_4 contains 8 Fe^{3+} ions in tetrahedral positions and 8 Fe^{2+} + 8 Fe^{3+} ions in the octahedral positions. γ-Fe_2O_3 has 21.333 Fe^{3+} ions distributed statistically in 24 types of positions, available in Fe_3O_4. Efficient oxidation of Fe_3O_4 may produce γ-Fe_2O_3 which then transforms back to Fe_3O_4 on heating to 250 °C in vacuum. The stages of oxidation FeO to γ-Fe_2O_3 (shown in Fig.2.1) may be described as follows: oxygen is added to the FeO structure in the form of new layers of close-packed oxygen atoms whereas part of the iron is oxidised to Fe^{3+}; the cations diffuse as a result of this addition to the oxide lattice thus reducing the Fe concentration in the elementary cell.

Haematite and oxide–hydroxides other than γ- and β-FeOOH, have structures which may be described from the viewpoint of the added layers of the O or OH ions which are almost close-packed. The iron atoms fit into interstitial holes without any significant deformation of the oxide lattice. Topotactic transformations take place because the cubic and hexagonal close-packed structures of

Table 2.2. Crystal structures of iron oxides and hydroxides [2].

Composition	Mineralogical name	Crystallographic system	Lattice parameters (nm)	Notes
α-Fe_2O_3	haematite	trigonal (hexagonal unit cell)	a = 0.5035 c = 133720	hcp oxygen, Fe^{3+} in octahedral sites
γ-Fe_2O_3	maghemite	cubic (spinel)	a = 0.833–0.838	spinel, various degrees of arrangement
Fe_3O_4	magnetite	cubic(fcc inverse spinel)	a = 0.8397–0.8394	inverse spinel
α-FeOOH	goethite	orthorhombic	a = 1.000 b = 0.303 c = 0.464	based on hcp oxygen
β-FEOOH	akaganeite		a = 1.048 c = 0.306	α-MnO_2 structure
γ-FeOOH	lepidokrokite	orthorhombic	a = 0.306 b = 1.240 c = 0.387	based on distorted ccp oxygen
δ-FeOOH			a = 0.2941 c = 0.4580	hcp oxygen, disordered CdI2 structure

ccp – cubic close-packed
hcp – hexagonal close-packed
fcc – face-centred cubic lattice

the oxide lattice are very similar and the compounds can be transformed easily from one arrangement to another.

Some of the transformations, shown in Fig.2.1, are sometimes ignored because the conditions of these transformations are not well known. A typical example is the transformation of Fe_3O_4 to α-FeOOH. However, it has been found that this transformation also takes place in nature because pseudomorphism α-FeOOH to Fe_3O_4 were found. A large number of pseudomorphous transformations is known and some of them also relate to natural minerals. For example, the results obtained under certain oxidation conditions enable the description of the formation of maghemite from magnetite or also from limonite gel containing goethite by the process identical to that in the laboratory conditions in which lepidokrokite was slowly heated. Maghemite also forms by the dissociation of pyrite, FeS_2, by heating in slightly alkaline aqueous systems in oxidation conditions.

The Earth's crust is formed almost exclusively by silicates and silicate rock and, consequently, silicates are often present as an addition in many ores and concentrates. The structure of the silicate

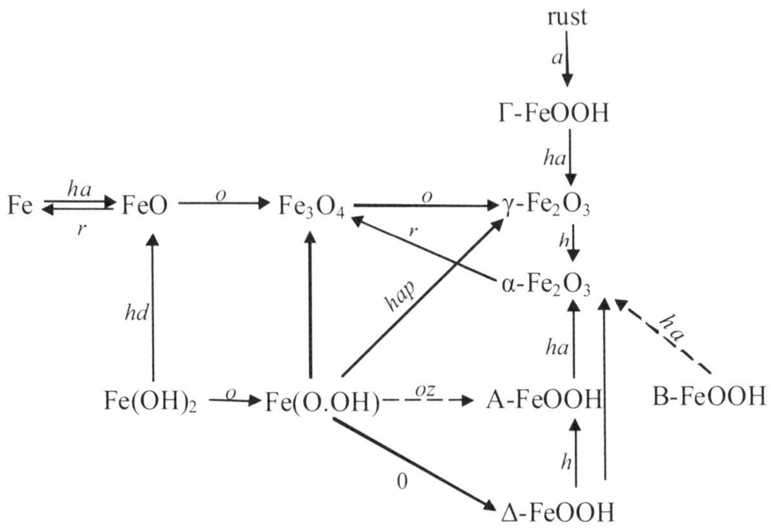

Fig. 2.1. Transformations between iron oxides and hydroxides, the solid arrows indicate topotactic transformation, the broken arrows non-topotactic transformations. Compounds in italics have the oxygen atoms in the cubic close-packed configuration, others in hexagonal close-packed configuration, with the exception of β-FeOOH which has a more complicated structure. Comments: *h* = heating, *a* – atmospheric oxidation, *z* – in alkali, *d* – in nitrogen or vacuum, *o* – after oxidation, *r* – after reduction, *p* – excess.

is based on the group SiO_4^{4-} containing four oxygen atoms in the tetrahedral positions around the silicate atoms. The Si–O distance is approximately 0.16 nm and the distance between the centres of the oxygen atoms is approximately 0.26 nm. There is a group of minerals referred to orthosilicates, whose lattice consists of simple SiO_4^{4-} ions and cations. The majority of silicates are, however, based on polymeric silicate groups of which $Si2O_7^{6-}$ is the simplest and which form rings, chains, strips and networks, some of which are shown in Fig.2.2.

In addition, there are three dimensional structures of different silicate forms. The tetrahedral group AlO_4^{3-} has almost the same size as the SiO_4^{4-} group and, consequently, aluminium may substitute silicon in the silicate anion structure thus enabling the existence of aluminosilicates.

The lattices of the silicates and aluminosilicates may be regarded as tetrahedrons for oxide ions containing silicon or aluminium ions in tetrahedral holes arranged in such a manner that they form polyhedrons containing holes of different co-ordination numbers and sizes. These contain cations with defined ion radii in a sufficient

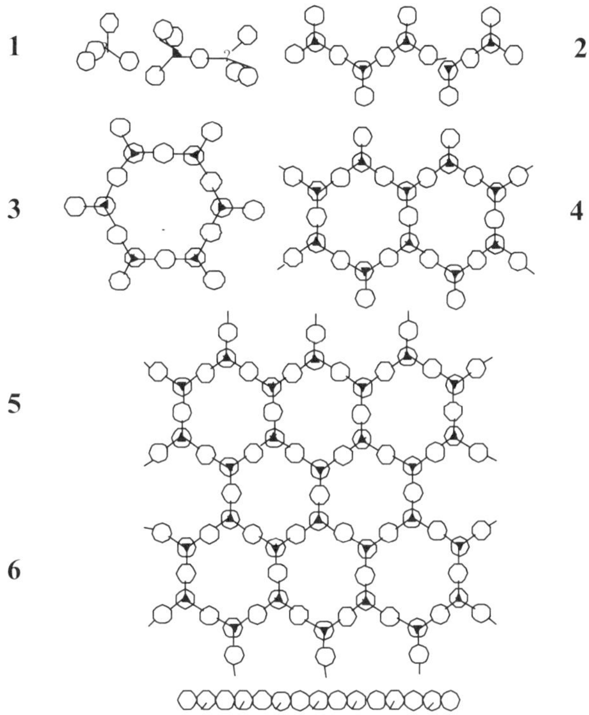

Fig. 2.2. Silicate structures, from top and left: 1) Orthosilicate ion $(SiO_4)^{4-}$; 2) Pyrosilicate ion $(Si_2O_7)^{6-}$; 3) Chain structure $(SiO_3)^{2n-}$; 4) Ring group structure $(Si_6O_{18})^{12-}$; 5) Banded structure based on groups $(Si_4O_{11})^{6-}$; 6) Section through the network structure.

number so that they neutralise the ion charge of the anion silicate structure. This means that if aluminium replaces part of silicon, a higher cation charge must be available. There is a large number of ions of similar sizes which may fill the octahedral holes in the silicate structure so that the isomorphous substitutions are very common in silicate minerals.

If aluminium replaces part of silicon in the three-dimensional network of the silicate structure and the composition $(Si,Al)O_2$ is produced, the substance has a negative charge and positive ions must be present in holes. Zeolites and quartz are suitable examples of these minerals. If the diffusion capacity of zeolite is sufficiently high for metal ions, zeolite can be used to produce these metals from solutions. On the other hand, in the production of metals present in the silicates, the silicate structure must be broken up in order to produce these metals efficiently. This also takes place in nature in erosion under certain conditions which may result, in a wet medium, in the transfer of a metal into a solution. This may be followed by repeated precipitation of the metal from the solution as a result of a change of pH or redox potential and by the formation of silicate intermediate products containing several metals. These may subsequently crystallise, usually with the formation of a poorly crystallised product with a wide range of composition because of the substitution of a metal to the 'ideal' composition of the mineral by other metals present with a suitable ion radius. This takes place in the laterization process resulting in the formation of, for example, nickel ores, processed by hydrometallurgical processes.

References

1. Vaughan, D.J., Craig J.R.: Mineral Chemistry of Metal Sulphides, Cambridge University Press, Cambridge, 1978.
2. Fasika, E.J.: *Corrosion Science*, 7, 1967, 833–839.

CHAPTER 3

PHASE EQUILIBRIUM OF COPPER AND IRON SULPHIDES

The metal sulphides are the most important group of ore minerals forming charge materials for the majority of processes for producing non-ferrous metals throughout the world. This is also the case in the production of copper by both pyrometallurgical or hydrometallurgical production method.

The metal sulphides in this case represent a group of natural crystal materials and their synthetic analogues which prevail in binary and ternary compounds of sulphur with copper and iron.

Although the sulphides of copper and copper and iron were known a long time ago as a source of copper, a more efficient systematic classification only appeared in the 19th century [1]. The rapid development and significant advances date back to the discovery and application of x-ray radiation for examining the structure of compounds [2] and this was followed immediately by examination and determination of the structures of the available sulphides of non-ferrous metals and iron, for example, Buerger [3-7].

The systematic study of the phase equilibrium of the sulphide phases in metallurgical laboratories greatly influenced the industrial production of copper and enabled introduction of advanced methods of production of copper including several types of continuous production and refining of copper and also hydrometallurgical methods of copper production. These results were also used for understanding the phase equilibrium of copper and iron sulphides.

3.1. One-component systems

3.1.1. Sulphur

The main components of the copper and iron sulphides are of course sulphur, copper and iron. Prior to analysing the phase equilibrium in the metal–sulphur systems, it would be useful to consider also the one-component system containing sulphur.

Sulphur cannot be removed from the processed raw materials of non-ferrous metals by dressing processes to avoid contamination of the environment (roasting, smelting). This is the reason why special attention should be paid to the behaviour of sulphur in hydrometallurgical methods of processing non-ferrous metals. In fact, elemental sulphur is a complex substance which has not as yet been completely understood and investigated. It is an element with the atomic number 16, the second element of group VI of the periodic table of elements so that it is a non-metal with the properties similar to those of oxygen and selenium. As an oxidation agent, the valency of sulphur is –2 and forms sulphides in combination with many other elements. The compounds of sulphur with the negative valency cannot of course act in chemical reactions as oxidation agents regardless of whether they have the ion or covalent bond and, on the contrary, they are more-or-less strong reduction agents.

At normal temperature pure sulphur is a light yellow solid substance without any smell. When the temperature is reduced, sulphur becomes progressively brighter and at the liquid air temperature it is almost white. On the market, sulphur is available in many different commercial forms with different physical and chemical properties and also purity.

The specific density of sulphur is approximately twice the density of water in which sulphur does not dissolve. It is slightly soluble in the majority of conventional solvents and more soluble in carbon disulphide. The reactivity of sulphur is very high. In suitable conditions, sulphur can directly react with the majority of other elements, with the exception of inert gases, iodine and molecular nitrogen.

Sulphur reacts not only with the majority of elements but also with a large number of inorganic and organic compounds. Of course, when sulphur forms a large number of allotropic modifications, it must also have different chemical behaviour and properties because of different energy content of these atoms. Unfortunately, only a small number of studies have been published in the case of sulphur to enable the determination of the chemical properties of various allotropic forms of elemental sulphur. At room temperature, only one modification of sulphur is thermodynamically stable: cyclic chain sulphur, S_8, but the number of the sulphur atoms in the molecule S_x can change from 1 to approximately 10^6 when only the temperature changes. This means that it may be possible to prepare at least one million forms of sulphur, or even more if they really exist, but the majority of these forms can be produced in only very complicated equilibrium systems, which depend on temperatures [8].

Solid sulphur has two allotropic modification: the first, intramolecular allotropy, considered for different molecular forms formed by the chemical bonding of the sulphur atoms, and the other one, intermolecular, for different structural arrangements of the molecules in the crystals. In comparison with other elements, sulphur has a very large number of polymorphs; more than thirty modifications are mentioned in the literature. As a chalcogen element with the closed external layer (the main state is represented by the form $3s^2 3p_x^2 3p_y^1 ep_z^2$, indicating that two non-paired electrons are placed in different *p*-orbitals; this explains why sulphur is bivalent: the sulphur atoms form only two covalent bonds with adjacent sulphur atoms with zero valency) sulphur can in fact form a large number of apriori possible molecular forms. It can be formed from linear molecules of any length which may close thus forming cyclic, ring-shaped molecules, or otherwise infinitely long chains. The large number of these hypothetical molecular groupings is arranged regardless of the stereochemical relationships, and others cannot be attributed to any other crystalline arrangement. In reality, all cyclic molecules described on the basis of crystallographic considerations without mutual penetration may form. However, only the linear chains with direct co-ordinates (which are linear in the crystallographic sense of the word) may be arranged in crystallographic bonds. Therefore, the molecular structures, with the exception of indirect chains, may form in accordance with any molecular type. The orientation in the space of the covalent bond depends on the orientation of the nearest subsequent bond of the chain, the free molecules are divided uniformly. In other words, the expected molecular formation is characterised by a low degree of symmetry. Of course, this has a strong effect on the possibility of identifying the individual allotropic forms of elemental sulphur.

It has already been mentioned that sulphur is present in many different forms. Some of them represent variations of the molecular configurations, others the variations of the crystallographic structures. In this case, only several forms of elemental sulphur are mentioned. They form or could form in leaching processes and, consequently, influence the course of hydrometallurgical reactions.

At the beginning, it should be stated that the terminology and description of various forms of elemental sulphur is not very efficient and to some extent confusing. Therefore, the terms and names of significant allotropic modifications of elemental sulphur are given prior to describing the individual allotropic forms of sulphur in Table 3.1.

3.1.1.1. Allotropic modifications of solid sulphur

Cyclooctasulphur, S_8

***Orthorhombic sulphur,* S_α:** The most significant form of sulphur is orthorhombic, S_α. It is the conventional form stable at room temperature and atmospheric pressure. This allotrope is described by many terms, such as rhombic sulphur, Muthmann sulphur I, α-sulphur and orthorhombic sulphur; this name is used most widely. The term α-sulphur is also used often because it is short. The term recommended by the IUPAC terminology is cyclooctasulphur and is used in most cases as a scientific and accurate term [9].

***Monoclinic sulphur,* S_β:** Crystallisation of a sulphur melt results in the formation of a monoclinic crystal form. Below the temperature of 95.4 °C the crystals transfer to the orthorhombic (α) form but rapidly cooled crystals may remain at room temperature in the monoclinic form for approximately one month. The structure is similar to orthorhombic sulphur S_α.

***Monoclinic sulphur,* S_γ:** The second form of monoclinic sulphur is referred to as Muthmann sulphur III, pearl-like sulphur, or γ-sulphur. Monoclinic prismatic crystals are formed by slow cooling of a sulphur melt heated to temperatures above 150 °C, or by cooling a hot solution of sulphur in alcohol, hydrocarbonates or carbon disulphide. Mineral rosickite can also be found. Its stability is still the subject of discussions. The melting point of S_γ is 106.8 °C and S_γ transforms to S_β and/or S_α.

In addition to these forms, a large number of other insufficiently identified allotropes of cyclooctasulphur has been described. Their identification is either incomplete or doubtful [11].

***Cyclohexasulphur,* S_6:** Thermodynamically unstable cyclohexasulphur may consist of pure crystals or be in a pure solution for a long period of time. However, in the presence of a small amount of impurities this form of sulphur breaks down very rapidly. It is sensitive to visible light and its chemical reactivity is alsoconsiderably higher than that of S_8. The cyclohexasulphur

Table 3.1. Names and synonyms of allotropic modification of elemental sulphur

Name	Synonym	Molecular type	Specification
α (alpha)	rhombic, orthorhombic, Muthmann I	cycloocta-S	orthorhombic - α
β (beta)	monoclinic I, prismatic, Muthmann II	cycloocta-S	monoclinic - β
γ (gamma)	monoclinic II, Gerne-Muthmann III, pearl-like, mother-of-pearl	cycloocta-S	monoclinic - γ
δ (delta)	monoclinic III, -monoclinic, Muthmann IV	cycloocta-S	allotrope S_8
ϵ (epsilon)	Engel, rhombohedral, Aten, monoclinic Engel	cyclohexa-S	rhombohedral
ζ (zeta)	5. monoclinic, Corinth	cycloocta-S	allotrope S_8
η (eta)	4. monoclinic, Corinth	cycloocta-S	allotrope S_8
θ(theta)	tetragonal, Corinth	cycloocta-S	allotrope S_8
ι (iota)	Erämetsä	cycloocta-S	allotrope S_8
κ (kappa)	Erämetsä	cycloocta-S	allotrope S_8
λ (lambda)		cycloocta-S	cycloocta S_8
μ (mi)	(a) insoluble (b) polymeric	polychain-S	solid/liquid polymer
ν (ny)	m	mixture	solid polymer
ξ (xi)	triclinic, Corinth	cycloocta-S	allotrope S_8
ο (omicron)	Erämetsä	cycloocta-S	allotrope S_8
π (pi)	(a) Aten, Erämetsä, (b) octa-S chain	ring mixture	frozen liquid
ρ (rho)	Aten, Engel	cyclohexa-S	cyclohexa-S_8
τ (tau)	Erämetsä	cyclohexa-S	allotrope S_8
ϕ (phi)	fibrous	mixture	fibrous
φ (ph)	fibrous, plastic	polychain-S	fibrous
χ (chi)	plastic	mixture	polymer
ψ (psi)	fibrous	mixture	fibrous

ω (omega)	insoluble, white, Das, supersublimation	mixture	polymer
m	triclinic	cycloocta-S	allotrope S8, solid, polymer
n	μ		
Athen	see ϵ, ρ	cyclohexa-S	rhombohedral
Braun	see μ	mixture	solid, polymer
Engel	see ϵ, ρ	cyclohexa-S	rhombohedral
Korinth	see ξ, η, θ, ζ	cycloocta-S	solid, polymer
Muthmann	see α, β, γ, δ	cycloocta-S	high pressure form
Schmidt	see orthrombic-S_{12}	cyclododeca-S	
amorphous	ω, μ	mixture	fibrous
cubic	high pressure cubic plasticϕ, φ, phase II		insoluble
fibrous	'Crystex', supersublimated	mixture	laminar
insoluble	phase I, white, ω, μ, χ	polychain-S	high pressure form
laminar	high pressure metallic	?	photosulphur
metallic	insoluble	?	cooled liquid
photosulphur	(a) Skjerven (b) Rice, Schenk	?	separated vapours
black	Malcev	mixture	separated vapours
brown	Rice	mixture	separated vapours
green	Erämetsä	mixture	separated vapours
orange	Rice	mixture	separated vapours
purple	(a) Rice (b) Erämetsä	mixture	separated vapours
red	Rice	mixture	separated vapours
violet	Erämetsä red	mixture	allotrope S_8
E,F,G,I,K,L,M	orange	mixture	

molecules are efficiently arranged. The crystals have an unit cell with 18 atoms and a specific density of 2.21 g·m^{-3}. It is the highest specific density of all known modifications, including the thermodynamically stable orthorhombic form S_8 [12].

Cycloheptasulphur, γ-S_7, and δ-S_7: S_7 may crystallise in the form of very long needles. Their bright yellow colour, in contrast to S_8, does not disappear by cooling to the temperature of liquid air. S_7 melts reversibly at 39 °C, undergoes polymer transformation at 45 °C, shows again a low viscosity at approximately 115 °C and polymerises again at 159 °C. Visible light results in rapid transformation to S_8 by polymerisation. At low temperatures this form of sulphur is stable for several weeks. The experimental specific density is 2.09 g·m^{-3} in comparison with 2.144 g·m^{-3} calculated for the 16 molecule unit cell of S_7 [13, 14].

The results of a large number of investigations of solid and dissolved S_7 show that S_7 crystallises in four different allotropes (α, β, γ, δ-S_7) whose thermodynamic stability is still unknown [14]. The structure of the γ and δ forms is known.

Cyclododecasulphur, S_{12}: This allotrope is the subject of special attention not only because of its surprisingly high stability. Bright yellow needles show the highest melting point (with breakdown) of all known sulphur modifications, 148 °C. Its solubility in conventional solvents is unexpectedly low. The reactivity of cyclododeca sulphur is between S_8 and S_6 but more on the side of S_8 [14].

Allotropes of polymeric sulphur: If the viscous sulphur melt (T > 160 °C) is rapidly cooled, this produces a plastic material with two easily detectable phases. All the allotropes are formed by sulphur chains. The chain molecule forms long helices. Three terms of a helix contain 10 atoms. Two values of the binding characteristics are very similar to those for S_{20} and S_{12} and are between the data for S_8 and S_6.

The solid polymeric chain sulphur exists in many forms [15–19], for example, rubber sulphur, plastic sulphur (χ), laminar sulphur, fibrous (Y, F), η, μ, and so on. All these forms are metastable mixtures of the allotropes containing more or less known defined numbers of helices, cyclooctasulphur and other forms which depend on the genesis of their formation. Their composition changes with time. If impurities are present, they slowly change to α-S_8 during a period of up to one month.

Liquid sulphur: The melting points of different crystalline allotropes, often accompanied by dissociation, are in Table 3.2.

Table 3.2. The melting points of sulphur allotropes [10]

Allotrope	Melting point (°C)	Allotrope	Melting point (°C)
S_6	~90	S_{10}	80–105
S_7	39	S_6–S_{10}	92
α-S_8	115.1	S_{11}	74
β-S_8	120.1	S_{12}	148
β-S_8	119.6	S_{13}	114
γ-S_8	108.6	α-S_{18}	126–128
S_9	> 50	S_{20}	121

The physical properties of liquid sulphur are very unusual. At a transformation temperature of 159 °C liquid sulphur unexpectedly changes to a highly viscous material which does not flow [20]. This exceptional 2000-fold increase of viscosity is accompanied by the change of colour from light grey to dark red. Almost all physical properties (specific heat, specific density, electrical conductivity, etc.) show a discontinuity at this transformation temperature. This unusual behaviour of sulphur is caused by polymerisation. Polymerisation is represented by two reaction steps; the initial reaction

$$S_8 \text{ ring} \rightarrow S_8 \text{ chain and } [\text{ring}]/[\text{chain}] = K_1$$

and the advancing reaction

$$(S_8 \text{ chain})_n + S_8 \text{ ring} \rightarrow (S_8)_{n+1} = K_3.$$

If K_1 and K_3 are known for two temperatures, their enthalpy and entropy are obtained from the van-Hoff equation.

This hypothesis describes accurately many properties of liquid sulphur below and above the viscosity maximum assuming at that the melting point the liquid consists of three units S_8 which at 159 °C polymerise to an average chain approximately 10^5 units of S_8 long. Continuous depolymerisation is expected to take place at higher temperatures. On the other hand, many properties show that liquid sulphur is in fact a far more complicated system.

In addition to the polymers and rings S_8, the presence of S_3, S_4,

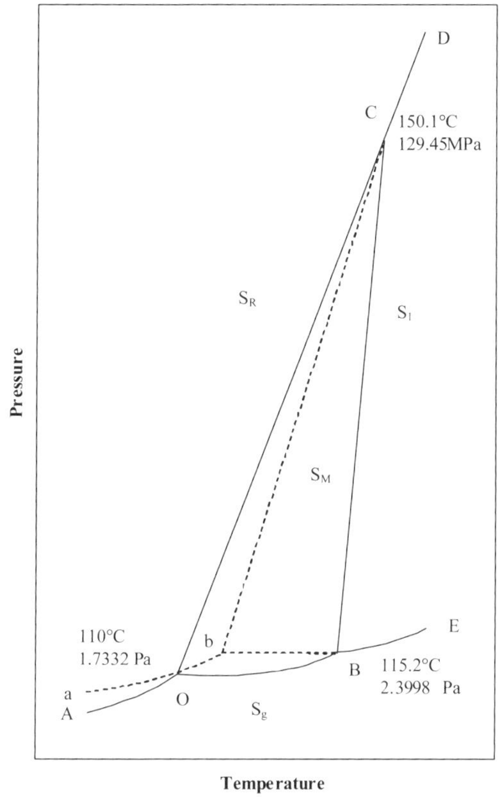

Fig. 3.1. Phase diagram of elemental sulphur.

S_5 and S_6 in liquid sulphur was also confirmed [21]. The conventional boiling point of sulphur is, according to IUPAC, a temperature of 444 6 °C.

3.1.1.2. Phase diagram

Figure 3.1 shows the phase diagram illustrating qualitative relationships between the orthorhombic (S_R), monoclinic (S_M), liquid (S_l) and gaseous sulphur (S_g) as a function of pressure and temperature [22, 23].

The area denoted by S_R is the region of stability of orthorhombic sulphur and S_M is the region of stability of monoclinic sulphur. The liquid sulphur exists in region S_l and sulphur vapours in S_g.

The line *AO* is the curve of the pressure of vapours of orthorhombic sulphur. If temperature increases quite rapidly, the curve of the pressure of orthorhombic sulphur vapours can be extended to the region of monoclinic sulphur in accordance with the dashed line *Ob*. This is possible because the transformation of orthorhombic to monoclinic sulphur is very long.

The *OC* is the curve of the crystallisation transformation from the orthorhombic to monoclinic form. The line *BC* is the curve of the melting point of monoclinic sulphur. However, if orthorhombic sulphur is rapidly heated, it may melt without the formation of monoclinic sulphur. In these conditions, the *bC* dashed line is the curve of melting of orthorhombic sulphur. It should be mentioned that orthorhombic sulphur melts at a lower temperature than monoclinic sulphur.

On the diagram, the line *bC* continues to point *D*. Section *CD* is the curve of the melting point of rhombic sulphur in the high pressure range in which monoclinic sulphur is not stable at any temperature.

The *OB* curve is the curve of the pressure of monoclinic sulphur vapours. In rapid cooling, monoclinic sulphur may exist temporarily in the region of orthorhombic sulphur. The dashed line *aO* being the continuation of the line *OB*, indicates the metastable pressure of the vapours of monoclinic sulphur.

The line *BE* is the curve of the melting point of liquid sulphur. Its continuation downwards to the point *b*, line *bB*, presents the curve of the boiling point of liquid orthorhombic sulphur which is a strictly metastable form.

Point *O* is the triple equilibrium point of the orthorhombic, monoclinic and gaseous sulphur. Point *C* is the triple equilibrium point of the orthorhombic, monoclinic and liquid sulphur. Point *b* is the triple 'equilibrium' point of orthorhombic, liquid and gaseous sulphur, although it is not the equilibrium point in the conventional sense of the word because it represents metastable conditions.

The diagram show that monoclinic sulphur is stable only under certain conditions. At a low temperature it changes to orthorhombic and melts at a very high temperature. At too low a pressure monoclinic sulphur evaporates and at too high a pressure it changes to orthorhombic sulphur.

3.2. Multi-component systems

3.2.1. The copper–sulphur equilibrium phase system

The copper–sulphur equilibrium phase system is relatively complicated. Before 1940, only two binary sulphide minerals of copper were known, chalcocite, Cu_2S, and covellite, CuS. Although it was known that chalcocite forms solid solutions up to the composition $Cu_{1.8}S$, the mineral digenite of the same composition was only described in 1942. Later, a further three minerals were identified: djurleite $Cu_{1.97}S$, anilite Cu_7S_5 and 'blue' covellite, $Cu_{1+x}S$. Only recently, this group of sulphides were supplemented by non-stoichiometric sulphides geerite, $Cu_{1.6}S$, spionokopite, $Cu_{39}S_{28}$, yarrowite, Cu_9S_8. In addition, some of these sulphides form polymorphous minerals. In any case, research in this area has not been completed.

The copper–sulphur equilibrium diagram in the temperature-composition co-ordinates is shown in Figs. 3.2 and 3.3. Figure 3.2 shows the temperature scale up to the region of existence of melts, Figure 3.3 shows the detail of the low temperature part of the diagram in a narrow range of the composition containing non-stoichiometric sulphides interesting from the viewpoint of direct leaching of copper sulphides. Table 3.3 gives the currently known minerals with a Cu_xS.

The following phases may exist in the system:

Chalcocite, Cu_2S: The approximate composition of chalcocite is Cu_2S. At 103.5±1.5 °C α-Cu_2S changes to the high temperature form β-Cu_2S with the sulphur atoms in close hexagonal arrangement. β-chalcocite exists in the composition range from Cu_2S up to approximately $Cu_{1.988}S$ at 105 °C but the size of the range decreases with increasing temperature. At 180 °C there are no longer any deviations from the composition Cu_2S. β-chalcocite cannot be cooled down to room temperature (~25 °C) and has not been found as a mineral. It is stable up to approximately 435 °C where it changes to the high-temperature digenite with a composition Cu_2S.

Djurleite, $Cu_{1.96}S$: Djurle [37] published a composition of djurleite $Cu_{1.96}S$ and described three forms: the low-symmetry form referred to as djurleite, the tetragonal form which is probably unstable, and the high-temperature form. Roseboom [24] found djurleite to be stable only below 932 °C and observed the tetragonal form only as metastable. Between 93 and 350 °C the phase with the composition $Cu_{1.96}S$ produced only a mixture of digenite and hexagonal chalcocite.

At present, the actual composition of djurleite is approximately $Cu_{1.93}S$ [38]. The high temperature tetragonal phase with composition $Cu_{1.96}S$ is also stable in the range between $Cu_{1.95}S$ to Cu_2S in the temperature 90–140 °C but the transition to the tetragonal phase is extremely slow.

Table 3.3. Minerals and phase in the Cu–S system

Mineralogical term	Composition	Thermal stability [°C] min	Thermal stability [°C] max	Year	Source
Chalcocite	Cu_2S	–	103	1966	Roseboom [24]
–	Cu_2S	103	~435	1971	Evans [25]
–	Cu_2S	~435	1129	1944 Cu_2S	Buerger [7] Roseboom [24]
–	Cu_2S	–	500	1963	Morimoto, Kullerud [26]
Djurleite	$Cu_{1.97}S$	–	93	1964	Morimoto [27]
Digenite	Cu_9S_5	–	83	1962 1966	Roseboom [24] Morimoto, Kullerud [26]
–	$Cu_{9+x}S$	83	1129	1962 1966	Roseboom [24] Morimoto, Kullerud [26]
Anilite	Cu_7S_5	–	70	1963	Morimoto, Koto [29]
Geerite	$Cu_{1.6}S$			1980	Goble, Robinson [30]
Spionokopite	$Cu_{39}S_{28}$			1980	Goble [31]
Yarrowite	Cu_9S_8			1980	Goble [31]
Blue covellite	$Cu_{1+x}S$	–	157	1964	Moh [32]
Covelitte	CuS	–	507	1954 1972	Berry [33] Rickard [34]
–	CuS_2	–	550	1966	Munson [35] Taylor, Kullerud [36]

Digenite, Cu_9S_5: The composition of digenite at room temperature is in the range $Cu_{1.765}S$ to $Cu_{1.79}S$. The edge, with a low copper content, is not affected by temperature whereas the copper-rich edge changes at 83 °C to $Cu_{1.83}S$. At this temperature it also changes to high-temperature digenite. The digenite with composition $Cu_{1.83}S$ is unstable below approximately 50 °C. Digenite forms three polymorphs: high-temperature, unstable and low-temperature digenite [26].

The digenite whose composition $Cu_{9-x}S_5$ ($x > 0.05$) crystallises in the cubic structure and the lattice parameters as well as the Cu:S

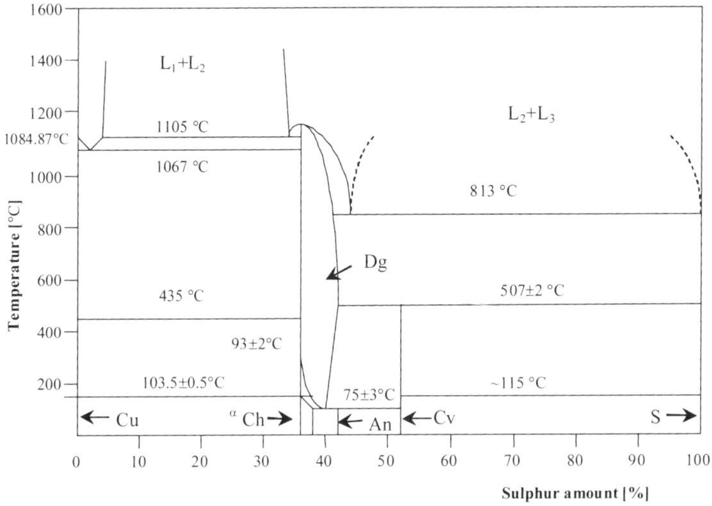

Fig. 3.2. Cu–S equilibrium phase system.

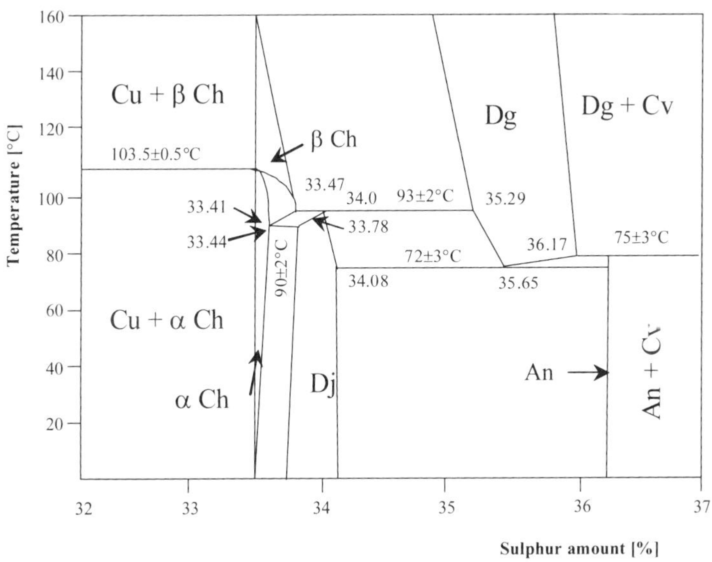

Fig. 3.3. Low temperature section of the Cu–S equilibrium phase system. *Cu – copper Cu; Ch – chalcocite* Cu_2S; *Dj – djurleite* $Cu_{1.93}S$; *Dg – digenite* $Cu_{1.8}S$; *An – anilite* Cu_7S_4; *Cv – covellite CuS.*

ratio depend on the genesis of formation of the mineral. However, a relatively wide range of non-stoichiometry for the mineral referred to as digenite is considered.

Crystal structures of metastable and low-temperature digenite are similar although the metastable form is more complex. The structure is derived from the basic rhombohedral sublattice. It contains one entity Cu_9S_5. The sulphur atoms occupy the nodes of the cubic phase-centred lattice and every tetrahedron contains 9/10 of the copper atom which is statistically distributed in 24 equivalent positions. If one layer of sulphur between two layers of copper, CuSCu, is regarded as a cell, the structure may be described as layered. Since copper is statistically distributed between the layers, some of the copper atoms are characterised by weaker bonds than others and may be removed with a smaller energy required for this purpose.

Anilite, Cu_7S_5: Later, in 1969, Morimoto [28] described another mineral with a composition Cu_7S_5. This mineral is thermally stable up to approximately 70 °C and breaks down into digenite and covellite above this temperature. Like djurleite, anilite is similar to digenite and it is quite difficult to identify by x-ray diffraction analysis.

Geerite, $Cu_{1.6}S$: Goble and Robinson described a new mineral with a composition $Cu_{1.6}S$ [30] with an efficiently crystallised pseudo-cubic substructure, similar to the structure of sphalerite with four main particles in the unit cell. Later, the structure of geerite was described as rhombohedral with a structure similar to digenite [39]. The temperature range of stability has not been specified but in addition to occurrence in the nature, this mineral is also identified as a product of reaction in leaching of anilite [40].

Spionokopite, $Cu_{39}S_{28}$: Frenzel [41], found that the blue covellite, formed by leaching of chalcocite, may have an excess of copper up to the composition $Cu_{1.4}S$. However, at a later stage Goble [31] defined new minerals, spionokopite and yarrowite.

Spionokopite, found in nature as a mineral, has the composition in the range between $Cu_{1.32}S$ to $Cu_{1.53}S$. Its idealised composition is given as $Cu_{39}S_{28}$ with the hexagonal, close-packed structure and covalent bonding of the sulphur atoms with 14 structural particles in the elementary cell. Its stability in relation to temperature has not been described.

Yarrowite, Cu_9S_8: Like spionokopite, yarrowite [31] was found as a natural mineral. Its composition changes in the range from $Cu_{1.12-1.32}S$ to the idealised composition Cu_9S_8. The structure was later clarified as hexagonal with 24 constructional particles in an unit cell. As in the previous case, its temperature stability has not been determined.

'Blue' covellite, $Cu_{1.05-1.1}S$: This phase is well known in mineralogy and according to Moh [32] is best represented by the composition $Cu_{1+x}S$, where x is in the range between 0.05 and 0.10. According to Moh, this mineral should be stable up to 157 °C and changes to a mixture of normal covellite and digenite above this temperature.

Covellite, CuS: The composition of covellite is CuS and does not show any deviation from the composition and is stable up to a temperature of 507 °C at which it changes to high-temperature digenite and a liquid phase rich in sulphur. Covellite, CuS, is one of the minerals whose simple composition does not indicate any unexpectedly complex structure. The structure contains one type of copper in the tetrahedral co-ordination. Each tetrahedral covers the corners forming a continuous layer. The second type of copper is placed in the trigonal interstitials along the ridge formed by two continuous layers, thus forming the planar layer CuS. The resultant structure consists of the plane of the triangle CuS_3, inserted between the double layer of the tetrahedral CuS_4. The sulphur-sulphur bond links these layers.

The above considerations confirm the complex nature of the problem and some uncertainties regarding the structure of copper sulphides. Nevertheless, the currently available data indicate that in the structure of sulphides with a high copper content, copper is bonded by two methods, in the bond with sulphur which results in the final CuS and the statistical distribution in the structure. As indicated by the binding relationships and distances, part of the copper atoms is not in such a strong bond as others. This also predicts the mechanism of leaching of chalcocite in two stages.

CuS_2: Munson [35] published the results of synthesis of copper disulphide with a pyrite-type structure. Later it was clarified that this symmetry may be only pseudo-cubic.

3.2.2. The iron–sulphur equilibrium phase system

The phase equilibrium in the Fe–S binary system has been known quite well at temperatures above 300 °C, although at lower temperatures there are still a number of uncertainties, especially in the regions of stability of pyrrhotite [80]. It may appear at first sight that the examination of high-temperature phases for hydrometallurgy is irrelevant. However, this is not so. In many cases, leaching is preceded by the so-called sulphation roasting by means of which of the low solubility components change to soluble.

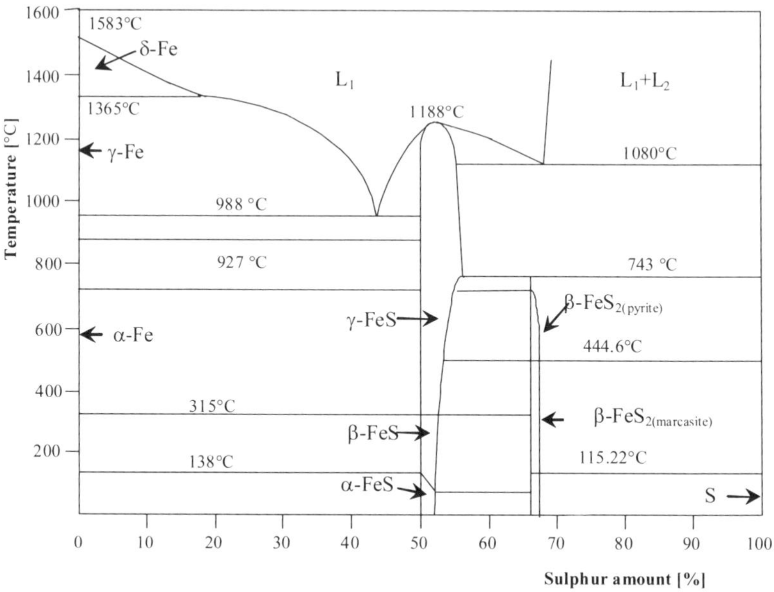

Fig. 3.4. Diagram of the iron–sulphur equilibrium phase system [41].

Of course, understanding these components is a pre-requisite for designing an effective process.

At present, the Fe–S phase systems is represented by the phase diagram shown in Fig. 3.4 [42]. The FeS–FeS_2 boundary composition is a very important region for the primary description of the stoichiometry and phase stability for the system in the solid state. The solid phases of the Fe–S system and their properties are summarised in Table 3.4.

The system contains the following phases:

Troilite, FeS: Strictly speaking, the term troilite may be applied only to the polymorphs of stoichiometric composition FeS, which are stable at 140 °C. Above 140 °C FeS has the structure of the high-temperature hexagonal pyrrhotite of the type NiAs (1C) whose composition range is greatly widened. Troilite appears only seldom in nature, and usually accompanies hexagonal pyrrhotite.

Mackinawite, FeS_{1-x}: This mineral is always found in nature either with troilite or low-temperature pyrrhotite. Mackinawite always contains small quantities of Co and Ni, although these elements are not important for the precipitation of this mineral from the solution in the temperature range 20–95 °C [53]. The ratio of metal to sulphur in mackinawite is slightly higher than 1 (1.04–1.07) [44] and, therefore, its chemical formula is FeS_{1-x}. However, on the

Table 3.4. Minerals and phases of the Fe-S system

Mineralogical term	Composition	Thermal stability [°C]		Year	Source
		min.	max.		
Troilite	FeS	–	140	1969	Yund, Hall [43]
Mackinawite	FeS_{1-x}	?	?	1970	Takeno et. al. [44]
Hexagonal pyrrhotite	$Fe_{1-x}S$	100	1190	1972	Kissin, Scott [45]
Pyrrhotite type MC	$Fe_{1-x}S$	262	308	1972	Kissin, Scott [45]
Pyrrhotite type NA	$Fe_{1-x}S$	~266	209	1972	Kissin, Scott [45]
Pyrrhotite type NC	$Fe_{1-x}S$	~100	~213	1972	Kissin, Scott [45]
Pyrrhotite type 5C	Fe_9S_{10}		~100	1971	Nakazawa, Morimoto [46]
Pyrrhotite type 11C	$Fe_{10}S_{11}$		~100	1971	Nakazawa, Morimoto [46]
Pyrrhotite type 6C	$Fe_{11}S_{12}$		~100	1971	Nakazawa, Morimoto [46]
Metastable pyrrhotite	$Fe_{1-x}S$		metastab.	1971	Nakazawa, Morimoto [46]
Pyrrhotite type 4C	$Fe_{7-x}S_8$		254	1972	Kissin, Scott [45]
Anomalous pyrrhotite	$Fe_{7-x}S_8$		?	1966	Clark [47]
γ-iron sulphide	Fe_2S_3		?	1973	Yamaguchi, Wada [48]
Mnythite	Fe_9S_{11}		~75	1972	Taylor, Willliams [49]
Greigite	Fe_3S_4		metastab.	1974	Scott [50]
Pyrite	FeS_2		743	1959	Kullerud, Yoder [51]
Marcasite	FeS_2		metastab.	1973	Rising [52]

other hand, Taylor and Finger [54] found in many cases a shortage of sulphur in this structure rather than an excess of the metal and, consequently, the formula $Fe_{1+x}S$ is more likely to be accurate. There are only a few data on the thermal stability of mackinawite but Zoka, et. al. [55] determined, for mackinawite from different deposits, the temperature of transformation to pyrrhotite as 120–153 °C.

Although the amount of information published on this mineral is relatively large, there are a number of uncertainties, especially for the phases coexisting at low temperatures when the very slow reaction kinetics leads to the formation of a large number of phases with ambiguous identification and description. The phase relationships have been explained more efficiently by, in particular, Nakazawa and Morimoto [46] Kissin and Scott [45] and Scott and Kissin [56].

Between the maximum melting point of 1188 °C and 315 °C there is an entire range of the pyrrhotite phase represented by the solid solution $Fe_{1-x}S$ in which iron and vacancies are randomly distributed in the cation sublattice of the structure NiAs (1C). The existence of this phase also spreads to low temperatures but with a limited non-stoichiometry range. Thc non-ordered pyrrhotite 1C cannot be retained even after rapid cooling and the crystals, produced by cooling this phase acquire one or several from a large number of superstructures. This transition, referred to as β-transition represents a discontinuity of the magnetic susceptibility in transition from the anti-ferromagnetic to paramagnetic state.

Nakazawa and Morimoto [46] found that if one of the ordered vacancies propagates through the pyrrhotite region with a decrease of temperature, the number of various of superstructures increases. The pyrrhotite phases 5C, 11C and 6C, found at room temperature, are stoichiometric phases with a composition $Fe_{n-1}S_n$ (n = 10, 11, 12) and are determined by the two-phase region [57]. In the upper limit of stability of the three stoichiometric types which is, however, unknown but is probably below 100 °C and 315 °C there are a further three phases with a substructure NiAs whose symmetry is not known. The end type, 'metastable' pyrrhotite, is produced if the pyrrhotite rich in iron is rapidly cooled. This results in the formation of the troilite–pyrrhotite two-phase range.

Monoclinic pyrrhotite, Fe_7S_8: The ferromagnetic pyrrhotite with the monoclinic superstructure and the composition close to Fe_7S_8 has been known for some time as a natural and also synthetic mineral. The equilibrium between the changes of the composition pyrite FeS_2, and pyrrhotite FeS in relation to temperature was studied by Arnold [58]. Toulmin and Barton [59] confirmed the form of the solidus line for pyrrhotite and obtained more accurate diffraction data for the pyrrhotite measured by Arnold. Yund and Hall [43] compiled the available diffraction data and proposed a relationship for calculating the amount of hexagonal pyrrhotite in relation to the atomic percent of iron. Monoclinic pyrrhotite has, however, a different composition, with the nominal composition being Fe_7S_8.

'Anomalous' pyrrhotite, Fe_7S_8: Clark [47] described a pyrrhotite containing 46.6% Fe with anomalous diffractions in the diffraction pattern in comparison with normal monoclinic pyrrhotite. This 'anomalous' pyrrhotite is found mostly in low-temperature sedimentary deposits. In contrast to the normal monoclinic pyrrhotite, which is ferromagnetic, this pyrrhotite is anti-ferromagnetic as hexagonal pyrrhotite. Its stability range has not been determined, although Taylor [60] has shown that the only method of formation of this mineral is the oxidation of hexagonal pyrrhotite.

γ-iron sulphide, Fe_2S_3: This phase has not been found in nature but was prepared by precipitation from an aqueous sulphide solution at 60 °C by Yamaguchi and Wade [48]. The spinel structure of this mineral was determined by electron diffraction and is similar to greigite, Fe_3S_4. However, like greigite, this mineral is magnetic. No information is available on its thermal stability.

Smythite, Fe_9S_{11}: Smythite was generally described by Erd, et. al. [61] as having a rhombohedral structure and composition Fe_3S_4. It was therefore concluded that it is a polymorph of greigite. After extensive study of natural smythite by Taylor and Williams [49], it structure was determined as pseudo-rhombohedral, similar to monoclinic pyrrhotite with a composition $(Fe,Ni)_9S_{11}$. Smythite is not a polymorph of greigite but it may be a pyrrhotite with a different arrangement of the group $Fe_{n-1}S_n$.

Only a small amount of information is available on the stability of smythite. It is present in the laminar form in monoclinic pyrrhotite and geodets which contain liquid inclusions at temperatures of 25–40 °C indicating that it is a phase stable only at low temperatures. Smythite has not been found in its hydrothermal recrystallisation above a temperature of 115 °C.

Greigite, Fe_3S_4: The thiospinel greigite is another problematic phase. It is found in low-temperature conditions. According to Berner [53], greigite is metastable in relation to FeS and FeS_2 at 25 °C, and according to Scott and Kissin [56] at temperatures between 115 and 350 °C.

Pyrite, FeS_2: Pyrite is stable up to a temperature of 743 °C at which it transforms by peritectic transformation to hexagonal pyrrhotite 1C + sulphur [51]. The relationship between pyrite and its polymorph marcasite is relatively unclear, despite extensive research. Buerger [3] found on the basis of high sensitivity chemical analysis that the composition of the minerals slightly differs; marcasite was characterised by a small deficit of sulphur (i.e. FeS_{2-x}) in comparison with pyrite which was close to the stoichiometry FeS_2. Kullerud found [62] that marcasite may be transformed to pyrite at temperatures below 150 °C in the presence of a surplus of sulphur, and not below 400 °C in the absence of sulphur. This indicates that for the Fe:S ratio with marcasite a sulphur excess is essential for the formation of pyrite. Kullerud [62] also described experiments with synthetic preparation of a mixture of marcasite and pyrite up to a temperature of 432 °C in the presence of water but not in its absence. This indicates that the H–S bond may stabilise marcasite. Later, it was found that the reverse rate of transformation of marcasite to pyrite at temperatures above 157 °C is directly proportional to temperature and indirectly proportional to the grain size. Marcasite was metastable in relation to pyrite and pyrrhotite in this temperature range. At temperatures below 157 °C marcasite is also metastable. Experiments have not confirmed the presence of marcasite at a temperature of 115 °C and higher temperatures.

3.2.3. The copper–iron–sulphur equilibrium phase system

Although special attention is paid to the examination of the copper-iron–sulphur equilibrium phase system by experts in mineralogy, pyrometallurgy and hydrometallurgy, the regions of stability of the individual possible phases and the coexistence conditions of these phases are not yet completely clear. Initial studies of this phase system were carried our by Mervin and Lombard [63]. The system was gradually studied by Schlegel and Schüler [64], Hiller and Probsthain [65], Yund and Kullerud [66], Kullerud, et. al. [67], Mukaiyama and Izawa [68], Cabri [69] and Barton [70]. The minerals and phases present in this system are summarised in Table 3.5.

The best known three-component sulphides include those present in the Cu–Fe–S system. Despite extensive research, there are many relationships, especially those relating to low temperatures, have not been completely explained. This is because the system contains many phases, solid solutions with a wide solubility range, high temperature phases and metastable phases with a long life. Figure 3.5 shows the occurrence of the phases in the Cu–Fe–S system. The abbreviations in Fig.3.5 are explained in Table 3.6.

The areas of stability of the individual phases of the system at

Table 3.5. Minerals and phases in the Cu–Fe–S system

Mineralogical term	Composition	Thermal stability [°C] min.	Thermal stability [°C] max.	Year	Source
Bornite	Cu_5FeS_4	–	228	1966	Morimoto,Kullerud [71]
–	Cu_5FeS_4			1966	Morimoto,Kullerud [71]
Bornite	Cu_5FeS_4	228	~1100	1966	Morimoto,Kullerud [71]
x-bornite	$Cu_5FeS_{4.05}$	–	125	1966	Yund, Kullerud [66]
Idaite (nukundamite)	$Cu_{5.5}FeS_{6.6}$	–	501	1966	Yund, Kullerud [66]
Fukuchilite	Cu_3FeS_8	–	~200	1969 1970	Kajiwara [78] Shimazaki, Clark [72]
Chalcopyrite	$CuFeS_2$	–	557	1972	Cabri, Hall [73]
Cubanite	$CuFe_2S_3$	–	200±210	1947	Buerger [5]
Talnakhite	$Cu_9Fe_8S_{16}$	–	~186	1973	Cabri [69]
Transition phase I	$Cu_9Fe_8S_{16}$	186	230	1974	Scott [50]
Transition phase II	$Cu_9Fe_8S_{16}$	230	520	1974	Scott [50]
Mooihoekite	$Cu_9Fe_9S_{16}$	~	167	1972	Cabri, Hall [73]
Transition phase A	$Cu_9Fe_9S_{16}$	167	236	1974	Scott [50]
Haycockite	$Cu_4Fe_5S_8$	–	?	1972	Cabri, Hall [73]
Cubic phase (pc)	wide range	20	200	1973	Cabri [69]
–	$Cu_{0.12}Fe_{0.94}S$	–	?	1970	Clark [74]

a higher temperature are quite well known. An isothermal section through part of the phase diagram in the Cu–Fe–S co-ordinates is shown in Fig.3.5. It is characterised by three extensive large solutions: 1) chalcocite–digenite–bornite (cc-dg); 2) transition solid solution (*iss*); 3) pyrrhotite (po).

The phase *iss* placed in the centre has a cubic fcc structure, the sphalerite type, includes a wide range of composition with a small shortage of sulphur. Cabri [69] states that the region of the phase

Table 3.6. Clarification and abbreviations used for the Cu–Fe–S system

Abbreviation	Mineral	Abbreviation	Mineral
cc	chalcocite	hc	haycockite
cdj	djurleite	cp	chalcopyrite
di	digenite	cb	cubanite
al	anilite	py	pyrite
cv	covelitte	mc	marcasite
bcv	blue covelitte	gr	greigite
bn	bornite	sm	smythite
x-bn	x-bornite	mpo	monoclinic pyrrhotite
a-bn	anomalous bornite	hpo	hexagonal pyrrhotine
id-I	idaite	tr	troilite
id-II	idaite	mk	mackinawite
tal	talnakhite	???	Cu-mackinawite
mh	mooiheokite	fk	fukuchilite

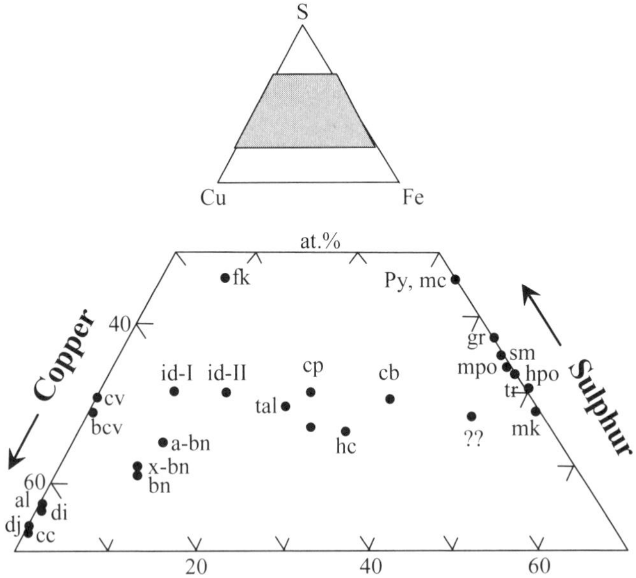

Fig.3.5. Minerals present in the Cu–Fe–S system.

iss may be divided into three zones in which each zone is characterised by different behaviour during rapid cooling. The first zone includes the section *iss* enriched in sulphur from the ratio Cu:Fe = 1 up to the part extremely enriched with iron. After rapid cooling from 600 °C, this part produces a mixture of chalcopyrite and *iss*. The second zone is depleted in sulphur and is located at the end of the iron-rich region. Rapid cooling results in the formation of a phase with a primitive cubic structure. The third zone, which separates the first and second zones and includes all central parts *iss* enriched with copper, produces on cooling a primitive cooling phase in a mixture with chalcopyrite or mooihoekite. These complications indicate the problems with the system at low temperatures. The solid solution, containing bornite, extends to the Cu–S bond and this continues in a solid solution 'high temperature chalcocite–high temperature digenite' up to the composition of the solid solution richer in iron and sulphur than stoichiometric bornite, Cu_5FeS_4. The pyrrhotite solution may contain in its structure copper up to 4.5 wt.%.

When the temperature is reduced below 600 °C, the simple, relatively well defined equilibrium states transform gradually to less defined and, in some regions, only estimated low stability phases. The most important process is the formation of chalcopyrite, a phase stable below 557 °C. It forms in the pyrite region *iss* and remains isolated from other Cu–Fe sulphides even with a gradually decreasing temperature. Other phases, gradually appearing in the system, include covellite, CuS, at 507 °C, and idaite, $Cu_{5+x}FeS_{6+x}$, at 501 °C. The phase relationships formed below 400 °C are shown in Fig.3.6 and with a further decrease of temperature they probably transform into the regions shown in Fig.3.7

Although the appearance of the diagram of the Cu–Fe–S ternary system does not change greatly with a further decrease of temperature, the relationships between these phases, similar to chalcopyrite, talnakhite, mooihoekite and haycockite are not well known. Special attention is given to examination of these phases but there are still a number of uncertainties regarding the low temperature phases of the Cu–Fe–S system and their polymorphs. As a result of compiling the data obtained by different authors, the isotherm was sectioned through the Cu–Fe–S diagram was constructed at room temperature, 25 °C, which is shown in Fig. 3.7.

The diagram shows the regions of stability of chalcopyrite and the evident mixture of chalcopyrite with pyrite up to the completely speculative mixture of bornite and mooihoekite. The obvious phases,

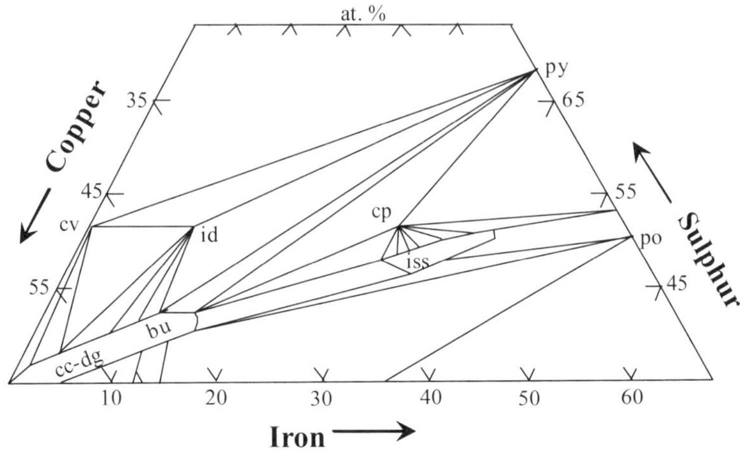

Fig. 3.6. Central part of the diagram of the Cu–Fe–S equilibrium phase system at 300 °C.

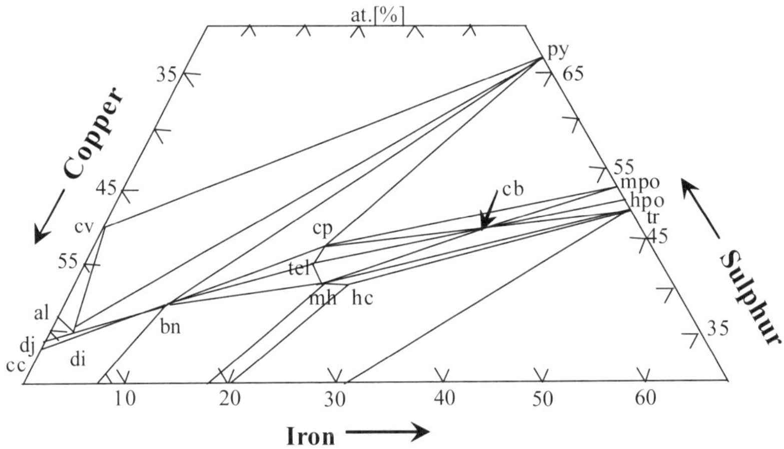

Fig. 3.7. Possible phase composition in the central part of the Cu–Fe–S phase system at 25 °C.

mooihoekite and haycockite are quite common, but they are generally overlooked because of their great similarity with chalcopyrite. Blue covellite and x-bornite were left out of the low temperature diagram because doubts regarding their existence in the stable state. Some doubts also remain even for the Cu–Fe conventional sulphides. Kullerud, et. al. found [68] that pyrite and chalcocite do not exist in stable coexistence and that the bornite-pyrite mixture is unstable below 228 °C. Barton and Skinner [75] claim however that the resistance of the pyrite–chalcopyrite interface is highly probable. The bornite–pyrite coexistence, recommended by Kullerud, et. al. [67] as metastable below 228 °C, is found quite often in many ores. In other words, the result of disappearance of anomalous bornite at temperatures close to 100 °C is the formation of chalcopyrite and chalcocite.

Therefore, the system contains the following phases:

Digenite, Cu_9S_5: Digenite, Cu_9S_5, which was regarded for a long time as part of the copper-sulphide group, was gradually also found as the Cu–Fe sulphide by Morimoto and Koto [29] who found that the natural digenites contain small amounts (~1%) of iron. Subsequently, they found that at temperatures below 70 °C the Cu–S pure system is characterised by the formation of a mixture of anilite and chalcocite and not of digenite. Increase of temperature above 25 °C leads gradually to the formation of digenite with a stable phase formed at 70 °C. The solid solution, formed at higher temperatures, have a composition rich in sulphur and iron. At temperatures approximately 335 °C, the digenite solid solution has a composition of the bornite solid solution and forms a single phase which transforms from Cu–S to the composition containing more than 15 at.% Fe. The maximum melting point of the digenite solid solution (1129 °C) is situated on the Cu–S line at a composition similar to Cu_2S depleted in copper.

Bornite, Cu_5FeS_4: Bornite, one of the best well-known sulphides, exists in several polymorphous modifications based on the sublattice with a parameter a = 0.55 nm. The low temperature form is found in ores and crystallises in the tetragonal system with a = 1.094 nm, c = 2.188 nm. As a result of heating to 228 °C it changes to the cubic form with a parameter a − · 0.55 nm. Morimoto and Kullerud [71] found that rapid cooling of high temperature bornite results in the formation of a metastable cubic structure with a = ~1.094 nm. The bornite, like digenite, is characterised by a wide range of the solid solutions with the Cu–Fe–S composition. The solid solution with digenite phase becomes complete above approximately 335 °C. Rapid cooling of the bornite–digenite transitional phase results in the formation of metastable polymorphs whose lattice parameters are integral multiples (2, 3, 4 or 5), or sometimes non-integral multiples (4.7 and 5.7). In addition to this, they repeat the main parameter of the sublattice of 0.55 nm. This depends on composition. The high temperature bornite phase is also soluble in the solid state in the direction to chalcopyrite. The breakdown of this composition results in extensive precipitation of texture usually found in ores. Brett [76, 77] found the formation of bornite, digenite, bornite–chalcocite and bornite–chalcopyrite bonds but the formation of this texture was never identified at the initial temperature of formation or cooling rate.

'Anomalous' X- or sulphur-enriched bornite: Heating of natural bornites [76, 77] results in the formation in some cases of chalcopyrite together with chalcocite, instead of a simple transformation to the high temperature polymorph. A number of authors consider these bornites as metastable sulphur-enriched bornites. Brett and Yund [76] believe that some of the bornites, formed at temperatures below 75 °C, have a sulphur to metal ratio higher than the values resulting from the stoichiometry Cu_5FeS_4. Yund and Kullerud [66] processed these results and found that *x*-bornite (as they referred to it) may be synthesised in a range with slight sulphur enrichment (0.5 wt.%) in relation to normal bornite and temperatures below 125 °C. It is not clear whether this phase is thermodynamically stable but its presence in nature has already been confirmed. On the other hand, the absence of *x*-bornite in the majority of ores indicates that this mineral has not formed by cooling by a solid state reaction but that it probably forms directly from low temperature solutions [76].

Cubanite, $CuFe_2S_3$: Cubanite, $CuFe_2S_3$, is often regarded as synonymous with chalcopyrite in which it is present in the form of high-anisotropy lamellae. The orthorhombic natural form is stable below 200–210 °C [69] and changes to cubic *iss* above this temperature. Heating of natural cubanite resulted in twinning and, consequently, the discovery of the high temperature hexagonal form of cubanite. It was also found that the precipitation of chalcopyrite from *iss* in the vicinity of the cubanite composition leads to the formation of a sulphide whose x-ray diffraction pattern indicates tetragonal composition.

The transition solid solution (iss) and the primitive cubic phase (pc): at temperatures above 500 °C, *iss* is the dominant ternary phase of the Cu–Fe–S system. *iss* was regarded for a long period of time as the higher temperature polymorph of chalcopyrite but *iss* is a different, although crystallographic similar phase. It has a disordered fcc structure, of the sphalerite type. The range of its chemical composition is wide, including talnakhite, mooihoekite, haycockite and cubanite. Many previously introduced interpretations were obviously based on examination of regions with a wide chemical composition. Low-temperature breakdown of *iss* has not as yet been completely explained, but it does form when the second zone in Fig. 3.6 inverts at the same temperature below 600 °C to a primitive cubic phase which is stable up to 20–200 °C. Two integrals of chalcopyrite with bornite, chalcopyrite with cubanite and natural clusters of talnakhite, mooihoekite and haycockite form by a mechanism similar to that of the products of breakdown of the original *iss*.

Idaite, $Cu_{5-x}Fe_xS_{6+x}$: Although the synthetic phase with the composition $Cu_{5-x}Fe_xS_{6+x}$ was described for the first time by Mervin and Lombard in 1937 [63], no natural analogues were found prior to 1959 when they were found by Frenzel [40] who also used for the first time the term idaite. The similarity of this phase with bornite (except for the fine orange shadow and strong anisotropy) at the previous authors to use the term 'orange bornite' for idaite. Idaite is stable below 501 °C and its unit cell was described by Frenzel [40] as hexagonal with the parameters a = 0.377 nm, c = 1.118 nm.

Fukuchilite, Cu_3FeS_8: Another contribution to examination of the Cu–Fe–S phase system was the identification of a new phase with a structure similar to pyrite, *i.e.* fukuchilite, Cu_3FeS_8 [74, 78]. Subsequent experimental study by hydrothermal synthesis in the temperature range 100–275 °C [72] shows that it is possible produce a compound of the pyrite type based on FeS_2–Cu_2S with the approximate ratio of FeS_2:Cu_2S = 3:7. However, Shimazaki and Clark [72] stressed that FeS_2–Cu_2S solid solutions are thermally unstable, despite the fact that they do form and remain in nature in the metastable state.

Chalcopyrite, $CuFeS_2$: Chalcopyrite, $CuFeS_2$, the most common sulphide in the Cu–Fe–S ternary system solidifies in the ordered tetragonal structure and is stable to 557 °C. Its composition slightly differs from the ideal $CuFeS_2$ in the direction to enrichment with a metal at high temperatures [70]. At temperatures higher than 557 °C chalcopyrite breaks down to pyrite + *iss*. This is not only a transition to high temperature polymorphs. There is also a close relationship between the disordered cubic structure *iss* (a = ~0.536 nm) and the ordered tetragonal structural chalcopyrite (a = 0.528 nm, c = 1.04 nm).

Talnakhite, $Cu_9Fe_8S_{16}$, *transition phase I, transition phase II:* Talnakhite, $Cu_9Fe_8S_{16}$, and its high temperature polymorphs I and II are examples of sulphides which were synthetically produced prior to determination in the mineral form. Talnakhite, observed for the first time by Hiller and Probsthain [65] regarded as the β-phase, was referred to as 'cubic chalcopyrite'. However, later Cabri [79] recommended the term talnakhite. In heating, synthetic talnakhite undergoes a transformation: at 186 °C to phase I, at 230 °C to phase II and to *iss* between 520 and 525 °C. Talnakhite has a cubic structure (a = ~1.059 nm), but the structures of the transition phases are not available.

Mooihoekite and transition phase A: Mooihoikite, $Cu_9Fe_9S_{16}$, represents another, chalcopyrite-like phase and like talnakhite, was found in nature [73] only after synthetic preparation of an identical phase described by Hiller and Probsthain [65]. Like other chalcopyrite-like phases, this phase is not found very frequently in nature, probably because of its similarity with chalcopyrite. Like talnakhite, mooihoekite is very similar to *iss* with a cubic structure, a = 0.53 nm, sphalerite type. Mooihoekite transforms to transition phase *A* with the unknown structure in heating at 167 °C. This phase changes to *iss* at temperatures higher than 236 °C.

Haycockite, $Cu_4Fe_5S_8$: haycockite, the third in the group of the chalcopyrite-like minerals was discovered by Cabri and Hall [73]. Despite the fact that it is an orthorhombic mineral, its structure may be regarded as that of sphalerite, such as talnakhite, mooihoekite, chalcopyrite, and *iss*. The amount of information on its stability is small and it has not been synthetically produced. It has not as yet been shown whether it transforms to *iss* at high temperatures.

$Cu_{0.12}Fe_{0.94}S_{1.00}$ *phase*: Clark [74] published data on a new, unnamed sulphide mineral, with a composition $Cu_{0.12}Fe_{0.94}S_{1.00}$. There is hardly anything known about this mineral, except for the fact that it is found together with chalcopyrite and cubanite. According to Clark's hypothesis, this may be a new mineral or may be one of the members of the series of solid solutions of copper mackinawite.

References

1. Dana E.S.: A Textbook of Mineralogy, Wiley, New York, 1877.
2. Bragg W.L.: *Proc. R. Soc. London*, 89A, 1913, 468–489.
3. Buerger M.J.: *American Mineralogist*, 16, 1931, 361–395.
4. Buerger M.J.: The pyrite–marcasite relation, American Mineralogist, 19, 1934, 32–61.
5. Buerger M.J.: *Zeitschrifts fuer Kristallographie*, 95, 1936, 83–113.
6. Buerger M.J., Buerger N.W.: *American Mineralogist*, 29, 1944, 55–65.
7. Buerger M.J.: *American Mineralogist*, 32, 1947, 415–425.
8. Meyer B. ed.: Elemental Sulfur – Chemistry and Physics, Interscience Publ., New York, 1965.
9. Steudel R., Reinhardt R., Schuster F.: *Angew.Chem.*, 89, 10, 1977, 756–

758.
10. Donohue J.: The Structures of the Elements, J.Wiley & Sons, New York, 1974, 324–369.
11. Donohue J., Caron A., Goldish E.: *J.Am.Chem.Soc.*, 83, 1961, 3748–3751
12. Steidel J., Steudel R.: *Z. Anorg. Allg. Chem.*, 476, 1981, 171–178.
13. Steudel R., Steidel, Pickardt J., Schuster F.: *Z.Naturforsch.*, *35B*, 1980, 1378–1383.
14. Schmidt M., Siebert W.: Sulphur, In: Comprehensive Inorganic Chemistry, Bailar et al eds., Pergamon Press, Oxford, 1973, 795–933.
15. Karchmer J. H., ed: The Analytical Chemistry of Sulfur and its Compounds, part 1., Wiley-Interscience, New York, 1970, 1–23.
16. Tuller W. N., ed.: The Sulphur Data Book, McGraw-Hill Book Comp., Inc., New York, 1954.
17. Müller A., Krebs B., eds: Sulfur, Its Significance for Chemistry, for the Geo-, Bio- and Cosmosphere and Technology, Elsevier, Amsterdam, 1984.
18. Meyer B.: Sulfur, Energy, and Environment, Elsevier, Amsterdam, 1977.
19. Thackray M: Phase Transition Rate Measurements, In: Meyer B., ed.: Elemental Sulfur – Chemistry and Physics, Interscience Publ., New York, 1965, 45–71.
20. Donohue J.: The Structure of the Allotropes of Solid Sulfur, In: Meyer B. ed., Elemental Sulfur – Chemistry and Physics, Interscience Publ., New York, 1965, 13–43.
21. Emstey J.: The Elements, Clarendon Press, Oxford, 1989, 180–181.
22. Krebs H., Weber E. F.: *Z. Anorg. Allg. Chemie*, 272, 1953, 289–296.
23. Roseboom E. H., jr: *Economic Geology*, 61, 4, June–July 1966, 641–674.
24. Evans H. T.: Crystal structure of low chalcocite, Nature Physical Science, 232, 1971, 69–70.
25. Morimoto N., Kullerud G.: *American Mineralogist*, 48, January–February 1963, 110–123.
26. Morimoto N.: *Mineralogical Journal* (Tokyo), 3, 1962, 338–344.
27. Morimoto N., Koto K., Shimazaki Y.: *American Mineralogist*, 54, September–October, 1969, 1256–1268.
28. Morimoto N., Koto K.: *American Mineralogist*, 55, 1970, 106–117.
29. Goble R. J., Robinson G.: *Canadian Mineralogist*, 18, 1980, 519–523.
30. Goble R. J.: *Canadian Mineralogist*, 18, 1980, 511–518.
31. Moh G. H.: Blaubleibender covellite, Carnegie Inst. Wash., Year Book, 63, 1964, 208–209.
32. Berry L. G.: *American Mineralogist*, 39, 1954, 504–509.
33. Rickard D. T.: *Mineral Deposita*, 7, 1972, 180–188.
34. Munson R. A.: *Inorganic Chemistry*, 5, 1966, 1296–1297.
35. Taylor L. A., Kullerud G.: Phase equilibria associated with the stability of copper disulfide, Neues Jahrb. Mineral. Monatshefte, 1972, 458–464.
36. Djurle S.: *Acta Chemica Scandinavica*, 12, 1958, 1415–1426.
37. Cook W. R., jr.: Phase changes in copper(I) sulfide as a function of temperature, Spec. Publ. U.S. National Bur. Stand., No. 364, 1972, 703–712.
38. Goble R. J.: *Canadian Mineralogist*, 23, 1985, 61–76.
39. Goble R. J.: *Canadian Mineralogist*, 19, 1981, 583–591.
40. Frenzel G.: *Neues Jb. Miner. Abh.*, 93, 1959, 87–132.
41. Massalski T.B. et al: Binary Alloy Phase Diagrams, 2nd ed., ASM Int: Materials Park, OH, 1990.
42. Yund R.A., Hall H.T.: *Economic Geology*, 64, 1969, 420–423.

43. Takeno S., Zoka H., Niihara T.: *American Mineralogist*, 55, 1970, 1639–1649.
44. Kissin S.A., Scott S.D.: *Economic Geology*, 67, 1972, 1007.
45. Nakazawa H., Morimoto N.: *Materials Res. Bulletin*, 6, 1971, 345–358.
46. Clark A.H.: *Trans. Inst. Min. Metall*, 75B, 1966, 232–235.
47. Yamaguchi S., Wada H.: *Kristall. Tech.*, 8, 1973, 1017–1019.
48. Taylor L.A., Willliams K.L.: *American Mineralogist*, 57, 1972, 1571–1577.
49. Scott S.D.: In: Sulfide Mineralogy, Ribbe P.H. Ed., Min. Soc. of America, November 1974, 41–110.
50. Kullerud G., Yoder H.S.: *Economic Geology*, 54, 1959, 533–572.
51. Rising B.A.: Phase relations among pyrite, marcasite and pyrrhotite below 300 °C, Ph.D. thesis, Penn. State University, 1973.
52. Berner R.A.: *Journal of Geology*, 72, 1964, 293–306.
53. Taylor L. A., Finger L.W.: Structural refinement and composition of mackinawite, Carnegie Institute of Washington, Geophys. Lab. Ann. Rept, 69, 1970, 318–322.
54. Zoka H., Taylor L.A., Takeno S.: *Journal Sci. Hiroshima University, Ser. C*, 7, 1973, 37–53.
55. Scott S.D., Kissin S.A.: *Economic Geology*, 68, 1973, 475–479.
56. Morimoto N., Gyobu A., Tsukuma K., Koto K.: *American Mineralogist*, 60, 1975, 240–248.
57. Arnold R.G.: *Economic Geology*, 57, 1962, 72–90.
58. Toulmin P., Barton P.B.: *Geochimica et Cosmochimica Acta*, 28, 1964, 641–671.
59. Taylor L.A.: Oxidation of pyrrhotites and the formation of anomalous pyrrhotite, Carnegie Institute of Washington, Year Book, 70, 1971, 287–289.
60. Erd R.C., Evans H.T., Richter D.H.: *American Mineralogist*, 42, 1957, 309–333.
61. Kullerud G.: Sulfide studies, In: Research in Geochemistry, Abelson P.H., ed., vol. II, Wiley, New York, 1967, 286–321.
62. Merwin H.E., Lombard R.H.: *Economic Geology*, 32, 1937, 203–284.
63. Schlegel H., Schüler A.: *Freiberger Forschungshefte*, *Sect. B*, 2, 1952, 1–32.
64. Hiller J.E., Probsthaim K.: *Zeitschrift für Kristallographie*, 108, 1956, 108–129.
65. Yund R.A., Kullerud G.: *Journal of Petrology*, 7, 1966, 454–488.
66. Kullerud G., Yund R.A., Moh G.H.: Phase relations in the Cu–Fe–S, Cu–Ni–S and Fe–Ni–S system, In: Magmatic Ore Deposits, Wilson H.D.B. ed, Economic Geology Monograph, 1969, 323–343.
67. Mukaiyama H., Izawa E.: Phase relations in the Cu–Fe–S system: the copper deficient part, In: Volcanism and Ore Genesis, Tatsumi T. ed., University of Tokyo Press, 1970, 339–355.
68. Cabri L.J.: *Economic Geology*, 68, 1973, 443–454.
69. Barton P.B.: *Economic Geology*, 68, 1973, 455–463.
70. Morimoto N., Kullerud G.: *Zeitschrift für Kristallographie*, 123, 1966, 235–254.
71. Shimazaki H., Clark L.A.: *Canadian Mineralogist*, 1970, 648–664.
72. Cabri L.J., Hall S.R.: *American Mineralogist*, 57, 1972, 689–708.
73. Clark A.H.: *Economic Geology*, 65, 1970, 590–591.
74. Barton P.B., Skinner B.J.: Sulfide mineral stabilities, In: Geochemistry of Hydrothermal Ore Deposites, Barnes H.L. ed., Holt, Rinehart & Winston,

New York, 1967, 236–333.

75. Brett P.R., Yund R.A.: *American Mineralogist*, 49, 1964, 1084–1098.
76. Brett P.R.: *Economic Geology*, 59, 1964, 1241–1269.
77. Kajiwara Y.: *Mineralogical Journal*, 5, 6, 1969, 399–416.
78. Cabri L.J.: A new copper–iron sulfide, *Economic Geology*, 62, 1967, 910–925.
79. Vaughan D.J., Craig J.R.: Mineral Chemistry of Metal Sulfides, Cambridge University Press, Cambridge, 1978.

Chapter 4

EQUILIBRIUM IN AQUEOUS SOLUTIONS

The molecules and ions in aqueous solutions interact by different mechanisms. These mechanisms are difficult to determine and there are many theories trying to explain this problem. The ions and molecules in the aqueous solution interact by means of interaction between ions of opposite sign or between ions and neutral molecules. This interaction may be described either on the basis of the Debye–Hückel theory [1] or on the basis of chemical equilibrium.

The original Debye–Hückel theory is valid only for highly diluted solutions of electrolytes and is based on a model in which the ions are considered simply as charged dimensionless particles. The change of the Gibbs energy of the system is calculated considering the electrostatic interaction between adjacent ions.

The validity of chemical equilibrium is based on the existence of the experimentally measured equilibrium constants. The change of the Gibbs energy of the system is determined from the values of these constants. In this method, the substances entering the reaction area are regarded as chemical units and the products are often regarded as complex substances.

Chemical equilibria are described by the relationships of the thermodynamic activity of substances taking part in the reaction. However, these activities are not directly proportional to the concentrations of the substances taking part because of the deviations from the ideal behaviour of the system and, in fact, these deviations are a source of the change of Gibbs energy and are included in the so-called activity coefficients. For every specific concentration the activity coefficient of each dissolved substance in the solution differs. The amount of a substance in chemical processes is described in most cases by the molar fraction x_i, which is the ratio of the moles of the component i to the total number of the moles of all components present:

$$x_i = \frac{m_i}{\sum_j n_j} \tag{4.1}$$

However, this relationship is never used for aqueous solutions or thermodynamic description of these solutions.

In practice, the concentration in the solution is expressed most frequently in the units of molarity c_i, mol of substance in 1 litre. The advantage of these units is that the water content in the sample of the solution does not have to be determined. Its disadvantage is that it depends on temperature and to a minor extent on pressure. The composition is independent of these factors, for example, in examination of the thermodynamics of the solution, we can use molality m_i, mol of substance in 1 litre, for expressing the concentration. In the case of water this means 1000/18.0152 = 55.51 mol of water per 1 kg. The molar fraction of the solution is:

$$x_i = \frac{m_i}{(m_i + 55.51)} \tag{4.2}$$

Activity a_i is given by its concentration in some units multiplied by the appropriate activity coefficient:

$$a_i = m_i\gamma_i = x_i f_i = c_i\gamma_i \tag{4.3}$$

If component i is, for example, sodium chloride, its molar activity coefficient in the appropriate solution would be γ_{NaCl}. Assuming that we consider a reaction in which the Cl^- ion plays some role, but the Na^+ ion does not, for example

$$Co^{2+} + 4Cl^- \rightarrow CoCl_4^{2-} \tag{4.4}$$

it would be interesting to examine the behaviour of cobalt from the viewpoint of the activity of substances taking part in the reaction. From this viewpoint, the activity coefficients of the substances taking part may be expressed by γ_{Cl-}. This expression has no thermodynamic meaning because according to the definition of the component, the latter is an independent variable component of the solution. This shows that the NaCl–H_2O two-phase system contains only two components, Na^+ and Cl^-, without any complexes. The advanced methods of calculating the activity coefficients consider, however, the activity coefficients of the single ions in accordance with the original Debye–Hückel theory. The activity of the single ions can be treated as the activity coefficients of the salt, $\gamma_{\pm}$.

In the case of dilute solutions of electrolytes in water, it is often assumed that the activity water is unity. This is not correct from the thermodynamic viewpoint and may cause problems in calculations. Generally, if a solution is in equilibrium with its vapour, the chemical potential of each component in the solution must be equal to the potential of its component in the gas phase:

$$\mu_i^{\text{solution}} = \mu_i^{\text{vapour}} = \mu_i^0 + RT \ln p_i \tag{4.5}$$

where p_i is the fugacity of the component i above the solution. This is usually taken as the partial pressure of vapour of the component i.

If the solution obeys Rault's law

$$p_i = p^0{}_i\, x_i \tag{4.6}$$

where p^0_i is the vapour pressure of the pure component i at the temperature of the solution. Then, equation (4.5) can be expressed as follows:

$$\mu_i^{\text{solution}} = \mu_i^0 + RT \ln p_i^0 + RT \ln x_i \tag{4.7}$$

Then

$$d\mu_i = RTd \ln f_i x_i \tag{4.8}$$

All very diluted solutions behave ideally, especially aqueous electrolytes which behave in this manner only at very high dilutions, as discussed later. In more concentrated solutions, the deviations from the ideality are caused by multiplication of the molar fractions by activity coefficients f_i. The activity coefficient of component 1 in a solution can be described as follows:

$$\mu_1^{\text{solution}} = \mu_1^0 + RT \ln p_1^0 + RT \ln f_1 x_1 \tag{4.9}$$

and

$$d\mu_1 = RTd \ln f_1 x_1 \tag{4.10}$$

In the case of the aqueous solution of an electrolyte this definition of the activity coefficient is not convenient for the solute which is not volatile so that the system does not have the equilibrium of the vapour pressure with the solution. Rault's law is not valid in this case and Henry's law can be applied. For the solute, component 2 of the solution, this can be written as

$$p_2 = k_2 x_2 \quad \text{or} \quad p_2 = k_2' m_2$$

and

$$\mu_2^{\text{solution}} = \mu_2^0 + RT \ln k_2 + RT \ln f_2' x_2 \tag{4.11}$$

$$d\mu_2 = RTd \ln f_2' x_2 \tag{4.12}$$

and

$$\mu_2^{\text{solution}} \mu_2^0 + RT \ln k_2' + RT \ln \gamma_2 m_2 \tag{4.13}$$

$$d\mu_2 = RTd \ln \gamma_2 m_2 \tag{4.14}$$

This is a theoretical background for using the activity coefficients for determining the activity of the substances on the basis of their concentration. These values are required for thermodynamic calculations.

An alternative method of processing the deviations from the ideal state is to use the Bjerrum 'osmotic' coefficient ϕ in the equation

$$\mu_1^{\text{solution}} = \mu_1^0 + RT \ln p_1^0 + \phi RT \ln x_1 \tag{4.15}$$

Equation (4.9) shows that

$$\phi \ln x_1 = \ln f_1 x_1 \tag{4.16}$$

Comparison of the equations (4.5) and (4.9) shows that

$$p_1 = p_1^0 f_1 x_1 \tag{4.17}$$

The activity coefficient is a unit used to multiply the vapour pressure of the solvent above the ideal solution. This gives the vapour pressure above the new solution of the same concentration. The resultant value is then used in physical chemistry in the relationships describing the increase of the boiling point, the decrease of the freezing point and osmotic pressure of liquids.

The molal osmotic coefficient ϕ can be defined as follows:

$$\ln a_s = -\frac{vmW_s}{1000}\phi \tag{4.18}$$

where a_s is the activity of the solvent and W is mole weight, m is the molality of the dissolved substance and for an electrolyte, v is the number of ions formed by dissociation of substances in the solution.

For water it holds that:

$$\ln a_w = -\frac{18.0152vm}{1000}\phi$$

For example, for a 2M solution of potassium chloride KCl at 25°C the activity of water is equal to 0.9364; $v = 2$ and $m = 2$, so that

$$\ln 0.9364 = -0.06571 = -0.07206 \quad ; \qquad \phi = 0.912$$

If a solution contains more than one component, we may use the equation for ln a_w and expression vm is replaced by the sum of all components present.

The activity of a solvent may be determined by measuring its partial pressure above the solution in comparison with partial pressure above the ideal solutions with the same concentration. This may be realised in experiments using the isopiestic method. In this method, we examine the solutions of salts that are in equilibrium with the solution as a reference substance for which the accurate values of the osmotic coefficients in a wide concentration range are available. The reference substances are represented by NaCl, KCl, $CaCl_2$, H_2SO_4, and saccharosis is used for non-electrolytes. Known amounts of these substances in defined concentrations are placed in a suitable vessel and equilibrium is established at a constant temperature with the salt solution. Equilibrium can be established more rapidly by evacuation. At equilibrium, the pressure of the vapours of both solutions is identical. The vessels are weighed to obtain the concentration of the solutions. The activity of a solvent may be calculated using the osmotic activity coefficient of water by many analytical or graphical methods. Equation (4.13) gives the chemical potential of dissolved substances in a solution, k' is the proportionality constant for the case in which the partial pressure of a volatile component is related to the molality of the non-volatile substance. If the dissolved substance is an electrolyte, equation (4.13), ignoring k', may be presented in the form

$$\mu_2 = \mu_2^0 \nu RT \ln \gamma_2 m_2$$

The changes of the chemical potential of the dissolved substance in relation to the concentration are described by the Gibbs–Duhem equation. Integration of this equation gives:

$$\int_{m_1}^{m_2} \frac{\phi - 1}{m} \mathrm{d}m + \phi m_2 - \phi m_1 = \ln \gamma m_2 - \ln \gamma m_1 \tag{4.19}$$

If the polynomial expression of the activity of water as a function of the molality of the electrolyte is available, numerical integration can be carried out. The isopiestic method is suitable for solution concentrationd from 0.1 M to the saturation state.

4.1. Ionic activities

4.1.1. Debye–Hückel theory

The activity coefficient of three electrolytes of different valences are presented in Fig.4.1 as a function of concentration. It may be seen that the curves have an infinite negative gradient after reaching zero concentration. The non-electrolytes, such as glycol or saccharosis, may show either an increase or decrease of the activity coefficients after reaching the zero concentration but the values are usually close to unity and are approximately distributed on straight lines.

The solutions of the non-electrolytes differ from the ideal state because of low binding forces, for example, van der Waals bonds, etc., whilst the form of the curves of the electrolytes is the result of strong bonding. Debye and Hückel [1] recommended that these forces are the result of interionic attraction and developed an equation describing the activity coefficient of salts as a function of their concentration. This equation is valid only for highly diluted solutions. The theory was summarised previously and explained in detail in [2, 3]. Prior to this, many authors had believed that ionic interaction is a case of strong bonding but none of the authors were able to support claims by convincing mathematical considerations.

The Debye–Hückel theory is based on the fact that the electrolytes dissociate in a solution producing ions which may be regarded as point charges. The chemical nature of the ions is without mass, but the charge in the unit volume of the solution controls the range of interaction between the charges. The salts

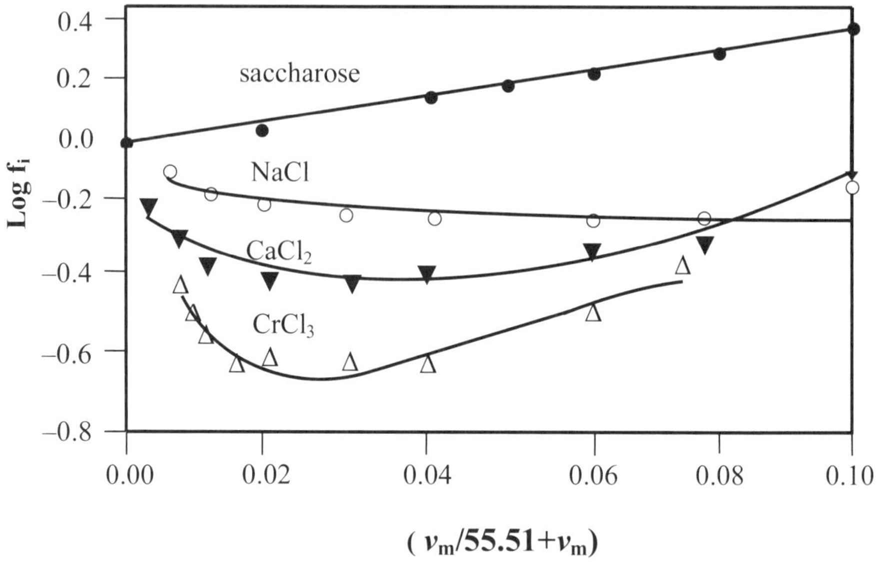

Fig. 4.1. Concentration dependence of the activity coefficient of molar fractions.

with different valences produce different amounts of ions in the solution. Electrolytes such as NaCl with a ratio of 1:1 produce 2 ions, electrolytes with a ratio 2:1 or 1:2 such as $CaCl_2$ or Na_2SO_4 produce 3 ions, electrolytes with a ratio 3:1 or 1:3 such as $LaCl_3$ or Na_3PO_4 produce 4 ions, etc. The ionic strength of the solution I may be defined as

$$I = \frac{1}{2}\sum_i m_i z_1^2 \tag{4.20}$$

where m_i and z_i are the molal concentration and valency of the ion i.

Every ion is considered to be surrounded by an ionic atmosphere with the opposite sign. The single positive ion in a dielectric solution of a binary electrolyte is surrounded by both negatively and positively charged ions. However, because of the existence of electrostatic attractive and repulsive forces, the negative ions are located closer to the central positive ion than other positive ions. This non-equilibrium of the charges is the ionic atmosphere which is the reason for the attractive force of the central ion.

Since the entire solution contains the same number of positive and negative charges, the total charge of the ionic atmosphere surrounding the central ion is equal to the size *a* and the opposite sign of the charge of the central ion. It was found that the effect of the ionic atmosphere of the ion is the same as the effect of the point charge of the same size as the ion located at the distance I/K from it

$$K = \left(\frac{4\pi e^2}{\varepsilon_r kT} \sum_i n_i z_i^2 \right)^{\frac{1}{2}} \tag{4.21}$$

where e is the charge of the electron, ε_r is the dielectric constant (relative permittivity) of the solution, k is the Boltzmann constant, T is absolute temperature, n_i is the number and z_i the valency of the ions i. If the concentration of these ions is equal to m_i in gram-ions, then

$$n_i = \frac{m_i N}{1000}$$

where N is the Avogadro number. Substituting $\varepsilon_r = 78.6$ for the value for water at 25°C $T = 298$ K we obtain

$$\frac{1}{K} = \frac{4.31 \cdot 10^{-8}}{\left(\sum_i m_i z_i^2 \right)^{\frac{1}{2}}} [\mathrm{cm}]. \tag{4.22}$$

This distance is known as the thickness of the ionic atmosphere and depends on the ionic strength of the solution, equation (4.20)

$$\frac{1}{K} = \frac{4.31 \cdot 10^{-8}}{\sqrt{2I}} [\mathrm{cm}] \tag{4.23}$$

The ionic atmosphere surrounds the central ion with potential ψ_i. If the charge of this ion is $z_i e$, then

$$\psi_i = \frac{z_i e \mathrm{K}}{\varepsilon_r}$$

The central ion possesses a certain amount of energy from the excess work in charging to the potential equal to half the product of its charge and potential. For 1 g-ion of ions which act as the centres of the ionic atmosphere, with N units charge, the energy is equal to

$$E_i = -\frac{N z_{\mathrm{i}}^2 e^2 K}{2\varepsilon_r} \tag{4.24}$$

It is assumed that this excess energy which the ion receives as a result of being surrounded by the ionic atmosphere, is the result of deviation of the electrolyte solutions from the ideal behaviour.

The chemical potential of ion i in the ideal solution is

$$\mu_i - \mu_i^0 + RT \ln x_i$$

where R is the gas constant. For a non-ideal solution

$$\mu_i - \mu_i^0 + RT \ln a_i = \mu_i^0 + RT \ln x_i + RT \ln f_i$$

The difference between the last two relationships, $RT \ln f_i$, is the difference of the free energy accompanying the addition or loss of 1 g-ion of the given substance from the large volume of the ideal and real solutionS. After substituting for E_i

$$\ln f_i = -\frac{N z_1^2 e^2 K}{2RT\varepsilon_r} \tag{4.25}$$

The result of substitution for K, using the molality of the ions and transformation to the decadic logarithm, one obtains

$$-\log \gamma_{\mathrm{i}} = \left[\frac{N^2 e^3 \sqrt{\dfrac{\pi}{1000}}}{2.303 R^{\frac{3}{2}}}\right]^{z_i^2} \frac{\sqrt{\sum_i m_i z_i^2}}{(\varepsilon_r T)^{\frac{3}{2}}} \tag{4.26}$$

The expression in the square brackets consists of universal constants and can be replaced by expression A'. Using the definition of the ionic force (4.30) gives

$$-\log \gamma_i = A''(\varepsilon_r T)^{-\frac{3}{2}} z_i^2 \sqrt{I} \quad (4.27)$$

where $A'' = \dfrac{A'}{\sqrt{2}}$

For water at 25°C with ε_r = 78.6 $A = A'(\varepsilon_r)^{-3/2}$ = 0.509.

The Debye–Hückel theory and the equations (4.27) and (4.30) represent the general form of the Debye-Hückel law applied to diluted solutions. This shows that the extent of deviation of the ion from the ideal behaviour in the given solution is controlled by the density of the charge in the solution in accordance with its ionic strength, and is independent of the chemical nature of the ions. Theoretically, we can determine directly the activity coefficient of the single ion in equation (4.27).

The individual activities of the ions and, therefore, the activity coefficients of the individual ions cannot be measured and therefore, as already mentioned, do not have any thermodynamic meaning. However, they may relate to the measurable mean activity of the ions. The main unit or the 'molecule' of a binary electrolyte dissociates into the total number of the ions v in which v_+ are cations and v_- are anions, the mean activity coefficient $\gamma\pm$ can be related to the activity coefficient of the single ion using the equation

$$\gamma_{\pm} = \sqrt[\nu]{\gamma_+^{\nu} \gamma_-^{\nu-}} \quad (4.28)$$

or

$$\log \gamma_{\pm} = \frac{\nu + \log \gamma_+ + \nu_- \log \gamma_-}{\nu_+ + \nu_-}$$

If the valencies of the ions are equal to z_+ and z_-:

$$\log \gamma_{\pm} = \frac{z + \log \gamma_+ + z_+ \log \gamma_-}{z_+ + z_-} \quad (4.29)$$

This gives

$$-\log \gamma_{\pm} = A z_+ z_- \sqrt{I} \quad (4.30)$$

4.1.1.1. Extension to more concentrated solutions

The above theoretical considerations valid for electrolytes are based on the concept of the ionic atmosphere and provide a basis for defining the law of ionic attraction in the solution. Harned and Owen [2] stressed that the concepts of electrostatics, hydrodynamics and statistical mechanics have been used to develop an accurate theory enabling description of the properties of the solutions of electrolytes with preferential attraction forces between the ions. They also determine the concentration limit of approximately 10^{-3} M, from which the electrolytes can be regarded as diluted.

In the majority of processes when using more concentrated solutions than 10^{-3} M, it is important to know the values of activity coefficients. There are a large number of studies which are concerned with the possibilities of applying the limiting law to more concentrated solutions. Shortly after publishing the main principles of the limiting law by Debye and Hückel, Bjerrum published his contribution to theory in which he defined several factors increasing ionic force, especially the formation of ion pairs due to interionic attraction. He assumed the simplest model in which the ions were supposed to be rigid unpolarisable spheres in a medium with a fixed dielectric constant and the formation of chemical bonds by electron sharing was ignored. Assuming that the sum of the radii of the two electrons of the electrolyte is greater than the minimum value defined by the equation $r_{\min} = z_+ z_- e^2/2\varepsilon_r kT$, it is possible to use thc limiting law, but if this value is lower the ions should be treated as forming pairs. In this paired state they do not contribute to the electrical energy of the central ion by the form of its ionic atmosphere. The rate of formation of ion pairs rapidly decreases with a decrease of r, and the equilibrium constant can be calculated. The final result is the expanded Debye–Hückel equation in which the size of the charge is z_+ and z_- are ignored, as indicated previously by the vertical bars in the equation. This leads to the equation in the form

$$-\log\gamma_\pm = Az_+z_-a^0\sqrt{1/(1+B/1} \tag{4.31}$$

where B is a constant. If the solution contains more than one electrolyte, it is thermodynamically unstable and the activity coefficients cannot be calculated so that the validity of equation (4.31) is limited. In addition, the closest distances between the ions are not known and a_0 is then sometimes regarded as the parameter of the size of the ion and not as the sum of two ion radii.

The possibility of eliminating these pseudo-basic parameters is offered by the equation in the form

$$-\log\gamma_{\pm} = \frac{Az^2\sqrt{I}}{1+\sqrt{I}} - \beta I \qquad (4.32)$$

where $z_+ = z_-$ and β is the empirical constant for a partial system selected for the best approximation from the available data. The equation with the parameters which cannot be set was recommended for

$$-\log\gamma_{\pm} = Az^2\sqrt{I/(1+I)} \qquad (4.33)$$

and although this equation is not very accurate, it may be used for solutions containing several electrolytes. Its generalisation is Davis's equation [4] which is in fact equation (4.32) with parameter β= 0.2:

$$-\log\gamma_{\pm} = Az^2\left(\frac{I}{1+\sqrt{I}} - 0.21\right) \qquad (4.34)$$

which results in good agreement with the measured values of the mean activity coefficients in diluted solutions with the mean deviation of approximately 2% in 0.1 M solutions and a smaller deviation in more diluted solutions. This is an empirical relationship based on processing the published results. Assuming that the amount of the experimental results is sufficiently large, the values of $\gamma_{\pm}$ of salts are also available so that one can determine the values of the parameters a^0 and β, so that the accuracy of calculation may be considerably higher or interpolation of $\gamma_{\pm}$ for the concentrations of electrolytes of lower valencies of up to approximately 1M is possible. However, generally none of these equations can be regarded as satisfactory.

4.1.1.2. Extension to mixed electrolytes

Due to increased demands on the availability of data on the thermodynamic properties of the aqueous solutions of electrolytes, more and more published studies appeared concerned with this problem in the sixties and seventies of the previous century. More and more data were published on different aqueous systems of the electrolytes and activity and osmotic coefficients in aqueous solutions, and the accuracy of the published data was improved [5-7].

The activities of osmotic coefficients of the majority of pure solutions of the electrolytes were measured at 25 °C, but there are various suitable interpolation procedures for processing these data. However, on the other hand, there is only a relatively small amount of data measured in the solutions of mixed electrolytes, although the demand for these data is strong. If it were possible to predict the activity or osmotic coefficients in the solutions of multicomponent electrolytes from the data obtained for simple electrolytes, considerable advances would be possible. This may be possible using a procedure developed by Pitzer, et. al., as one of the several existing methods.

4.1.2. Pitzer method

The Pitzer method described in several publications by Pitzer [8-27] is used widely. In these studies, Pitzer describes gradually the derivation of equations and parameters required for calculations for different systems of electrolyte solutions.

The Pitzer equations are based on the Debye–Hückel limiting law which is valid if ions are bonded by long-range interactions. It also gives equations for explaining the effect of bonding by short-range interactions without considering the formation of discrete chemical species, for example, non-charged ion pairs in electrolyte solutions. As expected, there are strong interactions between the cations and anions with higher valencies, and the behaviour of the electrolytes may be processed either by the Pitzer method of using equilibrium constants describing the resistance of discrete complex species.

Pitzer [8] proposed a general equation for efficient application of the Gibbs energy for a solution containing n_w kg of solvent and n_1, n_2, ... n_j moles of the dissolved substance i, j, ... in the following form:

$$\frac{G^{ex}}{RT} = n_w f(I) \frac{1}{n_w} \sum_{ij} \lambda_{ij}(I) n_i n_j + \frac{1}{n_w^2} \sum_{i,j,k} \mu_{ijk} n_i n_j n_k \qquad (4.35)$$

where G^{ex} is the difference between the real Gibbs energy of the solution and its ideal Gibbs energy.

In this equation, $f(I)$ is the function of the ionic strength, the properties of the solvent and temperature and expresses the long range effect of electrostatic forces. This way it is linked with the Debye–Hückel theory. $\lambda_{ij}(I)$ represents the effect of short-range forces between substances i and j and is a function of the ionic strength and this is again linked with the expression βI in equation (4.31). Equation (4.35) also includes the expression for the interaction of a triple ion, but it is assumed that μ_{ijk} is independent of the ionic strength.

The relationships for the osmotic and activity coefficients were then transformed using G^{ex}

$$\varphi - 1 = -\frac{\dfrac{\delta G^{ex}}{\delta n_w}}{RT\sum_i m_i} = \frac{(If') + \sum_{i,j}(\lambda_{ij} + I\lambda'_{ij})m_i m_j + 2\sum_{i,j,k}\mu_{ijk} m_i m_j m_k}{\sum_i m_i} \tag{4.36}$$

$$\ln \gamma i = \frac{1}{\mathrm{RT}}\frac{\delta G^{ex}}{\delta n_i} = \frac{z_i^2}{2} f' + 2\sum_j \lambda_{ij} m_j + \frac{z_i^2}{2}\sum_{i,j} \mu_{ijk} m_j m_k \tag{4.37}$$

where

$$f' = \frac{\mathrm{d}f}{\mathrm{d}I};\ \lambda'_{ij} = \frac{\mathrm{d}\lambda}{\mathrm{d}I} a m_i = \frac{n_i}{n_w};\quad \text{etc.}$$

Further development of the equations taking into account the valencies of the ions and the number of ions formed in dissolution of electrolytes of different types shows that every type can be solved independently and the mixed electrolyte would be regarded as a different case. This research would consist of processing the available data of osmotic coefficients of the individual classes of simple electrolytes and by evaluating them and optimising using equation (4.36) and (4.37), which are most accurate in this case. In most cases, the values of the osmotic coefficients were used and were subjected to critical analysis by Robinson and Stokes [3].

The results obtained for many simple electrolytes were published in [9]. These salts have one or both ions univalent. A suitable example are in this case the equations used for calculating G^{ex}, ϕ, γ:

$$\frac{G^{ex}}{n_w RT} = f^{Gx} + m^2(2\nu_M \nu_X)B_{MX}^{Gx} + m^3\left[2(\nu_M \nu_X)^{\frac{3}{2}}\right]C_{MX}^{Gx} \tag{4.38}$$

$$\varphi - 1 = |z_M z_X| f^{\varphi} + m\left(\frac{2\nu_M \nu_X}{\nu}\right) B_{MX}^{\varphi} + m^2 \frac{2(\nu_M \nu_X)^{\frac{3}{2}}}{\nu} C_{MX}^{\varphi} \quad (4.39)$$

$$\ln\gamma = z_M z_X f^{\gamma} = m\left(\frac{2\nu_M \nu_X}{\nu}\right) B_{MX}^{\varphi} + m^2 \frac{2(\nu_M \nu_X)^{\frac{3}{2}}}{\nu} C_{MX}^{\gamma} \quad (4.40)$$

$$f^{Gx} = -A_{\varphi} \frac{4I}{b} \ln(1 + bI^{\frac{1}{2}}) \quad (4.41)$$

$$f^{\varphi} = -A_{\varphi} \frac{I^{\frac{1}{2}}}{1 + bI^{\frac{1}{2}}} \quad (4.42)$$

$$f^{\gamma} = -A_{\varphi}\left[\frac{I^{\frac{1}{2}}}{1 + bI^{\frac{1}{2}}} + \frac{2}{b}(1 + bI^{\frac{1}{2}})\right] \quad (4.43)$$

$$B_{MX}^{Gx} = \beta_{MX}^{(0)} + \frac{2\beta_{MX}^{(1)}}{\alpha^2 I}\left[1 - e^{\alpha I^{\frac{1}{2}}}\left(1 + \alpha I^{\frac{1}{2}}\right)\right] \quad (4.44)$$

$$B_{MX}^{Gx} = \beta_{MX}^{(0)} + \beta_{MX}^{(1)} e^{e-\alpha I^{\frac{1}{2}}} \quad (4.45)$$

$$B_{MX}^{\gamma} = 2\beta_{MX}^{(0)} + \frac{2\beta_{MX}^{(1)}}{\alpha^2 I}\left[1 - e^{-\alpha I \frac{1}{2}}\left(1 + \alpha I^{\frac{1}{2}} - \frac{1}{2}\alpha^2 I\right)\right] \quad (4.46)$$

$$C_{MX}^{Gx} = \frac{1}{2} C_{MX}^{\varphi} \quad (4.47)$$

$$C_{MX}^{\gamma} = \frac{3}{2} C_{MX}^{\varphi} \quad (4.48)$$

where v_M and v_X are the numbers of M and X ions in the formula of the salt, z_M and z_X are the valencies M and X; $v = v_M + v_X$, n_w is the amount of the solvent in kg; m is the conventional molality; b and α are the numbers used for improving the accuracy of the equations for the data for calculating ϕ and γ.

f^{Gx} is an expression in the equation for $\frac{G^{ex}}{n_w RT}$ which includes long-range electrostatic forces. As indicated by equation (4.41), it is determined by the Debye–Hückel coefficient A in equation (4.27), but better results were obtained using the coefficients for the osmotic function. This is defined [9] as follows:

$$A_\phi = \frac{1}{3}\left(\frac{2\pi N d}{1000}\right)^{\frac{1}{2}}\left(\frac{e^2}{\varepsilon^r kT}\right)^{\frac{3}{2}} = \frac{1}{3}A_\gamma \tag{4.48}$$

Expression f^{Gx} also includes the values of the ionic strength of the solution and constant b. To obtain a simple form of the equation for mixed electrolytes, the values of b must remain the same for all solvents and, according to Pitzer [8], this value is equal to 1.2.

Coefficient λ_{ij} in equation (4.35), which also takes into account the effect of short-range interactions between the substances i and j, is represented by B^{Gx}_{MX}, in equation (4.38) for $\frac{G^{ex}}{n_w RT}$, and exponent GX again shows that the value B for the salt MX corresponds to the equation for $\frac{G^{ex}}{n_w RT}$. The form of the equation for deriving the values of B is clearly indicated from equation (4.45). The value B relates to the second problematic coefficient in equation (4.32), and its value MX is defined by two parameters $\beta^{(0)}_{MX}$ and $\beta^{(1)}_{MX}$ obtained by determining more accurate values of the osmotic coefficients MX. The values of $\beta^{(0)}_{MX}$ and $\beta^{(1)}_{MX}$ are characteristics of MX and may be used to calculate $(\phi-1)$ and $\ln\gamma$, using the equations (4.45) and (4.46) for determining the values B^{ϕ}_{MX} and B^{γ}_{MX}. B is a function of the ionic strength and also the numerical parameter α. Its value $\alpha = 2$ is satisfactory for all salts, considered in [2].

The third problematic coefficient is defined by C^{ϕ}_{MX} and the values required for equations (4.38) and (4.40) can be calculated from the equations (4.47) and (4.48). This third problematic coefficient is usually small and, in some cases, negligible.

Numerical factors such as $(2v_M v_X)$ multiplying second and third problematic coefficients in equation (4.38)–(4.40) are required for ensuring that the problematic coefficients represent the interaction of arrangement of the pairs and triple ions over short distances

According to Pitzer [9], there is a total of 227 strong electrolytes and they include inorganic compounds of the types 1:1, 2:1, 3:1, 4:1 and 5:1, the salts of carboxylic acid of the type 1:1 some halides, sulphonic acids and sulphonates of the types 1:1 and 2:1, and some organic salts of the type 1:1. The tabulated values give the values of $\beta^{(0)}$, $\beta^{(1)}$ and C^{ϕ} for each electrolyte, the maximum molality for f and the standard deviation together from the source of these data.

When considering the electrolyte of the type 2:2, it was found [10] that another term should be added to the equation (4.45)

$$B^{\phi} = \beta^{(0)} + \beta^{(1)} \exp(-\alpha_1 I^{\frac{1}{2}} + \beta^{(2)} \exp(-\alpha_2 I^{\frac{1}{2}}) \tag{4.50}$$

The added constants have the values $\alpha_1 = 1.4$, $\alpha_2 = 12.0$. Parameter $\beta^{(2)}_{MX}$ can be used to obtain more accurate data for the electrolytes with high valencies without considering the presence of discrete particles, complexes, other substances, etc.

4.1.2.1. Calculation of ϕ for salts in mixed electrolytes

The equations (4.36) and (4.37) for solutions of simple electrolytes indicate that three expressions are important. The first one is associated with long-range electrostatic interaction as a function of the ionic strength, the second one includes the short-range interactions between the anions and cations and the third one the same but amongst three ions. If two electrolytes are present at the same time, it is expected that the equations for $(\phi-1)$ and ln γ should contain the same first expression since the Debye–Hückel equation does not take into account the chemical nature of the ions, only their charges and concentration. As a first approximation, the interactions between the ions present in a mixed electrolyte solution would be the same as those in a solution containing only one salt in each, resulting in a summation of effects in the mixed solution. However, if equations (4.36) and (4.37) are used for calculating $(\phi-1)$ or ln and γ_i for a mixed electrolyte solution, it is then necessary to compare the results with the experimentally determined data and, if necessary, add further parameters. The accuracy of the functions should then be improved.

This was carried out Pitzer and Kim [11] when deriving an expression for the osmotic coefficient of a mixed electrolyte solution:

$$\phi-1=\left(\sum_i m_i\right)^{-1}\left\{2If^{\phi}+2\sum_c\sum_a m_c c_a\left[B^{\phi}_{ca}+\frac{\left(\sum m_z\right)}{(z_c z_a)^{\frac{1}{2}}}C^{\phi}_{ca}\right]+\right.$$
$$\left.+\sum_c\sum_{c'} m_c m_{c'}\left[\theta'_{cc}+I\theta'_{cc'}+\sum_a m_a\Psi_{cc'a}\right]+\sum_a\sum_{a'} m_a m_{a'}\left[\theta_{aa'}+I\theta'_{aa'}+\sum_c m_c\Psi_{caa'}\right]\right\}\qquad(4.51)$$

where f^{ϕ} was defined in equation (4.42); B^{ϕ}_{MX} from equation (4.45) is written here as B^{ϕ}_{ca}; c and c' are exponents of all the cations, and a and a' cover all the anions, μ is the expression for the interaction of a triple ion in equation (4.35) and λ represents the effect of the short-range forces:

$$C^{\phi}_{MX}=3\left[\frac{z_X^{\frac{1}{2}}}{z_m}\mu_{MMX}+\frac{z_M^{\frac{1}{2}}}{z_X}\mu_{MXX}\right]\qquad(4.52)$$

$$\theta_{MN}=\lambda_{MN}-\left(\frac{z_N}{2z_M}\right)\lambda_{MM}-\left(\frac{z_M}{2z_M}\right)\lambda_{NN}\qquad(4.53)$$

$$\psi_{MNX}=6\mu_{MNX}-\left(\frac{3z_N}{z_M}\right)\mu_{MMX}-\left(\frac{3z_M}{z_N}\right)\mu_{NNX}\qquad(4.54)$$

In equation (4.51): $\left(\sum mz\right)=\sum_c m_c z_c=\sum_a m_a z_a$

The first expression in the (4.51) in the combined brackets, $2If^{\phi}$, is the Debye–Hückel equation, the second expression includes the double sum of the molalities and second and third ones the problematic coefficient for pure electrolytes, as indicated by the previously discussed matters. The last two expressions include the differences between and second and third problematic coefficient for the different ions of the same sign, M and N are obtained from the estimates of the diameters of similar ions and it is expected that these values are small. The main effects of the mixed electrolytes then originate from the differences of the parameters $\beta^{(0)}$, $\beta^{(1)}$ and C^{ϕ} of the electrolytes. The parameters θ and ψ are small values with a negligible effect.

4.1.2.2. Calculation of γ for salts in mixed electrolytes

Pitzer and Kim proposed the following relationship for calculating the activity coefficient of the salt MX in the solution of a mixed electrolyte:

$$\ln\gamma_{MX} = |z_M z_X| f^{\gamma} + \left(\frac{2\nu_M}{\nu}\right)\sum_a m_a\left[B_{Ma}\left(\sum mz\right)C_{Ma} + \left(\frac{\nu_X}{\nu_M}\right)\theta_{Xa}\right] +$$

$$+\left(\frac{2\nu_X}{\nu}\right)\sum_c m_c\left[B_{cX} + \left(\sum mz\right)C_{cX} + \left(\frac{\nu_M}{\nu_X}\right)\theta_{Mc}\right] + \sum_c\sum_a m_c m_a\{z_M z_X B'_{ca} +$$

$$+\nu^{-1}\left[2\nu_M z_M C_{ca} + \nu_M\psi_{Mca} + \nu_X\psi_{caX}\right]\} +$$

$$+\frac{1}{2}\sum_c\sum_{c'} m_c m_{c'}\left[\left(\frac{\nu_X}{\nu}\right)\psi_{cc'X} + |z_M z_X|\theta'_{cc'}\right] +$$

$$+\frac{1}{2}\sum_a\sum_{a'} m_a m_{a'}\left[\left(\frac{\nu_M}{\nu}\right)\psi_{Maa'} + z_M z_X\theta'_{aa'}\right] \qquad (4.55)$$

Term f^{γ} is defined by the relationship (4.43)

$$B_{MX} = \beta_{MX}^{(0)} + \frac{2\beta_{MX}^{(1)}}{\alpha^2 I}\left[1 - \left(1 + \alpha + I^{\frac{1}{2}}\right)e^{-\alpha I^{\frac{1}{2}}}\right] \qquad (4.56)$$

$$B'_{MX} = \frac{2\beta^{(1)}_{MX}}{\alpha^2 I^2}\left[-1+\left(1+\alpha I^{\frac{1}{2}}+\frac{1}{2}\alpha^2 I\right)e^{-\alpha I^{\frac{1}{2}}}\right] \tag{4.57}$$

$$C_{MX} = \frac{C^{\phi}_{MX}}{2z_M z_X^{\frac{1}{2}}} \tag{4.58}$$

The expressions in the equations (4.51) and (4.55) which include θ or ψ have only a slight effect on the calculated values of $\phi-1$ or $\ln\gamma_{\pm}$, with thc exception of salts in which the short-range interactions are especially strong. They usually include tri- or multivalent ions, and only seldom bivalent ones. These expressions may be generally ignored, at least in initial calculations.

4.1.2.3. Calculation of γ for single ions

The Pitzer approach to calculating the activity coefficients is based on the Debye–Hückel equation (4.27) which gives the values for the single ion *i*. If the ionic activity coefficients for *M* and *X* are not determined by combining the data, but are calculated by separately, the following equations are used

$$\begin{aligned}\ln\gamma_M &= z_M^2 f^{\gamma} + \sum_a m_a\left[2B_{Ma}+\left(2\sum_c m_c z_c\right)C_{Ma}\right]+\\ &\sum_c\sum_a m_c m_a\left(z_M^2 B'_{ca}+|z_M|C_{ca}\right)+\\ &+\sum_c m_c\left(2\theta_{Mc}+\sum_a m_a\psi_{Mca}\right)+\frac{1}{2}\sum_a\sum_{a'} m_a m_{a'}\psi_{Maa'}\end{aligned} \tag{4.59}$$

$$\begin{aligned}\ln\gamma_X &= z_X^2 f^{\gamma} + \sum_c m_c\left[2B_{cX}+\left(2\sum_a m_a z_a\right)C_{cX}\right]+\sum_c\sum_a m_c m_a\left(z_X^2 B'_{ca}+z_X C_{ca}\right)+\\ &\sum_a m_a\left(2\theta_{Xa}+\sum_a m_c\psi_{Xac}\right)+\frac{1}{2}\sum_c\sum_{c'} m_c m_{c'}\psi_{Xcc'}\end{aligned} \tag{4.60}$$

The expressions f^{γ}, B, B' and C are identical with the expressions given in equations (4.43) and (4.56)–(4.58). The activity coefficient of salt *MX* is determined from the values for the single ions using equation (4.28).

Pitzer included the appropriate values of the stoichiometric coefficient $\left(\frac{2\nu_M\nu_X}{\nu}\right)$ of the parameters B^{ϕ}_{MX} and B^{γ}_{MX}, $\frac{2(\nu_M\nu_X)^{\frac{2}{3}}}{\nu}$ for C^{ϕ}_{MX} and C^{γ}_{MX} in the equations for ϕ and *ln* (4.39) and (4.40). The values $\beta^{(0)}$, $\beta^{(1)}$ and C^{ϕ} included by Pitzer in tables contain these coefficients and must be eliminated prior to using the parameters in the equations for the activity coefficients of the single ions.

4.1.2.4. Studies of complex systems

Pitzer and Kim calculated the values of ϕ or ln $\gamma_{\pm}$ for 52 binary mixed electrolytes using the values of the parameters of conventional ions for pure electrolytes and found differences between these and experimental values of ϕ or ln γ. Subsequently, the values of ϕ and ψ were determined from these differences. They used these values as further data for determining more accurate values of the parameters of solutions of mixed electrolytes. Similar calculations were also carried out for 11 binary mixtures without conventional ions. The results confirmed that all the values of θ and ψ are relatively low.

The majority of calculations were repeated [12] with two added expressions E_{θ} and $E_{\theta'}$, including more accurate equations. In some systems, it was shown that improvement of the accuracy greatly helped, in the cases of HCl–$SrCl_2$, HCl–$BaCl_2$ and HCl–$MnCl_2$ this was even more so, whereas less marked in the case of HCl–$AlCl_3$. In these systems, calculations were carried out using the measured value of the activity coefficient of HCl. The large number of the osmotic coefficients of the chlorides of rare-earth elements, nitrates and perchlorates were also determined more accurately using equation (30). The results were in good agreement.

Similarly, a relatively large number of data were published on geochemical systems [18] containing the ions Na^+, Mg^{2+}, Cl^- and SO_4^{2-} and these data were used for the binary mixture formed by $NaCl$, Na_2SO_4, $CuCl_2$ and $CuSO_4$. Further geochemical systems and studies, simulating sea water were published in [7]. The properties of other solutions of simple substances are also important for practical application, such as sulphuric and phosphoric acid, sodium chloride and sodium sulphate which were expressed by means of the Pitzer equations [13–15,19,20]. Detailed analysis of the published data for the NaCl–H_2O system transformed to 300°C at a pressure of 100 MPa required 28 parameters, including parameters for the standard state in water [21]. Calculations for temperature of up to 100°C require a different set of parameters. The temperature dependence of the activity coefficient depends on the enthalpy of compounds deviating from the ideal behaviour. When using the Pitzer equations, the temperature coefficients must be used in the parameters. The existing restrictions are based on the fact that the data for the thermal capacity of the majority of solvents are available only for temperatures close to 25°C, and these data are required for calculating the parameters.

Pitzer derived his equations for the short- and long-rane arrangement of ionic strength on the basis of the fact that these forces do not lead to the formation of discrete chemical species in the solution. This means that there are no interactions between the electrons of the ions forming the chemical bonds between them. Therefore, the Pitzer equation shows that the original form of the Pitzer equation cannot be used for describing the behaviour of a system containing non-ionic substances.

Bromley [22] used a different equation for the correlation of the mean activity coefficient of strong electrolytes in aqueous solutions in the form:

$$\log\gamma_{\pm} = \frac{A\,|\,z_+ z_-\,|\,I^{\frac{1}{2}}}{1+I^{\frac{1}{2}}} + \frac{(0.06+0.6B_m)\,|\,Z_+ z_-\,|\,I}{\left[1+(1.5I)\left(|\,z_+ z_-\,|^{-1}\right)\right]^2} + B_m I \qquad (4.61)$$

The value of the parameter B_m is a function of temperature.

Several examples of the application of the Pitzer equations with specific data obtained from the calculation of the parameters ϕ, a_w, γ, for the solutions NaCl, $MnCl_2$, the mixed electrolytes and NaCl-KCl, KBr–NaCl, etc., are presented in [23].

4.2. Formation of metallic complexes and equilibrium constant

The deviations from the Debye–Hückel limiting law, valid in solutions of strong electrolytes with a concentration higher than 10^{-3} M indicate that in these solutions the electrostatic attractive forces between the ions do not play any significant role in the determination of the values of G^{ex}. Further, it is assumed that they form chemical bonds by the effect of electron interactions between the ions and not discrete chemical species. This assumption is essential because at present that are no mathematical procedure which would make it possible to determine the type of interaction calculating the values of G^{ex} in the solutions. This means that it is not possible to calculate the Gibbs energy of the formation of discrete chemical species formed by the reaction between the components of the solution. This holds if ions react with ions or ions react with neutral molecules. However, these reactions are very important for hydrometallurgists. Therefore, the reactions are treated as equilibrium and are described quantitatively by experimentally determined equilibrium constants.

Metal Me with the valency z^+ in a solution containing single-valency anion L will be considered In the presence of ionic interaction, L is a ligand and the product of the interaction is a complex. Complexes form gradually and each step is controlled by the equilibrium constant

$$Me^{z+} + L^- \rightarrow MeL^{(z-1)+} \qquad K_1 = \frac{\{MeL^{(z-1)+}\}}{\{Me^{z+}\}\{L^-\}}$$

$$MeL^{(z-1)+} + L^- \rightarrow MeL_2{}^{(z-2)+} \qquad K_2 = \frac{\{MeL_2{}^{(z-2)+}\}}{\{MeL^{(z-1)+}\}\{L^-\}}$$

$$MeL_{n-1}^{(z-n+1)+} + L^- \rightarrow MeL_n^{(z-n)+} \qquad K_n = \frac{\{MeL_n{}^{(z-n)+}\}}{\{MeL_{n-1}^{(z-n+1)+}\}\{L^-\}}$$

The maximum number of the ions L which may form on complexes with Me^{z+} is n and n is also the co-ordination number Me^{2+}. The combined equilibrium constant β is defined as follows

$$Me^{z+} + 2L^- \rightarrow MeL_2^{(z-2)+}$$

$$\beta_2 = \frac{\{MeL_2^{(z-2)+}\}}{\{Me^{z+}\}\{L^-\}^2}$$

$$\beta_2 = K_1 K_2$$

Generally, it holds that $\beta_m = K_1K_2K_3 \ldots K_m$.

If the ligand is a non-charged molecule, for example, ammonia, equilibria can be explained in the same manner but the charge of each complex is z^+.

The factors controlling the absolute and relative amounts of each specie, containing a metal and free ligand in the solution are:

- the values of each equilibrium constant;
- the total concentration $[Me_t]$ in any form;
- total concentration of the ligand $[L_t]$;
- the ratio of these two concentrations;
- activity coefficients of each species present in the solution.

Depending on the total concentration of the metal in the solution, complex MeL is the first to form with the increase of the concentration of the ligand from zero. Its concentration increases and then decreases with the formation of MeL_2 then increases and again decreases while higher complexes form, Fig.4.2.

The extent of formation of the complex MeL_m is defined by the equation

$$\alpha_{\mathrm{Me}L_m} = \frac{[\mathrm{Me}L_m]}{[\mathrm{Me_t}]}$$

but if multi-atomic complexes are present, the concentration of the complexes containing more than one atom of the metal in each ion the molecule must be multiplied by the number of metal atoms.

The average number of the ligand is given by the equation

$$\bar{n}_L = \frac{([L_t]-[L])}{\mathrm{Me_t}}$$

In this case, the concentration of the ligand in the complex species is divided by the total concentration of the metal and the relationship is quite important, especially if the equilibrium constants are measured.

When writing the chemical equation representing equilibrium, the solvation of the species present is usually ignored. The metallic ions in the aqueous solutions are highly hydrated and in many cases it may be assumed that the ligand displaces the water molecule in the co-ordination site around the metal atom. For example, the ion $Cu(NH_3)_4^{2+}$ has three amino groups arranged in the four corners of the square with the copper atom in the centre, and the molecule is planar. The amino groups gradually displace the water molecules in the same positions.

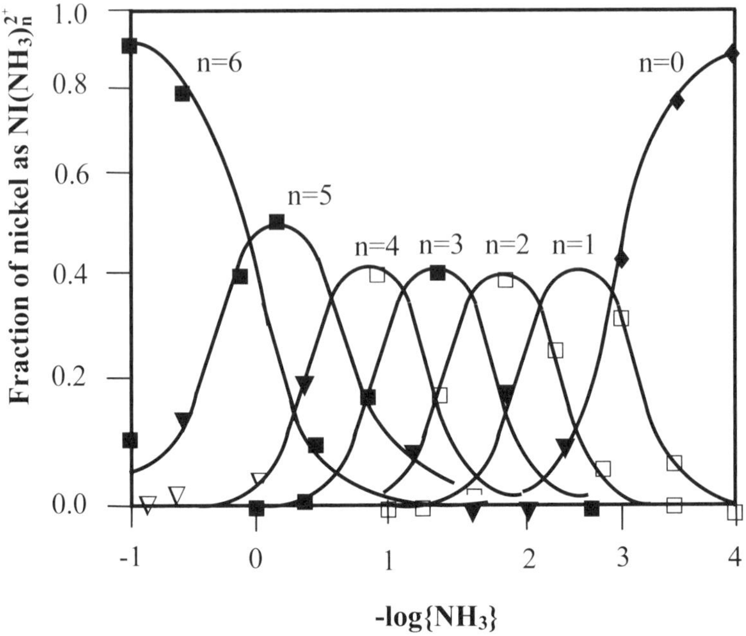

Fig. 4.2. Gradual formation of complexes of NH_3 with Ni^{2+} depending on the increase of the activity of NH_3.

Like the other bivalent and trivalent ions of the metals of the first transition group of the periodic table of elements, the simple hydrated ions Cu^{2+} has six directed molecules of water arranged in the octahedral positions. Under the influence of the Jahn–Teller effect in the case of the Cu^{2+} octahedron the ion is deformed and this is the reason why the metallic ions bond five and six ligands (including hydration water) only slightly. In the case of amino complexes, the equilibrium constants are as follows

$\log K_1$	$\log K_2$	$\log K_3$	$\log K_4$	$\log K_5$
4.15	3.50	2.89	2.13	−0.52

The ion $Cu(NH_3)_5^{2+}$ may form in strong ammonia aqueous solutions but the sixth molecule of ammonia must be added as the ammonium hydroxide.

Bjerrum studied the decrease of the values of *K* in relation to the mechanism by which the groups NH_3 obtained the Cu^+ ion. In Bjerrum's terminology, the logarithm of the ratio of two successful equilibrium constants is referred to as the 'sum effect' $T_{(m-1),m}$. This ratio was also divided into two quantities, 'the statistical effect', and $S_{(m-1),m}$ 'effect on the ligand' $L_{(m-1),m}$. The statistical effect was explained on the basis of the fact that the tendency to the loss of a group of ligands L with a species MeL_m is proportional to the value m, whereas the tendency of these species were reach a different group of the ligand is proportional to $(n–m)$. The stability constant n is then proportional to

$$\frac{n}{1}, \frac{(n-1)}{2}, \ldots \frac{(n-m+1)}{m}, \frac{(n-m)}{(m+1)}, \ldots \frac{2}{(n-1)}, \frac{1}{n}.$$

This shows that:

$$S_{m,(m+1)} = \log K_m - \log K_{(m+1)} = \log \frac{(n-m+1)(m+1)}{m(n-m)}$$

This equation is used if every ligand group occupies only one co-ordination position and if n identical co-ordination positions are around the metallic ion. The first four values of K for the system Cu^{2+}–NH_3 have the same order of magnitude and adjustment to statistical factors gives the modified values which are even quite close

$\log K_{1(cor)}$	$\log K_{2(cor)}$	$\log K_{3(cor)}$	$\log K_{4(cor)}$
3.55	3.32	3.07	2.73

Thus, it may be assumed that the majority of the experimental data have a statistical effect. The effect of the ligand was divided by Bjerrum into the 'electrostatic effect' and the 'residual effect'. The electrostatic effect is determined by the charge of the ligand and species containing the metallic ion. The ion of the ligand L^- is attracted in the direction to Me^{2+} or MeL^+, but is repulsed from MeL_3^-. Bjerrum derived an equation determining the strength of the electrostatic effect but since there are uncertainties in application of this equation, it is preferred to use the residual effect for the cases of non-charged ligand.

4.3. Thermodynamics of equilibrium constants

The value of the equilibrium constant by the formation of a metallic complexes may be determined by comparison with the values of the equilibrium constant of the individual hydrated metallic ions and molecules and the ions of the ligand in the solution. The forces of the chemical bonds between the metal and ligand play a role in the determination of the stability of the complex but other factors must also be included.

The equilibrium constant depends on the change of the standard Gibbs energy of the reaction ΔG^0, through the van't-Hoff isotherm:

$$\Delta G^0 = -RT \ln K$$

this is controlled by the change of the standard enthalpy ΔH^0 and entropy ΔS^0

$$\Delta G^0 = H^0 - TS^0$$

The change of the standard enthalpy is measured by the change of the thermal volume of the reactants and products in the formation of the complex and is determined by the type of chemical bond formed between the metallic ion and the ligand. In the case of single valency ligands the value ΔH^0 is usually between +20 and –20 kJ/mol for each step, but if stronger covalent bonds form, it is approximately –80 kJ/mol.

The change of the standard entropy in the formation of complexes is sensitive to the nature of the medium in which the complex is located. In aqueous solutions, ΔS^0 is usually positive. This unexpected value is determined by the disordered structure of water around the complex. The positive change of the entropy as a result of disordering is usually considerably greater than the negative value of the change of entropy as a result of the transitional entropy of the individual mechanic ions and ligands to the vibration and rotation contribution of the complex. If the ligand has a negative charge, the neutralisation of the charge in the formation of the complex reduces the number of the ions in the system that cause a change of entropy, and also reduces the number of co-ordinated water molecules. This results in a large positive change of entropy and destabilises the complex.

Bonding of the metallic ion with the non-charged ligand does not reduce the number of ions present in the system and the water molecules are slightly reoriented. In the formation of a complex this results in a relatively small positive or even negative change of entropy. ΔS^0 is generally the most important factor controlling the stability of complex compounds. The standard entropies of the formation of cations in the aqueous solutions tend to more positive values at higher temperatures and for anions these values are more negative. Generally, it appears that as the temperature increases, the change of the entropy of the formation of complexes become more positive and the complex because more stable.

At different temperatures from 25 °C, the change of the standard enthalpy of the reaction, ΔH_T^0, can be written in the following form:

$$\Delta H_T^0 = \Delta H_{298}^0 + \int_{298}^{T} \left(\frac{\delta \Delta H^0}{\delta T} \right)_p dT. \tag{4.62}$$

Similarly, the change of the standard entropy may be expressed by the equation

$$\Delta S_T^0 = \Delta S_{298}^0 + \int_{298}^{T} \left(\frac{\delta \Delta S^0}{\delta T} \right)_p dT. \tag{4.63}$$

Calculation of ΔG^0 at higher temperatures requires information on the temperature dependence of the changes of standard enthalpy and entropy. In this case it is considered that the pressure does not change, but at pressures other than normal pressure the effect of pressure on the values of $\Delta H^0{}_T$ and $\Delta S^0{}_T$ must taken into account.

The change of the reaction enthalpy depends on the enthalpies of all reactants taking part. The thermal content, i.e., enthalpy is defined by the equation:

$$H = U + pV$$

were U is the total internal energy, p is pressure and V is the volume of the system.

In other words, the thermal content must depend on the amount of heat required to increase the temperature of the system by one degree which is the heat capacity c_v at constant volume and c_p is constant pressure. To increase temperature dT at a constant volume the required amount of heat d_{qv} is $d_{qv} = \mathrm{d}U = c_v \mathrm{d}T$ and consequently:

$$c_v = \left(\frac{\delta U}{\delta T} \right)$$

At constant pressure $d_{qv} = dH - c_p dT$ and consequently $c_p = \left(\frac{\delta H}{\delta T} \right)_p$.

These definitions of the heat capacities can be used to determine the relationship between the change of the total heat capacity of the reaction and the calorific effect of the reaction in relation to temperature. If the initial state is I and the final state II, then

$$\Delta c_v = (c_v)_{II} - (c_v)_I$$
$$\Delta U = U_{II} - U_I$$
$$\Delta H = H_{II} - H_I$$

and subtracting the relevant equations for the initial and final states one obtains

$$\Delta c_v = \left(\frac{\delta \Delta U}{\delta T}\right)_v \tag{4.64}$$

$$\Delta c_p = \left(\frac{\delta \Delta H}{\delta T}\right)_p . \tag{4.65}$$

These are Kirchhoff laws. The heat capacity of 1 g of substance of specie is specific heat. For theoretical assumptions we consider heat capacity of 1 gram-molecule and the values c_v and c_p describe these molar capacities. The molar capacity is linked with a change of entropy by the following relation:

$$\Delta S = \frac{q}{T}$$

where q is the amount of absorbed heat. At a constant pressure, this amount of heat is equal to ΔH so that

$$\Delta S = \frac{\Delta H}{T}.$$

Equation (4.65) shows that

$$\frac{\delta c_p}{T} \mathrm{d}T = \left(\frac{\delta \Delta H}{T \delta T}\right)_P \mathrm{d}T = \left(\frac{\delta \Delta S}{\delta T}\right) \mathrm{d}T.$$

The measurable values of Δc_p are substituted into equations (4.62) and (4.63) relating to the considered chemical equations. The following equations hold for the values ΔH_T^0 and ΔS_T^0 at any temperature:

$$\Delta H_T^0 = \Delta H_{298}^0 + \int_{298}^{T} \Delta c_p^0 \mathrm{d}T \tag{4.66}$$

$$\Delta S_T^0 = \Delta S_{298}^0 + \int_{298}^{T} \frac{\Delta c_p^0 \mathrm{d}T}{T} \tag{4.67}$$

After substituting into the equation $\Delta G^0 = \Delta H^0 - T\Delta S^0$

$$\Delta G_T^0 = \Delta H_{298}^0 + \int_{298}^{T} \Delta c_p^0 \mathrm{d}T - T\Delta S_{298}^0 - T\int_{298}^{T} \frac{\Delta c_p^0}{T} \mathrm{d}T \tag{4.68}$$

The expression

$$\frac{1}{T}\int_{298}^{T} \Delta c_p^0 \mathrm{d}T - \int_{0}^{T} \frac{\Delta c_p^0}{T} \mathrm{d}T$$

is known as the function of Gibbs free energy, Δfef_T^0 which may be expressed as follows

$$\Delta fef_T^0 = \frac{1}{T}\int_{298}^{T} \Delta c_p^0 \mathrm{d}T - \int_{298}^{T} \frac{\Delta c_p^0}{T} \mathrm{d}T - \Delta S_{298}^0 \tag{4.69}$$

Substitution into (4.67) gives

$$\Delta G_T^0 = \Delta H_{298}^0 + T\Delta fef_T^0 \tag{4.70}$$

$$\log K_T = -\frac{\Delta H_{298}^0}{2.303\mathrm{RT}} - \frac{\Delta fef_T^0}{2.303R} \tag{4.71}$$

The equilibrium constant of the reactions is determined from equation (4.71) for any temperature assuming that the temperature function of the standard Gibbs energy is available which then assumes that the change of the standard heat capacity if the reaction at constant pressure ΔC_p^0 is known.

Although the free energy functions are available for many chemical reactions, they are not available for the majority of equilibria in aqueous solutions. For these purposes it is then necessary to make a number of assumptions in order to be able to estimate the approximate values of log K_T of the equations of hydrometallurgical systems.

One of the approximation methods is the van't-Hoff reaction isochore. The Gibbs–Helmholtz equation may be written in the form

$$\Delta G^0 - \Delta H = T\left(\frac{\delta \Delta G^0}{\delta T}\right)_p$$

and the temperature change of the equilibrium constant is obtained by substituting $-RT \ln K$ for ΔG^0. The van't-Hoff reaction isochore expressed in this form

$$\frac{\delta \ln K}{\delta T} = \frac{\Delta H}{RT^2} \tag{4.72}$$

has the integral form

$$\log K_T = -\frac{\Delta H}{2.303R}\left(\frac{1}{\mathrm{T}} - \frac{1}{298}\right) + \log K_{298} \tag{4.73}$$

To solve this equation it is necessary to consider ΔH as a constant between the temperatures ΔT and 298 K because if the change ΔH with temperature is known, Δc_p^0 may be calculated from equation (4.66). Assuming that ΔH may be replaced by ΔH_{298}^0 at all temperatures is identical with the assumption that Δc_p^0 is equal to zero.

The values ΔH^0, ΔS^0 and ΔC^0_p are obtained by the conventional procedure from the values of the individual chemical reactions of each product and the reactant, they all can change with temperature. Because of the shortage of heat capacity data, it may be rational to assume that ΔC^0_p is a constant and is equal to zero. Consequently, equation (4.69) changes to

$$\Delta fef_T^0 = \frac{\Delta c_p^0}{T}\int_{298}^{T} \mathrm{d}T - \Delta c_p^0 \int_{298}^{T} \frac{\mathrm{d}T}{T} - \Delta S_{298}^0 \tag{4.74}$$

and after integration

$$\Delta fef_T^0 = \frac{\Delta c_p^0}{T}(T - 298) - \Delta c_p^0(\ln T - \ln 298) - \Delta S_{298}^0 \tag{4.74}$$

Combining this relationship with (4.71) gives

$$\log K_T = -\frac{\Delta H^0_{298}}{2.303\mathrm{R}}\left(\frac{1}{T}-\frac{1}{298}\right)+\frac{\Delta c^0_p}{R}\left[\frac{1}{2.303}\left(\frac{298}{T}-1\right)-\log\frac{298}{T}\right]+\log K_{298} \quad (4.75)$$

This is equivalent to the assumption that ΔH^0 of the reaction changes in a linear manner with temperature.

In other words, the errors of the values of log K_T, caused by the assumption that $\Delta c^0_p = 0$ and by the use of the van't-Hoff reaction isochore, depend on the relative values of the expression for enthalpy and heat capacity in the equation (4.76). If the value of ΔH^0_{298} is high, the error is small but, as already mentioned, the change of the enthalpy of the formation of many equilibrium complexes is small. The change of the heat capacity of these reactions is generally in the range +200 to –200 J/mol deg at 25°C, and if ΔH^0 is of the order of 8 kJ, the resultant error in the values of log K_T may be of the order of 1–2 units at higher temperatures.

There is insufficient information on the values of Δc^0_p, ΔH^0 and ΔS^0 for complex reactions. There are only several cases of the available values of the equilibrium constants in a small temperature range which means that Δc^0_p can be calculated by assuming that ΔH^0 is a linear function of temperature. Of course, the resultant values are highly sensitive to even small errors of the equilibrium constants. Later, we shall describe the method of calculating the thermodynamic data for higher temperatures.

The values of the equilibrium constants depend on pressure at constant temperature in accordance with the equation:

$$\left(\frac{\delta \ln K}{\delta p}\right)_T = -\frac{\Delta V^0}{RT} \quad (4.77)$$

where ΔV^0 is the change of the partial molal volume in the reaction. Its value and also the volume changes of the dissolved ions and molecules in transition from the standard state to the given solution at higher pressures are low. This means that in the case of subcritical pressures the effect of pressure on the equilibrium constants in the aqueous solutions is small.

4.3.1. Selection of the values of equilibrium constants

In the beginning of the section it was stated that the equilibrium constants of the complex species are based on their formation from simple hydrated metallic ions and ligands. Equations were also presented for determining the constants through the expressions of the activity of specie, taking part in the equilibrium reactions. The extent of the formations is complex and the average number of the ligands was defined using concentration values. The methods of measuring the equilibrium constants usually include the change of the ratio of the ligand to the concentration of the metallic ion with determined constant, and also the measurement of the amount of ligands in the complex or of the number of metallic ions in the complex for every ratio. However, since this is a very complicated system, sensitive to many parameters, there are relatively small number of the values of the equilibrium constants of specific systems. The Pitzer methods cannot be used because the interaction coefficients of the complex ions of the salts are not available.

In order to avoid these deviations, the salt with a relatively high concentration of 0.2–2 M is usually added to the solutions. The salt supports the electrolyte properties so that the change of the total ionic strength during the experiment is small. If measurements are taken of the equilibrium constants at several values of ion strength, the resultant values can be extrapolated in such a manner as to obtain the value of the constant at the total zero ionic strength which is identical for all activity coefficients.

The additional electrolyte should be inert so that its ions should not interfere with any investigated equilibrium reaction. Usually, they include the perchlorates of alkaline metals or nitrates, assuming that none of these ions forms complexes with the majority of metals. Of course, this may not always be the case and if the investigated ligands form relatively brittle complexes with a metal, a competing reaction may also take place. For example, if it is required to determine the value β_4 for the chloride complex $ZnCl_4^{2-}$ for calculations to determine the separation of zinc and nickel by liquid extraction, it is necessary to consider whether the strong solution of sodium nitrate can be used as an additional electrolyte because the following reaction may take place:

$$ZnNO_3^+ + 4Cl^- \rightarrow ZnCl_4^{2-} + NO_3^-$$

If the stability of $ZnNO_3^+$ is sufficiently high taking into account the simple hydrated ion Zn^{2+}, the exchange reaction will influence greatly the measured value of β_4. The values of log β_3 and log β_4 at I = 4 and 25°C at +1 and –1 for Cl^- and for the nitrate the value of log K_1 in the same conditions is equal to 0.11 [24].

In fact, there are two sources of basic information on the stability constants. Sillen et. al. [25, 26] took into account the published data on the equilibria of metallic complexes, heats, entropies and free energies. Another source [24] is published in four volumes, with the inorganic compounds discussed in the fourth volume which also contains the tables of free energies (values of K), heat and entropies determined by at least two authors. This source is suitable for many applications because it gives the individual values of the constants for every ionic strength from 1 to 3, including zero, together with the appropriate values of ΔH and ΔS. Although the calculation methods used at present are very effective, these sources are exceptionally valuable from the viewpoint of the content of the experimental measured data.

In conclusion, if we are concerned with the relatively concentrated solutions, the two main approaches, i.e. the formation of complexes and grouping of the atoms cannot be mixed together. The equation derived by Pitzer on the basis of an interaction between the ions does not consider the formation of new specific species. Consequently, for example, in the diluted solution $CoCl_2$ the value G^{ex} and then γ_{CoCl2} is regarded as the result of the interaction between Co^{2+} and Cl^- and this is described by the values of $\beta^{(0)}$, $\beta^{(1)}$ and $\beta C^{(\phi)}$ in the equations (4.39) and (4.44)–(4.46). This solution is light pink and contains simple hydrated ions Co^{2+}. After adding HCl or highly soluble chlorides, for example, LiCl or $CaCl_2$ and the concentration of Cl^- reaches the values of approximately 6 M, the colour changes to blue because of the formation of $CoCl_4^{2-}$, containing the tetragonal (sp^3) configuration of Cl^- ions around the Co^{2+} ion with covalent bonding. Of course, it is then not rational to use the Pitzer equation for obtaining more accurate data for the activity in the $CoCl_2$–HCl system for high chloride concentrations without using many parameters. Although these high concentrations of the chlorides are essential for the formation of species, no experimental data have been published on the values of β_4 for $CoCl_4^{2-}$ and, therefore, the values of ΔG^0 of its formation.

Many bivalent metallic sulphates have activity coefficients in the range 0.7–0.035 for the concentration in the range 10^{-3} to 2 M. The low values of $\gamma_{\pm}$ are usually interpreted as the consequence of the presence of ionic associations of various specie, such as $NiSO_{4(aq)}$, $CoSO_{4(aq)}$, $ZnSO_{4(aq)}$ and so on. Of course, the activity data of these 2:2 electrolytes may be made more accurate by adding other parameters to the Pitzer equation [10]. The behaviour of the metallic salt-water systems can be described by interaction coefficients for Me^{2+}, SO_4^{2-} or equilibrium constants providing the thermodynamic data ΔG^0 and ΔS^0 for $MeSO_{4(aq)}$ compounds assuming that the equilibrium constants for a suitable temperature range were measured. However, if the data of Me^{2+}, SO_4^{2-} and $MeSO_{4(aq)}$ are used together with the interaction coefficients for $Me^{2+}SO_4^{2-}$, serious errors may appear because these data were combined and two independent methods were used.

The thermodynamics of the behaviour transition metals in aqueous solutions is almost always described by the model of grouping of the ions and equilibrium constants were discrete species. In most cases, these constants are determined from the measurement of the number of ligands or metal complexes at different metal–ligand ratios using several metal concentrations. The measurements are evaluated on basis of the presence of certain species in the solution. For example, in the Cu^{2+}–Cl^-–H_2O system at different pH conditions and chloride concentration the presence of the following species is considered: Cu^{2+}, H^+, OH^-, Cl^-, $CuO^{2-}_{2(aq)}$, $HCuO^{2-}_{(aq)}$, $Cu(OH)^+_{(aq)}$, $Cu(OH)_{2(aq)}$, $Cu(OH)^{2-}_{4(aq)}$, $Cu_2(OH)^{2+}_{2(aq)}$, $CuCl^+_{(aq)}$, $CuCl_{2(aq)}$, $CuCl_3^-$, $CuCl_4^{2-}$. The experimental measurements have confirmed or rejected the presence of some species under certain conditions and also the values of the appropriate equilibrium constants. However, to confirm the presence of specific species in a solution it is necessary to provide further information. It may be obtained by special analytical methods, such as the measurement of spectral certain species, ultrasound methods, measurements of electrical characteristics, etc.

The model of the ionic interaction for this system requires the cation–anion interaction between Cu^{2+}–OH^- and Cu^{2+}–Cl^-, or the anion–anion interaction OH^-–Cl^- and the triple interaction Cu^{2+}–OH^-–Cl. An advantage of the model of ion association for hydrometallurgical conditions is that it permits the formation of diagrams depicting the regions of stability of the individual species in the system under the given conditions. This will be discussed later.

References

1. Debye P., Hückel E.: *Physik. Z.*, 24, 185–206, 1923.
2. Harned H.S., Owen B.B.: The Physical Chemistry of Electrolytic Solutions, 3rd edition, Reinhold Publ. Corp., New York, 1958
3. Robinson R.A., Stokes R.H.: Electrolyte Solutions, 2nd.edition, Butterworths,

London, 1965
4. Davies C.W.: *J. Chem. Soc.*, London, 1938, 2093–2098.
5. Zemaitis J.F., Jr., Clark D.M., Rafal M., Scrivners N.C.: Handbook of Aqueous Electrolyte Thermodynamics, Design Inst. for Physical Property Data, AIChE, New York, 1966.
6. Horvath A.L.: Handbook of Aqueous Electrolyte Solutions, Physical Properties, Estimation and Correlation Methods, Ellis Horwood Ltd., Chichester, 1985
7. Davies R.H.: Aqueous Solutions of Dilute and Concentrated Electrolytes, Chemical Thermodynamics in Industry, Barry T.I. ed., Blackwell Sci. Publ., Oxford, 1985.
8. Pitzer K.S.: *J. Phys. Chem.*, 77, 1973, 268–277.
9. Pitzer K.S., Mayorga G.: *J. Phys. Chem.*, 77, 1973, 2300–2308.
10. Pitzer K.S., Mayorga G.: *J. Solution Chem.*, 3, 1974, 539–546.
11. Pitzer K.S., Kim J.J.: *J. Am. Chem. Soc.*, 96, 1974, 5701 – 5707.
12. Pitzer K.S.: *J. Solution Chem.*, 4, 1975, 249–265.
13. Pitzer K.S., Silvester L.F.: *J. Solution Chem*, 5, 1976, 269–278.
14. Pitzer K.S., Rabindra N.R., Silvester L.F.: *J. Am. Chem. Soc.*, 99, 1977, 4930–4936.
15. Silvester L.F., Pitzer K.S.: *J. Phys. Chem.*, 81, 1977, 1822–1828.
16. Pitzer K.S., Peterson J.R., Silvester L.F.: *J. Solution Chem.*, 7, 1978, 45–56.
17. Silvester L.F., Pitzer K.S.: *J. Solution Chem.*, 7, 1978, 327–337.
18. Wu Y.C., Rush R.M., Scatchard G.: *J. Phys. Chem.*, 72, 1968, 4048–4057.
19. Pitzer K.S., Murdzek J.S.: *J. Solution Chem.*, 11, 1982, 409–413.
20. Rogers P.S.Z., Pitzer K.S.: *J. Phys. Chem.*, 85, 1981, 2886–2895.
21. Rogers P.S.Z., Pitzer K.S.: *J. Phys. Chem.*, Ref. Data, 11, 1982, 15–81.
22. Bromley L.A.: *AIChE J.*, 19, 1973, 313–320.
23. Burkin .R.: Chemical Hydrometallurgy, Theory and Principles, Imperial College Press, 2001.
24. Smith R.M., Martell A.E.: Critical Stability Constants, Plenum Press, New York, volumes 1–5, 1974–1982.
25. Sillén L.G., Martell A.E.: Stability Constants of Metal–Ion Complexes, 2nd edition, Special Publication No. 17, The Chemical Society, London, 1964.
26. Sillén L.G., Högfeld E., Martell A.E.: Stability Constants of Metal–Ion Complexes, Supplement No.1, Special Publication No. 25, The Chemical Society, London, 1971.

Chapter 5

THERMODYNAMIC STUDIES OF HETEROGENEOUS SYSTEMS IN AN AQUEOUS MEDIUM

The thermodynamic studies of hydrometallurgical systems are based on calculating the conditions of chemical equilibrium between species. They may be in the solid, liquid or gaseous state. For different conditions and circumstances it is therefore necessary to obtain different types of information and data which may be obtained and processed by different methods. Three methods are used in most cases:

- diagrams of the areas of stability of system component;
- species diagram;
- determination of equilibrium in concentrated solutions

The diagrams of the areas of stability show the conditions in which the components of the system in the solid, liquid and gaseous state take part and dissolved species are stable in aqueous solutions, owing to the fact that they are hydrometallurgical systems. The systems are simple, of the type Me–H_2O or more complex, Me–S–H_2O, Me_1–Me_2–S–H_2O, Me–Cl_2–H_2O, Me–S–Cl_2–H_2O, etc., and are usually depicted in the co-ordinates pH and the oxide–reduction potential *E*. These are diagrams are identical with the so-called Pourbaix diagram used in corrosion studies. If the system contains also a reaction gas in water, for example, CO_2, the variables used may include pH and the logarithm of partial pressure log p_{CO2}. These diagrams help to understand the chemism of the system at individual temperatures and activities of dissolved species, using fugacities of gaseous reactants.

In many cases it is important to know the concentration or, at least, activity of the individual species in a solution and the proportions of the individual species present; there may be many of these species in complex solutions, and it is also important to know their behaviour with a change of chemical conditions. The usual types of diagrams of species show how the individual fractions of the total amount of the metal present in the form of individual species change with a change of the conditions in the system, for example, change of pH log E, log $\{Cl^-\}$, etc. Complex solutions contain many metals and ions. If the total activity is given by the sum of partial activities of each species present, or if the activity of each species is taken into account, these diagrams (either simple or more complicated) may be constructed. Since this is preceded by relatively complicated calculations, this work should be carried out using computers for both calculations and graphical images.

The phase equilibrium in the concentrated solutions is not utilised so often as the construction of diagrams. Instead of this, the model of the Gibbs energy in aqueous solutions is used for calculating solubility in the system. In all these methods one can obtain at least the Gibbs energy on reaching equilibrium. However, in a real system, the establishment of equilibrium is often inhibited by kinetic factors.

The information on the chemistry of aqueous solutions has been obtained together with a large number of published and available theoretical and experimental thermodynamic data, and the large number of numerical data was gradually made more accurate and 'denser' using diagrams and graphs. For the systems including metallic and non-metallic compounds in the aqueous medium, typical example, hydrometallurgy, the most widely used and obviously more suitable procedure for presenting the data is the so-called potential–pH or *E*–pH diagram showing the areas of stability of the individual considered compounds in the co-ordinates pH and the oxidation potential *E* in relation to the available data, i.e. in the so-called normal conditions (environment temperature 25 °C, total pressure 0.1 MPa) or also at higher temperatures.

This problem has been studied by many researchers and their studies have been based on original studies by Latimer [1] and Pourbaix and Valensi [2, 5]. Initial studies were published for the room temperature of 25 °C, depending on the type of thermodynamic data available [6–10]. However, later it was found that temperature has a strong effect on the course of hydrometallurgical processes and, therefore, *E*–pH diagrams were published gradually for higher temperatures [11–18].

A fundamental study in this area is that of Criss and Coble [19] which laid foundations for the extrapolation of thermodynamic values required for the calculation of the *E*–pH diagram at higher temperatures. Gradually, computers were used for calculation and graphic presentation of the *E*–pH diagrams. Several programs have been made available for thermodynamic calculations of hydrometallurgical processes [16, 20–24]. One of the most universal system is the HSC system developed at Outokumpu Oy, Finland [25] with a huge database containing more than 9000 standard data. After supplementing the user standard database, this program was also used for this work. The study [26] provides information on the user software, for pyro- and hydrometallurgical processes and thermodynamic parameters.

The *E*–pH diagrams indicate the areas of stability of the individual compounds in relation to the pH value and the potential. The diagrams can be used to predict the essential conditions for the existence of these compounds and the tendency for their behaviour in the hydrometallurgical process. The above considerations already indicate the most significant disadvantage of the currently available diagram, i.e., they do not include all existing compounds, either in the solid or ion state, although their existence has been described many times in the literature. In the case of sulphur, it is a large number of allotropes of elemental sulphur or ions of sulphur compounds. As regards the sulphides of copper and iron, it is the existence of non-stoichiometric sulphides of copper $Cu_{2-x}S$ and $Cu_xFe_yS_z$, which have been confirmed to exist as minerals or even as leaching products.

The actual leaching process takes place in the solid phase (s) – liquid phase (l) or gas component (g) heterogeneous system. The required components are transferred from ores and concentrates to the solution by leaching mainly in electrolyte solutions (solutions of acids, alkali and salts). To release the required ions from the leached material into the solution, it is initially necessary to weaken the binding forces in the main matrix. This takes place mainly by the effect of the leaching agent with a sufficiently high permitivity leading to weakening of the attracting electrostatic forces between the ions to such an extent that thermal motion forces are already sufficient to release them. High permitivity is typical of water which, in addition to this effect, also influences the transition of ions into the solution by the effect of dipole moments. This is one of the main theoretical reasons why the majority of leaching processes are performed in aqueous solutions.

The potential–pH diagrams as presented by Pourbaix are the diagrams stating the equilibrium conditions of existence of the stable phases of the individual elements in water under equilibrium conditions. These diagrams may theoretically describe reactions of hydrolysis, oxidation and reduction taking place in aqueous solutions. However, since the hydrometallurgical process, such as leaching, electrolysis, precipitation, etc., take place in more complicated systems containing several components, the studies by Pourbaix were used gradually for examining and publishing special studies concerned with three- and multi-component systems found in hydrometallurgical practice. Typical systems are: $Me–S–H_2O$, $Me–Cl–H_2O$, $Me–NH_4–H_2O$, $Me–N_2–C–H_2O$, etc., where Me is the examined metal in the solution. Quaternary or polycomponent systems have of course been also described with varying success.

E–pH diagrams are not the only type of diagram used in the thermodynamic studies of heterogeneous equilibria. These are a special case of the thermodynamic representation, also available in pyrometallurgy, where it is realised by a means of the potential–potential diagrams, i.e. the chemical potential of the substances taking part. In comparison with these diagrams, the *E*–pH diagrams have an advantage that the scale of the co-ordinates *E* and pH is used in the real range, i.e. $0 < pH < 14$ and the values of the real potentials in the area of stability of water, for example $-1\ V < E < +1.4\ V$. In most cases, the potential–potential diagrams show only the tendencies of variation of the composition of the participating substances in the equilibrium system, and many values indicated by the diagram, cannot be reached in reality. A typical example are partial pressures lower than 10^{-10} and higher than 10^{+10} MPa.

A considerable disadvantage of the *E*–pH diagrams is the fact that the substances are in the dissociated state in the solutions and their activity is not equal to 1. For these compounds there is only a restricted set of the thermodynamic values and, consequently, the *E*–pH diagrams cannot be used in such a wide range as the potential–potential diagrams for which the number of data is considerably greater, since these are pure substances with unit activity.

Anyway, it must be remembered that the hydrometallurgical processes are relatively far away from the equilibrium conditions (= thermodynamic) and even a small amount of the impurities which are always present in the real material, can greatly change the final behaviour of the system. For example, the presence of sodium or potassium ions in ferrous sulphate at low values of pH (<3) results in the formation of more suitable compounds of the yarosite type than oxides or hydroxides of iron.

5.1. Theoretical principle of the *E*–pH diagrams

The detailed description and application of the *E*–pH diagrams were presented in fundamental studies [27, 28] and, therefore, in this case, we only describe the system briefly which is essential for construction and understanding *E*–pH diagrams. It is assumed that the reader has basic knowledge of chemical thermodynamics and thermal chemistry.

The criterion for the feasibility of the chemical process is the maximum use for work which is equal to the decrease of the free enthalpy at constant pressure and temperature. The reaction takes place in the given direction if its maximum use for work, i.e. the decrease of the free enthalpy ΔG, is positive. The maximum use for work can be regarded as a measure of the chemical affinity of the reacting substances. Chemical reactions are irreversible processes taking place spontaneously in a specific direction until thermodynamic equilibrium is reached.

In a closed system consisting of several substances, the latter may react mutually in accordance with the equation

$$aA + bB + \ldots \rightarrow pP + rR \ldots \tag{5.1}$$

At a constant pressure and temperature the reaction takes place spontaneously in the directly indicated by the arrow if the system doesn't carry out any work, i.e., if the Gibbs energy of the system decreases ($\Delta G < 0$). If a mols of substance A, b mols of substance B, etc., interact, resulting in the formation of p mols of substance P, r mols of substance R, i.e., the following equation holds for the change of the free enthalpy of the system:

$$\Delta G = p\mu_p + r\mu_r + \ldots - a\mu_A - b\mu_B - \tag{5.2}$$

where μ indicates the chemical potentials of the individual substances – partial molar Gibbs energies. If all the reacting substances are in the standard state, the appropriate change of the Gibbs energy ΔG^0 is expressed by

$$\Delta G^0 = p\mu_p^{\circ} + r\mu_r^{\circ} + \ldots - a\mu_A^{\circ} - b\mu_B^{\circ} - \tag{5.3}$$

The difference between the equations (5.2) and (5.3) is therefore

$$\Delta G - \Delta G^0 = p\left(\mu_p - \mu_p^{\circ}\right) + r\left(\mu_R - \mu_R^{\circ}\right) + \ldots - a\left(\mu_A - \mu_A^{\circ}\right) - b\left(\mu_B - \mu_B^{\circ}\right) - \ldots \tag{5.4}$$

This equation holds for all systems, homogeneous and heterogeneous. The following equation holds for the dependence of the chemical potential on the activity of the component

$$\mu_i - \mu_i^\circ = RT \ln a_i \tag{5.5}$$

which after substituting into (5.4) and modification gives

$$\Delta G - \Delta G^0 = RT \ln \frac{a_p^p \cdot a_R^r \ldots}{a_A^a \cdot a_B^b \ldots} \tag{5.6}$$

For a system to reach equilibrium, it must apply that $\Delta G = 0$ and then

$$-\Delta G^0 = RT \ln \left[\frac{a_p^p \cdot a_R^r \ldots}{a_A^a \cdot a_B^b \ldots} \right]_{equilib} \tag{5.7}$$

the values in the square brackets indicate the equilibrium activities of the reacting component, in contrast to the values of a in equation (5.6) which represent the activity of the components at the given moment of the reaction (for example, initial activity). Since ΔG^0 is the standard change of the Gibbs energy and is constant for a specific system and depends only on the selection of the standard state of temperature, then the value

$$K_a = \left[\frac{a_p^p \cdot a_R^r \ldots}{a_A^a \cdot a_B^b \ldots} \right]_{equilib} \tag{5.8}$$

is also constant. This value which depends only on temperature and pressure and is independent of the activity of the reacting substances, is referred to as the equilibrium constant of the appropriate reaction. After substituting reaction (5.8) and (5.7) into (5.6) we obtain the general equation of the reaction isotherm

$$\Delta G = -RT \left(\ln K_a - \ln \frac{a_p^p \cdot a_R^r \ldots}{a_A^a \cdot a_B^b \ldots} \right) \tag{5.9}$$

This equation is valid equally for the reactions taking place in a gas mixture, a liquid solution and also for heterogeneous reactions. The general form of the reaction isotherm is often presented in the shortened form

$$\Delta G = -RT(\ln K_a - \Delta \ln a) \tag{5.10}$$

where $\Delta\ln a$ is the stoichiometric sum of the logarithms of the activities of the reacting components.

If the activities of the reacting components are equal to equilibrium activities, then $\Delta\ln a = \Delta\ln K_a$ and therefore $\Delta G = 0$. This means that the reaction does not take place, or the amount of the substance reacting in a specific direction during a specific period is the same as in the reverse direction. The difference between the activities of the reacting substances and the equilibrium activity increases, the value of ΔG also increases. For every spontaneous phenomenon $\Delta G < 0$, i.e. takes place in the direction in which the Gibbs energy decreases and the system carries out work. The reaction takes place in the direction given by its equation when $\ln K_a > \Delta\ln a$ and in the reverse direction when $\ln K_a < \Delta\ln a$.

The standard change of Gibbs energy described by the equation

$$\Delta G^{\circ} = -RT \ln K_a \tag{5.11}$$

is regarded as the numerical measure of the chemical affinity of the reacting substances. This is of course the maximum work of the reaction, and the initial activities of all substances taking part in the reaction are equal to one, i.e., in the standard state. When comparing the possible reactions in a system under the same conditions, the reaction whose ΔG value is more negative is thermodynamically more likely to take place.

The potential–pH diagrams, as already mentioned, are defined for examining the heterogeneous equilibrium of the considered substances in the aqueous medium. The co-ordinates E and pH result from the available and suitable data and quantities.

5.1.1. pH and its principle

A water molecule may be slightly dissociated in accordance with the equation

$$H_2O \Leftrightarrow H^+ + OH^- \tag{5.12}$$

The equilibrium between the molecules and the ions is characterised by the dissociation constant of water

$$K_a = \frac{a_{H^+} a_{OH^-}}{a_{H_2O}} \tag{5.13}$$

or after substituting the concentration

$$K_a = \frac{\gamma_{\mp}^2}{\gamma_{H_2O}} \frac{c_{H^+} c_{OH^-}}{c_{H_2O}} \tag{5.14}$$

where γ_{H2O} is the activity coefficient of the electrically neutral non-dissociated molecules of water which may be regarded as equal to one, i.e. $\gamma_{H2O} = 1$, and $\gamma_{\pm}$ is the mean activity coefficient of ionised water. Since the dissociation degree of water is very low, the concentration, and consequently, activity of non-dissociated water may be regarded as constant which means that dissociation has no significant effect on its value. The product $K_a \cdot a_{H2O}$ can be replaced by a new constant K_V:

$$K_v = a_{H^+} \cdot a_{OH^-} \tag{5.15}$$

or

$$K_v = \gamma_{\pm}^2 c_{H^+} \cdot c_{OH^-} \tag{5.16}$$

Constant K_V is given by the product of the activities of the hydrogen and hydroxide ions and is referred to as the ion product of water.

Since pure water is a very weak electrolyte ($\mu \to 0$) it may be assumed that $\gamma_{\pm} = 1$. Therefore, the ion product of pure water can be described by the equation

$$K_v = c_{H^+} \cdot c_{OH} \tag{5.17}$$

The dissociation of molecules of water in pure water results in the formation of the same number of hydrogen and hydroxide ions, so that:

$$c_{H^+} = c_{OH^+} = \sqrt{K_v} \tag{5.18}$$

The value of the ion product of water depends on temperature. Table 5.1 gives the values of K_V for several temperatures.

Table 5.1. Values of the ion product of water

t [°C]	0	10	20	25	30	50	100
$K_v \cdot 10^{14}$[mol^2 l^{-2}]	0.1139	0.292	0.6809	1.008	1.469	5.474	59
pH	7.472	7.268	7.084	6.999	6.917	6.631	6.12

For 25 °C it applies with sufficient accuracy that $K_v = 10^{-14}$ mol^2 l^{-2}, in pure water $c_{H+} = 10^{-7}$ mol l^{-1}.

The equilibrium constant expressed by the activity depends only on temperature and not on the impurities of other substances. Therefore, the ion product of water is constant not only in pure water but also in every aqueous solution of acids, alkali and salts, of course, in this case $a_{H+} \neq a_{OH-}$, or $c_{H+} \neq c_{OH-}$. Since in pure water $c_{H+} = c_{OH-}$, the solution is neutral, and if in this solution $a_{H+} > a_{OH-}$ ($c_{H+} > c_{OH-}$) it is acid if hydrogen ions prevail in it, i.e. $a_H > a_{OH}$, ($c_H > c_{OH-}$) and basic if $a_{H+} < a_{OH-}$, ($c_{H+} < c_{OH-}$). To characterise the reaction of the solution, it is sufficient to know a_{H+} (or c_{H+}), since a_{OH-} (or c_{OH-}) can be easily calculated from the equations (5.15) and (5.17). The reaction may be characterised efficiently by the negative decadic logarithm of the activity of hydrogen ions which describes the pH of the solution as

$$\text{pH} = -\log a_{H^+} \tag{5.19}$$

In pure water at 25 °C, pH = 7, in acid solutions pH < 7 and in alkali solutions pH > 7.

5.1.2. Potential E and its principle

In specific suitably selected conditions the work of a chemical reaction can be transformed to the work of electrical current in a device referred to as the galvanic cell. Each galvanic cell consists of two electrodes mutually conductively connected, and a specific potential difference is established between them. Chemical phenomena – electrode processes take place on the individual electrodes, and the sum of these processes represents the total chemical reaction of the given cell.

If a pure metal Me is immersed in pure water, the effort to equalise the chemical potentials of the metal in the solid and liquid phases results in a phenomenon similar to dissolution of a salt, i.e. there is a method for obtaining equilibrium between the ions in the metal and the ions in the liquid. The difference in comparison with the equilibrium between the saturated solutions and the solid phase in dissolution of salt and other crystalline substances is that the appropriate number of the electrons cannot be transferred into the solution together with the ions. The electrons remain in the metal and inhibit, by electrostatic forces, the transfer of other ions into the solution. An electric double layer forms at the interface between the metal and the solution. This layer may also be treated as a planar condenser whose negative plate is the surface of metal, the positive plate is the layer of cations adjacent to the surface. If a small amount of ions is dissolved, the charge of the double layer increases to such an extent that the effort of the metal to transfer the ions into the solution is no longer sufficient for overcoming the attraction force between the metal and the ions. Further dissolution ceases and equilibrium, characterised by a specific difference of the potentials between the metal and the solution, is established. In comparison with the solution, the potential of the metal is more negative and, consequently, its effort to transfer the ions into the solution is greater.

The galvanic cell is formed by two electrodes immersed in a solution, for example, Daniel's cell is formed by a positive copper electrode immersed in a solution of copper sulphate and a negative zinc electrode immersed in a zinc sulphate solution. Both solutions are separated by a diaphragm transmitting the ions. It may be assumed that the electrical potential inside every solution is the same everywhere. A potential difference forms between the electrodes and is referred to as electromotive voltage *E* of the galvanic cell. The zinc electrode is charged more negatively in relation to the solution than the copper electrode and has a higher negative potential. Connecting both electrodes by a conductor through a galvanometer results in an effort to equalise the charges and the potential difference leads to the transfer of the electrons from the more negative electrode to the more positive one. This disrupts the equilibrium at the interface between the electrodes and the solution. A decrease of the number of the electrons in metallic zinc is equalised by further dissolution of zinc so that the oxidation reaction will take place:

$$Zn^0 \rightarrow Zn^{2+} + 2e^-$$

On the other hand, the increase of the number of the electrons in copper is compensated by further precipitation of copper – by reduction – from the solution

$$Cu^{2+} + 2e^- \rightarrow Cu^0$$

Overall, the following reaction takes place in this cell:

$$Zn^0 + Cu^{2+} \rightarrow Zn^{2+} + Cu^0$$

Electric current is taken from the cell at the expense of the useful work of the cell. Zinc dissolves on the negative electrode and copper then precipitates on the positive electrode until the electrodes are conductively connected. The electromotoric force (EMF) voltage is equal to the voltage between the electrodes in the condition with no current.

The transfer of electric current in the galvanic cell is accompanied by the transfer of ions from one electrode to the other. The potential difference between the electrodes U indicates the work in transferring the unit amount of electricity from one electrode to the other. If the charge Q is transferred, the work carried out A' will be

$$A' = QU \tag{5.20}$$

The galvanic cell is in fact electrolytic equipment working in the reverse direction. If electric energy in electrolysis is supplied from the outside resulting in the transfer of ions and electrochemical reactions, in the galvanic cell the situation is completely reversed. The energy of the electrochemical reaction forms a potential difference at the electrodes. An electric field forms in which the ions move from one electrode to the other and carry out work. If one mole of ions of the given electrolyte reacts in the galvanic cell, the charge transfer according to the Farraday law is $Q = zF$, where z is the number of elementary charges and F is Farraday's constant. The work carried out will be:

$$A' = zFU \tag{5.21}$$

If the galvanic cell operates by a thermodynamically reversible mechanism, the useful work of the reaction is maximal. At constant pressure and temperature, this work is equal to the decrease of the Gibbs energy:

$$A' = -\Delta G \qquad (5.22)$$

Since the maximum voltage, equal to the EMF voltage, is linked with the maximum work in accordance with (5.21),

$$U_{\max} = E \qquad (5.23)$$

then

$$-\Delta G = zFE \qquad (5.24)$$

Using these assumptions, we can find the relationship between the thermal effects of the reaction taking place in the cell and its electromotive voltage. According to the Gibbs–Helmholtz equation:

$$\Delta G = -\Delta H = T\left(\frac{\delta \Delta G}{\delta T}\right)_p \qquad (5.25)$$

Substituting the partial derivation by the total derivation and after substitution:

$$zFE + \Delta H = T\left(\frac{\delta \Delta G}{\delta T}\right)_p \qquad (5.26)$$

or

$$E - T\frac{\mathrm{d}E}{\mathrm{d}T} = -\frac{\Delta H}{zF} \qquad (5.27)$$

where ΔH is the thermal effect of the reaction, and $\mathrm{d}E/\mathrm{d}T$ is the temperature coefficient of the electromotive (EMF) voltage at constant pressure.

EMF voltage can be accurately measured. Equation (5.24) gives the maximum useful work of the reaction taking place in the cell. Measuring the temperature coefficient of EMF voltage it is possible to determine the thermal effect of the reaction using (5.27).

At constant pressure

$$\frac{d\Delta G}{dT} = -\Delta S \tag{5.28}$$

After substituting for ΔG from (5.24):

$$\Delta S = zF\frac{dE}{dT} \tag{5.29}$$

Measurement of the temperature dependence E makes it possible to calculate the change of entropy during the reaction taking place in the cell.

If the reaction took place irreversibly away from the galvanic cell, the heat which would be generated by the reacting system will be $-\Delta H$. If the reaction takes place in the galvanic cell and is reversible, part of the energy changes to the electric work $A' = -\Delta G$, and part remains in the form of heat Q'_{min}, which cannot be converted to work, i.e., bound energy

$$-\Delta H - \Delta G + Q'_{min} \tag{5.30}$$

For the isothermal phenomenon:

$$-\Delta H - \Delta G - T\Delta S \tag{5.31}$$

and comparing these two relationships gives the following equation for the heat Q'_{min} released in the reversible reaction

$$Q'_{min} = -T\Delta S \tag{5.32}$$

or according to (5.29)

$$Q' = -TzF\frac{dE}{dT} \tag{5.33}$$

If the cell generates heat during a reaction, i.e. $Q'_{min} > 0$, the temperature coefficient E is negative. This is the most frequent case, at $Q'_{min} < 0$, E increases with increasing temperature.

The previously mentioned relationships and assumptions can also be applied to the reactions of a heterogeneous system taking place in aqueous solutions, i.e. in a conducting medium. The relationship between the value of E of the cell and the activity of the reacting substances is determined using the reaction isotherm. If a reaction

$$aA + bB + \ldots + ze^- = pP + rR + \ldots ze^-$$

takes place in a general galvanic cell, its maximum useful work can be described by the equation:

$$-\Delta G = RT\left(\ln K_a - \ln \frac{a_p^p \cdot a_R^r \ldots}{a_A^a \cdot a_B^b \ldots}\right) \tag{5.34}$$

where *Ka* is the equilibrium constant of the reaction, and a_A, ... a_p ... are the activities of the reacting substances. Substituting this equation into (5.24) gives

$$zFE = RT\left(\ln K_a - \ln \frac{a_p^p \cdot a_R^r \ldots}{a_A^a \cdot a_B^b \ldots}\right) \tag{5.35}$$

where z is the number of electrons exchanged in the given reaction between the reacting substances. Consequently, the EMF voltage of the cell may be described by the equation

$$E = \frac{RT}{zF}\left(\ln K_a - \ln \frac{a_p^p \cdot a_R^r \ldots}{a_A^a \cdot a_B^b \ldots}\right) \tag{5.36}$$

If chemical equilibrium is established in the galvanic cell, i.e., when the activities are equilibrium, the following equation holds

$$K_a = \frac{a_p^p \cdot a_R^r \ldots}{a_A^a \cdot a_B^b \ldots} \tag{5.37}$$

so that $E = 0$ and the cell does not generate electric current. According to (5.36) from the available value of E and activities we can determine the equilibrium constant of the reaction.

Equation (5.36) can be presented in the form

$$E = \frac{RT}{zF}\ln K_a + \frac{RT}{zF}\ln \frac{a_p^p \cdot a_R^r \ldots}{a_A^a \cdot a_B^b \ldots} \tag{5.38}$$

The first member on the right hand side of the equation contains constant quantities at constant temperature. It can be replaced by a single constant E^{o} so that the following relationship is obtained:

$$E = E^{o} + \frac{RT}{zF}\ln\frac{a_P^p \cdot a_R^r \ldots}{a_A^a \cdot a_B^b \ldots} \tag{5.39}$$

referred to as the Nernst equation.

E^{o} is the EMF voltage of the cell in the case in which the activities of all the reacting substances are equal to unity and is therefore referred to as the standard electromotoric force of the given cell. Expressing activities using the concentrations ($a_i = \mu_1\gamma_1$) gives

$$E = E^{\circ} + \frac{RT}{zF}\ln\frac{\mu_A^a \cdot \mu_B^b \ldots a_P^p \cdot a_R^r \ldots}{\mu_P^p \cdot \mu_R^r \ldots a_A^a \cdot a_B^a \ldots} \tag{5.40}$$

For the diluted solutions, the ratio of the activity coefficients on the right and left hand sides of the reaction equation may be regarded as equal to approximately one, so that

$$E = E^{\circ} + \frac{RT}{zF}\ln\frac{\mu_A^a \cdot \mu_B^b \ldots}{\mu_P^p \cdot \mu_R^r \ldots} \tag{5.41}$$

The EMF of the galvanic cell is equal to the difference of the potentials between the electrodes in the currentless state. The values of the absolute electrode potentials cannot be determined. Therefore, the potentials of the individual electrodes are determined in relation to the potential of a conventionally selected standard electrode. This is based on the standard hydrogen electrode (SHE) on which the following reaction takes place:

$$2H^+ + 2e^- \underset{\text{oxidation}}{\overset{\text{reduction}}{\Leftrightarrow}} H_2$$

The direction of the process is given by the polarity of the hydrogen electrode. Consequently, the electrode potential is given by the EMF voltage of the cell consisting of the given electrode and the standard hydrogen electrode, and the potential of the standard hydrogen electrode is equal to zero. Generally, it applies that

$$E_i = E_{oi} = \Delta V_i - \Delta V_{\circ} \tag{5.42}$$

where E_i is the electrode potential of the i-th electrode, E_{oi} is the

EMF voltage of the cell consisting of the standard hydrogen electrode and the given electrode, ΔV_i is the absolute potential of the *i*-th electrode, ΔV_a is the absolute potential of the standard hydrogen electrode. The following equation holds for the two electrodes of any galvanic cell:

$$E = \left(\Delta V_i - \Delta V_\circ\right) - \left(\Delta V_2 - \Delta V_\circ\right) \tag{5.43}$$

and therefore

$$E = E_1 - E_2 \tag{5.44}$$

The EMF of a galvanic cell is equal to the difference of the potentials of both electrodes. The electrode reaction in the given system can be expressed by the following general scheme

$$aOx_1 + bOx_2 + \ldots + \mathrm{ze}^- \rightarrow p\,\mathrm{Red}_1 + r\,\mathrm{Red}_2 + \ldots$$

The left hand side of the reaction contains the substances (*Ox*) with a high valency which are reduced by accepting electrons, and the right hand side contains the reduced substances (*Red*). The symbols *a*, *b*, ... *p*, *r*, ... give the appropriate number of mols. Since the oxidation reaction

$$\frac{\mathrm{z}}{2}\mathrm{H}_2 \rightarrow \mathrm{zH}^+ + \mathrm{ze}^-$$

also takes place on the hydrogen electrode, the overall reaction of the galvanic cell consisting of the given and hydrogen electrode is:

$$aOx_1 + bOx_2 + \ldots + \frac{\mathrm{z}}{2}\mathrm{H}_2 \rightarrow p\,\mathrm{Red}_1 + r\,\mathrm{Red}_2 + \ldots + \mathrm{zH}$$

According to (5.39), the electromotoric force of this cell can be described by the equation

$$E_i = E_\circ + \frac{RT}{zF}\ln\frac{a_{Ox_1}^{a}.a_{Ox_2}^{b}\ldots a_{\mathrm{H}_2}^{\frac{z}{2}}}{a_{\mathrm{Red}_1}^{p}.a_{\mathrm{Red}_2}^{r}\ldots a_{\mathrm{H}^+}^{z}} \tag{5.45}$$

or

$$E_i = E_\circ + \frac{RT}{zF}\ln\frac{a_{\mathrm{H}_2}^{\frac{z}{2}}}{a_{\mathrm{H}^+}^{z}} + \frac{RT}{zF}\ln\frac{a_{Ox_1}^{a}.a_{Ox_2}^{b}\ldots}{a_{\mathrm{Red}_1}^{p}.a_{\mathrm{Red}_2}^{r}\ldots} \tag{5.46}$$

Since the activity of gaseous hydrogen and the activity of hydrogen ions for the standard hydrogen electrode is accurately defined and is constant, the first two members on the right hand side of the equation (5.46) can be combined and the generally valid equation can be written for the potential of the electrode on which the reaction takes place:

$$E_i = E_i^{\circ} + \frac{RT}{zF} \ln \frac{a_{Ox_1}^{a} . a_{Ox_2}^{b}}{a_{Red_1}^{p} . a_{Red_2}^{r}} \tag{5.47}$$

Value E° is the potential of the electrode at which the activities of all the reacting substances are equal to one and is referred to as the standard electrode potential.

In the solution which contains the ions of the given substance in the reduced and oxidised forms, oxidation–reduction reaction or redox reactions take place to establish equilibrium. For example, in a solution containing $FeCl_2$ and $FeCl_3$ the following oxidation–reduction reaction takes place:

$$Fe^{3+} + e^{-} = Fe^{2+} \tag{5.48}$$

In principle, these electrodes do not differ, others or, more accurately, all the electrodes are of the oxidation–reduction type, but this term is used generally only for the electrodes in which the reaction takes place between the substances in the liquid solution. There are many types of oxidation–reduction electrodes and many of them play a very important role in fundamental metallurgical processes. Table 5.2 gives the standard potentials of the conventional redox system at 25 °C.

The potential of the oxidation–reduction electrodes is described by the general equation (5.47). Applying this equation to a specific case, for example, equation (5.48) gives

$$E_{Fe^{2+},Fe^{3+}} = E^{0}_{Fe^{2+},Fe^{3+}} + \frac{RT}{F} \ln \frac{a_{Fe^{3+}}}{a_{Fe^{2+}}} \tag{5.49}$$

The standard oxidation–reduction electrode potential is equal to the electrode potential at the same activity of the oxidised and reduced form. If $a_{Fe}^{3+} = a_{Fe}^{2+}$, then the ratio of the activities is equal to one and consequently, the second member in the equation is equal to zero.

Table 5.2. Standard potentials of conventional redox systems

Redox system	**Redox potential [V]**
$F_{2(g)}$ / 2 F^-	+ 2.85
O_3 + 2 H^+ / O_2 + H_2O	+ 2.07
Co^{3+} / Co^{2+}	+ 1.84
H_2O_2 + 2 H^+ / 2 H_2O	+ 1.77
MnO_4^- + 4 H^+ / MnO_2 + 2 H_2O	+ 1.69
Mn^{3+} / Mn^{2+}	+ 1.50
$Cl_{2(g)}$ / 2 Cl^-	+ 1.36
$Cr_2O_7^{2-}$ + 14 H^+ / 2 Cr^{3+} + 7 H_2O	+ 1.33
ClO_2^- + 4 H^+ / Cl^- + 2 H_2O	+ 1.25
MnO_2 + 4 H^+ / Mn^{2+} + 2 H_2O	+ 1.23
O_2 + 4 H^+ / 2 H_2O	+ 1.229
$Br_{(l)}$ / 2 Br^-	+ 1.07
2 Hg^{2+}/ Hg_2^{2+}	+ 0.907
Fe^{3+} / Fe^{2+}	+ 0.77
O_2 + 2 H^+ / H_2O_2	+ 0.69
MnO_4^- / MnO_4^{2-}	+ 0.558
J_2 / 2 J^-	+ 0.535
O_2 + 2 H^+ / 2 OH^-	+ 0.40
Cu^{2+} / Cu^+	+ 0.153
2 H^+ / H_2	0.00

5.2. Calculation and construction of *E*–pH diagrams

In the form published by Pourbaix [2], the potential–pH diagram indicates the thermodynamic equilibrium in the Me–H_2O system in standard conditions, i.e. at 25 °C and the unit total pressure where Me is the generally studied element. The diagrams can be used, for example, for hydrolysis processes or corrosion of metals in aqueous media. The diagrams plotted in this manner have, however, only a small information content if they are used for examining the conditions in hydrometallurgical processes, for example, leaching or precipitation. This is due to the fact that these processes are characterised by the interaction of several components of the heterogeneous system, such as the leached material, mostly sulphides or sulphides of metals, the leaching agent based on chlorides, sulphates, nitrates, etc., and of course water. Therefore, it is necessary to add further components to the original Pourbaix diagram.

In calculation and graphical representation of the *E*–pH diagrams in the standard conditions (25 °C, p = 0.1 MPa) it is necessary to formulate all combinations of all theoretically possible chemical equations, potentially present in the system. No error is made if one of the equations is ignored because in gradual representation of the interface in the system every further equation follows from the context.

Basically, there are only three types of reaction in the system. The first type are the non-redox reactions which take place with participation of hydrogen ions. These reactions are represented graphically in the diagram by the vertical line at constant pH. The relationship between the change of the Gibbs energy and the equilibrium constant is given by the equation

$$\ln K_T = -\frac{\Delta G_T^\circ}{RT} \tag{5.50}$$

so that for standard conditions

$$\ln K_{298} = -\frac{\Delta G_{298}^\circ}{2477.69}$$

Another type of reaction taking place in the system are the reactions during the exchange of electrons, or redox reactions, and hydrogen or hydrogen ions take part in the reaction. Generally, this reaction may be written in the form of the reaction of the total cell:

$$aA + bB - z\mathrm{H}^+ \rightarrow pP + rR - \frac{z}{2}\mathrm{H}_2$$

and for the reaction of a half-cell

$$aA + bB - ze^- \rightarrow pP + rR$$

The reduction potential of the half-cell, E_T, is given by the Nerns equation

$$E_T = E_T^\circ + \frac{RT}{zF}\ln\frac{a_P^p.a_R^r\ldots}{a_A^a.a_B^b\ldots} \quad (5.51)$$

and for the standard conditions

$$E_{298} = E_{298}^\circ + \frac{0.02568}{z}\ln\frac{a_P^p.a_R^r\ldots}{a_A^a.a_B^b\ldots}$$

$$E^\circ = -\frac{\Delta G_{298}^\circ}{zF}$$

where F is the Faraday constant equal to 96484.56 $C.mol^{-1}$.

In calculating the potentials the value pH is substituted for the logarithms of the hydrogen ions in the second member of equation (5.49). Since such a reaction depends on both pH and E, it will have the form of inclined lines, with the angle of inclination determined by the number of mols taking part in the reaction.

The third type of reaction is represented by horizontal lines. They correspond to reactions at electron exchange, i.e. oxidation and reduction, but without hydrogen ions taking part. These types of reaction are, of course, independent of pH and only take place at a specific value of potential E, as indicated by equation (5.51).

The line, representing equilibrium between two areas of stability in this system is obtained by substituting into the equation any collected value of pH, for example, pH = 1, followed by calculating the potential at pH = 1. Similarly, a different value is substituted, for example, pH = 10 and this gives the potential at this value of pH. Connecting both points gives the line of the investigated interface. For the equations taking place only at constant pH, or only at constant potential, the calculation using the appropriate equation is of course sufficient.

The resultant lines are straight lines and are located in the entire range of pH at/or potential. Their validity is limited by the point of intersection with another equilibrium line which restricts the area of stability of another substance in the system which satisfies, in the given conditions of pH and potential, the stability condition, i.e. $\Delta G = 0$ and the substance whose standard change of Gibbs energy is more negative is also thermodynamically more stable. Using this method, unlikely reactions are eliminated from the system and, at the same time, other possible reactions which were not considered at the beginning, appear. For the substances in the ionic state, containing the studied metal it is necessary to determine activity values. These values, equal to one in the limiting case, are usually in the range between 10^{-4} and 10^{-1} M and are depicted in the diagrams by lines of different type. This representation indicates the accurate tendencies of increase or decrease of the size of the regions of stability of the individual substances in the system in relation to the change of their activity in the solution. In a number of cases, one of the regions of stability appears or disappears in this fashion.

The main problems in constructing *E*–pH diagrams include a shortage of thermodynamic data. As indicated by (5.51), to calculate the lines indicating the stability interface in the *E*–pH diagram it is necessary to determine the values ΔG^0 (or ΔH^0 and ΔS^0) of the formation of individual compounds. The number of thermodynamic data is restricted mainly to the fact that the required values include the values of ΔG^0 of the formation of not only all substances present in the system and of their allotropic modifications, but also the values of ΔG^0 of the formation of individual ions in the solution. These values are not available in sufficient amount and, with the exception of the basic set of data [1], further data have been published only very seldom.

The *E*–pH diagram of the investigated system cannot be calculated if some of the thermodynamic data are not available. The area of stability of such a substance is not present in the diagram which may cause problems in evaluation. On the other hand, it may be that this substance does not have the equilibrium state.

5.2.1. E–pH diagrams at elevated temperatures

The majority of hydrometallurgical processes take place efficiently at elevated temperatures and/or pressures which has resulted in the development of the method of the so-called high pressure leaching. However, the hydrometallurgical processes, taking place in the so-called normal conditions, i.e. at temperatures of up to 100 °C and the total pressure of 0.1 MPa, are effective mostly at temperatures close to the boiling point. This shows that it is very important to carry out theoretical examination of the hydrometallurgical systems at high temperatures and, therefore, the extent of application of the *E*–pH diagrams in the standard conditions (25 °C) is limited.

As already mentioned, to calculate the equilibrium in the system it is necessary to know the change of the standard Gibbs energy ΔG^0 of the substances in the pure and ionic state. Since attention is given here to the higher temperature *E*–pH diagrams, the main prerequisite for constructing these diagrams is the availability of the value of the standard Gibbs energy of the formation of substances in the pure and ionic state not only at the standard but also selected higher temperatures. It is not difficult to determine the values ΔG°_{T} because they have already been calculated, tabulated and published in a number of well known tables [29–34], or are available in the data form or a computer net.

It is far more difficult to determine the values of ΔG°_{T} for the substances in the ionic state, as discussed previously. There are several procedures for this which can be applied to obtain the required values. Fundamental studies in this area are those of Criss and Coble and Coble [19, 35] which laid foundations for the extrapolation of the thermodynamic values required for calculating the *E*–pH diagram at elevated temperatures.

For the reactions between the substances in which there is no phase transformation in a specific temperature range, ΔC°_{p} is usually small and, therefore, the so-called Ulich approximation may be used. The values of the standard change of the Gibbs energy ΔG°_{T} at temperature *T* are then calculated from the equation

$$\Delta G^{0}_{T_2} = \Delta H^{\circ}_{T_2} - T_2 \Delta S^{\circ}_{T_2} \tag{5.52}$$

and the values of ΔH° and ΔS° for elevated temperatures are determined from the equations

$$\Delta H^{\circ}_{T_2} = \Delta H^{\circ}_{T_1} + \int_{T_1}^{T_2} \Delta c^{\circ}_{p} \mathrm{d}T \tag{5.53}$$

and

$$\Delta S^{\circ}_{T_2} = \Delta S^{\circ}_{T_1} + \int_{T_1}^{T_2} \frac{\Delta c^{\circ}_{p}}{T} \mathrm{d}T \tag{5.54}$$

Function (5.52) is not of course linear when calculating ΔG^{0}_{T2}

therefore, as the average value ΔC_p° the expression $\left.\overline{c}_p^{o}\right]_{T_1}^{T2}$ is therefore used as the average value between the considered temperatures.

Consequently, the equation (5.52) has the form

$$\Delta G_{T_2}^0 = \Delta G_{T_1}^0 - \Delta S_{T_1}^{\circ}\left(T_2 - T_1\right) + \left.\overline{c}_p^{\circ}\right]_{T_1}^{T_2}\left(T_2 - T_1\right) - T_2 \left.\overline{c}_p^{\circ}\right]_{T_1}^{T_2} \ln \frac{T_1}{T_2} \quad (5.55)$$

Temperature T_1 is in this case equal to the standard temperature, i.e. 298 K.

The average values of the heat capacities for non-ionic substances are determined from the well known published tables. One of the well known procedures is used for the substances in the ionic state. The Criss and Coble's corresponding principle is one of the most efficiently developed procedures. In this procedure, if the standard state is selected efficiently by defining entropy $H^{+}_{(aq)}$ for every temperature, the partial molar ion entropies at this temperature depend in a linear manner on the corresponding entropy values at the reference temperate. In this case, the temperature of 298 K is regarded as the standard temperature and corresponds to the entropy of the hydrogen ion equal to –20.92 $J.mol^{-1}deg^{-1}$ in accordance with the values for the 'absolute' ion entropy $H^{+}_{(aq)}$ recommended by other authors.

The corresponding principle may be described generally by the following equation

$$\overline{S}_{T_2}^{\circ}(abs) = a_{T_2} + b_{T_2}.\overline{S}_{298}^{\circ}(abs) \quad (5.56)$$

where a_{T2} and b_{T2} are the constants which depend on the type of ion, i.e. whether they are cations, anions, oxianions and acid oxianions and whether they also depend on the temperature considered in this case. The value $\overline{S}_{T2}^{o}(\text{abs})$ refers to the ion partial molar entropies on the 'absolute' scale which of course in reality is not absolute because the standard state is the one at 298 K:

$$\overline{S}_{298}^{\circ}(\text{abs}) = \overline{S}_{298}^{\circ}(\text{conv}) - 20.97z \quad (5.57)$$

where z is the ion charge.

Table 5.3 summarises the constants a_{T2} and b_{T2} of equation (5.36 [35] recalculated to the permitted units.

Table 5.3. Entropy constants a_{T2} and b_{T2} of equation (5.56) in $Jmol^{-1}deg^{-1}$

Temperature [K]	Simple cations		Simple anions and OH^-		Oxianions MeO_n^{-m}		Acid oxianions $MeO_n(OH)^{-m}$		Standard state entropy $H^+_{(aq)}$
	a_T	b_T	a_T	b_T	a_T	b_T	a_T	b_T	
298	0	4.184	0	4.184	0	4.184	0	4.184	–20.92
333	16.318	3.996	–21.338	4.054	–58.576	5.092	–56.484	5.774	–10.46
398	43.095	3.665	–54.392	14.184	129.704	6.176	126.775	7.924	8.368
423	67.78	3.314	–89.119	4.138	–194.137	7.058	–209.2	9.962	27.196
473	97.487	2.975	126.357	4.104	–280.328	8.452	–292.88	12.385	53.724

Applying these considerations to the average value of the partial molar heat capacity between 298 K and temperature T_2 gives

$$\bar{c}_p^\circ\Big]_{298}^{T_2} = \frac{\bar{S}_{T_2}^\circ(\text{abs}) - \bar{S}_{298}^\circ(\text{abs})}{\ln\dfrac{T}{298}} \qquad (5.58)$$

and from equation (5.56) it follows

$$\bar{c}_p^\circ\Big]_{298}^{T_2} = \alpha_{T_2} + \beta_{T_2}\bar{S}_{298}^\circ(abs) \qquad (5.59)$$

where α_{T2} and β_{T2} are constants which depend on the type of ion and the considered temperature and also on the previously mentioned constants a_{T2} and b_{T2}.

Under normal conditions when constructing E–pH diagrams at 298 K it is necessary to consider the reaction of the half-cell in relation to the standard hydrogen electrode (SHE) whose G_{298}° value is equal to zero. The EMF voltage of the total cell is defined as the equation (5.44)

$$E_{298}\ (\text{cell}) = E_{298}\ (\text{half-cell}) - E_{298}\ (\text{SHE})$$

which gives

$$E_{298}\ (\text{cell}) = E_{298}\ (\text{half-cell})$$

assuming the already mentioned conventions.

Using the equation (5.55) in the form

$$\Delta G_T^0 = \Delta G_{298}^0 - (T-298)\Delta S_{298}^\circ + (T-298)\Delta \overline{c}_p^\circ]_{298}^T \ln \frac{T}{298} \tag{5.60}$$

the E–pH diagrams can be extrapolated to higher temperatures. When calculating the values of ΔS°_{298} it may be seen that the relationship

$$\Delta G = \Delta H - T\Delta S \tag{5.61}$$

is not applicable to the reactions of the half-cell when ignoring the thermodynamic values for the electron. However, these values can be easily obtained, even at any temperature, if we consider the reaction of the half-cell of the standard hydrogen electrode. If the temperature dependence of E° for SHE is not available, the values at all temperatures for the hydrogen half-cell are usually determined from the following equations:

(*i*) $\Delta G^0_T\,(\mathrm{SHE}) = 0;\ E^\circ_T = 0;\ a_H = 1,\ p_{H2} = 0.1$ MPa

Using the Gibbs–Helmholtz equation ΔH°_T is determined from

$$\Delta H_T^\circ = zFT\left(T\frac{\mathrm{d}E^\circ}{\mathrm{d}T} - E^\circ\right) \tag{5.62}$$

and ΔS°_T is determined from

$$\Delta S_T^\circ = zF\frac{\mathrm{d}E^\circ}{\mathrm{d}T} \tag{5.63}$$

and Δc°_{pT} is determined from

$$\Delta c_{pT}^\circ = \left(\frac{\delta \Delta H^\circ}{\delta T}\right)_p \tag{5.64}$$

For all temperatures it holds that:

(*ii*) ΔH°_T (SHE) = 0
(*iii*) ΔS°_T (SHE) = 0
(*iv*) Δc°_{pT} (SHE) = 0

From (*i*), (*ii*), (*iii*), (*iv*) one can obtain the thermodynamic values of the electron for all temperatures as follows:

$$\Delta\bar{G}_T^0[e]=0$$

$$\Delta\bar{H}_T^\circ[e]=0$$

$$\bar{S}_T^\circ[e]=\frac{1}{2}S_T^\circ[H_{2(g)}]-\bar{S}_T^\circ[H_{(aq)}^+]$$

$$\bar{c}_{pT}^\circ[e]=\frac{1}{2}\bar{c}_{pt}^\circ[H_{2(g)}]-\bar{c}_{pT}^\circ[H_{(aq)}^+]$$

It may be seen that different values are obtained for $\Delta\bar{S}_{298}^{\text{o}}$ and $\Delta\bar{c}_{298}^{\text{o}}$ (as well as for other temperatures) if the reaction of the half-cell including or not including the values of $\bar{S}^{\text{o}}$ and $\bar{c}_p^{\text{o}}$ for the electron is taken into account. Even if the ‘absolute’ entropies are used in the form defined by Criss and Coble instead of conventional definitions, it is still necessary to consider the entropy of the electron.

If calculations are carried out taking into account conventional entropies, then usually at all temperatures $\Delta\bar{S}_{298}^{\text{o}}[H^+]$ = 0. At 298 K using the entropy value for gaseous hydrogen $\Delta\bar{S}_{298}^{\text{o}}[H^+]$ = 130.587 Jmol^{-1}deg^{-1} gives the entropy of the electron equal to $\Delta\bar{S}_{298}^{\text{o}}[H^+]$ = 65.296 Jmol^{-1}deg^{-1}.

On the absolute scale $\bar{S}_T^{\text{o}}[H^+]$ changes with temperature. If calculations are carried out taking into account the values determined by Criss and Coble and equal to $\bar{S}_{298}^{\text{o}}[e]$= 20.92 Jmol^{-1}deg^{-1}, the calculated electron entropy will be $\bar{S}_{298}^{\text{o}}[e]$ = 86.215 Jmol^{-1}deg^{-1}.

Problems with the calculations of the properties of the electron are easily eliminated by considering the reaction of the total cell

$$aA+bB-zH^+=pP+rR-\frac{z}{2}H_2$$

because the electrons are eliminated by the previously considered reaction of the half-cell

$$aA+bH-ze=pP+rR$$

and the reaction of hydrogen half-cell

$$H^+_{(aq)} = e^- = \frac{1}{2}H_{2(g)^+}$$

Calculations of the Gibbs energy, enthalpy, entropy or heat capacities can therefore be carried out directly for any temperature, for example, for the reaction of the total cell assuming that it is extrapolated using a suitable method.

Introducing the ΔG^0_T obtained in this manner into the Nernst equation valid for the reaction of the total cell gives

$$E_T = -\frac{\Delta G^0_T}{zF} - \frac{RT}{zF}\ln\frac{a^P_p.a^r_R.a^z_{H^+}}{a^a_A.a^b_B.a^2_{H_2}} \tag{5.65}$$

and taking into account that (for the total cell) we have:

$$\Delta G^0_T = p\Delta\bar{G}^0_T(P) + r\Delta\bar{G}^0_T(R) - a\Delta G^0_T(A) - b\Delta G^0_T(B) + zG^0_T(H^+) - \frac{z}{2}\Delta G^0_T(H_2) \tag{5.66}$$

the half-cell $aA + bB + ze = pP + rR$ is regenerated in the normal conditions ΔG°_T(SHE) = 0 and a_{H+}, p_{H2} = 0.1 MPa, and the Nernst equation has the following form

$$E_{T(\text{complete cell})} = E_{T(\text{half cell})} = E^0_T - RT\ln\frac{a^p_P \cdot a^r_R}{a^a_A \cdot a^b_B} \tag{5.67}$$

Another feature which makes the reaction of the total cell interesting is that these reactions have the same resultant values of ΔS°_{298} when using conventional or 'absolute' values.

Although the approximation according to Criss and Coble is used quite frequently and has also been integrated in a computer program for calculating E–pH diagrams [23, 25], there are other methods of approximating the data for calculating high temperature E–pH diagrams.

Barnes and Kullerud [7] calculated and constructed high temperature E–pH diagrams of the Fe–S–H_2O system. This method uses the van't Hoff relationship which contains constant values for the change of enthalpy ΔH in the considered temperature range. This method permits extrapolation of the thermodynamic data to 250 °C.

Djackova and Khodakovskii [12] based their calculations of the sulphur–water system in the temperature range 25–300 °C on extrapolation of ΔG° published by Khodakovskii, et al. [11]. The proposed relationship is suitable for the association of weak electrolytes, redox reactions, solutions of gases and low solubility substances. This relation

$$\Delta G^{\circ}_{\text{reaction}} = A' - D'T + C'T^2 \tag{5.68}$$

was determined assuming the linear change of Δc^{0}_{p} in relation to temperature

$$\Delta c^{\circ}_{p} = -2C'T \tag{5.69}$$

and assuming the validity of the relationship

$$\Delta G^{0}_{T} = -RT \ln K^{0} \tag{5.70}$$

gave the following final relationship

$$\log K^{\circ}_{\text{reaction}} = -\frac{A^{*}}{T} + D^{*} - C^{*}T \tag{5.71}$$

$$\Delta H^{\circ}_{\text{reaction}} = A' - CT^2 \tag{5.72}$$

$$\Delta S^{\circ}_{\text{reaction}} = D' - 2C'T \tag{5.73}$$

Using this method, Djackova and Khodakovskii calculated [12] diagrams for the activity of sulphur-containing substances as equal to 10^{-1} M, at a temperature of 25 °C and 0.1 MPa, and also at 150 and 300 °C at 0.5 and 8.5 MPa.

Michard and Allegre [36] proposed a substitution of E–pH diagrams by log S–pH diagrams in which the total concentration of sulphur was in the oxidised state –2. To support their concept, the authors considered the sulphate–sulphide reaction (the first step in the formation of metallic sulphides) and the rate of the reaction was very low and the reaction was far away from the equilibrium state. The reaction

$$E_{298} = +0.148 - 0.0591\text{pH} - 0.0074 \log \frac{\left(\text{S}^{2-}\right)}{\left(\text{SO}_4^{2-}\right)} \tag{5.74}$$

was regarded as very important (the initially published equation with the values given in calories) assuming that equilibrium was reached between $SO^{2-}_4 \Leftrightarrow S^{2-}$, although the measurements of the actual potentials in the sulphide solutions depends in reality on the sulphide–polysulphide couple which reaches equilibrium at a considerably higher rate. The measured potentials did not represent the sulphide–sulphate ratio, as expected by the authors. Since the equilibrium redox potential was not reached, the measured values corresponded basically to the mixed potential for which the Nerns equation is not valid. Therefore, the authors proposed to solve this problem by using Ag/Ag_2S electrodes because they are efficiently defined and measurable, even at very low concentrations. Of course, it is then no longer necessary to reach the sulphate–sulphide equilibrium since the amount of the measured sulphide gives accurately the extent of the reaction.

The usefulness and practical importance of the E–pH diagrams led to more extensive studies and applications. In addition to proposing procedures in which several components of the system were considered in interpretation, either reactive gases such as CO_2 and SO_2 or complex-forming anions as NH_4 and Cl^- [9], a very interesting variant of the thermodynamic study of phase equilibrium in aqueous solutions was proposed. In this variant, the E–pH diagrams are transformed to (chemical) potential–electron number diagrams, z-pH [37]. This diagram retains the information content of the E–pH diagram but also acquires advantages of the conventional metallurgical phase diagrams. The E–pH diagrams depict basically the three-dimensional space represented by the co-ordinates E–pH log a_Me. The z–pH diagram is transformed to the space in which the potential E is replaced by the electron number z. Since potential E is a measure of the chemical potential of the electrons and z is a measure of the number of the electrons. In E and z they summarise thermodynamic variables such as pressure and volume. This shows that the transformation from the E–pH–log a_Me is spaced to the z–pH–log a_Me space and acquires special properties, especially if one point in the E–pH–log a_Me space which represents tw (or three) phases in equilibrium, is transformed to two (or three) individual points in the z–pH–log a_Me space. This is due to the fact that the chemical potential of the electrons E is the same in the phase in equilibrium but in general the concentration of the electrons in individual phases differs.

Figure 5.1 shows the interpretation of the E–pH diagram of the $Cu–H_2O$ equilibrium system. The same system transformed to the

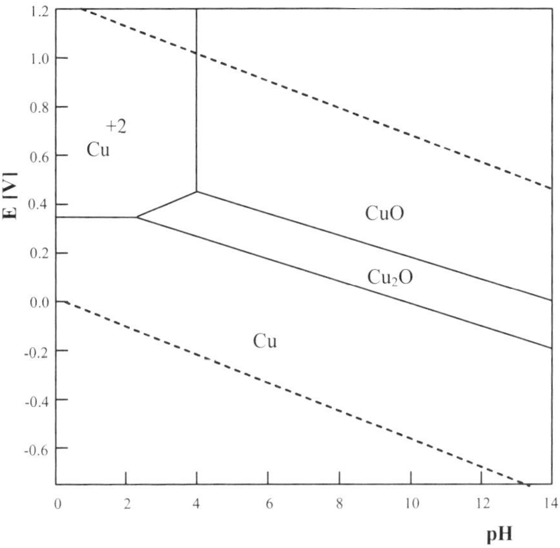

Fig. 5.1. E–pH diagram of the Cu–H_2O system at 25 °C and the activity of copper-containing ions of 1 M and a total pressure of 0.1 MPa.

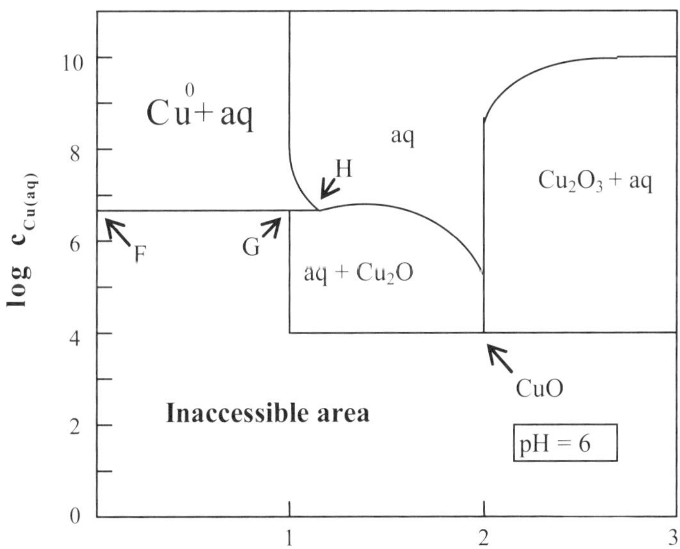

Fig. 5.2. z–pH diagram of the Cu-H_2O system. Intersection of the three-dimensional image by the pH = 6 plane. The reaction at the interface between Cu, Cu_2O and the water phase takes place along the line FGH.

z–pH diagram in shown in Fig.5.2.

Of course, a number of studies have been published gradually describing the thermodynamic studies of equilibrium systems of metals in different aqueous solutions. It is difficult to name all of them here. In addition to the already sited studies, historically interesting or advanced (at the time) studies includes those of Garrels and Naeser [6], Biernat and Robins [13], Murray and Cubicciotti [38] into the S–H_2O system, and the studies by Barnes and Kullerud [7], Asworth and Boden [14], Biernat and Robins [39], Yokokawa, et. al. [40], and Wadsley [41] concerned with the thermodynamic studies of the Fe–S–H_2O system.

The equilibrium of other metals in an aqueous solution has been studied in, for example, investigations carried out by Woods, et. al. [10, Havlik and Kmetova [15], Cubicciotti [42] and Havlik, et. al. [43] concerned with the Cu–S–H_2O system, Marcus and Protopopoffova [44] for the Ni–S–H_2O system, Larsen and Linkson [18] for the Zn–S–H_2O system, El-Raghy and El-Demerdash [45] for the Fe–S–H_2O, Cu–S–H_2O, Ag–S–H_2O, Mo–S–H_2O, Zn–S–H_2O and Pb–S–H_2O systems, as well as studies by Muir and Senanayake [46] for the Cu–Fe–Ag–As–Sb–Bi–S–H_2O complex system.

It may be seen that these studies have been concerned with the behaviour of metals in the Me–S–H_2O system, de facto in the aqueous medium of sulphuric acid and the majority of the studies have also been carried out in this direction. However, since actual leaching systems use media other than sulphuric acid – for example chloride or ammonia environment, these systems have also been examined from the thermodynamic viewpoint. Suitable examples include studies by Froning, et. al., and Davis, et. al. [47, 48], concerned with the Fe–Cl–H_2O and l–Cl–H_2O systems. The cyanide system was studied by Adams [49], Araia and Toguri [50], and Allabergin, et. al. [51].

The thermodynamic studies of ammonia systems were conducted in, for example, investigations by Acharya, et. al., and Osseo-Asare [52–54] for systems containing magnesium and iron applied to leaching of deep sea manganese concretions. Further studies concerned with the ammonia systems are those by Bhuntumkomol, et. al. [55] for a system containing nickel, and by Vu, et. al. [56] for a cobalt-containing system.

The thermodynamics of leaching of gold and silver in thiourea solutions was also studied by Gaspar, et. al. [57].

As already mentioned, one of the most important problems in the calculation and design of E–pH diagrams is a shortage of thermodynamic data or differences in these data. For example, for one of the main chemical equilibria in the S–H_2O system

$$HS^{-} \rightarrow S^{2-} + H^{-}$$

the data published by Latimer [1] and National Bureau of Standards [58] differ quite greatly, as shown in Table 5.4.

Table 5.4. Thermodynamic data for HS^- and S^{2-}

	ΔG°_{298}[kJ/mol]	ΔS°_{298}[J/mol · deg]	ΔG°_{298}[kJ/mol]	ΔS°_{298}[J/mol·deg]
	Latimer	**NBS**	**Latimer**	**NBS**
HS^-	+12.602	+12.06	+61.13	+62.802
S^{2-}	+95.53	+85.83	–26.79	–14.65

Using these data gives of course different results; the pH values of equilibrium for different temperatures are as follows:

		pH_{298}	pH_{373}	pH_{423}
HS^-/S^{2-}	Latimer	14.00	11.95	10.81
	NBS	12.92	10.98	9.89

This shows unambiguously that when calculating and constructing the E–pH diagrams it is necessary to use carefully the data that are available, especially if it is necessary to compile these data. Practice shows that it is more advantageous to use data from a single source, even if these data are not completely accurate, in preference to the compilation of data from different sources.

5.3. Potential–pH diagrams in leaching of copper sulphides

The E–pH diagrams considered here depict the thermodynamic equilibrium of substances in aqueous solutions. It is therefore necessary to define initially the area of stability of water indicating the range of validity of these diagrams.

The association of water is described by the reaction

$$H_2O \rightarrow H^{+} + OH^{-}$$

which consists of the reactions of half-cells

Reduction: $$2H^+ + 2e^- \rightarrow H_2 \qquad (5.75)$$

and oxidation: $$2OH^- \rightarrow O_2 + 2H^+ + 4e^- \qquad (5.76)$$

Combining the reactions (5.75) and (5.76) gives the overall reaction

$$2H_2O \rightarrow 2H_2 + O_2 \qquad (5.77)$$

Generally, reaction (5.77) is expressed in the form

$$2H_2O \rightarrow O_2 + 4H^+ = 4e^- \qquad (5.78)$$

and both reactions of half-cells are directly expressed in the form of H^+ ions. The potentials of these reactions for 25 °C are expressed by the equations

$$E = 0.00 - 0.0591 \text{ pH} - 0.0295 \log P_{H2}$$

and

$$E = 1.228 - 0.0591 \text{ pH} - 0.0147 \log P_{O2}$$

for reduction (5.75) and oxidation (5.78).

For the pressure p_{H2} = 0.1 MPa and P_{O2} = 0.1 MPa these equations may be expressed as follows:

$$E = 0.00 - 0.0591 \text{ pH}$$

and

$$E = +1.28 - 0.0591 \text{ pH}$$

The equilibrium pressures of hydrogen and oxygen are found between these two lines below 0.1 MPa. Between these lines there is the region thermodynamic stability of water at a pressure of 0.1 MPa and a temperature of 25 °C. To provide more accurate information, these lines (the range of stability of water) are placed in every *E*–pH diagram depicting the region of the realistically possible existence of the thermodynamically stable substances under the given conditions in the investigated *E*–pH diagram.

The thermodynamically possible reactions of the H–O system are represented by the equations

$$H_2O \rightarrow OH^- + H^+ \qquad (5.79)$$

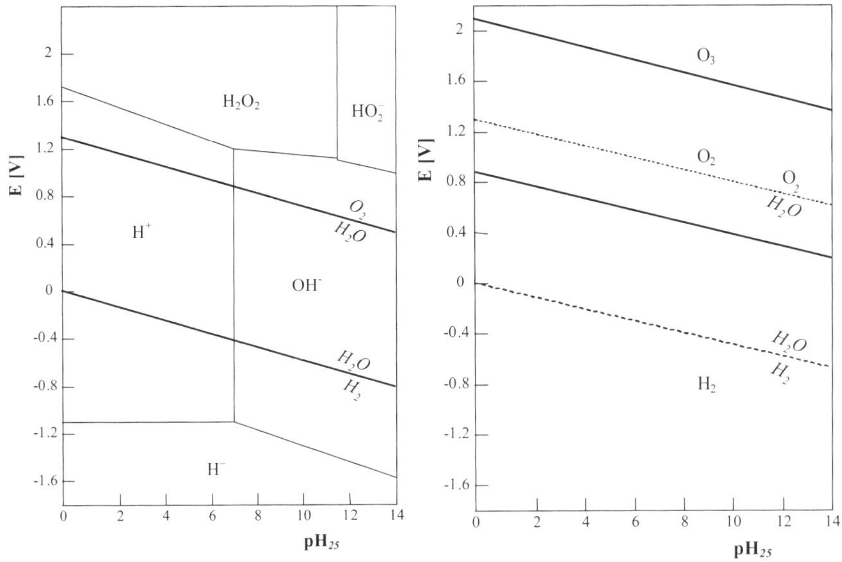

Fig. 5.3. *E*–pH diagram of the H–O system. a) the area of stability of individual ions, b) the area of stability of gas components.

$$H_2O_2 \rightarrow HO_2^- + H^+ \quad (5.80)$$

$$H^- \rightarrow H^+ + 2e^- \quad (5.81)$$

$$H_2O + H^- \rightarrow OH^- + 2H^+ + 2e^- \quad (5.82)$$

$$2H_2O \rightarrow H_2O_2 + 2H^+ + 2e^- \quad (5.83)$$

$$2H_2O \rightarrow HO_2^- + 3H^+ + 2e^- \quad (5.84)$$

$$H_2O + OH^- \rightarrow H_2O_2 + H^+ + 2e^- \quad (5.85)$$

$$H_2O + OH \rightarrow HO_2^- + 2H^+ + 2e^- \quad (5.86)$$

The area of stability of the thermodynamically possible phases of the H–O system are shown in Fig. 5.3*a*. Figure 5.3*b* shows the area of stability of gaseous components in this system. This is important information for later application of reactions between the solid, liquid and gas phases in leaching processes.

5.3.1. S–H_2O equilibrium system

In the investigated equilibrium S–H_2O system it is important to consider not only the stable substances like sulphur, sulphides and sulphates, but also metastable substances such as sulphites, hydrosulphites, di-, tri-, tetra- and pentathionates. 35 substances were considered, and the values of the standard Gibbs energy of their formation for temperatures in the range 25–150 °C were determined from several published sources [4, 59–61].

The following reactions between the potentially existing phases are considered in the system:

$$H_2S \rightarrow HS^- + H^+ \tag{5.87}$$

$$HS^- \rightarrow S^{2-} + H^+ \tag{5.88}$$

$$H_2S_2O_3 \rightarrow HS_2O_3^- + H^+ \tag{5.89}$$

$$HS_2O_3^- \rightarrow S_2O_3^{2-} + H^+ \tag{5.90}$$

$$2S_5O_6^{2-} + 3H_2O \rightarrow HS_2O_3^- + H^+ \tag{5.91}$$

$$2S_5O_6^{2-} + 3H_2O \rightarrow 5S_2O_3^{2-} + 6H^+ \tag{5.92}$$

$$5H_2S_5O_3 \rightarrow 2S_5O_6^{2-} + 3H_2O + 4H^+ \tag{5.93}$$

$$HS_2O_4^- \rightarrow S_2O_4^{2-} + H^+ \tag{5.94}$$

$$H_2SO_3 \rightarrow HSO_3^- + H^+ \tag{5.95}$$

$$HSO_3^- \rightarrow SO_3^{2-} + H^+ \tag{5.96}$$

$$HSO_4^- \rightarrow SO_4^{2-} + H^+ \tag{5.97}$$

$$2S^{2-} \rightarrow S_2^{2-} + 2e^- \tag{5.98}$$

$$3S_2^{2-} \rightarrow 2S_3^{2-} + 2e^- \tag{5.99}$$

$$4S_3^{2-} \rightarrow 3S_4^{2-} + 2e^- \tag{5.100}$$

$$5S_4^{2-} \rightarrow 4S_5^{2-} + 2e^- \tag{5.101}$$

$$2HS^- \rightarrow S_2^{2-} + 2H^+ + 2e^- \tag{5.102}$$

$$3HS^- \rightarrow S_3^{2-} + 3H^+ + 4e^- \tag{5.103}$$

$$4HS^- \rightarrow 4S_4^{2-} + 4H^+ + 6e^- \tag{5.104}$$

$$5HS^- \rightarrow S_5^{2-} + 5H^+ + 8e^- \tag{5.105}$$

$$2S^{2-} + 3H_2O \rightarrow S_2O_3^{2-} + 6H^+ + 8e^- \tag{5.106}$$

$$2HS^- + 3H_2O \rightarrow S_2O_3^{2-} + 8H^+ + 8e^- \tag{5.107}$$

$$2H_2S + 3H_2O \rightarrow S_2O_3^{2-} + 10H^+ + 8e^- \tag{5.108}$$

$$2H_2S + 3H_2O \rightarrow HS_2O_3^- + 9H^+ + 8e^- \tag{5.109}$$

$$5H_2S + 6H_2O \rightarrow S_5O_6^{2-} + 22H^+ + 20e^- \tag{5.110}$$

$$4H_2S + 6H_2O \rightarrow S_4O_6^{2-} + 20H^+ + 18e^- \tag{5.111}$$

$$2S^{2-} + 4H_2O \rightarrow S_2O_4^- + 8H^+ + 10e^- \tag{5.112}$$

$$2S^{2-} + 4H_2O \rightarrow HS_2O_4^- + 7H^+ + 10e^- \tag{5.113}$$

$$S^{2-} + 3H_2O \rightarrow SO_3^{2-} + 6H^+ + 6e^- \tag{5.114}$$

$$S^{2-} + 3H_2O \rightarrow HSO_3^- + 5H^+ + 6e^- \tag{5.115}$$

$$HS^- + 3H_2O \rightarrow SO_3^{2-} + 7H^+ + 6e^- \tag{5.116}$$

$$HS^- + 3H_2O \rightarrow HSO_3^- + 6H^+ + 6e^- \tag{5.117}$$

$$H_2S + 3H_2O \rightarrow HSO_3^- + 7H^+ + 6e^- \tag{5.118}$$

$$S^{2-} + 4H_2O \rightarrow SO_4^{2-} + 8H^+ + 8e^- \tag{5.119}$$

$$HS^- + 4H_2O \rightarrow SO_4^{2-} + 9H^+ + 8e^- \tag{5.120}$$

$$H_2S + 4H_2O \rightarrow SO_4^{2-} + 10H^+ + 8e^- \tag{5.121}$$

$$H_2S + 4H_2O \rightarrow HSO_4^- + 9H^+ + 8e^- \tag{5.122}$$

$$S_2^{2-} + 3H_2O \rightarrow S_2O_3^{2-} + 6H^+ + 6e^- \tag{5.123}$$

$$2S_5^{2-} + 15H_2O \rightarrow 5S_2O_3^{2-} + 30H^+ + 24e^- \tag{5.124}$$

$$S_2O_3^- + H_2O \rightarrow S_2O_4^{2-} + 2H^+ + 2e^- \tag{5.125}$$

$$S_2O_3^{2-} + H_2O \rightarrow HS_2O_4^- + H^+ + 2e^- \tag{5.126}$$

$$2S_2O_3^{2-} \rightarrow S_4O_6^{2-} + 2e^- \tag{5.127}$$

$$S_2O_3^{2-} + 3H_2O \rightarrow 2SO_3^{2-} + 6H^+ + 4e \tag{5.128}$$

$$S_2O_3^{2-} + 3H_2O \rightarrow 2HSO_3^- + 8H^+ + 4e^- \tag{5.129}$$

$$S_4O_6^{2-} + 6H_2O \rightarrow 4HSO_3^- + 8H^+ + 6e^- \tag{5.130}$$

$$S_4O_6^{2-} + 6H_2O \rightarrow 4H_2SO_3 + 4H^+ + 6e^- \tag{5.131}$$

$$S_2O_4^{2-} + 2H_2O \rightarrow 2SO_3^{2-} + 4H^- + 2e^- \tag{5.132}$$

$$S_2O_4^{2-} + 2H_2O \rightarrow 2HSO_3^- + 2H^+ + 2e^- \tag{5.133}$$

$$HS_2O_4^- + 2H_2O \rightarrow 2HSO_3^- + 3H^+ + 2e^- \tag{5.134}$$

$$HS_2O_4^- + 2H_2O \rightarrow 2H_2SO_3 + H^+ + 2e^- \tag{5.135}$$

$$S_2O_4^{2-} + 2H_2O \rightarrow 2H_2SO_3 + 2e^- \quad (5.136)$$

$$2SO_3^{2-} \rightarrow S_2O_6^{2-} + 2e^- \quad (5.137)$$

$$2HSO_3 \rightarrow S_2O_6^{2-} + 2H^+ + 2e^- \quad (5.138)$$

$$2H_2SO_3 \rightarrow S_2O_6^{2-} + 4H^+ + 2e^- \quad (5.139)$$

$$SO_3^{2-} + H_2O \rightarrow SO_4^{2-} + 2H^+ + 2e^- \quad (5.140)$$

$$HSO_3^- + H_2O \rightarrow SO_4^- + 2H^+ + 2e^- \quad (5.141)$$

$$HSO_3^- + H_2O \rightarrow HSO_4^- + 2H^+ + 2e^- \quad (5.142)$$

$$2SO_4^{2-} \rightarrow S_2O_8^{2-} + 2e^- \quad (5.143)$$

$$2HSO_4^- \rightarrow S_2O_8^{2-} + 2H^+ + 2e^- \quad (5.144)$$

$$S_2O_8^{2-} + 2H_2O \rightarrow 2HSO_5^- + 2H^+ + 2e^- \quad (5.145)$$

$$S_5^{2-} \rightarrow 5S + 2e^- \quad (5.146)$$

$$HS^- \rightarrow S + H^+ + 2e^- \quad (5.147)$$

$$H_2S \rightarrow S + 2H^+ + 2e^- \quad (5.148)$$

$$5S + 6H_2O \rightarrow S_5O_6^{2-} + 12H^+ + 10e^- \quad (5.149)$$

$$2S + H_2O \rightarrow S_2O_3^{2-} + 6H^+ + 4e^- \quad (5.150)$$

$$4S + 6H_2O \rightarrow S_4O_6^{2-} + 12H^+ + 10e^- \quad (5.151)$$

$$S + 3H_2O \rightarrow H_2SO_3 + 4H^+ + 4e^- \quad (5.152)$$

$$2S + 6H_2O \rightarrow S_2O_6^{2-} + 12H^+ + 10e^- \quad (5.153)$$

$$S + 4H_2O \rightarrow SO_4^{2-} + 8H^+ + 6e^- \quad (5.154)$$

$$S + 4H_2O \rightarrow HSO_4^- + 7H^+ + 6e^- \quad (5.155)$$

$$H_2S_{(g)} + 2H_2O \rightarrow SO_2 + 6H^+ + 6e^- \quad (5.156)$$

$$SO_{2_{(g)}} + H_2O \rightarrow SO_3 + 2H^+ + 2e^- \quad (5.157)$$

$$H_2S_{(g)} \rightarrow HS^- + H^+ \quad (5.158)$$

$$H_2S_{(g)} \rightarrow H_2S_{(l)} \quad (5.159)$$

$$SO_{2_{(g)}} + H_2O \rightarrow SO_3^{2-} + 2H^+ \quad (5.160)$$

$$SO_{2_{(g)}} + H_2O \rightarrow HSO_3^- + H^+ \quad (5.161)$$

$$SO_{2_{(g)}} + H_2O \rightarrow H_2SO_3 \quad (5.162)$$

$$SO_{3(g)} + H_2O \rightarrow SO_4^{2-} + 2H^+ \quad (5.163)$$

$$SO_{3(g)} + H_2O \rightarrow HSO_4^- + H^- \quad (5.164)$$

$$5H_2S_{(g)} + \rightarrow S_5^{2-} + 10H^+ + 8e^- \quad (5.165)$$

$$5H_2S_{(g)} + 6H_2O \rightarrow S_5O_6^{2-} + 22H^+ + 20e^- \quad (5.166)$$

$$2H_2S_{(g)} + 3H_2O \rightarrow S_2O_3^{2-} + 10H^+ + 8e^- \quad (5.167)$$

$$2H_2S_{(g)} + 3H_2O \rightarrow HS_2O_3^- + 9H^+ + 8e^- \quad (5.168)$$

$$4H_2S_{(g)} + 6H_2O \rightarrow S_4O_6^{2-} + 20H^+ + 18e^- \quad (5.169)$$

$$2H_2S_{(g)} + 4H_2O \rightarrow S_2O_4^{2-} + 12H^+ + 10e^- \quad (5.170)$$

$$2H_2S_{(g)} + 4H_2O \rightarrow HS_2O_4^- + 11H^+ + 10e^- \quad (5.171)$$

$$3H_2S_{(g)} + 6H_2O \rightarrow S_3O_6^{2-} + 18H^+ + 16e^- \quad (5.172)$$

$$S_2O_6^{2-} + 4H_2O \rightarrow 5SO_{2(g)} + 8H^+ + 8e^- \quad (5.173)$$

$$S_2O_3^{2-} + H_2O \rightarrow 2SO_{2(g)} + 2H^+ + 4e^- \quad (5.174)$$

$$HS_2O_3^{2-} + H_2O \rightarrow 2SO_{2(g)} + 3H^+ + 4e^- \quad (5.175)$$

$$S_4O_6^{2-} + 2H_2O \rightarrow 4SO_{2(g)} + 4H^+ + 6e^- \quad (5.176)$$

$$S_2O_4^- \rightarrow 2SO_{2(g)} + 2e^- \quad (5.177)$$

$$HS_2O_4^{2-} \rightarrow 2SO_{2(g)} + H^+ + 2e^- \quad (5.178)$$

$$S_3O_6^{2-} \rightarrow 3SO_{2(g)} + 2e^- \quad (5.179)$$

$$H_2S_{(g)} \rightarrow S + 2H^+ + 2e^- \quad (5.180)$$

$$S + H_2O \rightarrow SO_{(g)} + 2H^+ + 2e^- \quad (5.181)$$

$$S + 2H_2O \rightarrow SO_{2(g)} + 4H^{|} + 4e^- \quad (5.182)$$

$$S + 3H_2O \rightarrow SO_{3(g)} + 6H^+ + 6e^- \quad (5.183)$$

The equilibrium conditions of the chemical and electrochemical reactions were calculated on the basis of the already considered facts, equations and procedures. The diagrams for which we consider the presence of substances stable in an aqueous solution at 25, 100 and 150 °C are shown in Fig. 5.4, and the total considered activity of the sulphur ions is 10^{-1} M at 0.1 MPa of total pressure.

On the basis of the diagram and for the considered conditions,

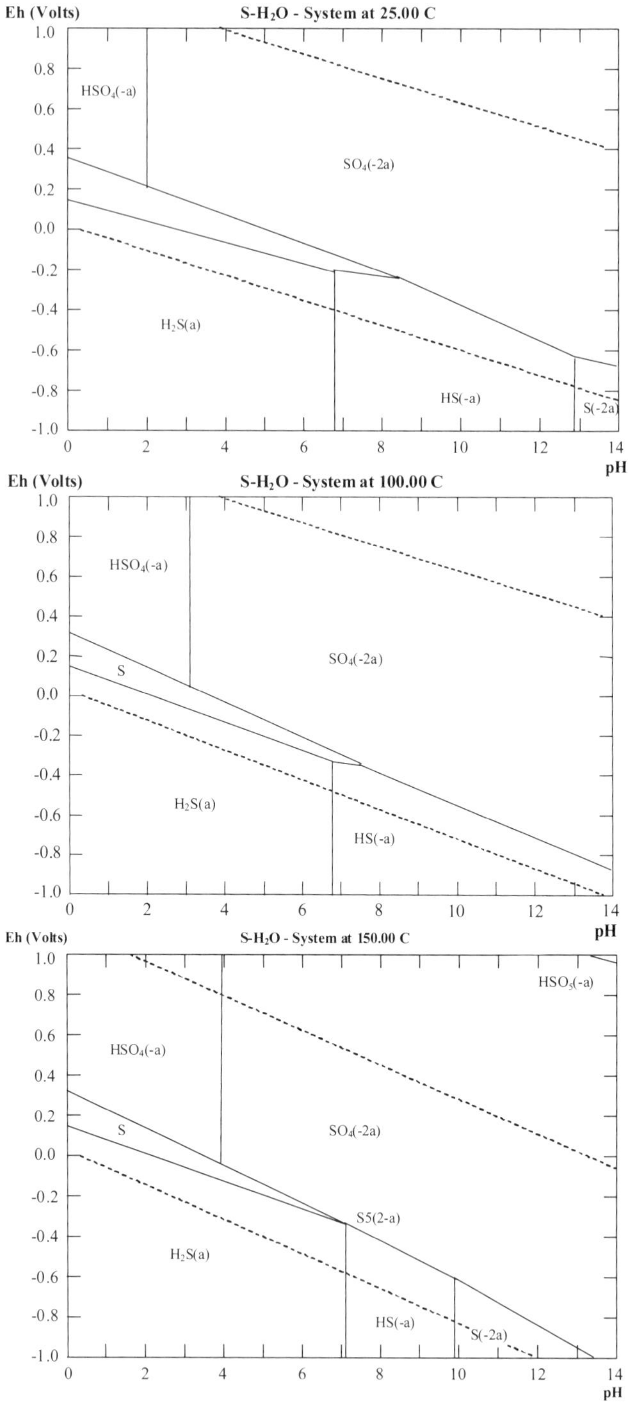

Fig. 5.4. *E*-pH diagram of the S-H_2O system for 25, 100 and 150 °C, the activity of sulphur-containing ions equal to 10^{-1} M at a total pressure of 0.1 MPa.

the given substances are stable in an aqueous solution in the region of their stability and, of course, in the region of the stability of water indicated by the broken lines. In this case, the presence of other metastable substances is not taken into account. The individual equilibrium relationships are represented by the lines between the individual regions of stability.

For a non-redox homogeneous reaction (5.97)

$$HSO_4^- \rightarrow SO_4^{2-} + H^+$$

the calculated values are pH_{25} = 1.91, pH_{100} = 3.02 and pH_{150} = 3.68. The equilibrium line increases with increase of temperature towards higher values of pH thus expanding the area of stability of HSO_4^-.

The calculated values of pH for the reaction (5.87) which is also a non-redox homogeneous reaction

$$H_2S \rightarrow HS^- + H^+$$

are: pH_{25} = 7.00, pH_{100} = 6.70 and pH_{150} = 6.95. The H_2S/HS^- equilibrium line is shifted, with increase in temperature, to lower pH values until it reaches a temperature of 373 K. Above this value, the calculated value is displaced to higher pH values taking into account the occurrence of the SO_4^{2-}/H_2S equilibrium.

For the same type of reaction (5.88)

$$HS^- \rightarrow S^{2-} + H^+$$

the values are: pH_{25} = 14.00, pH_{100} = 11.95 and pH_{150} = 10.81 which shows that for the range of the pH values from zero to 14 this equilibrium is not considered at temperatures below 25 °C. However, above this value, the presence of S^{2-} ions must be taken into consideration.

Assuming that the activity of all the substances in the solution is the same, these three equilibria are independent of ion activity and are represented by the equilibrium relationship:

$$-pH + \log\frac{(A)}{(B)} = \log K \qquad (5.184)$$

The remaining equilibrium reactions some are homogeneous others heterogeneous but they all depend on pH and the potential which means that they are graphically represented by inclined lines.

It should be mentioned that the reaction (5.147)

$$HS^- \rightarrow S + H^+ + 2e^-$$

for the activity of substances with a sulphur content of 10^{-1} M disappears at temperatures above 150 °C. At these temperatures sulphur cannot be reduced to HS^- because the area of stability of sulphur is suppressed below the lower limit of stability of HS^-. This also shows that the SO_4^{2-} ion in acid solutions can be directly reduced to H_2S. Similarly, at 25 °C a decrease of concentration results in the eventual disappearance of the area of stability of sulphur and the formation of the H_2S/SO_4^{2-} interface which is reflected at the activity of the sulphur-containing ions of 10^{-4} M.

Generally, it may be assumed that at temperatures of up to 150 °C, the ions of H_2S, HS^- and S^{2-} are stable in water and aqueous solutions without oxidation agents, H_2S in acid solutions, HS^- mainly in alkaline solutions and S^{2-} at very high values of pH. The stability of HSO_4^- is dominant at low pH values. A decrease of the activity of the ions in the solution reduces the widths of the area of stability of elemental sulphur and leads to the disappearance of the reaction between the hydrogen sulphide ions and sulphur.

The area of stability of sulphur is located completely in the area of stability of water and is stable in solutions without any oxidation agents. The upper boundary of pH is influenced by the increase of temperature. This increase of temperature results in a change of the potential of the equilibrium reaction (5.154) and (5.155) in the direction to negative values, and the potential of the equilibrium (5.148) changes less markedly. Consequently, the area of stability of elemental sulphur decreases and, probably, completely disappears at a higher temperature. This is similar to the situation in which we consider the activity of sulphur-containing substances in the solution equal to 10^{-4} M which reduces the size of the area of stability of elemental sulphur, even already at room temperature and disappears at 150 °C.

Thermodynamics operate with substances in the equilibrium state. However, the hydrometallurgical processes are far away from the equilibrium state and, consequently, the conditions in the solutions may differ from those described previously. Therefore, attention will be given to *E*–pH diagrams of the S–H_2O system which contains less stable compounds of sulphur but which may be present as intermediate phases during leaching. Figure 5.5 shows the *E*–pH equilibrium diagrams which consider only hydrogen sulphates, sulphates and dithionates.

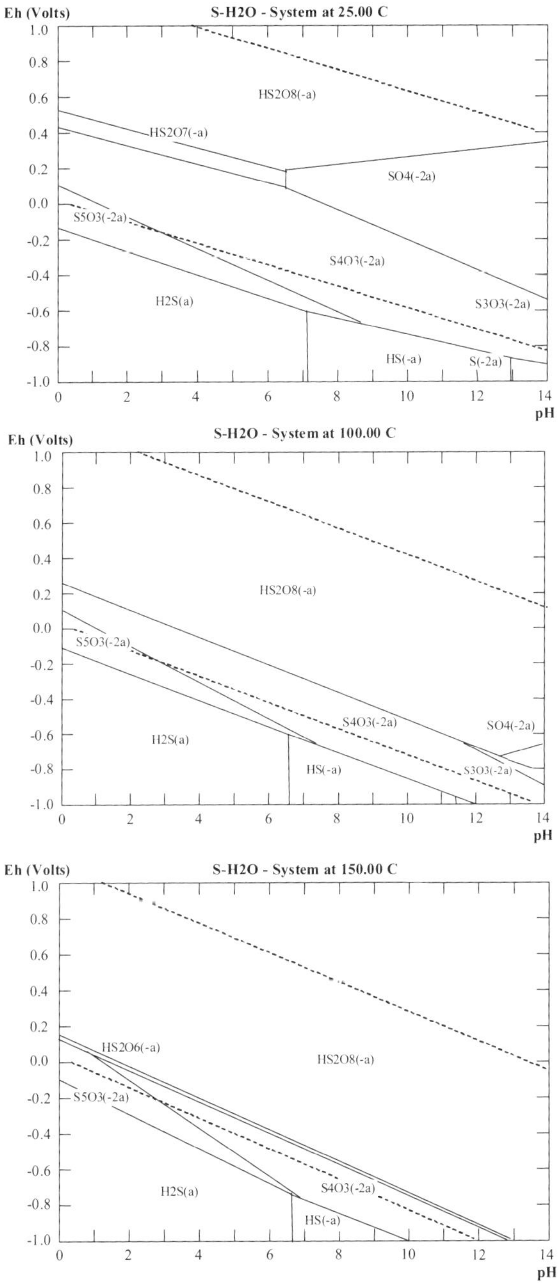

Fig. 5.5. *E*–pH diagram of the sulphur-water system taking into account hydrogen sulphates, sulphates and dithionates for 25, 100 and 150 °C.

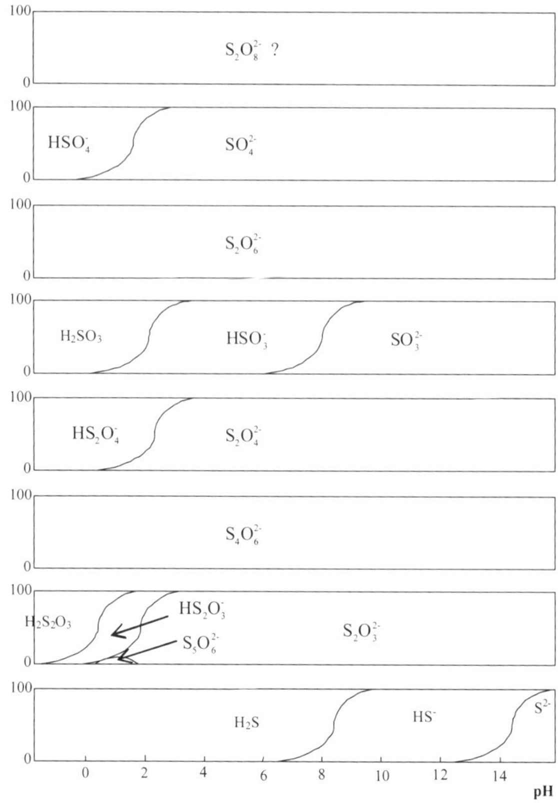

Fig. 5.6. Effect of pH on dissociation of dissolved substances at 25 °C for different valences of sulphur.

The effect of pH on the dissociation of dissolved substances for five valances of sulphur from –2 to +6 is shown in Fig. 5.6. However, in this case it should be stressed that no data are as yet available for the dissociation of $S_4O_6^{2-}$ (+2.5), $S_3O_6^{2-}$ (+3.3), $S_2O_6^{2-}$ (+5) and $S_2O_8^{2-}$ (+7).

The effect of pH and pressure on the solubility of $H_2S_{(g)}$ and $SO_{2(g)}$ is shown in Fig.5.7 for six values of the partial pressure $H_2S_{(g)}$ and $SO_{2(g)}$ from 10^{-7} to 10^{-2} MPa.

Figure 5.8 shows the *E*–pH diagram at 25 °C indicating the presence of stable and unstable substances in the presence of $H_2S_{(aq)}$ or $SO_{2(aq)}$ at a total pressure of 0.1 MPa.

These diagrams show the area of stability of coexisting substances in the equilibrium state in the S–H_2O system and it is also possible to estimate the behaviour of water which contains both H_2O and oxygen O_2. This simulated situation is of considerable

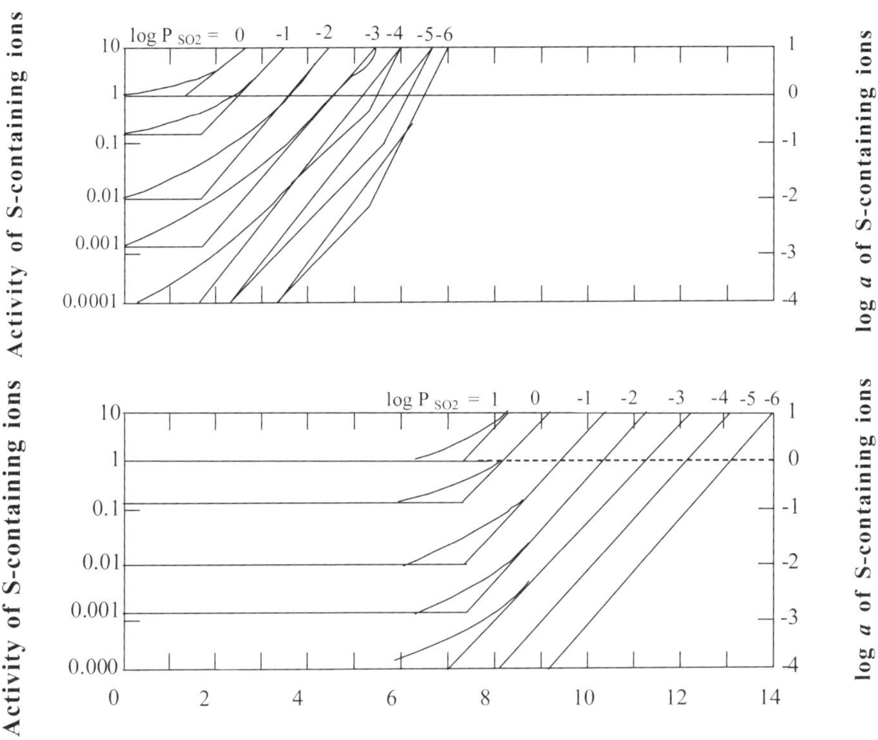

Fig .5.7. The effect of pH and pressure on the solubility of gaseous $H_2S_{(g)}$ $SO_{2(g)}$ at 25 °C. Equilibrium between the gaseous $H_2S_{(g)}$ and sulphide solutions between gaseous $SO_{2(g)}$ and sulphate solutions.

importance in the application of hydrometallurgical processes.

Figure 5.8 shows the situation in which the water is permanently saturated with H_2S at a pressure of 0.1 MPa, considering two extreme cases. In the first case, solution is in contact with oxygen at a partial pressure of 0.001 MPa only for a short period of time. In the second case, it is taken into account that the supply of oxygen is constant and water is in constant contact with the atmosphere containing H_2S at 0.1 MPa and O_2 at 0.001 MPa. Line (7) in Fig. 5.8 separates areas in which gaseous H_2S (below) and gaseous SO_2 (band line) prevail. Below the line (7) there are two groups of lines (4) and (5) indicating the conditions of metastable equilibrium of the solutions of thiosulphates $S_2O_3^{2-}$ and tetrathionates $S_4O_6^{2-}$ (for concentration between 10^{-4} and 10^0 g · atom S per litre) in the presence of gaseous H_2S at 0.1 MPa. As indicated in the upper part of Fig. 5.7, the three vertical lines in the lower part of the graph indicate the total sulphur concentration (in the form of H_2S

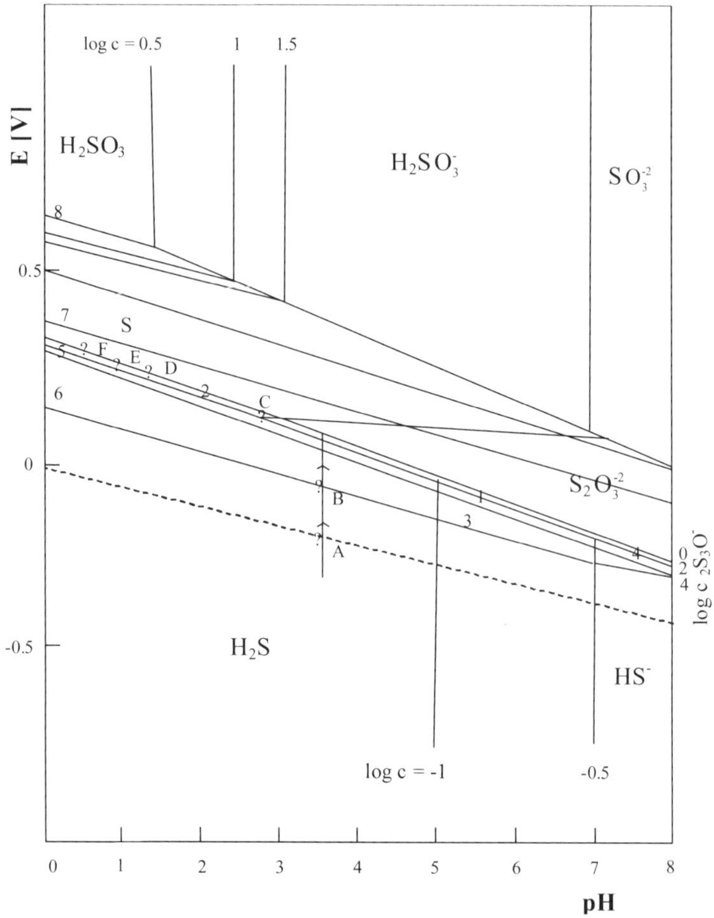

Fig. 5.8. Equilibrium *E*–pH diagram of the S–H_2O system containing sulphur, sulphides, thiosulphates, tetrathionates, and sulfites at 25 °C in the presence of gaseous H_2S and SO_2 at a pressure of 0.1 MPa.

and H^-) in a solution in equilibrium with gaseous H_2S at a total pressure of 0.1 MPa.

Above the line (7) there is a group of lines (8) showing the metastable equilibrium conditions of solutions of tetrathionates $S_4O_6^{2-}$ in the presence of gaseous SO_2 at a pressure of 0.1 MPa, indicating the total sulphur concentration ($10^{0.5}$, 10 and $10^{1.5}$ g · atom of sulphur per litre). The three vertical lines in the upper part of the diagram show, as indicated in the bottom part of Fig. 5.7, the total sulphur concentration (in the form of H_2SO_3 and HSO_3^-) in solutions in equilibrium with SO_2 at a pressure of 0.1 MPa. However, because of a high concentration, these values are approximate.

Effect of temperature and concentration on solubility of oxygen in water

The effect of temperature on the solubility of oxygen in pure water, at a partial oxygen pressure of 10^{-3} MPa, is shown in Table 5.5 and Fig. 5.9.

Table 5.5. Solubility of oxygen in pure water at a partial oxygen pressure of 10^{-3} MPa

Temperature [°C]	25	50	75	100	125	150
Dissolved $O_{2(g)}$ [mol/l·10^5]	1.26	0.93	0.8	0.76	0.78	0.87
Concentration [mg/l]	0.403	0.298	0.256	0.243	0.25	0.278

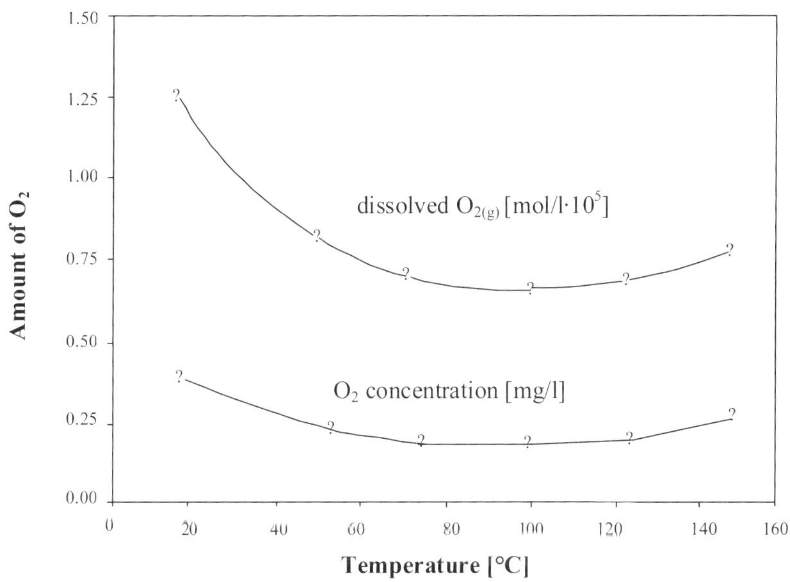

Fig. 5.9. Temperature dependence of the solubility of oxygen and water at a partial oxygen pressure of 10^{-3} MPa.

Formation of elemental sulphur

According to the equilibrium, shown in Figs. 5.1 and 5.2, oxidation of H_2S may result in the formation of elemental sulphur (solid below 119 °C and liquid above 119 °C) and, depending on the pH value, thiosulphate $S_2O_3^{2-}$ and tetrathionate $S_4O_6^{2-}$. Because of a shortage of data, substances such as $HS_2O_3^-$ or $H_2S_2O_3$ and pentathionate $S_5O_6^{2-}$ have not been considered.

If oxidation with oxygen takes place, the following reactions occur:

$$H_2S \rightarrow 2S + 4H^+ = 4e^-$$

and

$$O_2 + 4H^+ + 4e^- \rightarrow 2H_2O$$

or

$$2H_2S + O_2 \rightarrow 2S + 2H_2O$$

with the formation of two mols of sulphur per every mol of oxygen, i.e. at 25 °C of water, saturated with oxygen at a pressure 0.001 MPa with a formation of $2.52 \cdot 10^{-5}$ mols of sulphur (or 0.8 mg S per litre). This reaction does not change the pH value.

Thiosulphate $S_2O_3^{2-}$ forms by the reactions:

$$H_2S + 1.5H_2O \rightarrow 0.5S_2O_3^{2-} + 5H^+ + 4e^-$$

and

$$O_2 + 4H^+ + 4e^- \rightarrow 2H_2O$$

or

$$2H_2S + O_2 \rightarrow 0.5S_2O_3^{2-} + 0.5H_2O + H^+$$

with the formation of 0.5 mol of $S_2O_3^{2-}$ and 1 g · atom of H^+ per every mol of O_2 (or, at 25 °C, and for p_{O2} = 0.001 MPa we obtain $0.63 \cdot 10^{-5}$ $S_2O_3^{2-}$ and $1.26 \cdot 10^{-5}$ g atom of H^+ per litre) with a decrease of pH.

The formation of tetrathionate S_4O_6 may be described by the relationships:

$$4H_2S + 6H_2O \rightarrow S_4O_6^{2-} + 20\ H^+ + 18e^-$$

and

$$O_2 + 4H^+ + 4e^- \rightarrow 2H_2O$$

or

$$4H_2S + 4.5O_2 \rightarrow S_4O_6^{2-} + 3H_2O + 2H^+$$

with the formation of 0.22 mol $S_4O_6^{2-}$ and 0.44 g · atom of H^+ per mol of O_2 (i.e. at 25 °C and p_{O2} = 0.001 MPa gives $0.28 \cdot 10^{-5}$ mol of $S_4O_6^{2-}$ and $0.55 \cdot 10^{-5}$ g · of H^+ per litre). This reaction also reduces pH.

Characteristics of water in E–pH diagram at 25 °C

The pH value of pure water saturated with gaseous H_2S at a pressure of 0.1 MPa at 25 °C is equal to approximately 3.9. The characteristic of this water is located around the point denoted by *A* in Fig. 5.8. If more oxygen is added, this point is shifted upwards at constant pH until it reaches the line (6) representing the equilibrium between $H_2S_{(g)}$ and S and a pressure of 0.1 MPa. At this point (denoted by *B* in Fig. 5.8) oxidation of H_2S with a formation of elemental sulphur is thermodynamically possible according to reaction (3) or (6).

If the characteristic point reaches the lines (1) and (2) which correspond to the equilibrium $H_2S/S_2O_3^{2-}$ and $H_2S/S_4O_6^{2-}$, oxidation of H_2S may also lead to the formation of $S_2O_3^{2-}$ and $S_4O_6^{2-}$ from the thermodynamic viewpoint. Under these conditions, one may expect the simultaneous formation of elemental sulphur, thiosulphates and tetrathionates. Assuming that the previously considered initial amount of oxygen ($1.26 \cdot 10^{-5}$ mol/litre) is no longer renewed, the three previously mentioned reactions cease to occur since all the available oxygen has been consumed.

Detailed knowledge of the system in these conditions is based only on detailed understanding of the kinetics of the reactions (3), (1) and (2). If this is not so, two hypotheses may be proposed:

- the reaction H_2S/S (3) is reversible. In this case, only this reaction takes place and $S_2O_3^{2-}$ or $S_4O_6^{2-}$ form. The characteristic point is stabilised at point *B* in Fig. 5.8.
- the H_2S/S equilibrium is not reversible and the equilibria $H_2S/S_2O_3^{2-}$ and $H_2S/S_4O_6^{2-}$ (2) are reversible. The characteristic point is then displaced approximately along the lines (4) and (5) for the pressure of gaseous H_2S of 0.1 MPa. As already mentioned, both reactions of H_2S to $S_2O_3^{2-}$ and $S_4O_6^{2-}$ form H^+ (1 g · atom of H^+ mol of O_2 of reaction (4) and 0.44 g · atom of H^+ per mol O_2 for the reaction (5)). As indicated by Fig. 5.8, the tetrathionate $S_4O_6^{2-}$ is the dominant phase in comparison with thiosulphate $S_2O_3^{2-}$ and in this case the amount of produced H^+ is equivalent to the amount of oxygen present ($1.26 \cdot 10^{-5}$ mol O_2/litre) is approximately $1.26 \cdot 0.44 \cdot 10^{-5}$ gram-ions of H^+. The

presence of this amount there is no significant change of pH. If it is assumed that the total amount of dissolved sulphur (as $S_4O_6^{2-}$ + $S_2O_3^{2-}$) is approximately $10^{-4.9}$ mol/litre (i.e. approximately 0.3 mg S/L), the characteristic point is close to pH = 3.9 and E = 0.027 V (denoted by C in Fig. 5.8).

The validity of this hypothesis has not been proved, although it appears in practice that, under suitable conditions, the solution contains small quantities of suspended solid sulphur.

If the water is not in permanent contact with the atmosphere containing H_2S at a pressure of 0.1 MPa and O_2 at a pressure of 0.01 MPa, oxidation of H_2S will continue without breaks with the simultaneous formation of every or some of three oxidation products, i.e. S, $S_2O_3^{2-}$ and $S_4O_6^{2-}$. Formation $H_2S_2O_3$ and $S_5O_6^{2-}$ is also possible, as indicated by Fig. 5.8, at pH lower than 3 by hydrolysis of thiosulphate $S_2O_3^{2-}$ by a means of one or two mols of S (32 to 64 g) per mol of O_2 (32 g), i.e. 1–2 g of S per 1 g of O_2.

For example, the supply of 0.1, 1 or 10 g of oxygen into one liter of solution forms 0.1–0.2, 1–2 or 10–20 g of sulphur in a litre (in the form of elemental sulphur and/or dissolved in the form of $S_2O_3^{2-}$, $S_4O_6^{2-}$, $HS_2O_3^-$, $H_2S_2O_3$ and $S_5O_6^{2-}$). Assuming that the amount of elemental sulphur is negligible in comparison with the dissolved sulphur, the solution will be the solution of thiosulphate and polythionates with a concentration of completely dissolved sulphur of approximately 0.003–0.006, 0.03–0.06 or 0.3 0.6 mols of sulphur per litre (or log c = 2.5 to 2.2, –1.5 to 1.2 or –0.5 to –0.2). In this case, the characteristic point in Fig. 5.8 should be shifted from the points denoted by B and C to the left and along the lines (5) for log C = –2.5 to –0.2 to the point for which the concentration of sulphur, pH and potential E are approximately equal to the values given previously, as a function of the total amount of added oxygen.

The characteristic point of the water saturated with H_2S at a pressure of 0.1 MPa which is subjected to the temporary effect of oxygen at a pressure of 0.001 MPa is displaced along the points $A \rightarrow B \rightarrow C$. If the water is subjected to the constant effect of the same oxygen atmosphere, the characteristic point is displaced further along the points $C \rightarrow D \rightarrow E \rightarrow F$, with the formation of an acid solution formed by thiosulphate and polythionates $S_4O_6^{2-}$, $S_5O_6^{2-}$ and H_2S and probably also a certain amount of suspended elemental sulphur.

This interpretation of the diagram in Fig. 5.8 is based on several assumptions on the reaction substances and reaction kinetics of the considered reactions. However, the kinetics of the majority of these reactions is probably more complicated than assumed in this case on the basis of the available information.

5.3.2. Cu–S–H_2O equilibrium system

The published studies, concerned with the thermodynamic investigation of the Cu–S–H_2O system [28, 62] usually considered only the existence of the phases Cu_2S and CuS in the system. However, as indicated by the chapter concerned with mineralogy, the system also contains several stable phases at room temperature. The existence of djurleite, $Cu_{1.96}S$, and anilite $Cu_{1.75}S$, at low temperatures was described by Potter or Vaughan and Craig [63–64] and shows that these substances should be considered in the *E*–pH diagrams representing the equilibrium of this system [65].

It is possible that metastable copper sulphides form during leaching. They have already been identified in experiments during leaching of chalcosite in an acid medium [66–68], and the standard change of the Gibbs energy was identified for the phases $Cu_{1.95-1.91}S$, $Cu_{1.86-1.80}S$, $Cu_{1.68-1.65}S$, $Cu_{1.40-1.36}S$ and CuS. Since these are substances with a certain range of composition, the calculation and plotting of *E*–pH diagrams were carried out considering the average composition $Cu_{1.93}S$, $Cu_{1.83}S$, $Cu_{1.67}S$ and $Cu_{1.38}S$. The value ΔG^0 for CuS, formed by the electrochemical oxidation of chalcosite, –47.06 kJ mol^{-1} [67] is less negative than the same value for the mineral CuS published by Potter [63], i.e. –53.96 J mol^{-1} and indicates that CuS, formed during oxidation in the aqueous medium, is a metastable substance whose structure is not defined accurately as in the case of the stable form of the same sulphide.

The following main reactions between the potential existing phases have been considered in the system:

$$Cu^{2+} + H_2O \rightarrow CuO + 2H^+ \qquad (5.185)$$

$$CuO + H_2O \rightarrow CuO_2^{2-} + 2H^+ \qquad (5.186)$$

$$CuO_2^{2-} + H^+ \rightarrow HCuO_2^- \qquad (5.187)$$

$$CuO + H_2O \rightarrow H^+ + HCuO_2^- \qquad (5.188)$$

$$2CuO + H_2 \rightarrow Cu_2O + H_2O \qquad (5.189)$$

$$Cu_2O + H_2 \rightarrow 2Cu + H_2O \qquad (5.190)$$

$$Cu^{2+} + H_2 \rightarrow Cu + 2H^+ \qquad (5.191)$$

$$Cu^{+} + 1/2H_2 \rightarrow Cu + H^{+} \quad (5.192)$$

$$Cu^{2+} + 1/2H_2 \rightarrow Cu^{+} + H^{+} \quad (5.193)$$

$$2Cu^{2+} + H_2O + H_2 \rightarrow Cu_2O + 4H^{+} \quad (5.194)$$

$$2Cu^{2+} + HSO_4^{-} + 5H_2 \rightarrow Cu_2S + 4H_2O + 3H^{+} \quad (5.195)$$

$$2Cu^{2+} + SO_4^{2-} + 5H_2 \rightarrow Cu_2S + 4H_2O + 2H^{+} \quad (5.196)$$

$$2Cu^{2+} + H_2S_{(aq)} + H_2 \rightarrow Cu_2S + 4H^{+} \quad (5.197)$$

$$2Cu^{2+} + HS^{-} + H_2 \rightarrow Cu_2S + 3H^{+} \quad (5.198)$$

$$2CuS + H_2 \rightarrow Cu_2S + H_2S_{(aq)} \quad (5.199)$$

$$2CuS + H_2 \rightarrow Cu_2S + HS^{-} + H^{+} \quad (5.200)$$

$$Cu_2S + SO_4^{2-} + 2H^{+} + 3H_2 \rightarrow 2CuS + 4H_2O \quad (5.201)$$

$$Cu_2S + HSO_4^{-} + H^{+} + 3H_2 \rightarrow 2CuS + 4H_2O \quad (5.202)$$

$$2Cu + HSO_4^{-} + H^{+} + 3H_2 \rightarrow Cu_2S + 4H_2O \quad (5.203)$$

$$2Cu + SO_4^{2-} + 2H^{+} + 3H_2 \rightarrow Cu_2S + H_2O \quad (5.204)$$

$$2Cu_2S + H_2 \rightarrow H_2S_{(aq)} + 2Cu \quad (5.205)$$

$$2HCuO_2^{-} + 2H^{+} + H_2 \rightarrow 2Cu_2O + 3H_2O \quad (5.206)$$

$$2CuO_2^{2-} + 4H^{+} + H_2 \rightarrow Cu_2O + 3H_2O \quad (5.207)$$

$$CuO_2^{2-} + 2H^{+} + H_2 \rightarrow Cu + 2H_2O \quad (5.208)$$

In addition to these main equations, all the equations containing the above mentioned non-stoichiometric copper sulphides have been considered.

The diagrams of the Cu–S–H_2O system for temperatures of 25, 100 and 150 °C are shown in Fig. 5.10, and the total activity of the sulphur ions is assumed to be 10^{-1} M, the activity of the copper-containing substances is 10^{-6} M at 0.1 MPa of the total pressure, and only the presence of stoichiometric copper sulphides was considered. The temperature of 150 °C was considered taking into account that the melting point of elemental sulphur is approximately 120 °C which means that after reaching this temperature, molten sulphur will coat solid particles in the solution so that all the conditions will greatly change. The diagrams at 150 °C were used to illustrate the tendency for a change in the regions of stability of the individual phases present in the solution with increasing temperature.

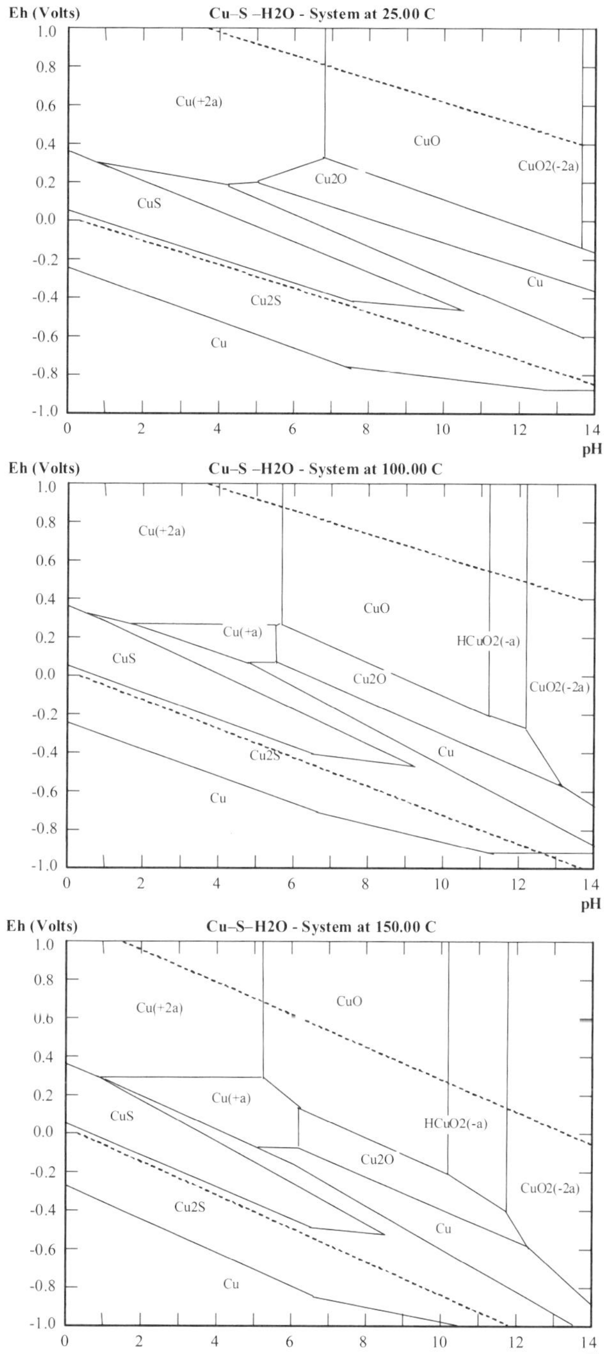

Fig. 5.10. *E*–pH diagram of the Cu–S–H_2O system for 25, 100 and 150 °C, for the activity of sulphur-containing substances of 10^{-1} M and the activity of copper-containing substances of 10^{-6} M and the total pressure of 0.1 MPa.

The reaction (5.186)

$$CuO + H_2O \rightarrow CuO_2^{2-} + 2H^+$$

is a non-redox reaction and the CuO/CuO^{2-} interface is represented by the vertical line. Comparison of the values of pH for this range and the interface is represented by the equations (3) and (4) and the activity of the substances containing copper equal to 10^{-6} M give

equation (5.186)	**equation (5.187)**	**equation (5.188)**
CuO / CuO^{2-}	CuO_2^{2-} / $HCuO_2^-$	CuO / $HCuO_2^-$
$pH_{298} = 12.97$	$pH_{298} = 13.14$	$pH_{298} = 12.8$

Schematically this may be depicted as follows

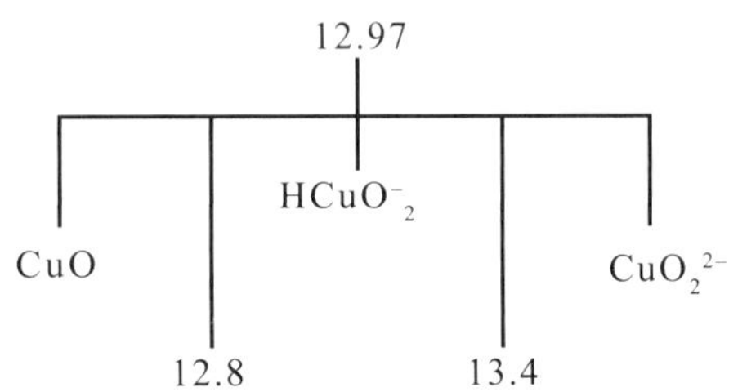

which means that there is no equilibrium between CuO and CuO_2^{2-} or, more accurately, the concentration of CuO_2^{2-} at pH below 13.4 is very low and, consequently, the CuO/CuO_2^{2-} equilibrium (equation 5.186) may be ignored. Similar considerations may also be applied to temperatures of 100 and 150 °C.

For the solutions with the activity of the copper-containing ions of approximately 10^{-3} M, the equilibrium reactions given by the equations (5.186)–(5.188) were determined for the following values of pH:

equation (5.186)	**equation (5.187)**	**equation (5.188)**
CuO / CuO_2^-	CuO_2^{2-} / $HCuO_2^-$	CuO / $HCuO_2^-$
$pH_{298} = 14.47$	$pH_{298} = 13.14$	$pH_{298} = 15.8$

and, consequently, the equilibria (5.186) and (5.187) may be ignored because of the previously mentioned reasons. In addition to this, equilibrium (5.188) is outside the range of the considered values of pH. At higher temperatures the equilibrium (5.188) is displaced to lower pH and (5.186) and (5.187) may still be ignored.

Comparison of the relationships (5.191)–(5.193) at different temperatures shows that they are independent of pH and, therefore, the interfaces are represented by lines at constant pH.

In the solutions in which the activity of the copper-containing ions reaches the values of the order of 10^{-6} M the equilibrium values are as follows:

equation (5.191)	**equation (5.192)**	**equation (5.193)**
Cu^{2+} / Cu	Cu^{+} / Cu	Cu^{2+} / Cu^{+}
E_{25} = + 0.160 V	+0.166 V	+0.153 V
E_{100} = + 0.115 V	+0.065 V	+0.221 V
E_{150} = + 0.085 V	–0.005 V	+0.266 V

At 25 °C, these equilibria may be described as follows:

Cu^{+}/Cu ——— + 0.166V		area of Cu^{2+}
Cu^{2+}/Cu ——— + 0.160V	———	
Cu^{2+}/Cu^{+} ——— + 0.153V		area of Cu

At potentials lower than +0.160 V metallic copper is stable and the Cu^{2+}/Cu^{+} equilibrium is not considered because the activity of Cu^{+} ions is negligible. Similar considerations also hold for potentials higher than 0.160 V when the area of Cu^{2+} and, therefore, the equilibrium Cu^{+}/Cu (5.192) may be ignored.

The results obtained for higher temperatures differ:

	100°C	150°C
Cu^{2+} / Cu^{+}	+0.165 V	+0.179 V
Cu^{2+} / Cu	+0.115 V	+0.085 V
Cu^{+} / Cu	+0.065 V	+0.008 V

At 150 °C and potential of +0.065 V metallic copper oxidises to the Cu^+ ion. The area of stability of metallic copper is below this value and direct oxidation of metallic copper to the Cu^+ ion is highly unlikely to take place. However, Cu^+ can be oxidised to Cu^{2+} at potentials higher than +0.165 V. Consequently, the Cu^{2+}/Cu equilibrium at 100 °C can be ignored which, however, does not apply to the equilibria (5.192) and (5.193) at 150 °C. Briefly speaking, at 25 °C the existence of Cu^+ ions may be ignored but at 100 and 150 °C there is already a significant region of the stability of Cu^+ ions. At these temperatures it is also necessary to consider the Cu^+/Cu_2S and Cu^+/Cu_2O equilibria. At 25 °C the area of stability of Cu^+ ions becomes significant only when the activity of copper-containing ions reaches the values of 10^{-8} M.

An interesting result is the one that shows the Cu^{2+}/CuS equilibrium which exists at 25 °C for negative values of pH transforms to positive values (close to zero) at 100 and 150 °C. On the other hand, reaction (5.200) tends to disappear with increasing temperature because the area of stability of CuS is displaced to the left away from the area of stability of the HS^- ions. Relationship (5.203) is relevant only in solution in which the activity is of the order of 10^{-3} at room temperature.

To summarise, it may be concluded that the equilibrium diagram of the copper-sulphur water system is situated inside an area of stability of water for the entire temperature range considered here. Covelin and sulphur may coexist in equilibrium in the entire temperature range. The Cu^+ ions, metastable at 25 °C, show a significant area of stability at elevated temperatures above 100 °C in solutions in which the activity of the copper-containing ions reaches the values of approximately 10^{-6} M. In this case, the hydroxide $Cu(OH)_2$ was not considered because it is less stable than CuO.

However, as already mentioned, the system contains a large number of non-stoichiometric sulphides, Cu_2S and, of course, other, often metastable forms of sulphur. After collecting the available thermodynamic values, especially for the copper sulphides, new calculations were carried out for the copper–sulphur–water system and the results are presented in Fig.5.11. The presence of metastable forms of sulphur had no effect on the resultant diagrams, but the presence of non-stoichiometric sulphides is clearly visible. The diagrams indicate the area of stability of the individual sulphides in the form of a thin edging of the lower sulphide. Table 5.6 shows the variation of the standard Gibbs energy of the

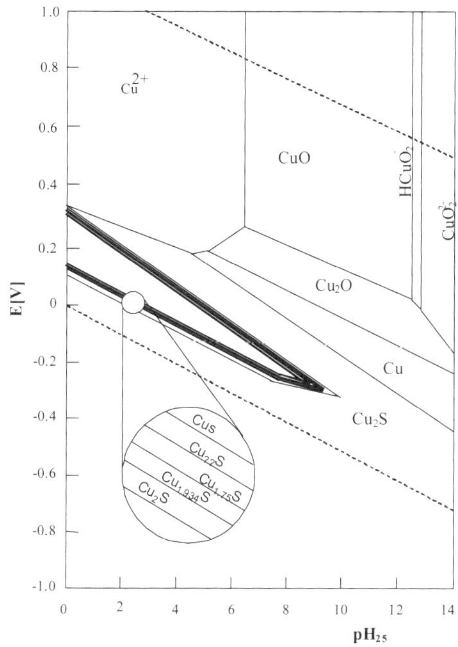

Fig. 5.11. *E*–pH diagram of the Cu–S–H_2O system at 25 °C.

Table 5.6. Values of standard Gibbs energy of formation of non-stoichiometric sulphides

Compound	**ΔG° [kJ/mol]**	
	25 °C	100 °C
$CuS_{metastable}$	–47.06	–47.526
CuS	–53.96	–54.48
$Cu_{1.1}S$	–56.98	–57.443
$Cu_{1.4}S$	–64.45	–65.558
$Cu_{1.75}S$	–78.58	–78.565
$Cu_{1.934}S$	–83.99	–85.476
$Cu_{1.965}S$	–84.78	–86.344
Cu_2S	–85.71	–87.46

formation of the individual non-stoichiometric sulphides.

A similar procedure is used in calculating diagrams at elevated temperatures. Their appearance did not change greatly and they copy the form shown in Fig.5.12, taking into account the fact that

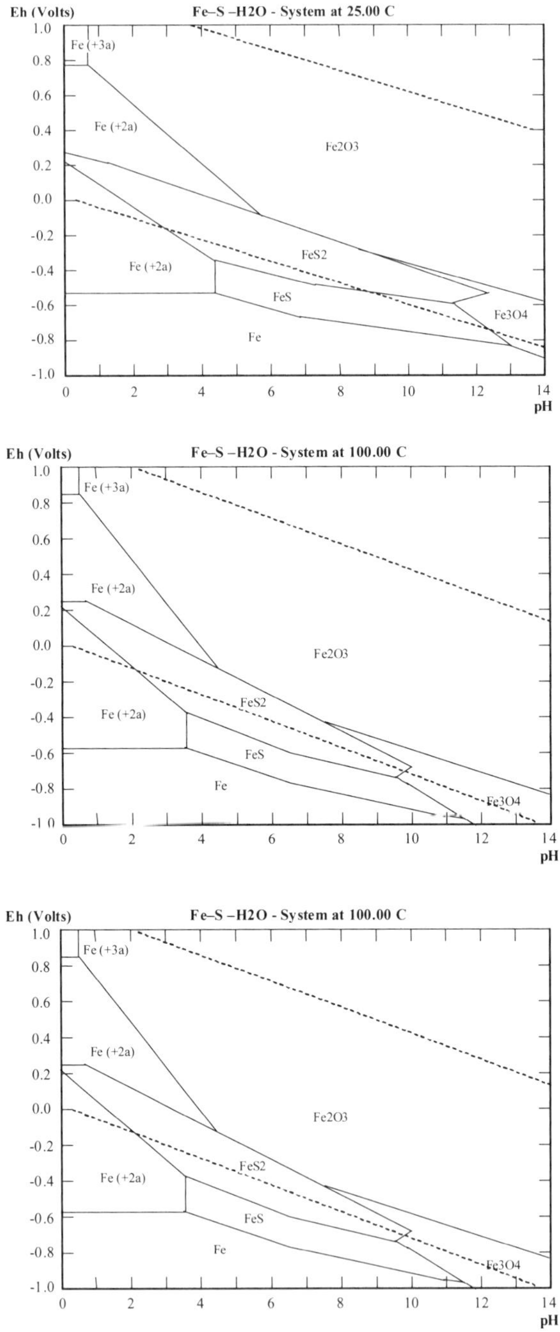

Fig. 5.12. *E*–pH diagram of the Fe–S–H_2O system at 25, 100 and 150 °C and a total pressure of 0.1 MPa. The activity of Fe-containing compounds is 10^{-6} M and that of the sulphur-containing compounds 10^{-1} M.

the CuS/Cu_2S interface contains non-stoichiometric sulphides of the type Cu_xS as already mentioned. An exception is $Cu_{1.75}S$ represented by anilite which is thermally stable up to approximately 70 °C and breaks down into digenite covellite above this temperature [69, 70]. Covellite is not found in diagrams for 100 and 150 °C.

5.3.3. The Fe–S–H_2O equilibrium system

The following reactions were taken into account in the system:

$$Fe^{2+} + 2HSO_4^- + 7H_2 \rightarrow FeS_2 + 8H_2O \tag{5.209}$$

$$Fe^{2+} + 2SO_4^{2-} + 2H^+ + 7H_2 \rightarrow FeS_2 + 8H_2O \tag{5.210}$$

$$FeS_2 + 2H^+ + H_2 \rightarrow Fe^{2+} + 2H_2S_{(aq)} \tag{5.211}$$

$$Fe_2O_3 + 4SO_4^{2-} + 8H^+ + 15H_2 \rightarrow 2FeS_2 + 19H_2O \tag{5.212}$$

$$Fe_2O_3 + 4HSO_4^- + 4H^+ + 17H_2 \rightarrow 2FeS_2 + 19H_2O \tag{5.213}$$

$$Fe_3O_4 + 6SO_4^{2-} + 12H^+ + 22H_2 \rightarrow 3FeS_2 + 28H_2O \tag{5.214}$$

$$3FeS_2 + 4H_2O + 2H_2 \rightarrow Fe_3O_4 + 6HS^- + 6H^+ \tag{5.215}$$

$$3FeS_2 + 4H_2O + 2H_2 \rightarrow Fe_3O_4 + 6S^{2-} + 12H^+ \tag{5.216}$$

$$FeS + 2H^+ \rightarrow Fe^{2+} + H_2S_{(aq)} \tag{5.217}$$

$$Fe_3O_4 + 3HS^- + 3H^+ + H_2 \rightarrow 3FeS + 4H_2O \tag{5.218}$$

$$Fe_3O_4 + 3S^{2-} + 6H^+ + H_2 \rightarrow 3FeS + 4H_2O \tag{5.219}$$

$$FeS_2 + H_2 \rightarrow FeS + HS^- + H^+ \tag{5.220}$$

$$FeS_2 + H_2 \rightarrow FeS + H_2S_{(aq)} \tag{5.221}$$

$$Fe^{2+} + 2S + H_2 \rightarrow FeS_2 + 2H^+ \tag{5.222}$$

$$3Fe_2O_3 + H_2 \rightarrow 2Fe_3O_4 + H_2O \tag{5.223}$$

$$Fe_2O_3 + 4H^+ + H_2 \rightarrow 2Fe^{2+} + 3H_2O \tag{5.224}$$

$$Fe_2O_3 + 6H^+ \rightarrow 2Fe^{3+} + 3H_2O \tag{5.225}$$

$$2Fe^{3+} + H_2 \rightarrow Fe^{2+} + H^+ \tag{5.226}$$

The diagrams of this system for temperatures of 25, 100 and 150 °C, the activity of iron-containing compounds in the solution of 10^{-6} M and the activity of sulphur-containing compounds of 10^{-1} M at a total pressure of 0.1 MPa are shown in Fig.5.12.

It may be seen that the form of the diagrams does not change greatly with increasing temperature. However, at 25 °C there are changes with a change of the activity of sulphur in the solution. In this case, the area of stability of pyrite decreases together with the increase of the area of stability of pyrrhotite FeS which is stable in strong reduction conditions at pH = ~7 to 9.5.

At temperatures of 100 and 150 °C sulphur and Fe^{2+} ions can still coexist but more concentrated solutions at the given values of pH and temperatures at equilibrium can be ignored.

The area of stability of pyrite is, at higher temperatures of up to 150 °C, partially overlapped by the area of stability of sulphur so that these compounds can coexist. However, this coexistence can also be ignored in the case of more concentrated solutions in the considered range of the values of pH and temperatures. The area of stability of pyrrhotite, which is quite small at 25 °C, tends to disappear with increase of temperature and no longer exists at 150 °C.

The area of stability of Fe^{2+} ions decreases in size with increasing temperature or with increasing activity of iron ions. At 25 °C, Fe_3O_4 is stable in strong reduction conditions at very high values of pH, but at higher temperatures the region of its stability widens in the direction to the lower values of pH.

The area of stability of the Fe^{3+} ions decreases with increasing temperature in solutions with a low concentration of ions with the iron content of 10^{-6} M and the activity of sulphur-containing ions of 10^{-1} M. In more concentrated solutions (10^{-1} M of iron and sulphur) this region can be ignored.

5.3.4. Cu–Fe–S–H_2O equilibrium system

The *E*–pH diagram of the Cu–Fe–S–H_2O system for a temperature of 25 °C is shown in Fig.5.13, for the activity of copper- and iron-containing substances in the solution of 10^{-3} M and for the activity of sulphur-containing substances of 10^{-1} M, at a total pressure of 0.1 MPa.

As in the case of non-stoichiometric copper sulphides of the type Cu_xS, a number of sulphides $Cu_xFe_yS_z$ also exists in this case. Taking into account the absence of thermodynamic data, only the

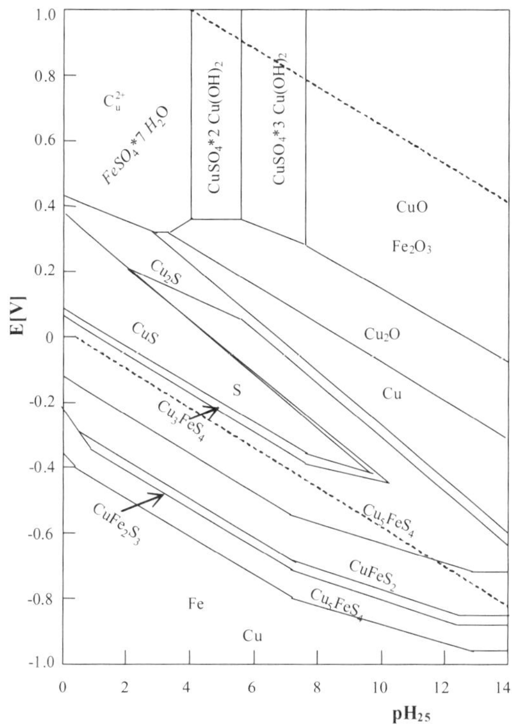

Fig. 5.13. *E*–pH diagram of the Cu–Fe–S–H_2O system at 25 °C, for the activity of copper- and iron-containing substances equal to 10^{-3} M, the activity of sulphur-containing substances of 10^{-1} M and the total pressure of 0.1 MPa.

Table 5.7. The values of standard Gibbs energy offormation of some sulphides $Cu_xFe_yS_z$

Compound	ΔG° [kJ/mol] at 25 °C
$CuFeS_2$	187.51
$CuFe_2S_3$	–287
Cu_3FeS_4	–314
Cu_5FeS_4	–386.39

basic sulphide chalcopyrite $CuFeS_2$ is usually found in the published diagrams [28, 71].

The potential presence of sulphides has been considered for this case; the thermodynamic data available for these sulphides are shown in Table 5.7.

The diagram does not show the presence of pyrite and pyrrhotite in order to provide better information. However, these are found in regions overlapping with covellite, CuS or bornite Cu_5FeS_4, as indicated by the diagram of the Fe–S–H_2O equilibrium system shown in Fig. 5.12.

The diagram shows that in the area of stability of water at low pH covellite, CuS and bornite, Cu_5FeS_4 are in equilibrium. The standard change of the Gibbs energy of bornite is considerably lower than that of the potentially present phases $CuFeS_2$, $CuFe_2S_3$, Cu_3FeS_4 or non-stoichiometric sulphides Cu_xS. In the range of higher values of the redox potential, there are soluble ions of copper Cu^{2+} and iron Fe^{2+} or Fe^{3+} and at higher values of pH, also the oxide CuO, Fe_2O_3 or complex $CuFeO_2$. The increase of the values of pH results, depending on the values of the redox potential, in the formation of chalcocite, Cu_2S, cuprite Cu_2O, hematite Fe_2O_3. The form of the diagram is very similar to that of the Cu–S–H_2O system shown in Fig.5.11.

The formation of copper and iron sulphides cubanite $CuFe_2S_3$ and chalcopyrite $CuFeS_2$ is hypothetically possible only in the area away from the area of stability of water and from the viewpoint of the considered system, it is irrelevant, with the exception of a small area at pH close to 14.

Similarly, the form of the diagrams does not change greatly with increasing temperature. In the area defined by the stability of water, the area of stability of Fe^{2+} decreases, like the area of stability of CuS. The areas of stability of non-stoichiometric sulphides of copper is reduced to the existence of only djurleite $Cu_{1.934}S$ and chalcocite Cu_2S.

At temperatures close to 100 °C, the line of transition of Fe^{2+} to Fe_2O_3 is shifted to lower values of pH. The area of stability of Fe^{3+} ions is shifted to the area of very acid solutions. The transfer of copper and iron to soluble forms is moved to the region of lower redox potentials. The coexistence of the soluble ion Cu^{2+} and insoluble oxides or iron hydroxides is still observed in a wide range of pH although the boundary of transformation of Cu^{2+} to CuO is displaced greatly to lower values of pH but does not exceed the value close to pH = 4.

The area of stability of elemental sulphur, overlapped by the area of stability of covellite, decreases with increasing temperature. Consequently, the size of the area of stability of Cu_5FeS_4 increases.

As in any other cases, the increase of the concentration of copper and iron in their compounds, present in the system, results in the shifting of the areas of stability in the direction of lower pH values. In practice, this means that in the entire considered temperature range the system will contain stable Cu_5FeS_4 already from the values of pH close to 1 and gradually with increasing redox potential it will be Cu_2S and from the values pH < 1.8 and E > 0.35 V the system will contain stable oxides of copper and iron CuO, Fe_2O_3 or $CuFeO_2$.

5.3.5. The Cu–S–Cl–H_2O equilibrium system

Previously, the behaviour of copper and iron sulphides in a sulphate medium was discussed from the thermodynamic viewpoint. From the practical viewpoint in leaching, these equilibrium systems are used for theoretical examination of leaching of copper sulphides or copper and iron in the solutions of sulphuric acid, or using iron-sulphide, this is a relatively frequent case and is also utilised in industry, either for pressure leaching or leaching in normal conditions. In addition to the iron sulphate, iron chloride is also used as an oxidation agent, especially because it is a far more efficient leaching agent than the sulphate. Leaching systems using copper chloride as an oxidation agent are also being introduced. Therefore, the Me–S–Cl–H_2O system is also a subject of theoretical studies, although not so frequently as the Me–S–H_2O systems [72–74].

The equilibrium diagrams of the Cu–S–Cl–H_2O system at temperatures of 25, 100 and 150 °C are shown in Fig. 5.14. The activity of copper-containing substances in the solution is 10^{-3} M and the activity of sulphur-containing substances is equal to 10^{-1} M at a total pressure of 0.1 MPa.

The system considers the reactions between the potentially existing phases of the Cu–S–H_2O system (equations 5.185–5.208) and ions and substances containing chlorine, Table 5.8.

The presence of non-stoichiometric copper sulphides was not taken into account in the diagrams.

In comparison with the previously discussed diagrams, the diagrams of the Cu–S–Cl–H_2O system are characterised by the presence of a large area of stability of univalent chlorine complexes of copper. With increasing temperature the amount of the copper complexes in the system varies. If at room temperature the highest stability was shown by $CuCl_3^{2-}{}_{(aq)}$, at higher temperatures it is $CuCl_2^{-}{}_{(aq)}$. This division of copper into species depends not only on temperature but also on the total activity of copper in the solution.

Figure 5.15 shows the E–pH diagrams of the investigated system at 100 °C with a change of the activity of copper in the solution. In solutions with a high copper content above 0.1 M Cu the copper

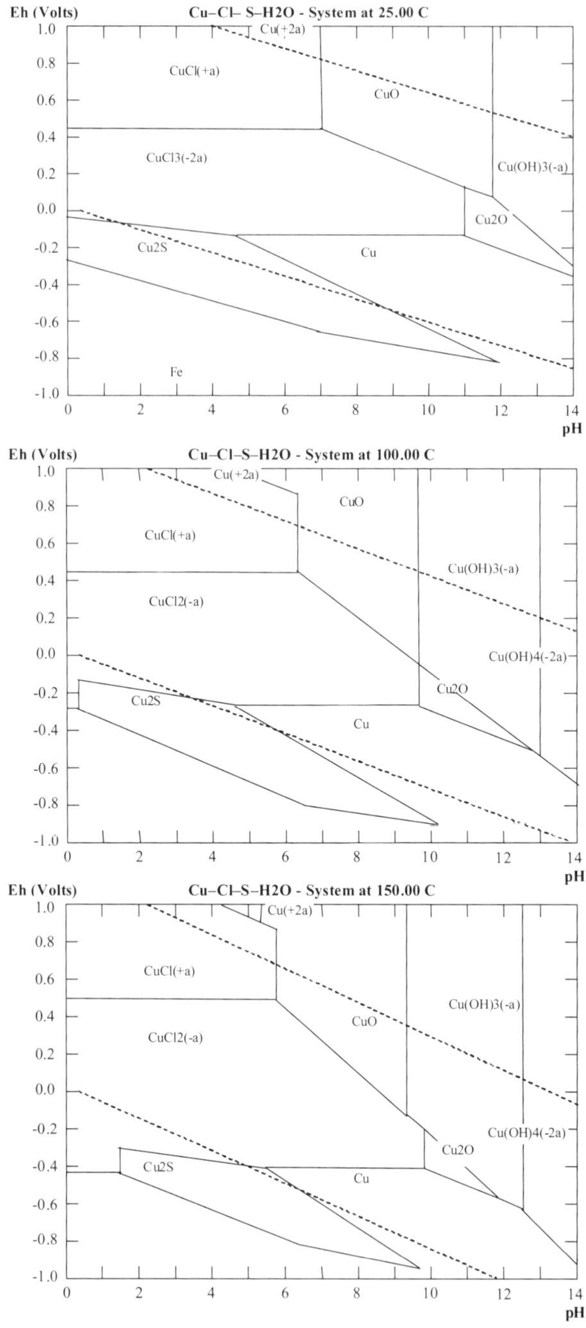

Fig. 5.14. *E*–pH diagram of the Cu–S–Cl–H_2O system at 25, 100 and 150 °C, the activity of sulphur-containing substances is equal to 10^{-1} M, the activity of copper-containing substances of 10^{-6} M and the unit activity of chlorides and the total pressure of 0.1 MPa.

chloride CuCl is stable. A decrease of the total concentration of copper below 0.01 M results in the area of stability of $CuCl_{2\ (aq)}^{-}$. The system shows this behaviour also with a decrease of the

Table 5.8. Substances containing chlorine in the equilibrium system Cu–S–Cl–H_2O

Soluble substances	Cu-containing substances
$Cl_{(aq)}^{-}$	CuCl
$Cl_{2(g)}$	$CuCl_2$
$Cl_{2(aq)}$	$CuCl_2·2\ H_2O$
$Cl^{-}_{3(aq)}$	$CuCl_2·Cu(OH)_2$
$ClO^{-}_{(aq)}$	$CuCl_2·2\ Cu(OH)_2$
$ClO^{-}_{2(aq)}$	$CuCl_2·3\ Cu(OH)_2$
$ClO^{-}_{3(aq)}$	$3CuCl_2·7\ Cu(OH)_2$
$ClO^{-}_{4(aq)}$	$CuCl^{-}_{2(aq)}$
$HCl_{(g)}$	$CuCl^{2-}_{3(aq)}$
$HClO_{(aq)}$	$CuCl^{+}_{(aq)}$
$HClO_{2(aq)}$	$CuCl_{2(aq)}$
	$CuCl^{-}_{3(aq)}$
	$CuCl^{2-}_{4(ag)}$

total concentration of chlorides in the solution.

Therefore, it appears that the application of E–pH diagrams to chloride systems is more complicated than in the case of the sulphate systems owing to the fact that the chloride form a large number of species, electropositive, electronegative and electroneutral. Species diagrams should be applied to these purposes, as discussed later.

5.3.6. The Fe–S–Cl–H_2O equilibrium system

The equilibrium diagrams of the Fe–S–Cl–H_2O system for temperatures of 25, 100 and 150 °C are shown in Fig. 5.16. The activity of substances containing iron, chlorides and sulphur is equal to 1 at a total pressure of 0.1 MPa. As in the previous case, the reactions between the potentially existing phases of the Fe–S–H_2O system (equations 5.209–5.226) and the ions and substances

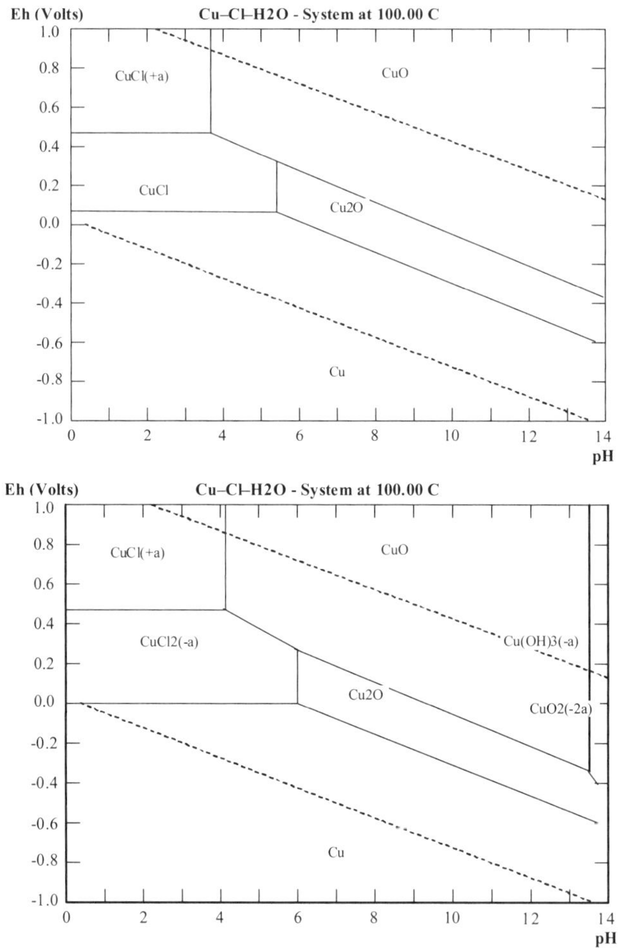

Fig. 5.15. *E*–pH diagram of the Cu–S–Cl–H_2O system for 100 °C, the activity of sulphur-containing substances of 10^{-1} M, the activity of copper-containing substances of 10^{-1} and 10^{-2} M at the unit activity of chlorides.

containing chlorine were taken into account in the system.

At room temperature, the system shows the area of stability of the $Fe^{2+}_{(aq)}$ ion. However, with increasing temperature this region disappears and the area of stability of $FeCl^{+}_{(aq)}$ at its expense. At the same time, the size of the areas of stability of $FeCl_2^{+}{}_{(aq)}$) as well as $FeCl^{+}_{(aq)}$ decreases with increasing temperature.

The reduction of the activity of iron-containing substances in the solution increases the size of the areas of stability of $FeCl_2^{+}{}_{(aq)}$ and $Fe^{2+}_{(aq)}$. The tendency in their behaviour is similar to that of more concentrated solutions, as shown in Fig. 5.17.

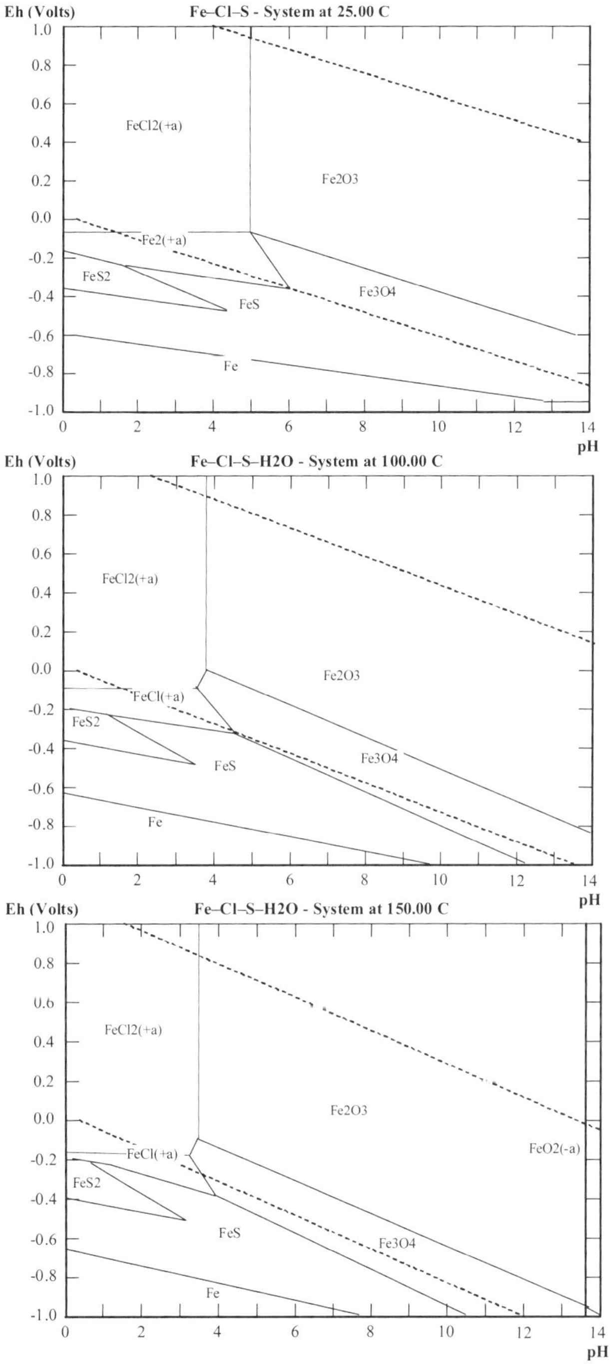

Fig. 5.16. *E*–pH diagram of the Fe–S–Cl–H_2O system at 25, 100 and 150 °C, unit activity of iron-, sulphur- and chloride-containing substances at a total pressure of 0.1 MPa.

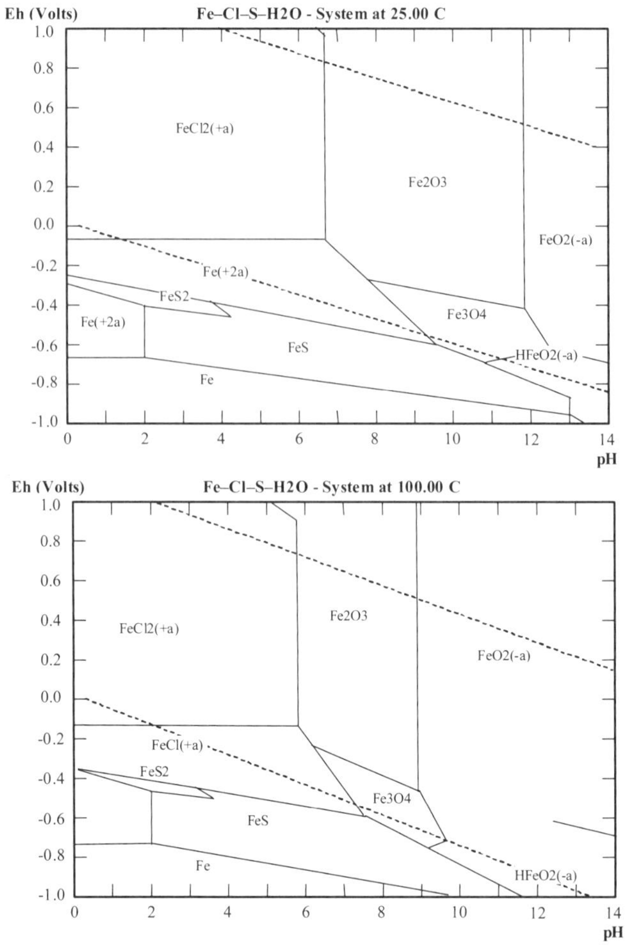

Fig. 5.17. *E*–pH diagram of the Fe-S-Cl-H_2O system at 25 and 100 °C, the activity of iron-containing substances of 10^{-6} M, the unit activity of chlorides and sulphur-containing substances at a total pressure of 0.1 MPa.

A similar trend is also detected when the total activity of the sulphur-containing substances is reduced, Fig. 5.18.

Investigations of the systems containing iron chlorides are very important for hydrometallurgy because the application of the trivalent iron ions as the oxidation agent in leaching of copper sulphides, either in the form of sulphate or chloride, is one of the main methods of oxidation leaching. This system is characterised by oxidation–reduction reaction in which Fe^{3+} is reduced and copper oxidised to the soluble form. Figure 5.19 shows the behaviour of Fe^{3+} ions in the Fe–S–Cl–H_2O system. With increase in

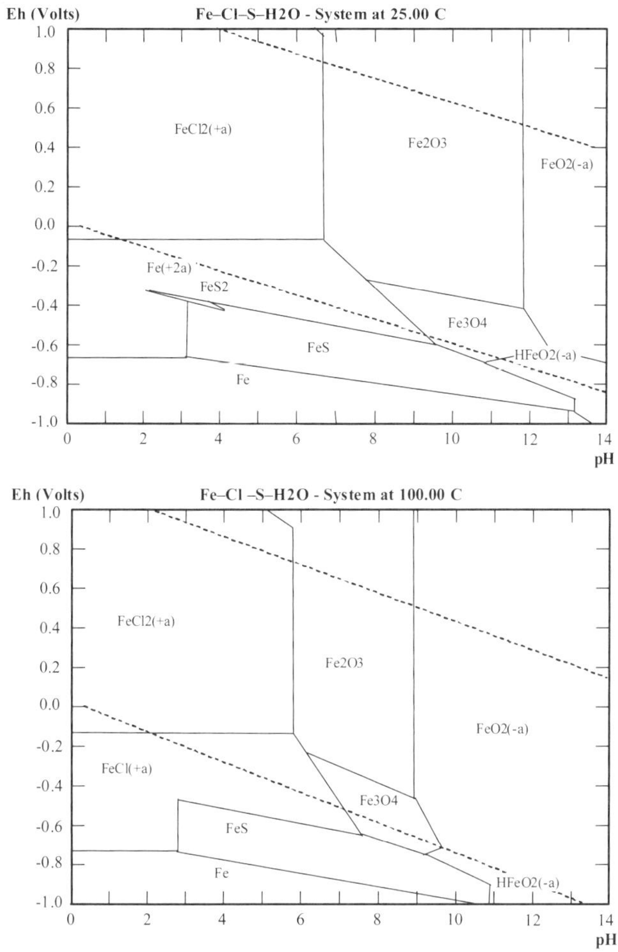

Fig. 5.18. *E*–pH diagram of the Fe–S–Cl–H_2O systems for 25 and 100 °C, the activity of iron-containing substances of 10^{-6} M, the unit activity of chloride- and sulphur-containing substances of 10^{-1} M at a total pressure of 0.1 MPa.

temperature the area of stability of the Fe^{3+} ions is displaced to lower values of pH, and at 150 °C it is even in the range of negative values of pH. For leaching in the normal conditions of pressure and temperature the Fe^{3+} ion shows stability in the region of acid solutions which may be used in practice.

5.3.7. Equilibrium diagram of the Cu–Fe–S–Cl–H_2O system

Using the previous considerations for calculations, the resultant *E*–pH diagram of the Cu–Fe–S–Cl–H_2O system is shown in Fig. 5.20 for 25 °C, the activity of copper- and iron-containing

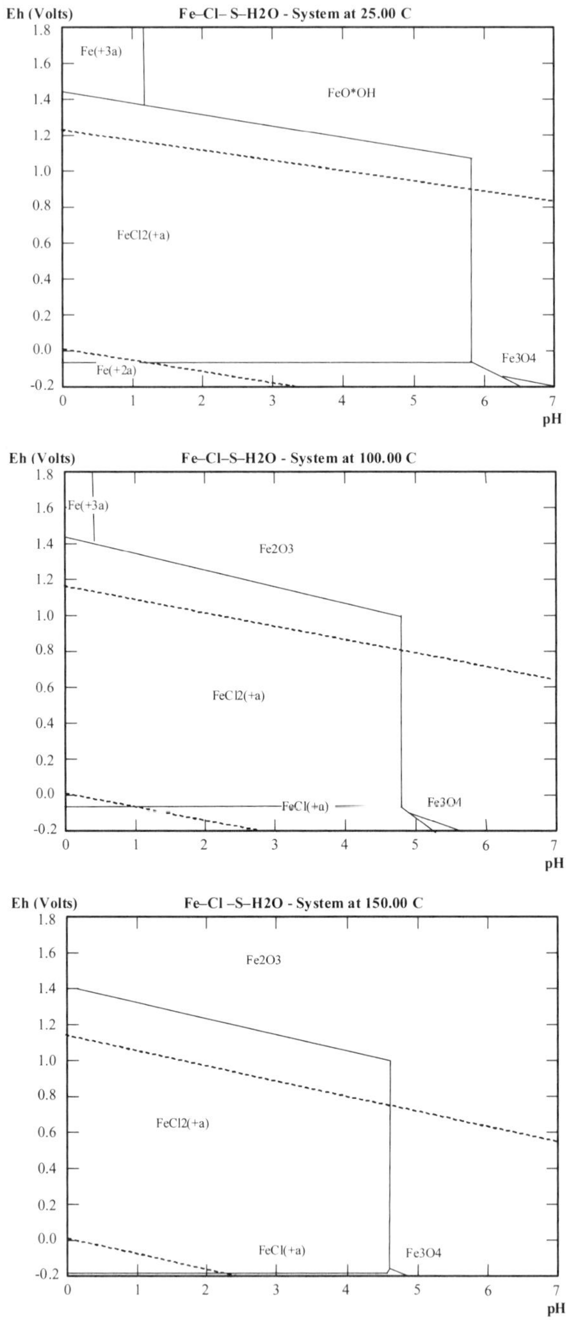

Fig.5.19. E–pH diagram of the Fe–S–Cl–H_2O system for 25 and 100 °C, the activity of iron-containing substances of 10^{-6} M, the unit activity of chloride- and sulphur-containing substances of 10^{-1} M at a total pressure of 0.1 MPa.

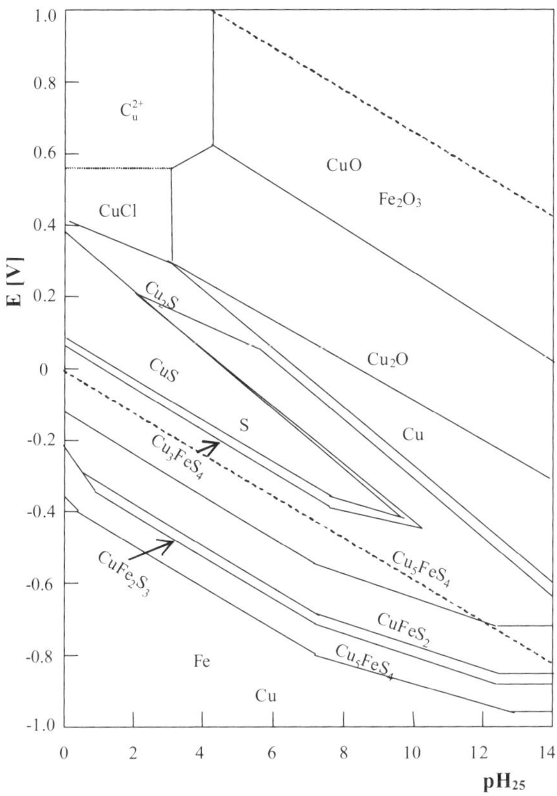

Fig. 5.20. *E*–pH diagram of the Cu–Fe–S–Cl–H_2O system for 25 °C, the activity of copper- and iron-containing substances equal to 10^{-3} M and the activity of sulphur-containing substances equal to 10^{-1} M at a total pressure of 0.1 MPa.

substances in the solution of 10^{-3} M and the activity of sulphur-containing substances of 10^{-1} M at a total pressure of 0.1 MPa. To simplify the diagram, the presence of non-stoichiometric copper sulphides Cu_xS and of the area of hydrated copper sulphates was not considered. The large area of stability of Cu^+ ions is clearly visible and the area of stability of $FeCl_2^+$ these activities are suppressed.

The potential–pH diagrams represent a suitable basis for selecting the leaching conditions, because they depict the course of reactions in aqueous solutions as a function of variable values of the potential and pH. Despite the fact that leaching is accompanied by a number of non-equilibrium reactions, and the thermodynamic considerations do not provide exhaustive information on these reactions, the diagram does indicate the areas of existence of the individual substances in the solution and the conditions, determined on the basis of the values of pH at the potential, of when and how the iron and copper sulphide or copper and iron are oxidised in the solution and what is the product of oxidation.

Oxidation leaching of sulphides of non-ferrous metals, present in complex sulphide ores and concentrates is aimed at application of acid solutions. Copper can be transferred from chalcopyrite to the soluble form of the bivalent ions by the formation of intermediate products of covellite, djurleite and anilite; at higher pH (~2.5) this is possible by the formation of bornite and chalcocite with the simultaneous oxidation of S^{2-} ions to elemental sulphur or SO_4^{2-} ion. Iron is oxidised to the soluble bivalent form with the formation of bornite. A further increase of the potential results in oxidation of iron to the trivalent form which is the essential oxidation agent and takes place in the leaching reaction. The increase of pH in the solution by a relatively small value (pH ~ 0.6) at room temperature causes the iron to transfer to the oxide form and this is followed by hydrolytic precipitation from the solution. However, copper remains in the form of the soluble bivalent ion in the solution in these conditions. This confirms thermodynamically the possibility of separating copper and iron in the solution by diluting the latter.

From this viewpoint, the presence and behaviour of species, especially in chloride solutions in which chlorine complexes form, is still a complicated problem. In this case, thermodynamic studies by a means of the *E*–pH diagrams do not explain efficiently the behaviour of the system in some cases regardless of the fact that there are still insufficient data for calculations. This problem may be solved using species diagrams which will be discussed later.

In most cases, an increase of temperature increases the leaching rate. Therefore, these systems were constructed not only for room temperature, 25 °C, but also for temperatures of 100 and 150 °C. The upper limit in temperature is limited by the fact that at temperatures of approximately 120 °C elemental sulphur melts, coats the sulphide particles so that almost all thermodynamic conditions are cancelled. Despite this, this temperature is given for more efficient description of the tendencies shown by the system with a change of temperature. The temperature of 100 °C is quoted as the limiting temperature for leaching in the so-called normal conditions, i.e., up to 100 °C and the unit total pressure.

The increase of temperature results in the transfer of copper from chalcopyrite to the soluble bivalent form by the formation of intermediate products of covellite and djurleite or bornite and chalcocite, and the equilibrium is displaced for lower values of pH.

Therefore, in the case of iron, the area of stability of Fe^{3+} ions no longer forms at a temperature of 100 °C in the realistic range of values of pH. As a result of dilution, Fe^{2+} changes directly to the insoluble oxide form.

A similar situation also exists when using a chloride medium for leaching. Copper is transferred to the univalent soluble form CuCl, which is more advantageous from the thermodynamic viewpoint than the bivalent form. This shows that it is more efficient to use the chloride medium for leaching than the sulphate medium which forms only the bivalent soluble form of copper.

5.4. Species diagrams

The *E*–pH diagrams describing the areas of stability of the individual substances in the solution do not fulfill efficiently the requirements on the presence of the components of the solution, if the molar ratio between the ligand and the metal changes. Species diagrams make it possible by a considerably simpler procedure.

It should be stressed that some of the published data on the stability constants of the individual species are contradicting. This is also in the case of chlorine complexes of copper and this is very important for hydrometallurgy. Assuming the sufficiently high concentration of the chloride, the solution is characterised by the formation of the thermodynamically stable Cu^{I} at room temperature, but the Cu^{+} ion does not form. In these solutions, the Cu^{I}–Cu^{II} pair can be used as the oxidation system and this is often utilised in copper hydrometallurgy. The possibilities of using the system have already been verified using the published data on the stability of Cu^{I} complexes by adapting the Pitzer method, as described previously. After deciding which of the possible species is indeed present, calculations are carried out to determine their equilibrium constants corresponding to the tabulated value [75] ΔG^0 of the formation of species, as shown in Fig. 5.21.

The characteristic feature of the hydrometallurgical processes is the regulation and control of the chemical conditions and obtaining the results. This often results in the formation of metallic anion species in the solution, although these metals should be produced from the cationic form.

An efficient method of extracting metals from the chloride solution by their mutual separation is liquid extraction. It is based on the fact that the substances $MeCl_z$ of metal Me^{z+} are not charged and if the metal is bonded with a large amount of chloride

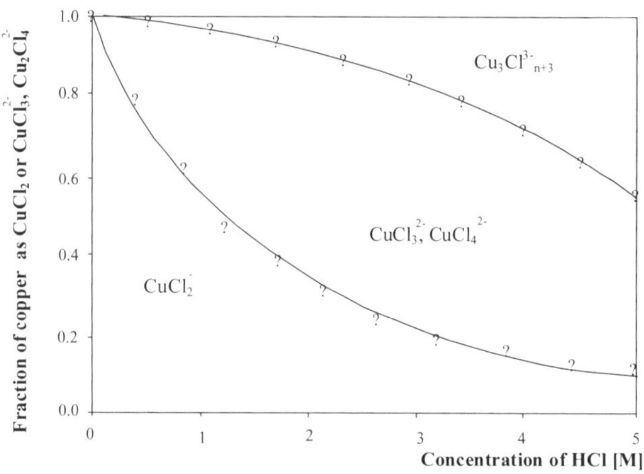

Fig. 5.21. Species containing Cu^{I} present in the solutions of CuCl in HCl with different concentration.

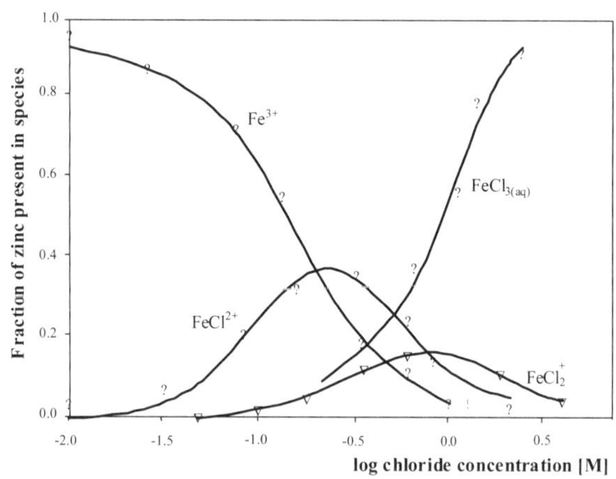

Fig. 5.22. Species diagram for Fe^{III} in chloride solutions.

ions, then *z* becomes anionic. Taking into account the fact that the values of the equilibrium constants of the formation of complex species are unique for individual elements, they form non-charged and anionic species at different concentrations of the chloride ions [76]. If it is necessary to determine the experimental conditions of the process in which two or more metals should be separated, it is rational to use the species diagram indicating the fractions of the metals present in its chlorine complex within the framework of the considered range of the total chloride concentration, which may be up to 10 M. Examples are shown in Figs. 5.22–5.24 for Fe^{III}, Cu^{II} and Zn^{II}.

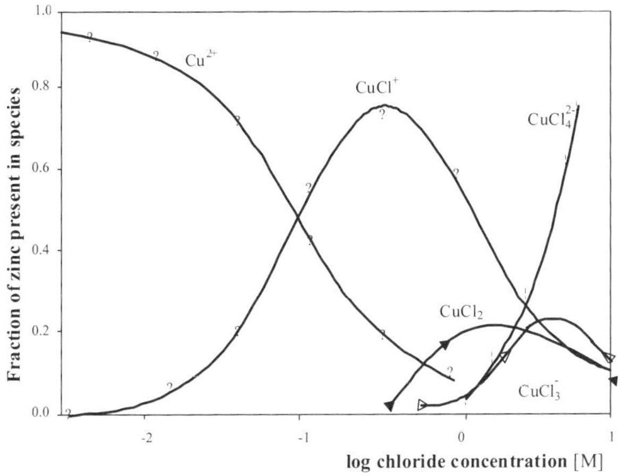

Fig. 5.23. Species diagram for Cu^{II} in chloride solutions.

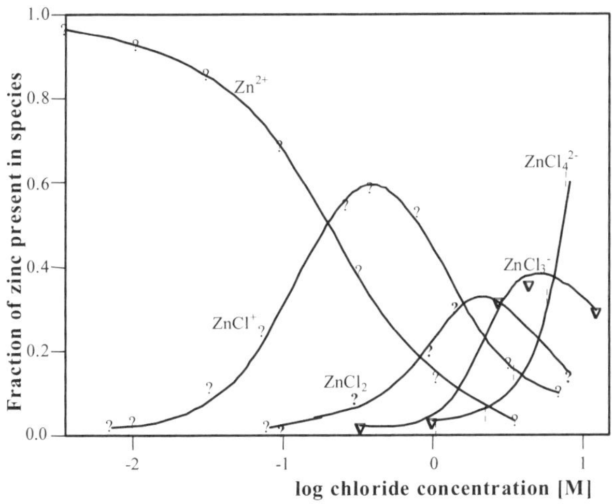

Fig. 5.24. Species diagram for Zn^{II} in chloride solutions.

For plotting species diagrams it is more efficient to use lower concentrations of metal. The values of thermodynamic equilibrium constants can be obtained by extrapolation to infinite dilution or by using the ΔG^0 values of the species, if these are available. Of course, it is also necessary to take into account the effect of the change of the activity coefficient with a change of concentration.

References

1. Latimer W.M.: Oxidation Potentials, Prentice-Hall, 1952, pp. 392.
2. Pourbaix M.: Thermodynamics of dilute aqueous solutions: Graphical Representation of the Role of pH and Potential, Cebelcor Publication F 227, 1963.
3. Valensi G., Van Muylder J.: Comportement electrochimique du soufre. Diagrames d'equilibres tension – pH du systeme S–H_2O, a 25 °C, 1 atm. Rapport Cebelcor n° 559, CEFA/R.17, 1958.
4. Valensi G., Van Muylder J., Pourbaix M.: Atlas d'equilibres electrochimiques, Gauthier-Villars ed., 1963.
5. Pourbaix M.: Atlas of Electrochemical Equilibria in Aqueous Solutions, Pergamon Press, New York, Rap. Cebelcor, Brussels, 1966.
6. Garrels R.M., Naeser C.R.: *Geochimica et Cosmochimica Acta*, 15, 1958, 113-130
7. Barnes H.L., Kullerud G.: *Economic Geology*, 56, 1961, 648-688.
8. Horvath J., Hackl L.: *Corrosion Science*, 5, 1965, 525-538.
9. Angus J.C., Lu B., Zappia M.J.: *Journal Applied Electrochemistry*, 17, 1987, 1-21.
10. Woods R.,Yoon R.H.,Young C.A.: *International Journal of Mineral Processing*, 20, 1987, 109-120.
11. Chodakovskij I.L., Ryzhenko B.N., Naumov G.B.: *Geokhimiya*, 12, 1968, 1486-1503.
12. Dyachkova I.B., Khodakovskij I.L.: *Geokhimiya*, 1, 1968, 1359-1375.
13. Biernat R.J., Robins R.G.: *Electrochimica Acta*, 14, 1969, 809-820.
14. Ashworth V., Boden P.J.: *Corrosion Science*, 10, 1970, 709-718.
15. Havlík T., Kme•ová D.: *Rudy*, 30, 5, 1982, 144-148.
16. Robins R.G.: Computer generated stability diagrams for the copper-sulfur -water System, Proc. Hydrometallurgy – Fundament., Technol., Innovat., Hiskey & Warren eds., 1993, 143-155.
17. Škrobian M., Havlík T.: *Metallurgical Letters*, 43, 2, 1989, 109-115.
18. Larsen D.M., Linkson P.B.: *Metallurgical Transactions B*, 23B, June 1993, 409-417.
19. Criss C.M., Cobble J.W.: *Journal of Amer. Chem. Soc.*, 86, 1964, 5385-5390.
20. Linkson P.B., Phillips B.D., Rowles C.D.: *Minerals Sci. Engng.*, 11, 2, 1979, 65-79.
21. Angus J.C.,Angus Ch.T.: *Journal of the Electrochemical Society*, 132, 5, 1985, 1014-1019.
22. Verink E.D. jr.: Adaptation of a computer program for Pourbaix diagrams to personal computers, Proc. Computer Usage in Mat. Education, Liedl ed., AIME, 1985, 123-134.
23. Havlík T., Škrobian M., Petrièko F.: E-pH '91: A computer program for calculating and plotting of high-temperature Pourbaix diagrams, Computer Software Conference, University Missouri-Rolla, June 1992.
24. Rutie L., Liang Y.: A single disk software for potential-pH diagrams with a large database, Proc. ICHM '92, vol 1, Jiayong ed., China, October 1992, 192-196.
25. Roine A.: HSC – A Windows version software for thermodynamic calculations, February 1994

26. Havlík T., Kammel R.: *Chemické listy*, 89, 1995, 603-610.
27. Garrels R. M.: Mineral Equlibria at Low Temperature and Pressure, Harper & Brothers Publ., New York, 1960.
28. Garrels R. M., Christ C. L.: Solutions, Minerals and Equilibria, Harper & Row Publ., 1965.
29. Barin I., Knacke O., Kubaschewski O.: Thermodynamic Properties of Inorganic Substances, Springer Verlag Berlin, 1977.
30. Barin I., Thermochemical Data of Pure Substances, Wiley-VCH, Weinheim, 1993.
31. Binnewies M., Milke E.: Thermochemical Data of Elements and Compounds, Wiley-VCH, Weinheim, 1999.
32. Pankratz L.B. et al.: Thermodynamic Properties of Elements and Oxides, US Bureau of Mines, Bulletin 672, 1982.
33. Chase M.W et al.: JANAF Thermochemical Tables, Third Edition, J. Phys. Chem. Ref. Data, vol. 14, 1985.
34. Kubaschewski O., Alcock C.B.: Metallurgical Thermochemistry, Pergamon Press, London, 1979.
35. Cobble J.W.: *Journal of American Chemical Society*, 86, 1964, 5394 – 5401.
36. Michard G., Allegre C.J.: Comportement geochimique des elements metalliques en milieu reducteur et diagrammes (log S, pH), Mineral. Deposita, 4, 1969, 1-17.
37. Angus J.C., Zappia M.J.: *Journal of Electrochemical Society*, June 1987, 1374-1383.
38. Murray R.C., jr., Cubicciotti D.: *Journal of Electrochemical Society*, 130, 4, April 1983, 868-869.
39. Biernat R.J., Robins R.G.: *Electrochimica Acta*, 17, 1972, 1261-1283.
40. Yokokawa H., Sakai N., Kawada T., Dokiya M.: *Journal of Electrochemical Society*, 137, 2, 1990, 388-398.
41. Wadsley M.: *Hydrometallurgy*, 29, 1992, 91-108.
42. Cubiccitti D.: *Corrosion Science*, 44, 12, 1988, 875-880.
43. Havlík T., Škrobian M., Malinowski Cz.: *Archiwum Hutnictwa*, 35, 1990, 3, 463-474.
44. Marcus P., Protopopoff E.: *Journal of Electrochemical Society*, 140, 6, 1993, 1571-1575.
45. El-Raghy S.M., El-Demerdash M.F.: *Journal of Electrochemical Society*, 136, 12, 1989.
46. Muir D.M.,Senanayake G.: *Hydrometallurgy*, 14, 1985, 279-293.
47. Froning M.H., Shanley M.E., Verink E.D., jr.: *Corrosion Science*, 16, 1976, 371–377.
48. Davis A.,Tran T.,Young D.R. *Hydrometallurgy*, 32, 1993, 143-159.
49. Adams M.D.: *Journal South African Inst. Min. Metall.*, 90, 2, 1990, 37-44.
50. Arai K.,Toguri J.M.: *Hydrometallurgy*, 12, 1984, 49-59.
51. Allaberginov R.D., Bogacheva L.M., Ismatov Kh. R.: *Zhurnal Prikladnoi Khimii*, 8, 1985, 1755-1759.
52. Acharya S., Anand S., Das S.C., Das R.P., Jena P.K.: *Erzmetall*, 42, 2, 1989, 66-73.
53. Osseo-Asare K.: *Trans. Instn. Min. Met., sect.C: Min. Process. Extr. Metall.*, 90, 1981, C152-158.
54. Osseo-Asare K.: *Trans. Instn. Min. Metall, sect.C: Miner. Proc. Extr. Met.*, 90, 1981, C159-163.

55. Bhuntumkomol K., Han K.N., Lawson F.: Leaching behaviour of metallic nickel in ammonia solutions, Inst. Min. and Metallurgy, March 1980, C7-13
56. Vu C., Han K.N., Lawson F.: *Hydrometallurgy*, 6, 1980, 75-87.
57. Gaspar V., Mejerovich A.S., Meretukov M.A., Schmiedl J.: *Hydrometallurgy*, 34, 1994, 369-381.
58. Wagman D. D. *et al.*: Selected Values of Chemical Thermodynamic Properties, Technical Notes National Bureau of Standards, 270/4, 1969, 43 – 55.
59. Zhang H.M., Zhang Z.C., Yang X.Z., Pourbaix M.: Chemical and Electrochemical Equilibria in the Presence of a Gaseous Phase. Sulphur, Rapports Techniques Cebelcor 149, RT.281, 1984.
60. Zhang H.M., Zhang Z.C., Yang X.Z., Pourbaix M.: Chemical and Electrochemical Equilibria in the Presence of a Gaseous Phase. Oxygen–Sulphur, Rapports Techniques Cebelcor 149, RT.282, 1984.
61. Pourbaix M., Pourbaix A.: Potential–pH Diagrams for the System S–H_2O, from 25 to 150 °C. Influence of Access of Oxygen in Sulphide Solutions Rapports Techniques Cebelcor 159, 1990.
62. Ferreira R.C.: High-temperature E-pH diagrams for the systems S-H_2O, Cu-S-H_2O and Fe-S-H_2O, In: Leaching and Reduction in Hydrometallurgy, IMM London, 1975, 67-84.
63. Potter R.W.: *Economic Geology*, 72, 1977, 1524–1542.

Chapter 6

SOFTWARE AND DATABASES FOR THERMODYNAMIC CALCULATIONS

As indicated by the previous chapters, for efficient thermodynamic calculation and plotting of different types of diagrams it is necessary to carry out complicated calculations but, in particular, have the required number of data of sufficient quality at our disposal. This is greatly helped by the application of powerful computers and the internet.

The literature describes a large number of algorithms including thermodynamic or thermochemical calculations, including databases for calculating phase composition, equilibrium diagrams, chemical reactions, thermal equilibria, solubilities, complex chemical equilibria, chemical modelling, etc. [1, 2]. However, many efficient algorithms operate with non-compatible databases and, therefore, in parallel. A problem in this case is obviously the protection of copyright and, on the other hand, the relatively high price of these program facilities. A quote for unification on the basis of compatible data sets was made by Bale and Eriksson [3] and at present there are several very large database systems enabling calculations and modelling of thermodynamic processes in hydrometallurgy, although each system has its specific features. The most important fact from this viewpoint is the existence of the concept of integrated databases brought into life by the Scientific Group Thermodata Europe SGTE. The integrated thermochemical database (ITD) is the defined as a system offering large thermochemical data banks and powerful programming enabling equilibrium thermodynamic calculations typical of complex chemical equilibria and calculations of phase diagrams in multicomponent multiphase systems.

SGTE is a consortium of seven west European organisations concerned with the development and operation of thermochemical databases in the form in which they can be used widely, although the majority of the existing program facilities are incompatible. At present, the integrated database SGTE contains the data on approximately 3000 compounds, of these 2000 condensed phases, 1000 gaseous compounds and 500 substances in the ion form for aqueous solutions [4].

It is not possible to list all the existing database systems nor the program facilities that are available. They are almost all of the commercial nature and their price very often controls their application and extent of use. On the other hand, new systems are being continuously developed. that is why as an example, attention will be given to several well known systems which may greatly help the thermodynamic calculation and other tasks in hydrometallurgy.

FactSage© [5] is one of the largest fully integrated database computer systems for chemical thermodynamics in the world. In this form, the system appeared in 2001 after merger with all the systems F*A*C*T and ChemSage/SolGasMix, which were developed over a period of almost 20 years in Umeå, Sweden Montreal, Canada and Aachen, Germany [6, 7].

The FactSage© program package operates on a PC under the Windows system and consists of a complex of information, databases, calculation and manipulation modules enabling work with a large number of pure substances and also substances in the ion state and species. At the present time, the system utilises several hundreds of industrial, governmental and academic institutions in the area of material science, pyrometallurgy, hydrometallurgy, electrometallurgy, corrosion, glass industry, ceramics, geology, combustion processes, etc. [8].

Figure 6.1 shows the areas of application of FactSage program package, version 5.3, in the year 2005.

From the viewpoint of hydrometallurgy, it is possible to calculate the equilibria in the solutions, *E*–pH and species diagrams, apply with the Debye–Hückel limiting law, model diluted and concentrated aqueous solutions on the basis of the Pitzer model, model aqueous solutions at high pressures and temperatures, apply the Helgeson model to diluted aqueous solutions, etc.,

In addition to the possibility of purchasing program package, either as a single or multiple licence, the researcher can work directly on the Internet where individual applications are free of charge otherwise a fee must be paid.

Figure 6.2 shows the possibilities of working with FactSage on the Internet, Fig. 6.3 and Fig. 6.4 show examples of the results.

Another relatively widely used system is HSC [10] which, however, is far smaller and designed for individual work on a PC. Its name comes from working with enthalpy (*H*), entropy (*S*) and

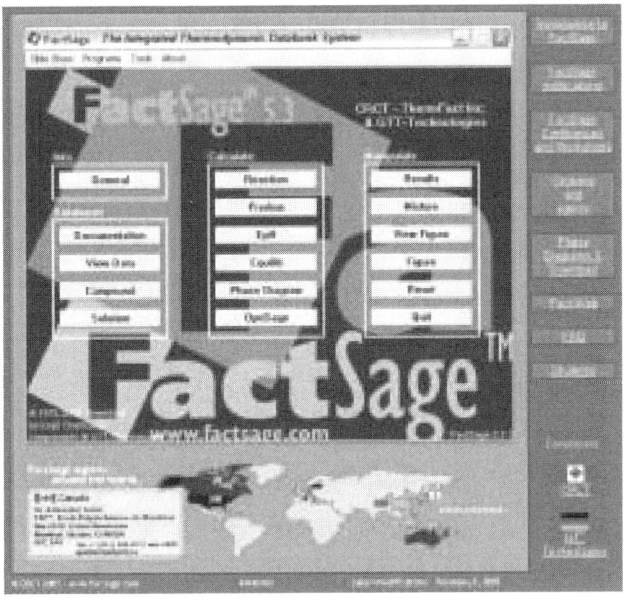

Fig. 6.1. Menu of the FactSage program package.

Fig. 6.2. Operation of the Fact-Web system on the Internet.

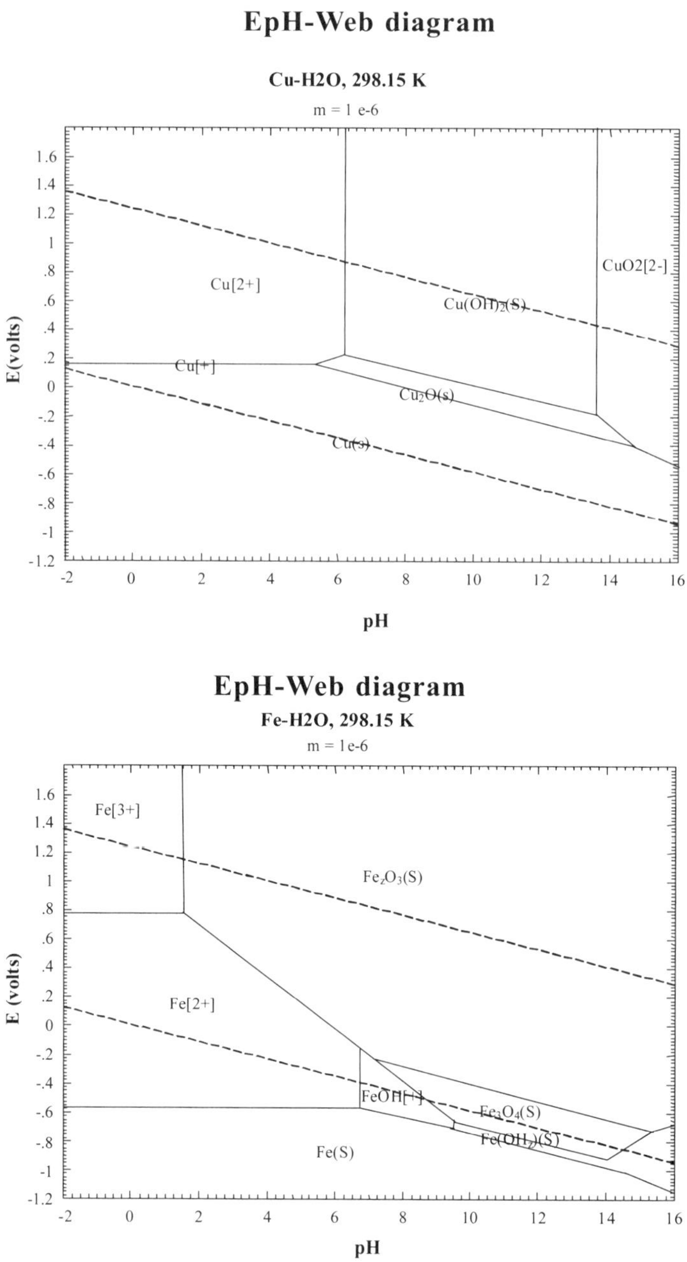

Fig. 6.3. Pourbaix diagrams of the Cu–H_2O and Fe–H_2O system obtained using the Fact-Web [9].

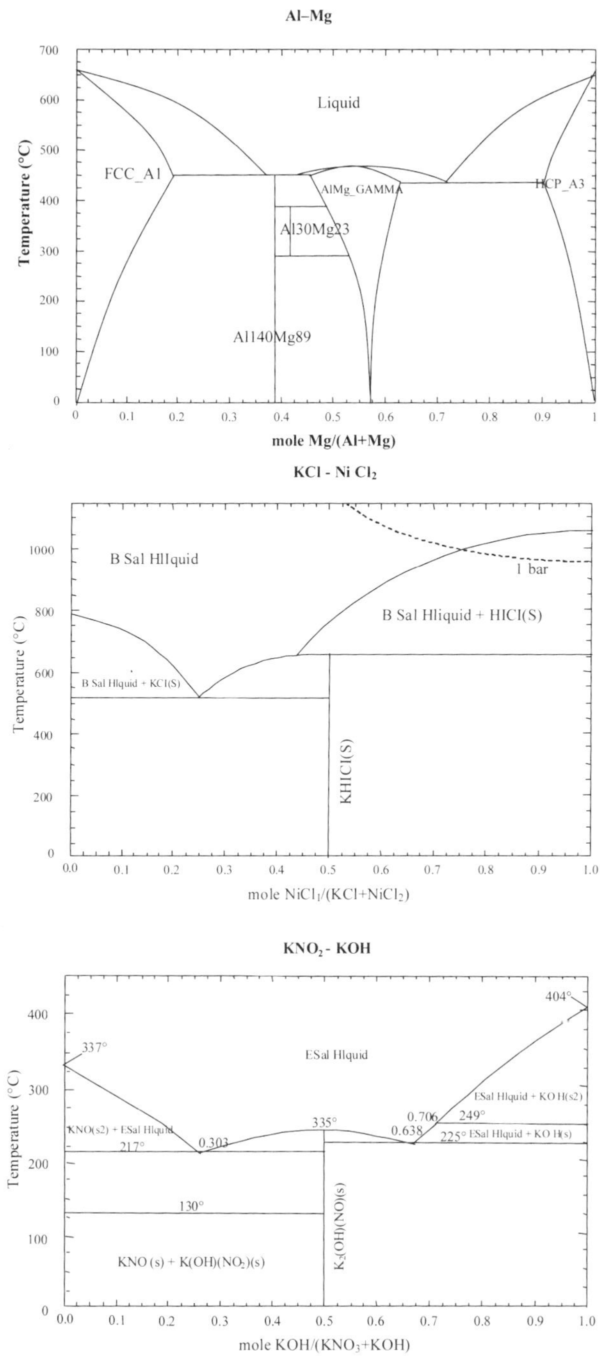

Fig. 6.4. Phase diagrams obtained using the Fact-Web [9].

thermal capacity (*C*). The current version HSC 6.1 operates under Windows®. The system was developed by Outokumpu Research Centre in Pori, Finland, in 1981 and has been greatly changed since then. The version of the program 5.11 has 14 application possibilities, Fig. 6.5, of which the application most important for hydrometallurgy is the calculation of reaction equations, material and heat balance, calculation of equilibrium compositions, electrochemical equilibria, plotting and displaying *E*–pH diagrams, plotting of species diagrams, etc.

At the present time, the database contains data on approximately 17,000 chemical substances. The entire system is organised in modules enabling independent activities. They are:

- equilibrium module;
- module of the mass/enthalpy equilibrium;
- Excel module enabling easy operations with calculations and imaging;
- tabulation and drawing thermodynamic parameters of the individual components;
- calculation of mineralogy and elemental composition;

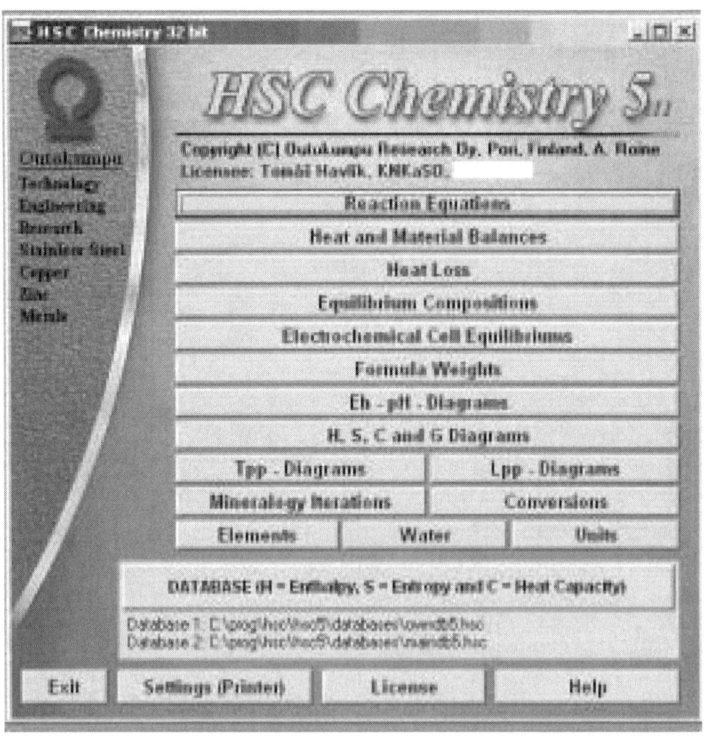

Fig. 6.5. Menu of HSC software, version 5.11.

- depicting the data from the database;
- conversions between the units.

The *E*–pH diagrams presented in Chapter 5 were designed and drawn by the HSC program, version 5.11.

The database system MTDATA – Metallurgical and Thermochemical Data Service [11] was developed and operated by the National Physics Laboratory, Harwell Laboratory, England, which is a member of SGTE. The development of the program started in 1971, with the most marked development possible only with the development of personal computers. Although this large system is designed mainly for high temperature systems, it may also be used for equilibrium calculations for multicomponent multiphase systems, calculation and graphical imaging of thermodynamic functions for pure substances, calculation and graphical imaging of thermodynamic functions in solutions, calculations and graphical imaging of binary and ternary phase diagrams, calculation of phase interfaces and crystallisation routes in multicomponent systems, calculation of the regions of stability and calculation and graphical imaging of *E*–pH diagrams, Fig. 6.6.

The METSIM software [12] (Metallurgical Simulations) is a system of programs for solving complex chemical, metallurgical and environmental processes, developed and distributed by the international consortium PROWARE [13]. The system is suitable for calculating the mass and energy equilibria and optimisation of engineering calculations, and is used for modelling and also production processes, such as waste processing, processing of ore minerals, dressing flowsheets, hydrometallurgy, pyrometallurgy, coal processing, heap leaching, etc.

A suitable example is the calculation of material equilibrium in heap leaching where it is necessary to consider the chemical reactions, precipitation and evaporation, the content of actual solid reagents and water, drainage and logistics. The model is not designed for the steady state and forms the time dependences of individual yield curves during leaching. It may also be used for comparing the control strategies, the effect of changes of the key parameters and the effect seasonal changes.

Figure 6.7 shows the simulated situation and Fig. 6.8 its output

GWB (Geochemist's Workbench®) [14] is a set of interactive programs for solving the equilibrium in the aqueous phase, and offers facilities for calculations in environmental protection and improvement of environment, oil industry and economical geology. The following tasks can be solved in metallurgy:

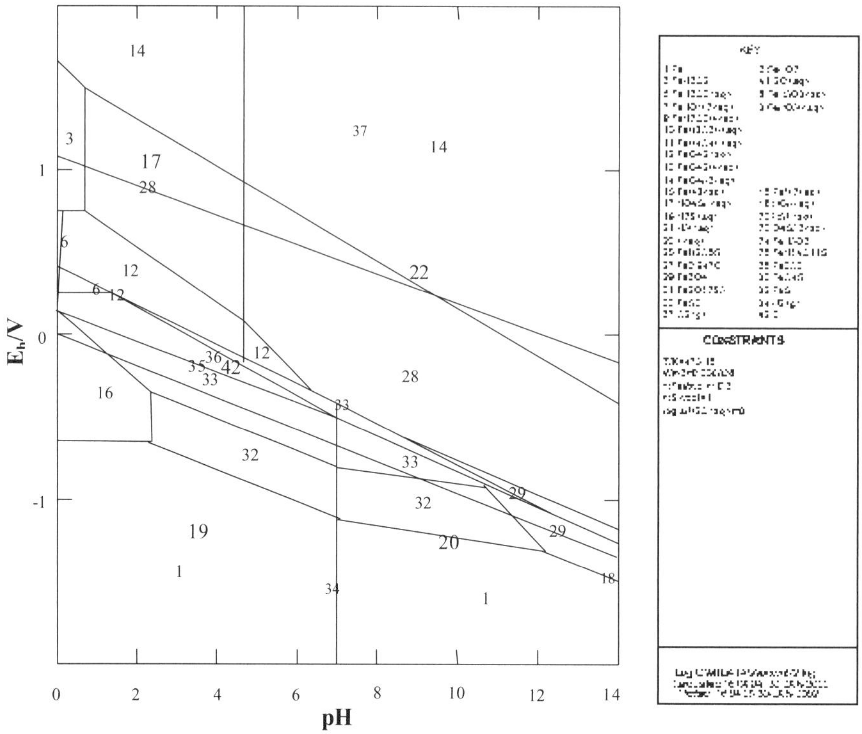

Fig. 6.6. *E*–pH diagram of the Fe–S–H_2O system at 200 °C, calculated by the MTDATA system.

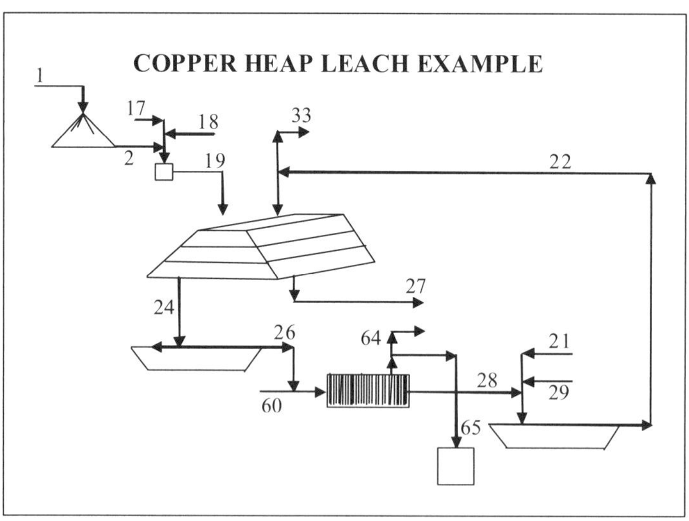

Fig. 6.7. Flow diagram of copper heap leaching in created by METSIM software.

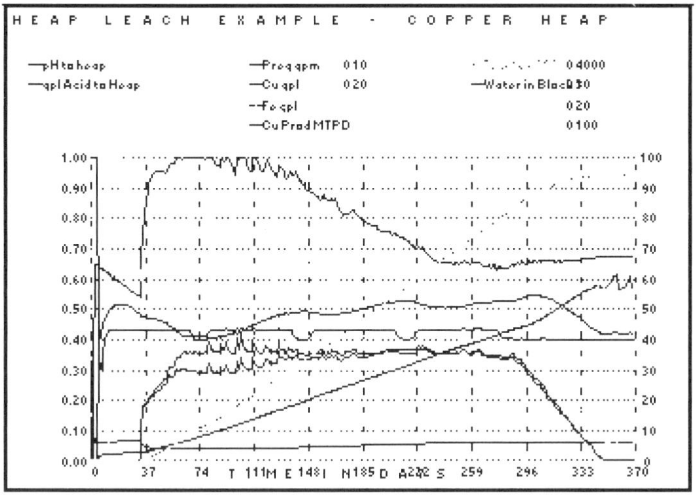

Fig. 6.8 Outputs of copper heat leaching simulation.

- chemical reaction equilibria;
- formation of *E*–pH and activity diagrams;
- calculation of species in the solution and gas fugacity;
- model sorption of ions and formation of complexes at the interface, etc.

Figures 6.9–6.11 show examples of the output of the GWB software for hydrometallurgy.

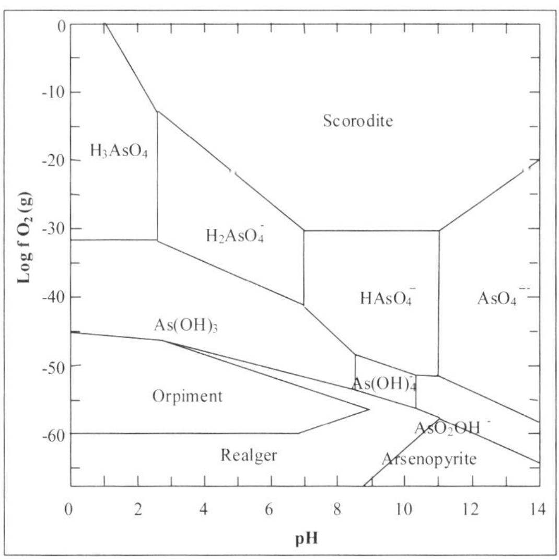

Fig. 6.9. Diagram of the As–Fe–S–H_2O system at 100°C generated by GWB software.

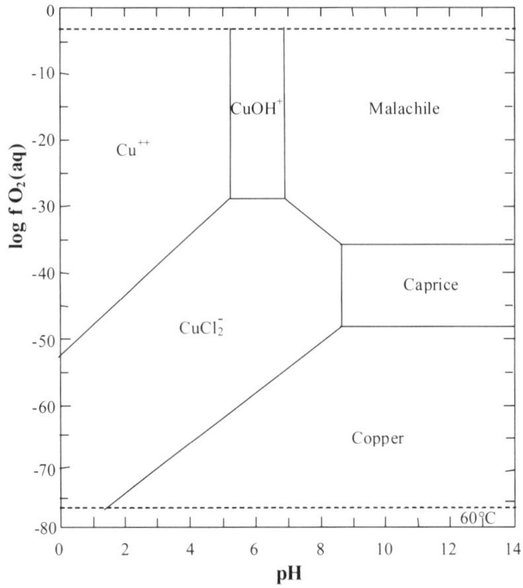

Fig. 6.10. Redox–pH diagram. Effect of chlorides on the formation of copper complexes at 60°C at $f_{CO2} = 1$.

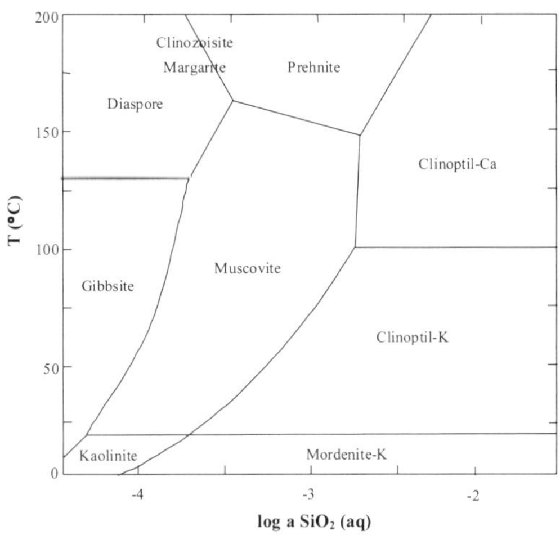

Fig. 6.11 Diagram of the stability of aluminosilicate in the presence of SiO_2 in the temperature range 0–200°C.

Of course, there are large numbers of other powerful or less powerful computer systems or software used for thermodynamic and thermochemical calculations in hydrometallurgy. The most important part of the discussed programs are the databases which of course determine the price of these programs. In many cases it is possible to obtain partially or even completely comparable results,

Table 6.1. Price list of several programs and databases for hydrometallurgical applications

System		Price		
Program	**Database**	**Single user**	**Multilicence**	**Universities**
			\$6000	
FACT	FACT		\$15000	
	SGTE		\$8500	
HSC		\$595	\$7000	\$3500
CHEMSHEET		\$2695	\$3995	\$1395 / 2095
MTDATA (for aqueous solutions)		£600		
METSIM		according to agreement	according to agreement	
GWB	\$99.00 handbook	\$799.00	+ \$399.00	\$599.00

may be less comfortably, also on considerably cheaper products. In any case, the application of integrated thermochemical databases ITD in connected with the computer network is most promising.

Table 6.1 shows the input data on the prices of the previously mentioned program facilities in spring of 2005.

References

1. Havlík T., Kammel R.: *Chemické listy*, 89, 1995, 603–610.
2. STN Database Catalog – the world premier on-line service for sci-tech information, FIZ Karlsruhe, January 2003.
3. Bale Ch.W., Eriksson G.: *Canadian Metallurgical Quarterly*, 29, 2, 1990, 105–132
4. http://www.sgte.org/
5. Bale Ch.W. a kol.: FactSage thermochemical software and databases, Calphad, 26, 2, 2002, 189–228
6. Thermfact/CRCT (Montreal, Canada) http://www.crct.polymtl.ca
7. GTT-Technologies (Aachen, Germany) http://www.gtt-technologies.de
8. http://www.esm-software.com/
9. http://www.factsage.com/
10. Roine A.: HSC Chemistry for Windows, Outokumpu, ver. 5.0, 2002
11. Davies R.H. a kol: MTDATA – Thermodynamic and phase equilibrium software from the National Physical Laboratory, Calphad, 26, 2, 2002, 229–271
12. Swanson V.N.: Modelling heap leach operations with METSIM, In: Recent Trends In Heap Leaching Conference, Bendigo, Victoria, Australia, 27–29 September 1994
13. http://members.ozemail.com.au/~ozmetsim/met2/http://www.rockware.com/catalog/pages/gwb

Chapter 7

KINETICS OF HETEROGENEOUS REACTIONS OF LEACHING PROCESSES

Thermodynamics provides information on the given system in the equilibrium state and makes it possible to forecast the behaviour of the equilibrium with the change of various external conditions, such as temperature, pressure, concentration of reagents, etc. The actual metallurgical systems, especially hydrometallurgical ones, greatly deviate from the equilibrium conditions and it is therefore necessary to know the actual behaviour with time, the factors controlling the rate of the process and parameters which must be ensured so that the reaction is ended in real time.

These questions are answered by chemical kinetics which studies the quantitative relationships of the course of a reaction in time. The kinetic studies usually include two stages: experimental determination of the degree or rate of transformation in relation to the conditions of the reaction and mathematical description of the determined relationship and the evaluation of the kinetic parameters of the reaction and their interpretation in relation to the nature of the processes taking place.

The hydrometallurgical processes are characterised mainly by heterogeneous reactions, for example, in leaching the solid phase interacts with the liquid or also gas phase, and the solid phase is partially or completely dissolved. Another type of reaction, the gas reacts with the solution and the overall of the process may be controlled by the rate of dissolution of the gas in the liquid in which the reactions take place. A typical example is pressure reduction of nickel or copper from a solution by hydrogen. Of great practical importance is also the kinetics of solvent extraction in which a metal passes through the interface between two liquid phases, one aqueous and the other one organic, and usually the ions of the metals and H^+ are exchanged. In the case of ion exchange, the rate of exchange of anions or cations depends on the diffusion of anions inside the ion exchanger.

A special feature of heterogeneous reactions is the location of the reaction zone at the interface. The general surface and thickness of the reaction zone may differ and are controlled by both the nature of the investigated process and the conditions of the process. The actual rate of the reaction may change from extremely slow to very high and is influenced by a number of factors.

These reactions greatly differ from the reactions generally considered in the basic principles of physical chemistry in which the problems of heterogeneous kinetics are solved on the basis of the main principles of heterogeneous reactions. The rate of a chemical reaction depends on the concentration of the reactants. If one of these reactants is a solid, its efficient concentration depends on the actual surface area and the roughness of the surface and internal surfaces must also be considered. The rate of the reaction rapidly decreases with the consumption of the reactants and, for this reason, it is necessary to consider the rate only on the time scale. However, if it is possible to prepare experiments in such a manner that it is possible to measure the kinetics of the reaction so that the concentrations of the reactants remain approximately constant during the measurements, then the measurements provide information on the reaction rate in the given conditions. If this is not the case, the measurements show the progressive deceleration of the reaction and the rates must be determined by a means of tangents to the kinetics curves at selected times. For mathematical processing of the results it is necessary to measure also the order of the reaction taking into account every reagent taking part in the reaction.

Although the majority of the reactions include a number of simple steps, which often require separate evaluation, there are certain general views on the overall reaction valid for a wide range of these reactions. Understanding of these general aspects is very important in solving the kinetics and mechanisms of the individual reactions in the investigated system and helps analysis of the complex systems represented in this case by leaching processes in hydrometallurgy.

Reaction rate

The rate of the reaction $A + B \rightarrow C + D$ is represented by the amount of the reactant A, or B, transformed to the reaction products in unit time, i.e.

$$v = -\frac{\Delta n}{\Delta t} \qquad (7.1)$$

where Δn is the change of the number of moles of the reacting substance in time period Δt.

This expression determines only the average rate of the reaction during the investigated period. If c_1 is the concentration of the substance formed at time t_1 and c_2 is its concentration at time t_2, then

$$v = -\frac{c_2 - c_1}{t_2 - t_1} \qquad (7.2)$$

In a limited case in which the differences $(c_2 - c_1)$ and $(t_2 - t_1)$ are very small it applies that

$$v = -\frac{\mathrm{d}c}{\mathrm{d}t} \qquad (7.3)$$

where dc is an infinitely small change of concentration at time dt.

The reaction rate may change from extremely slow to extremely fast. Of course, to determine the rate in the minimum time it is necessary to know the factors controlling these conditions.

The dependence of the reaction rate on concentration is expressed by the law of mass action (C. M. Guldberg, P. Waage, 1867) according to which the reaction rate at a constant temperature is proportional to the product of the concentrations of the substances taking part in the reaction. For a reaction to take place, the reacting molecules must meet. If the number of these meetings in unit volume per unit time increases, the reaction rate also increases. An increase of the number of molecules in the unit volume increases the number of these meetings. The number of meetings and, consequently, the reaction rate are therefore proportional to the concentration of the reacting molecules.

According to the law of mass action, the following equation holds for the reaction rate $A + B \rightarrow AB$

$$v = k\, c_A . c_B \qquad (7.4)$$

The proportionality constant k which is independent of concentration is referred to as the reaction rate constant. Numerically, it is equal to the reaction rate in a mixture in which the concentrations of all the reacting substances are unit concentrations. Its value depends on temperature and increases very rapidly with increasing temperature, or also depends on the presence of catalysts. For a general reaction

$$aA + bB + \ldots \rightarrow pP + rR + \ldots \tag{7.5}$$

we have

$$v = k.c_A^a.c_B^b \tag{7.6}$$

During a reaction, the concentrations of the initial components decrease and, consequently, the total reaction rate also decreases and after certain time drops to zero. At this moment, the reaction stops because the amount formed from each component is the same as the amount of the component which breaks down. The system is in chemical equilibrium, and:

$$\left[\frac{c_P^p.c_R^r}{c_A^a.c_B^b}\right]_{equilib} = \frac{\vec{k}}{\overleftarrow{k}} = K_c \tag{7.7}$$

where $\vec{k}$ and $\overleftarrow{k}$ are the rate constants of the direct and reversed reaction, and K_c is the equilibrium constant and is identical with the thermodynamic equilibrium constant. Its value is independent of concentration because the rate constants $\vec{k}$ and $\overleftarrow{k}$ do not depend on this constant. According to the kinetic concentrations, the reaction is not interrupted at the chemical equilibrium but takes place at the same rates in both directions – the system is in dynamic equilibrium.

Order of the reaction

The kinetic equation is a differential equation between the concentrations of the substances and time and, consequently, for equation (7.5) we can write:

$$v_A = -\frac{dc_A}{d\tau} = f(c_A, c_B, \ldots c_P, c_R \ldots) \tag{7.8}$$

where the left-hand side contains the increase of the product or the rate of the reaction.

As already mentioned (equation (7.6)), equation (7.8) may be expressed as follows

$$v_A = -\frac{dc_A}{d\tau} = k_A.c_A^a.c_B^b \ldots \qquad (7.9)$$

and this also applies to the reaction rate of the reactant B, etc.

Exponents a, b, ... are referred to as the order of the reaction in relation to its component A, B, ... The sum of the exponents $n = a + b + \ldots$ is referred to as the (total) order of the reaction.

In order to apply the kinetic equations, it is necessary to know the order of the investigated reaction, partial orders in relation to the reagents taking part and the values of the rate constants. The theoretical methods of determination are not yet sufficiently developed, and, therefore, the values are obtained by experiments, as described later. Generally, it is necessary to obtain the values of the concentration of the investigated substances in relation to the reaction or leaching time and, subsequently use the suitable numerical or graphic method.

Reaction interface

The surface, separating the individual phases is referred to as the reaction interface or interphase boundary. The reaction interface is characterised by the transfer of the mass of the reacting substances. Therefore, the interphase is the boundary between two phases, for example, in the reaction of the liquid–solid phase type, such as leaching, the interface is the outer surface of the solid substance in contact with the leaching agent. All the heterogeneous reactions may be divided into five categories, based on the nature of the interface, i.e.:

- solid substance – gas, (s) – (g);
- solid substance – liquid, (s) – (l);
- solid substance – solid substance, (s) – (s);
- liquid – gas, (l) – (g)
- liquid – liquid, (l) – (l)

In reality, only two phases take part always in the heterogeneous processes, although the number of the phases in a mixture may be greater. This is caused by the fact that the overall reaction rate is determined by the rate of a single elemental step – the rate–determining step which is the slowest of all the steps. For example, leaching of copper (solid substance) in a diluted acid (liquid) in the presence of oxygen (gas) is reduced to the (s)–(l) reaction because oxygen is transferred from the gas to liquid phase at a higher rate than other reactions taking place.

The rate controlling step of the reaction

The theory of chemical kinetics has been derived from experimental studies of homogeneous reactions in the liquid or gas state. In heterogeneous reactions in the liquid phase there are however also processes between one or several dissolved substances in the solution and the solid surface which cannot be described by the laws of homogeneous kinetics. A suitable simple example is the reaction between the solid surface and dissolved reactants. The surface is regarded as constant and for a reaction to take place, the dissolved reactant must reach this surface. This requirement can be described by relationships of simple or constrained diffusion. However, if ion is transferred from the solid to liquid phase during oxidation reactions on the surface, for example, Cu^{2+} from Cu_2S, an insoluble solid reaction product may form, in this case elemental sulphur and/or sulphate ions, if suitable oxidation conditions are ensured. Whilst the soluble product diffuses from the surface into the solution, sulphur tries to remain on this surface, thus inhibiting the access of the reactants to the surface of Cu_2S. In some cases, for example, in leaching nickel sulphides, sulphur may even form a continuous layer and the leaching reaction is stopped. In certain leaching reactions the oxidation of the solid substance produces species with very low solubility, like gold in acid leaching. But if there is a suitable ligand present in the solution, the formation of complexes like $Au(CN)_2^-$ in cyanide containing a leach solution makes gold highly soluble.

A large number of heterogeneous reactions take place through several elementary stages. According to [2], interaction between solid and liquid phases takes place through several steps:

(*i*) diffusion of reactants from the volume of the leaching agent to the interface (outer surface of the particles);

(*ii*) adsorption of the reactant on the surface of particles;

(*iii*) the chemical reaction of adsorbed reactants to the adsorbed products (surface reaction);

(*iv*) desorption of adsorbed products of the reaction;

(*v*) diffusion of products from the inner surfaces to the outer surface;

(*vi*) diffusion of products from the interface to the volume of the liquid.

If a layer of solid reaction products forms on the surface of the solid phase, other steps occur, namely:

(*ia*) the diffusion of reactants through the layer of the solid reaction product to the reaction interface;

(*va*) diffusion of reaction products through the layer of the solid reaction products from the reaction interface to the environment.

In some cases (relatively frequent) chemically controlled processes it is also necessary to take into consideration the transfer of electrons within the crystal structure as the rate-controlling step.

Depending on the conditions of the reaction, the relationships between the individual elementary stages change. All the stages, with the exception of the limiting stage, take place in equilibrium or quasi-equilibrium conditions. The overall rate of the process is controlled by its slowest stage. In most cases it is diffusion (either external or internal), the actual chemical reaction at the phase boundary, or nucleation. For this reason it is important to investigate initially the effect of all elementary stages of the reaction and determine the slowest stage, i.e. the rate controlling stage of the reaction.

Stages (*i*), (*v*) and (*vi*) are controlled by the diffusion rate of soluble reactants and reaction products and in systems with mixing of the solution also by the hydrodynamics of the system. Stages (ii), (iii) and (iv) may be regarded as chemically controlled processes.

7.1. Effect of variables on the kinetics of a heterogeneous reaction

The kinetics of the heterogeneous reaction investigates the rate of this reaction. However, this process is never simple and is affected by a large number of external factors so that generally it may be said that the rate is a function of many variables. From this the methods of experimental examination of the kinetics of heterogeneous processes result. Generally, it applies that the reaction rate is in fact the amount of the reacted initial substance per unit time. The dependences expressing the change of concentration are usually empirical but, in some cases, they may be forecast theoretically, if the reaction mechanism is known. If the kinetic equation of the reaction is known, it is possible to calculate the concentration of the reagent after a certain period of time under constant conditions at any reaction time. In this case, it is important to take into account all factors which may have a significant effect on the kinetics and mechanism of the reaction. The most important factors include:

- existence of the diffusion (boundary, Nernst layer);
- effect of the movement of the liquid phase;
- the interfacial area;
- geometry of the interface;
- formation of solidification nuclei;
- effect of temperature;
- effect of the reagent concentration;
- defects of the crystal structure;
- electrochemical leaching mechanism;
- autocatalytic reaction;
- effect of the solid product of the reaction

and many others.

Mass transfer in a heterogeneous reaction

Nernst's theory provides a significant contribution to mass transfer theory. Nernst assumed that the chemical processes taking place at the interface are always faster than the stage controlled mass transfer so that the overall reaction rate is controlled by mass transfer. In the simplest case, the reaction between one ion of the reactant and the solid is so fast that equilibrium is established almost instantaneously so that the concentration c_1 of the reactant at the interface remains negligible. The stage controlling the rate of the reaction is given by the rate with which the reactant reaches the reaction surface. If the solution is sufficiently mixed, concentration c in the volume of the solution is uniform and the concentration gradient between the values of c and c_1 is given by the difference of the concentrations at both sides of the thin stationary layer of the liquid with the thickness δ adjacent to the leached surface. If the area of the reaction surface is defined as A, then according to Fick's law the number of molecules of solution dN which transfer from the volume of the solution to the solid reaction surface during time dt is proportional to the concentration gradient normal to the surface, dc/dy:

$$\mathrm{d}N = -DA\left(\frac{\mathrm{d}c}{\mathrm{d}y}\right)\mathrm{d}t \tag{7.10}$$

The proportionality coefficient D is the diffusion coefficient with the dimension [(length)^2/time]. If the solid substance is in contact with the volume of the solution V then

$$-\frac{\mathrm{d}c}{\mathrm{d}t}=\left(\frac{DA}{V}\right)\left(\frac{\mathrm{d}c}{\mathrm{d}y}\right) \tag{7.11}$$

Nernst assumed that the concentration gradient is linear and expressed by $\frac{(c-c_1)}{\delta}$, so that

$$-\frac{\mathrm{d}c}{\mathrm{d}t}=\frac{DA(c-c_1)}{V\delta} \tag{7.12}$$

and the rate constant of the first order k is given by the expression $k=\frac{DA}{V\delta}$.

The rate constant through the unit surface in the unit volume k_T can be determined by experiments. Index T indicates that the rate constant belongs to the reactions controlling mass transfer. If the diffusion coefficient of the reactant is available, the thickness of the diffusion layer δ can be determined from the equation:

$$k_T=\frac{kV}{A}=\frac{D}{\delta} \tag{7.13}$$

The thickness of the diffusion layer δ for a large number of reactions in water at 20 °C is equal to approximately 0.03 mm. The fact that this value is the same for a large number of relatively different chemical reactions satisfy the assumption according to which the rates are controlled by diffusion processes. However, this value is physically unlikely because it includes a diffusion layer with a thickness of approximately 50,000 molecules. It is unlikely that the solid substance would greatly affect the molecule of water separated by many molecules of a different type. In reality, it is assumed that this layer is identical with a thin film of the liquid which remains at the surface of the solid substance washed by this liquid.

The effect of temperature on k_T may also be used to support the Nernst's view of heterogeneous reactions. If thickness δ is independent of temperature, then the ratio $\mathrm{d}k_T/\mathrm{d}T$ should be equal to the rate of the change of the diffusion coefficient with temperature. The activation energy of diffusion at 25 °C is usually between 12 and 27 kJ/mol, depending on the solvent and dissolved substance so that the observed critical increase of energy E_A for the processes, controlled by mass transfer, should be in the same range, approximately 17 kJ/mol. This is also in agreement with experimental results.

The critical increase of energy in a heterogeneous reaction is equal to the activation energy of a homogeneous reaction, as postulated by Arrhenius, equation (7.43). In examination of a homogeneous reaction the experimentally measured activation energy may be analysed by expressions for the free energy and entropy of the formation of a transitional state using the theory of the absolute reaction rate.

An increase in the mixing rate results in a decrease of the thickness of the Nernst layer, i.e. the value of δ decreases and the reaction rate increases. From the viewpoint of the mixing rate, the value δ depends on the dimensions and geometry of the system. If a powder solid substance is leached, it is then necessary to take into account the relative movement amongst the particles and the liquid and this is greatly influenced by turbulence.

In leaching of massive specimens the reaction rate increases with increasing mixing rate (i.e., rpm) by the value a which is lower than or equal to unity. The amount of solid substances insoluble in a pure solvent but reacting with the additions with the formation of soluble products may be dissolved at identical rates if the reactions are controlled only by the rate of transfer of the solution to the solid reaction surface. This has been confirmed in cases of dissolution of metallic mercury, cadmium, zinc, copper, silver, iron, nickel and cobalt in aqueous iodine solutions. In the case of zinc $a = 0.56$. Iodine acts as an oxidation agent and forms a metallic anion.

The literature describes a large number of examples of reactions between the liquid and solid phases controlled by the rate of the chemical reaction. Several reactions, described on the basis of Nernst's theory as controlled by mass transfer, are not considered completely correctly. The main objection is the assumption that the diffusion layer is stationary in relation to the solid reaction surface and also that the thickness of the Nernst layer is the same in many different reactions, approximately 0.03 mm. These differences are explained by the hydrodynamic theory of mass transfer.

It has been shown [3] that the movement of the liquid takes place very closely at the surface and was detected at a distance of 10^{-5} cm from the surface of the solid [4]. This means that the assumption that the concentration of the dissolved substance is a linear function of the distance y from the surface in the direction into the volume of the solution c, and $y = \delta$, is only suitable as an approximation. Of course, there are a large number of proofs of the existence of a concentration gradient between the solid surface and a point in the volume of the solution from which we can determine the distance δ in the systems in which the reaction is controlled by mass transfer. The properties and thickness of this region are determined by the values of the diffusion coefficient of the dissolved substance, the viscosity of the solution and also by the mechanism of the flow of the liquid around the solid reaction surface. This means that the degree of mixing is the most important variable for every reaction system, as described by the hydrodynamics expression.

The appropriate areas of the theories of hydrodynamics, analysed in the reactions controlled by mass transfer, have been developed to a sufficiently high level [5, 6] and this analysis shows that it is important to obtain mostly two main data:

(*i*) the mechanism of the change of the concentration of the dissolved substance with distance from the solid reaction surface and the size of the area through which the change is recorded;

(*ii*) the rate of transfer of the dissolved substance by forced diffusion from the volume of the solution to the solid reaction surface.

Of course, theory must indicate which factors should be taken into account and also their quantitative expression.

Convective diffusion equations are very complicated and in most cases they are solved using semi-empirical methods. When using the rotating disc method the high degree of symmetry is ensured and this enables a complete solution and also the uniform supply and flow of the entire volume of the liquid [6]. The disc rotates around the normal axis to the surface area and this eliminates the effect of edges. The volume of the solution is also sufficiently large to eliminate the effect of the Nernst diffusion layer.

In the volume of the solution away from the rotating disc, the liquid approaches the disc in the direction of the normal to the surface at the constant rate v_y:

$$v_y = -0.886\sqrt{(\nu\omega)} \tag{7.14}$$

where ν is the kinematic viscosity of the liquid and ω is the constant of the angular speed of the disc [radian·s^{-1}]. The liquid starts to rotate only if it is supplied very close to the disc but in this case the increase of the angular speed becomes greater with the decrease of the distance between the liquid and the surface of the disc, until the liquid reaches the disc completely. The centrifugal force, formed from the angular moment, also generates the radial component of the speed for the liquid and this component displaces the liquid away from the surface of the disc. Consequently, the liquid is continuously attracted in the direction to the surface of the disc and at the same time, pushed away from it until it comes very close to the surface (Fig. 7.1).

Thus, the liquid is formed by two types of flow, one normal to the direction of the surface with a constant speed v_y, the other one parallel to the surface. The transition from one type to another indicates the presence of a viscous boundary layer. The results show that when the distance from the surface is equal to approximately $2.8\sqrt{(\nu/\omega)}$ the rate of flow to the surface equals 80%

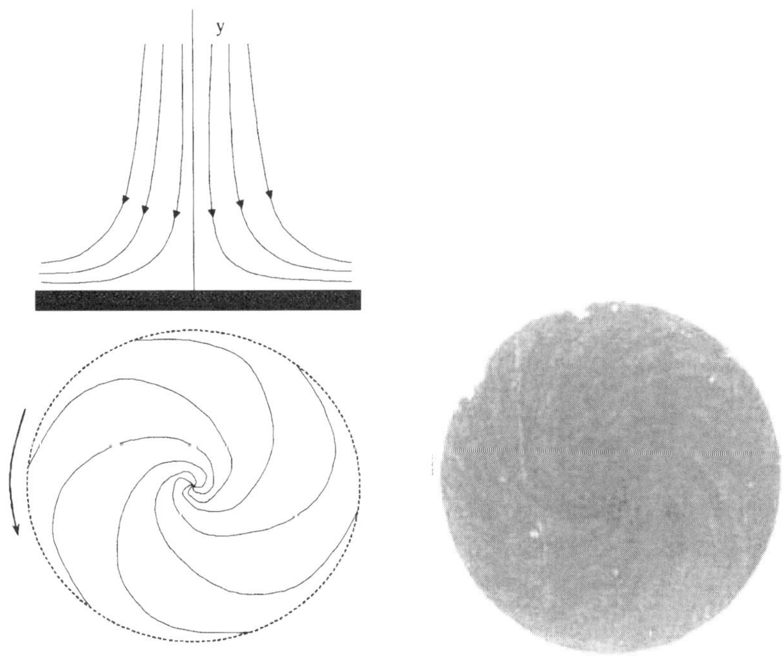

Fig. 7.1*a* (top) The flow of liquid to the disc rotating around the axis *y*. Close to the surface of the disc the liquid flows parallel to the surface. (bottom) The circle indicated by the interrupted line indicates the outer edge of the disc. The helical solid lines indicate the appearance of the liquid very close to the surface of the rotating disc. If the appropriate reaction takes place, this appearance is also typical of the surface of the disc either by etching away part of the surface or by forming a reaction product, Fig. 7.1*b* (right) [7].

of its maximum value, and the speed of the parallel component is 10% of the speed. The distance can be regarded as approximately equal to the thickness of the Nernst viscous boundary layer. In water at 25 °C this layer is approximately 0.5 mm thick at an angular rotating speed of the disc of 25 rad/s.

Taking into account the convective transfer conditions, the concentration of the dissolved substance in the liquid, which reacts with the surface of the rotating disc, should depend only on the distance from the surface and the distance from the axis of rotation, since the system is characterised by axial symmetry. Introducing the distance from the axis gives

$$D\frac{\mathrm{d}^2c}{\mathrm{d}y^2} = v_y\left(\frac{\mathrm{d}c}{\mathrm{d}y}\right) \tag{7.15}$$

If the value of y is high, v_y is constant and assuming that the angular speed of the disc ω is sufficiently high, so that the resultant value of v_y is sufficiently high, then the low rate of other processes, which may change concentration, for example, non-convective diffusion, does not contribute to the change of the concentration of the dissolved substance. Consequently, the concentration of the dissolved substance in the volume of the solution is constant, if the flow of the liquid in the distance from the disc is sufficiently high. The rate of diffusion to the layer depleted in the dissolved substance on the surface of the disc is too low to have any effect from a large distance from the disc.

In the area very close to the disc the value of v_y is considerably lower than its maximum value and the rate of transfer of the dissolved substance starts to be controlled by diffusion. Figure 7.2 shows the ratio of the concentrations at the distance y from the surface, c_y, to the concentration in the volume c [5, 6, 8]. This ratio is depicted as a function of the distance, expressed as the ratio of the thickness of the diffusion layer δ, defined by the Nernst equation (7.43). The broken line expresses this relationship.

The thickness of the diffusion boundary layer depends on the thickness of the hydrodynamic or viscous boundary layer which is regarded as constant over the entire disc surface. Of course, theory is valid only in the case of a disc diameter considerably larger than the thickness of the viscous layer. The distance of the walls of the vessel of the leaching system from the edge of the disc must also be sufficiently large so that these walls have no effect on the flow of the liquid which must be non-turbulent. Taking these conditions

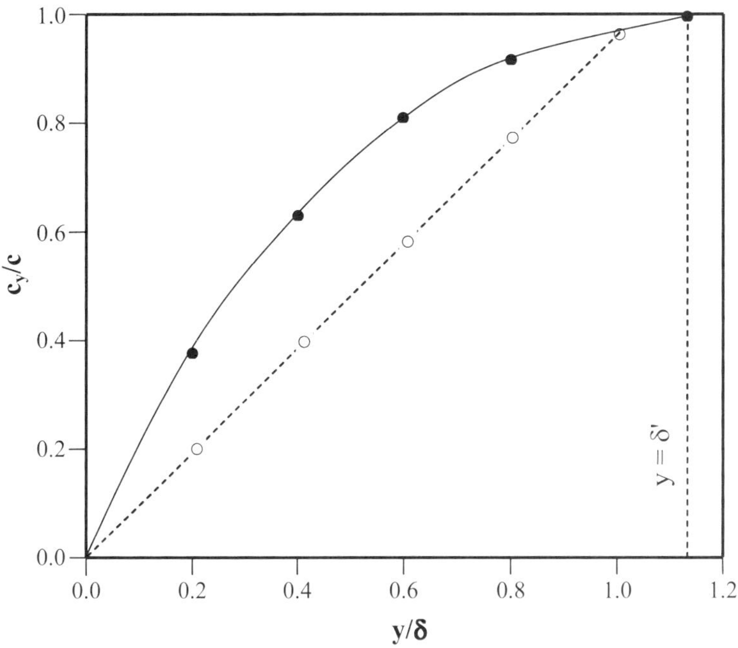

Fig. 7.2. (*solid line*) The ratio of the concentration of the dissolved substance at a distance y from the surface of the rotating disc c_y, to the volume concentration c as a function of the distance y, expressed as the ratio of the thickness of the diffusion layer (*broken line*), Nernst dependence according to equation (7.43).

into account, equation (7.15) may be used for finite size systems:

$$-\frac{\mathrm{d}c}{\mathrm{d}t} = \frac{DA(c - c_1)}{V(0.893\delta')} \tag{7.16}$$

which is equivalent to equation (7.43), and δ' is the thickness of the layer of the concentration gradient close to the surface of the disc. Comparison of the equations (7.43) and (7.38) shows that $d = 0.893\delta'$. Integration of equation (7.16) gives

$$\delta' = 1.805\left(\frac{D}{\nu}\right)^{\frac{1}{3}}\left(\frac{\nu}{\omega}\right)^{\frac{1}{2}} \tag{7.17}$$

so that

$$\delta = 0.893\,\delta' = 1.612\,D^{\frac{1}{3}}\nu^{\frac{1}{6}}\omega^{-\frac{1}{2}} \tag{7.18}$$

The flow rate of the dissolved substance from the volume of the solution to the solid surface is defined by the Nernst theory in the following form

$$I = DA\frac{c - c_1}{\delta} \tag{7.19}$$

so that in the simplest case, if the concentration on the surface is equal to zero,

$$I = \frac{D\pi r^2 c}{1.612\, D^{\frac{1}{3}} c^{\frac{1}{6}} w^{-\frac{1}{2}}} \approx 1.9 r^2 c D^{\frac{2}{3}} \nu^{-\frac{1}{6}} w^{\frac{1}{2}} \tag{7.20}$$

where r is the disc radius.

These considerations show that when using a rotating disc for leaching a solid substance, it is possible to calculate the maximum rate of reaction of the substance with a leaching agent, if the process is controlled completely by mass transfer.

The area and geometry of the interface

The rate of a given reaction and the shape of the kinetic curves are greatly influenced by the morphology of the surface of the solid phase entering the reaction. If the reaction surface does not change during the reaction and the reaction takes place through a constant area, the reaction rate remains constant so that

$$v = \frac{dn}{dt} = k\,Ac \tag{7.21}$$

where n is the amount of the substance reacting in time t through the area A, k is the rate constant, and c is the reagent concentration.

The area of the surface A remains constant in this case during the reaction

$$-\int_{n_0}^{n} dn = k\,A\,c \int_0^t dt \tag{7.22}$$

$$n_0 - n = k\,A\,c\,t \tag{7.23}$$

This means that the dependence $n_0 - n$ is a straight line with the angle of inclination $k \times A \times c$ from which we can calculate the value of k.

If the reaction area is not constant, but has a geometrical shape, for example, sphere, cube or some other body, the area of the surface and, consequently, the reaction rate will gradually change. A suitable example is the shape of the reaction area in the form of a sphere. The surface area A decreases with time according to equation (7.21) and the area is equal to $A = 4\pi r^2$ and the volume $m = \frac{4}{3}\pi r^3 \rho$, where r is radius and ρ density. Consequently

$r = \left(\frac{3m}{4\pi\rho}\right)^{\frac{1}{3}}$ and $A = 4\pi\left(\frac{3}{4\pi\rho}\right)^{\frac{2}{3}} m^{\frac{2}{3}}$.

$$\frac{dm}{dt} = k4\pi\left(\frac{3}{4\pi\rho}\right)^{\frac{2}{3}} m^{\frac{2}{3}} c = k' m^{\frac{2}{3}} \tag{7.24}$$

$$3\left(m_0^{\frac{2}{3}} - m^{\frac{1}{3}}\right) = k't \tag{7.25}$$

This means that the dependence $\left(m_0^{\frac{2}{3}} - m^{\frac{1}{3}}\right)$ in relation to t has the form of a straight line and the rate constant is calculated from the angle of inclination of the dependence.

A similar procedure is used to derive the rate equation for the cube and the disc which is the same as for the spherical shape.

The rate reactions may be efficiently expressed by a means of the value of conversion R. If only a part of the substance reacts, the following equation holds

$$R_i = \frac{n_i}{n_{i,\text{initial}}} \tag{7.26}$$

where n_i is the number of mols of the i-th reagent at time t from the start of the reaction, and $n_{i,\text{initial}}$ is the initial number of moles of the i-th reagent.

The value of R can be presented in the form

$$R = \frac{m_0 - m}{m_0} \tag{7.27}$$

For a partially reacted spherical specimen

$$R = \frac{\frac{4}{3}\pi r_0^3 \rho - \frac{4}{3}\pi r^3 \rho}{\frac{4}{3}\pi r_0^3 \rho} = 1 - \frac{r^3}{r_0^3} \tag{7.28}$$

$$\frac{r^3}{r_0^3} = 1 - R \tag{7.29}$$

$$r = r_0 \left(1 - R\right)^{\frac{1}{3}} \tag{7.30}$$

Substituting the values of the area and volume of the sphere into the general rate reaction (7.21)

$$-\frac{\mathrm{d}m}{\mathrm{d}t} = -4\pi\rho r^2 \frac{\mathrm{d}r}{\mathrm{d}t} \tag{7.31}$$

$$-\int_{r_0}^{r} \mathrm{d}r = \frac{kc}{\rho} \int_{0}^{t} \mathrm{d}t \tag{7.32}$$

$$r_0 - r = \frac{kc}{\rho} t \tag{7.33}$$

$$r_0 - r_0 \left(1 - R\right)^{\frac{1}{3}} = \frac{kc}{\rho} t \tag{7.34}$$

$$1 - (1 - R)^{\frac{1}{3}} = \frac{kc}{r_0 \rho} t \tag{7.35}$$

Similarly, the rate equation can be derived for a partially reacted cube-shaped specimen

$$1 - (1 - R)^{\frac{1}{3}} = \frac{2kc}{r_0 \rho} t \tag{7.36}$$

and the disc-shaped specimen

$$1 - (1 - R)^{\frac{1}{2}} = \frac{kc}{r_0 \rho} t \tag{7.37}$$

These relationships have the form of a straight line in relation to t and the rate constant can be determined from their angle of inclination.

In a special case in which $r = 1$, the final equation valid for the disc, sphere and cube has the form

$$1-(1-R)^{\frac{1}{3}}=\frac{4kc}{3r_0\rho}t \qquad (7.38)$$

The following experiments should be sufficient when investigating leaching: the required amount of the leaching agent with a predetermined concentration, for example 500 ml of 0.5 M solution of $FeCl_3$ in 0.5 M solution of HCl and a certain, predetermined amount of the leaching agent, for example, 5 g of chalcosite, Cu_2S, is placed in a reaction vessel. The temperature of the system is increased to the required level and a predetermined amount of the sample of the solution is taken after starting the experiment. In this sample, copper leached out from chalcosite is dissolved and the sample is used for chemical analysis of the amount of copper. This gives the kinetic dependence of the amount of dissolved copper. Of course, in analysis it is necessary to take into account all significant effects, otherwise the interpretation of the results may be loaded by an error. A number of experimental practices will be discussed in further chapters.

Therefore, if it is possible to prepare experiments in which the constant area of the reaction surface is ensured, it is possible to obtain by a simple procedure the values of the rate constants essential for deriving the rate equation and the effects of other parameters. In leaching, this method is represented by the rotating disc method which also efficiently eliminates diffusion through the boundary layer. Another method is the preparation of a massive sample of the defined size and dimensions, for example, a cube or plate, which is positioned in a stationary position in the leaching reactor. However, in this case, maximum attention should be given to ensure the hydrodynamics of the leaching system.

The problems are mostly of practical nature: area A in the laboratory experiments is small, maximum several cm^2, otherwise it would be difficult to fulfil other requirements on the distance from the walls of the reaction vessel, maintenance of the required number of revolutions, etc. This then means that the process will be relatively slow, equation (7.23) and the amount of the leached substance may be small, below the detection limit of the analytical method. This may greatly increase the experiment time (the value of t in equation (7.23)) and this may result in errors in measurements, as in analysis of very small quantities. However, since mass transfer takes place, the leached surface is 'attacked' and although its size may not change, the area may change as a result of wrinkling of the surface causing further measurement errors.

Another problem is the preparation of massive samples. These samples can be prepared by cutting out and treatment of large pieces or by pressing and/or sintering powder specimens or by casting and solidification of molten material. All these cases carry the risk of the anisotropy of the physical and mineralogical properties. In the case of pressed and non-sintered samples there is a risk of extensive porosity, i.e. a hidden surface which takes part in the reaction. The results may not correspond to the actual situation.

However, in some cases, for example, in the examination of leaching of pure metals, especially noble metals the rotating disc method has been used efficiently.

All the mentioned objections are also valid for the specimens of defined shapes – sphere, cube, disc, plate, etc. Taking into account other factors, it should not be expected that the original form will be also maintained when reducing the total volume which may again cause a number of possible errors in the system.

For these reasons, leaching experiments are carried out using powder specimens. These specimens have a sufficiently large surface for relatively rapid leaching but the largest problem is the change of the actual surface during leaching and also the actual reaction area of the sample. The individual grains are not only ragged, often covered with cracks, but in leaching some of the areas disappear more rapidly, others less rapidly. The leaching sludge must be mixed to remove the diffusion layer from the leached surface and, consequently, the individual grains come into contact and rub against each other, disrupt laminar flow-around, etc. The situation is even more complicated if the reaction product is some solid product, for example, elemental sulphur, covering partially or completely the leached surface. There are many other factors complicating the analysis of the results in experiments with the powder samples, but the changing active leached area is one of the most important ones. To solve the problem, it is necessary to use kinetic modelling equations, as discussed later.

Adsorption and desorption

Adsorption is the accumulation of the gas, liquid or dissolved substances on the surface of the solid. Desorption is the reverse process, i.e. elimination of the adsorbed substances from the surface. Adsorption is the result of the fact that the atomic forces acting on the atoms in the volume of the solid substance differ from those acting on the surface. Schematically this is shown in Fig. 7.3. In this case, the surface is slightly more active and will adsorb similar molecules or ions.

The nature of the adsorption processes greatly varies because they depend on the properties of the solid surface and adsorbed substances which may be of the ion nature or of the type of polar and non-polar molecules.

In principle, there is physical and chemical or ion adsorption. Physical adsorption may be characterised as follows:

- No chemical bonds form between the solid surface and the adsorbed substance;
- The process is accompanied by the generation of a small amount of heat;
- The process is reversible, i.e. the adsorbed ions or molecules may be desorbed by, for example, reducing pressure or increasing temperature;
- The process is not selective and, therefore, takes place on any surface.

Physical adsorption is an exothermic process and, therefore, the amount of the adsorbed substance will decrease with temperature according to van't Hoff reaction. This means that this type of adsorption is quite significant at slightly elevated temperatures.

Chemisorption may be characterised as follows:

- Electron exchange between the surface and the adsorbed substance contribute to the process. Electron transfer takes place in two directions – from the adsorbed substance to the solid

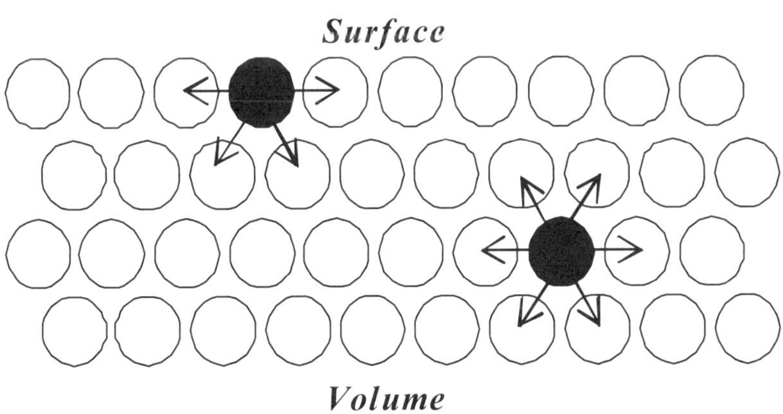

Fig. 7.3. Interatomic forces acting on the surface and in the volume of the solid substance.

surface and vice versa. The defects of the structure of the solid, impurities and non-stoichiometry increase adsorbtivity;

- The process is endothermic and this means that the degree of adsorption increases with increasing temperature. This type of adsorption is usually attributed to the association of molecules or interaction of the ions on the surface;
- The process is specific which means that it takes place only on certain surfaces;
- The process is irreversible and the adsorbed substances cannot be desorbed by a simple change of the external conditions.

Nucleation

The process of formation of solidification nuclei plays a significant role in heterogeneous reaction and, consequently, in the reactions between the solid and liquid phases. In hydrometallurgical processes, these reactions include mainly crystallisation, precipitation of the solid phase from the solution, the formation of gaseous products of the reaction, evaporation at temperatures close to the boiling point, and so on. In all these processes, the formation of nuclei at the start of the reaction may be the rate-determining step.

The creation of a solid from this viewpoint usually takes place in three steps:

Nucleation is an induction period in which nuclei form in certain areas of the surface. This stage depends greatly on the defectiveness of the structure of the material.

The formation of the reaction interface characterised by the formation of a certain amount of nuclei with the reaction taking place at a measurable rate. This period is referred to as acceleration.

Growth of the reaction interface. After starting the reaction, the rate of the reaction increases until the size of the reaction interface reaches the maximum value. Subsequently, the reaction rate decreases depending on the decrease of the reaction surface and the disappearance of the initial phase. This stage is referred to as the attenuation stage and is manifested as an inflection point on the kinetic curve.

Each of these stages is controlled independently by kinetic relationships and the range in which these relationships are valid greatly varies. This depends on the substance itself and, in some cases, on the particle size. For example, for very small particles, the induction period is difficult to detect.

If the nucleation products are in active areas and the rate at which of rate of growth of the solidification nuclei is the lowest and, therefore, the rate is controlling, the reaction rate will be proportional to the mass of the reacting substance in time

$$\frac{dn}{dt} = kn \tag{7.39}$$

$$-\int_{n_0}^{n} \frac{dn}{n} = \int_{0}^{t} dt \tag{7.40}$$

$$-\ln \frac{n}{n_0} = kt \tag{7.41}$$

After introducing the conversion expression, i.e. the reacted proportion of the substance, R

$$-\ln \frac{1}{1-R} = kt \tag{7.42}$$

The dependence ln $1/(1-R)$ should have the shape of a straight line in relation to t.

Figure 7.4 shows schematically the nucleation and form of the nucleation relationships.

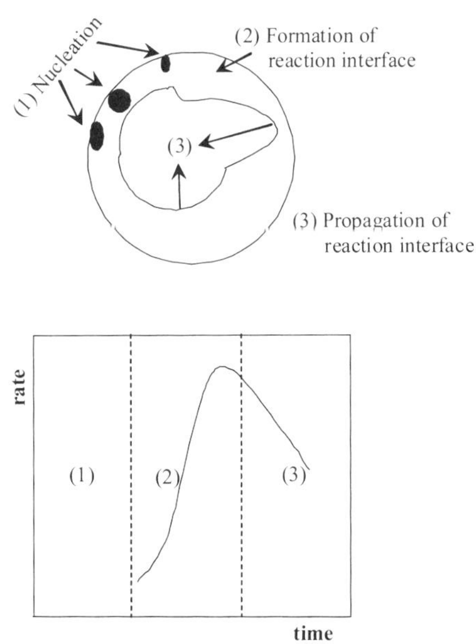

Fig. 7.4. Schematic representation of the nucleation process and the dependence of rate on time.

The shape of the kinetic curves depends on the nature of the process. The induction period characterises the initial changes of the system (defects of the structure, removal of easily soluble particles from the surface, subsurface diffusion, etc.) resulting in the concentration of the product and occurrence of interactions between the reagents leading to an increase of the size of the reaction zone. The specific proportion of the individual partial processes in the overall course of the reaction is then determined on the basis of the results of examination of the process kinetics.

The period of rapid increase of the reaction rate is clearly linked with the formation and growth of the nuclei of the product in the reaction zone. In a number of cases this process may be more complicated and includes mutual dissolution of the reagent and the formation of nuclei of the products, their transfer to the soluble form, and diffusion into the surrounding medium.

Effect of temperature on the reaction rate

The reaction rate depends on the temperature by a means of a rate constant. In simple reactions, both the rate constant and the reaction rate increase with increasing temperature. The rate of parallel and successive reactions also increases with increasing temperature. In the case of reactions characterised by a more complicated mechanism, the rate may decrease with increasing temperature. The increase of temperature in leaching processes usually greatly accelerates these processes. Arrhenius (*1889*) showed that the dependence of the rate constant k on absolute temperature is described by the equation

$$k = Ae^{-\frac{E}{RT}} \tag{7.43}$$

where A is the frequency factor, E is the so-called activation energy and R is the gas constant. Taking the logarithm of equation (7.43) gives

$$\ln k = \ln A - \frac{E}{RT} \tag{7.44}$$

which is the equation of the straight line. Therefore, if the rate constants for a least two temperatures are available, activation energy E can be determined by comparing these relationships in the form

$$E = R\frac{T_2 T_1}{T_2 - T_1}\ln\frac{k_{T_2}}{k_{T_1}} \tag{7.45}$$

The strong effect of temperature on the reaction rate explains why it is necessary usually to heat the reacting substances. The reaction taking part at room temperature at an unmeasurable rate may be very fast after heating.

Empirical equation (7.43) is referred to as the Arrhenius equation which can be used to determine the energy for the reaction, i.e., the activation energy. A more exact relationship was derived later [1]

$$k = \frac{k_B T}{h} e^{-\frac{\Delta G^{\#}}{RT}} = \frac{k_B T}{h} e^{\frac{\Delta S^{\#}}{R}} e^{-\Delta H^{\#}} \tag{7.46}$$

where k_B is the Boltzman constant, h is the Planck constant, $\Delta G^{\#}$ is Gibbs activation energy, $\Delta H^{\#}$ is activation enthalpy and $\Delta S^{\#}$ is activation entropy.

For examining the effect of temperature on the leaching process it is sufficient to use the Arrhenius equation which may characterise the reaction from the viewpoint of their rate-controlling stage. Strictly speaking, in the case of heterogeneous reactions, one should also take into account adsorption energies but this is difference is basically ignored in the published results.

The diffusion-controlled processes are not greatly affected by temperature, and the chemically controlled processes depends strongly on temperature. The reason for this is that the diffusion coefficient is, according to the Stokes–Einstein equation, linearly dependent on temperature, whereas the rate constant of the chemical reaction depends exponentially on temperature, as expressed by the Arrhenius dependence. In other words, with increase of temperature the value of coefficient D increases in units, but coefficient k increases by orders of magnitude. For this reason, the values of activation energy of the diffusion-controlled processes are low and vary in the range 4–13 kJ/mol and the values of the activation energy of the chemically controlled reactions are usually higher than 42 kJ/mol. The reactions controlled by a mixed mechanism have the values of the activation energy in the range 20–35 kJ/mol. The situation in the reactions taking place in the solid state is different because these diffusion coefficients depend on temperature and the activation energy and vary basically in the range 800–1600 kJ/mol.

Effect of the concentration of reactants

Usually, the reaction rate increases with increasing concentration of the reactants. However, if the system contains concentrations of more than one substance, the kinetic equation cannot be solved unambiguously, because it is a single equation with several dependent variables. To derive a differential equation for a single dependent variable, it is necessary to carry out mass balance.

The concentration of the substances reacting in accordance with the reaction (7.5) are linked by the mass balance equations

$$\begin{aligned} c_A &= c_{A0} - ax \\ c_B &= c_{B0} - bx \\ &\ \ \vdots \\ c_P &= c_{P0} - px \\ c_R &= c_{R0} - rx \end{aligned} \qquad (7.47)$$

where c_{i0} are the initial concentrations, i.e., concentrations at time $\tau = 0$.

For the chemical reaction (7.5) the equation (7.47) can be summarised in a single equation

$$c_i = c_{i0} - v_i x \qquad i = 1, 2, ..., n \qquad (7.48)$$

where n is the number of substances taking part in the reaction, and v_i are stoichiometric coefficients, negative for initial substances and positive for reaction products.

Using the mass balance equations, the kinetic equation of a simple reaction (7.9) may be written in a new form

$$a\frac{dx}{d\tau} = k_A(c_{A0} - ax)^a(c_{B0} - bx)^b..... \qquad (7.49)$$

This gives the differential equation between a single dependent variable x and an independent variable τ. The initial condition is $x = 0$ and time $\tau = 0$.

The uncontrollably increasing concentration of the reagent may be uneconomical or it may cause undesirable secondary reactions. For example, in leaching processes it is necessary to know the minimum required concentration of the leaching agents which does not react in the transfer of any part of the leaching agent to the waste at the optimum leaching rate of the given metals from their initial materials.

Effect of defect structure and impurities

The rate of the chemical reaction is influenced quite strongly by non-stoichiometry, the existence of the defect structure and the presence of impurities at the reaction interface. Non-stoichiometric compounds of copper and/or iron are often found in their sulphide minerals, and, therefore, the effect of non-stoichiometry is also reflected during leaching. Non-stoichiometric compounds as such in which certain areas in the structure are not occupied and atoms A are not in a stoichiometric ratio to atoms B. With further processing, either physical or physical–chemical, the degree of non-stoichiometry usually becomes greater, for example, by intensive milling of sulphides or by heating them. The chapter, concerned with copper and iron sulphides presents a list of these compounds indicating the actual existence of a specific non-stoichiometric sulphide. Only in simple sulphides of copper of the type Cu_xS, the composition changes from CuS to CuS_2. The composition of the iron sulphide changes in the range from $Fe_{0.88}S$ to FeS_2.

Basically, there are four types of non-stoichiometric compounds.

- The atoms of a non-metal, for example, oxygen or sulphur are not present in the crystal structure so that a metal surplus becomes evident. The electrons, belonging to the anions, remain trapped in the vacancies. The situation may be described as follows:

$$2B^- \rightarrow B_2 + 2e^- + \square$$

- The atoms of a non-metal also disappear from the crystal structure and the electrons with which these atoms were previously associated, are released. An excess of metallic ions rapidly takes up interstitial sites and the free electrons are trapped in the vicinity of these interstitial cations;
- The crystal structure obtains additionally the atoms of the non-metal which become anions after acquiring electrons as a result of oxidation of certain metal anions to a high valency:

$$B_2 + 2e^- \rightarrow 2B^-$$
$$A^+ \rightarrow A^{2+} + e^-$$

- The crystal structure receives further atoms of the non-metal which become anions after obtaining the electrons as a result of oxidation of certain metal anions, but the added anions occupy interstitial sites.

Figure 7.5 shows schematically the type of defects of the crystal structure.

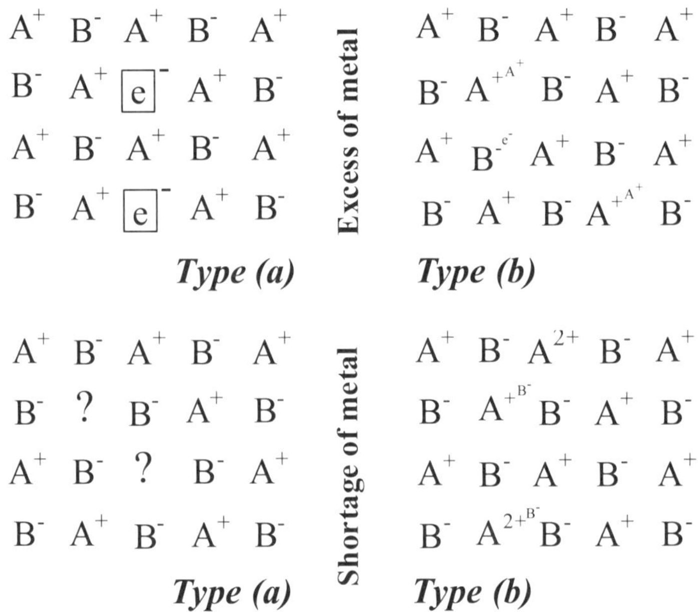

Fig. 7.5. Defects in a non-stoichiometric crystal structure.

It may be seen that the type of non-stoichiometry with an excess of metal contains free electrons, whereas the type with a metal shortage does not. The trapped electrons in the type with a metal excess may be excited to a higher energy level. They are also mobile and may migrate inside the entire structure so that they become type *n* semiconductors.

The type of non-stoichiometry with a metal shortage is found only in compounds showing different valences. These compounds are also semiconductors but with different conductivity mechanism because they do not contain free electrons. Conductivity occurs by a means of differences in the vacancies because at a potential difference the electron may jump from one anion with a lower valency to an anion with a higher valency. This type of semiconductivity is referred to as the *p*-type, or hole conductivity.

Impurities have a significant effect on the kinetics of heterogeneous reactions. The presence of foreign atoms may cause stresses in the crystal structure which will be sufficient to initiate the reaction. Similarly, the presence of impurities from the macro viewpoint may cause a change of the rate and/or mechanism of the reaction. The presence of foreign solid substances may lead to the formation of a local galvanic cell and, subsequently, an electrochemical reaction. In leaching of sulphides, it is well known that the presence of pyrite in a leaching mixture accelerates the entire process exactly for these reasons.

In addition to the crystal structure defects and the presence of impurities, the kinetics of the process is also greatly affected by macro defects represented by the morphology and shape of the particles and the quality of the leached surface. It is well known that the edges, projections and tips are characterised by accumulation of the energy which supports the leaching reaction and, on the other hand, smooth surfaces efficiently resists the effect of the reagents. Each of the points of the surface irregularities is the point of a nucleus. Therefore, this effect must be taken into account when investigating the leaching kinetics. Of course, the nature and type of the mineral is also very important in this case [12].

7.2. Elementary phenomena at the interface

Physical, chemical and electrochemical processes take place at the reaction interface. The dissolution of a solid substance in water, without any chemical reaction taking place, is a typical physical process, for example, dissolution of NaCl in water. A salt crystal immersed in a volume of water is immediately covered with a thin layer of saturated solution of NaCl. The ions will diffuse spontaneously into water in accordance with a Fick's law, until the volume is saturated as shown by the equation in the form

$$\frac{dc_{B(l)}}{dt}=\frac{DA}{yV}\left(c_{B(s)}-c_{B(l)}\right) \tag{7.50}$$

where $c_{B(l)}$ is the concentration B in the solution at time t, $c_{B(s)}$ is its solubility at the experimental temperature, i.e., the concentration in the saturated state, D is the diffusion coefficient, A is the reaction interface, V is the volume and y is the thickness of the boundary layer.

However, if the dissolved substance starts to react on the solid surface with the aqueous solution, the concentration of the solution at the interface is saturated $c_{B(s)}$ and in the volume it is equal to $c_{B(l)}$. In this case, there are three possibilities:

- If the rate of reaction of the reagent with the dissolved substances in the solution volume is very high, concentration $c_{B(l)}$ is equal to zero and the process is controlled by the rate of diffusion through the boundary layer:

$$v = \frac{D}{y} A\left(c_{B(s)} - c_{B(l)}\right) - k_1 A c_{B(s)} = \text{const} \qquad (7.51)$$

where $k_1 = \frac{D}{y}$

- If the rate of reaction of the reagent with the dissolved substances in the volume of the solution is very low, diffusion does not play a role and the dissolved substances build up in the solution, i.e. $c_{B(l)} \approx c_{B(s)}$. The reaction rate depends on the concentration of the reagent B because $c_{B(s)}$ is a constant and the process is controlled by the rate of the chemical reaction

$$v = k_2 A c_B \qquad (7.52)$$

- If the rate of the reaction of the reagent with the dissolved substances in the volume is equal to the diffusion rate, this is the case known as the reaction taking place in a mixed regime.

If a reaction is accompanied by electron transfer, i.e. it is an oxidation-reduction reaction, which often occurs in leaching, we are concerned with an electrochemical process. The reagent in a solution with concentration D diffuses through the boundary layer to obtain the electrons from the interface. In this case, there are also three possibilities:

- If the rate of the chemical reaction at the interface is considerably higher than the rate of diffusion of the reactants to the interface, then $c_{Di} = 0$ and the concentration of the reagent in the layer is equal to zero. These reactions are referred to as diffusion-controlled reactions and are described by (7.51).
- If the rate of the chemical reaction at the interface is considerably lower than the diffusion rate, the process is controlled by the rate of the chemical reaction and is described by the equation (7.52).
- If both rates are identical, the process takes place in the mixed regime and a concentration gradient is formed across the boundary layer

$$v = k_1 A(c_D - c_{Di}) = k_2 A c D_i \qquad (7.53)$$

and

$$c_{Di} = \frac{k_1}{k_1 + k_2} c_D$$

Substitution of the values of C_i to the above equations gives

$$v = \frac{k_1 k_2}{k_1 + k_2} A c_D = k A c_D \tag{7.54}$$

where $k = \frac{k_1 k_2}{k_1 + k_2}$

If $k_1 < k_2$, then $k = k_1 = D/y$, i.e. the process is diffusion controlled. If $k_1 > k_2$, then $k = k_2$, i.e. the process is controlled by the rate of the chemical reaction.

Figure 7.6 shows schematically the situation at the interface.

The reaction rate is determined by the slowest stage. The results show that the rate-controlling step may change depending on the reaction conditions and, therefore, information obtained on the rate for the given combination of the conditions cannot be applied to other combinations of the conditions. In many cases a single stage does not control the overall reaction rate because the overall rate is influenced to various degrees by several elementary steps. The change of the course of the reaction may also influence the effect of these steps. For this reason, it is important to understand the joint effect of individual elementary reaction steps not only when determining the rate-controlling step of the reaction in the given reaction conditions, but also when determining whether it will be necessary to take into account more than one stage when expressing the overall reaction rate.

In addition to the previously mentioned stages, including the

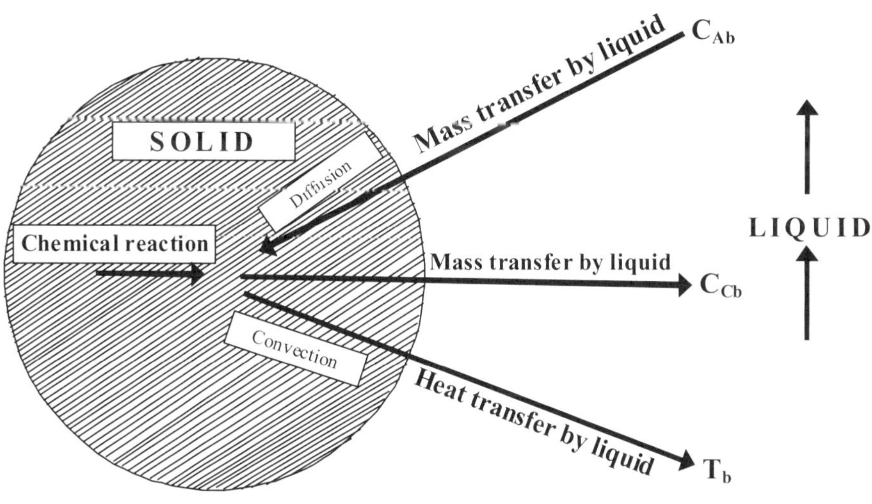

Fig. 7.6. Schematic diagram of the overall reaction process.

chemical changes of the reactants, the overall reaction rate may be greatly affected by the transfer of heat and structural changes in the solid phase. Many reactions of the liquid-solid phase type are of distinctively exothermal or endothermal nature. The reaction heat must be transferred either to the environment or supplied from the environment. Heat transfer includes conduction (convection) and/or radiation between the surrounding environment and the solid surface and conduction inside the solid. These considerations greatly depend on the heat capacities of the reactants taking part in the reaction.

Mass transfer between the solid surface and the liquid

The stage of external mass transfer has been studied quite extensively and has been investigated probably in the greatest detail of all the reaction stages. Although the rate of mass transfer between the moving liquid and solid surface can be calculated in certain cases by solving the problem of a flow and diffusion equation [5], better results are often obtained by approximation using the average values of the mass transfer coefficients from empirical correlations.

Taking into account the situation in which the substance A is transferred from the solid solution into the flow of the moving liquid, the concentration of substance A is equal to c_{As} in the vicinity of the solid surface and to c_{Ab} in the volume of the liquid. The rate of transfer of the mass through the unit surface of the solid is given by the equation

$$n_A = k_m (c_{Ab} - c_{As}) \tag{7.55}$$

where k_m is the mass transfer coefficient and n_A is the flow of substance A.

Of course, to estimate the mass transfer coefficients, it is necessary to carry out appropriate selection taking into account the natural conditions of the individual systems. To calculate the coefficient it is necessary to have relevant information on the diffusity and viscosity of the liquid. These values are usually presented in tables.

Diffusion of the liquid through pores in the solid

If the reacting solid is porous, diffusion through the pores in this substance becomes very important because the active surface for the potential reaction is considerably larger. In another important case the solid non-porous substance forms a porous reaction surface through which the reactants must diffuse. Of course, the rate of diffusion through the porous substance is considerably higher than that through the substance without pores.

However, diffusion through the pores is far more complicated than diffusion in the liquid. This is due to the fact that the diffusion paths through the porous solid substance are not straight and simple but twisted and is very difficult to determine this 'twisting' in relation to the structure of the pores. If the pores are small, the laws of molecular diffusion do not hold and Knudsen diffusion takes place in this case.

Diffusion through a porous solid substance is described by Fick's law in the form

$$n_A = -D_e \Delta c_A \tag{7.56}$$

where D_e is the effective diffusivity of substance A in the porous medium. D_e includes contribution of the cross section of the area occupied by the solid substance (and, therefore, not accessible to diffusion), and also the fact that the pores are not straight and this influences the increase of the diffusion paths. Δc_A is the concentration gradient and c_A is the actual concentration of the substance A in the space of the pore. Generally, the effective diffusivity of substance A in a porous medium may be estimated using Bosanquet's relationship

$$\frac{1}{D_{Ae}} = \frac{\tau}{\varepsilon}\left(\frac{1}{D_{AK}} + \frac{1}{D_{AB}}\right) \tag{7.57}$$

where D_{Ae} is the effective diffusivity of the substance A, D_{AK} is the Knudsen diffusivity, i.e. the single-capillary coefficient, D_{AB} is the molecular diffusivity of substance A in a mixture, τ is the tortuosity of the substance, and ε is the porosity of the solid substance.

The tortuosity factor, which is a measure of increasing diffusion distance in a porous medium, cannot be reliably estimated *a priori*. It must be determined by experiments and its value differs between 2 and 10. Porosity ε in equation (7.57) includes the transfer section of the area occupied by the solid and is therefore not accessible to diffusion.

7.2.1. Intrinsic kinetics of heterogeneous reactions on solid surfaces

7.2.1.1. Kinetics of processes of leaching a single particle

As already mentioned, certain generalisation can be made regarding the mass transfer by liquids and diffusion through the pores, but the stages of adsorption on the surface and the chemical reaction depends strongly on the nature of considered substances, to make them general. In published studies, authors often use a simple expression for the first rate order because of mathematical simplicity, although other expressions may also be used. However, in many cases in practice the reactions between the solid and liquid phases can indeed be approximated by first order reactions. Generally, the concentration dependence is far more complex.

There are two types of adsorption of liquid on the solid surface, physical adsorption and chemisorption. Chemisorption, which is a result of far more intensive interactions than physical adsorption, is responsible for the reactions of the liquid–solid phase type and catalytic reactions. The chemical kinetics of the liquid–solid phase reactions may be separated into several individual steps, i.e., adsorption of reactants, reactions between adsorbed substances and the solid surface, and desorption of the product. After taking into account the kinetics of each stage, an equation is obtained for the rate simplified by the approximation for the conditions of the steady state and the formation of the reaction equilibrium. The general expression for the rate-determining stage of the reaction after this simplification is:

$$\text{Speed} = \frac{k\left(c_R^n - \dfrac{c_p^m}{K_E}\right)}{1 + K_R c_R^r + K_p c_p^p} \qquad (7.58)$$

where c_R describes the concentration of the reactants and c_p is the concentration products in the liquid phase. If the concentration is low and $n = m = 1$, the rate becomes the rate of the first order. This is the reason why the first order of kinetics is used for describing the reactions of the liquid-solid phase type in addition to the fact that it is simpler from the mathematical viewpoint.

Reaction of a single non-porous particle

If the reacting solid substance initially non-porous, the reaction takes place a sharp interface between two phases, i.e., either liquid–solid or solid–solid, and this depends on whether or not the solid product is formed and if so, whether this product is porous or nonporous.

If no solid is formed, for example, in dissolution, or the solid product is removed from the surface immediately after formation, the solid reactant is always in contact with the bulk of the liquid and the size of the particle will diminish as the reaction progresses. However, if a coherent layer of the solid reaction product forms around the reacting solid substance, the reaction will take place at the interface between the unreacted and completely reacted zones. If the solid reaction product is porous, the liquid reactant can reach the reaction interface by diffusing through the pores of the solid product. If the product is non-porous, either the liquid species must diffuse into the solid either by solid state diffusion or a constituent species of the solid reactant must diffuse to the surface to react with the liquid reactant. The overall size of the solid substance will depend on whether the volume of the solid reaction product is larger or smaller than that of the reacting solid substance.

The chemical reaction and mass transfer take place in succession if the non-porous solid substance reacts with the liquid. These conditions are far simpler for analysis of the process than in the case of formation of the porous solid reaction product.

In certain reactions liquid–solid reactions, nucleation is a very important step. The growth of nuclei is a relatively complicated phenomenon. If the size of the solid substance increases or reaction temperature increases, the nucleation time occupies a small part of the total reaction time and, therefore, nucleation is less important from the viewpoint of analysis of the process rate.

Reactions in which no solid product layer is formed

A suitable example of such a reaction is leaching of a metal in an acid. This type of reaction may be generally described by the following scheme:

$$A_{(l)} + bB_{(s)} \rightarrow cC_{(l)} + dD_{\text{(removable s)}} \tag{7.59}$$

where b, c and d are stoichiometric coefficients.

It is assumed that a spherical particle of a non-porous solid

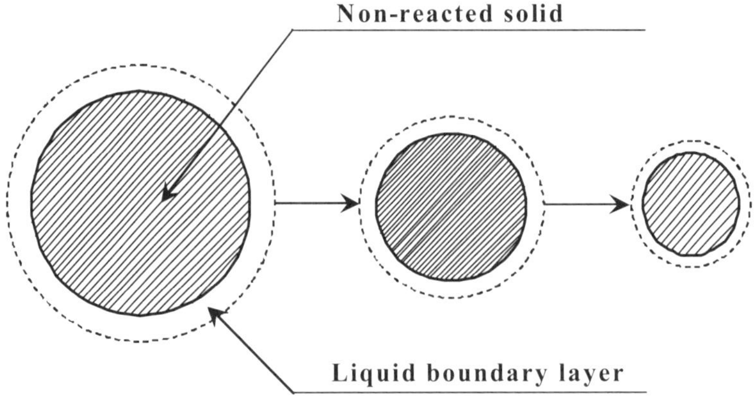

Fig. 7.7. Schematic diagram of a shrinking particle while reacting with the surrounding liquid.

substance reacts with the liquid without the formation of the solid product, as shown in Fig. 7.7.

The rate of consumption of the liquid substance A by the reaction on the solid surface is given by the equation

$$n_A = kf(c_{As}) \tag{7.60}$$

where n_A is the speed through the unit area, k is the rate constant of the reaction, and f determines the dependence of rate on concentration. If the build up in the boundary layer surrounding the solid reactant is ignored, the rate of the chemical reaction must be equal to the rate at which the liquid substances are transferred between the surface of the solid reactant and the volume of the liquid. According to the already mentioned equation, the rate of external mass transfer is described as follows:

$$n_A = k_m(c_{Ab} - c_{As}) \tag{7.61}$$

Combining equations (7.60) and (7.61) gives

$$kf(c_{As}) = hm(c_{Ab} - c_{As}) \tag{7.62}$$

The overall rate may be determined by solving this equation for unknown C_{As} followed by substituting back either into equation (7.60) or (7.61). Before finding the general solution, also including the effect of both chemical kinetcs and external mass transfer, it is necessary to examine asymptotic cases.

When $k \ll k_m$, equation (7.62) shows that $c_{As} \cong c_{Ab}$. This takes place when external mass transfer is without any problems and the overall reaction rate is controlled by the rate of the chemical reaction. In this case, the rate is given by the equation

$$n_A = kf(c_{Ab}) \qquad (7.63)$$

However, if $k \gg k_m$, the term $f(c_{As})$ tends to zero and this is found in cases in which concentration A approaches the equilibrium concentration in the conditions prevailing on the surface of the solid c^*_{As}. In this case, the chemical reaction takes place without any problems and external mass transfer controls the overall reaction rate. Substitution of c^*_{As} by c_{As} in equation (7.61) gives

$$n_A = km(c_{Ab} - c^*_{As}) \qquad (7.64)$$

For the intermediate regime in which the rate of both the chemical reaction and mass transfer is low, it applies that

$$n_A = kc_{As} \qquad (7.65)$$

and

$$c^*_{As} = 0 \qquad (7.66)$$

Solving equations (7.61) and (7.65) in order to eliminate c_{As} gives

$$n_A = \frac{c_{Ab}}{\frac{1}{k} + \frac{1}{k_m}} \qquad (7.67)$$

Equation (7.67) offers an easy solution for the first order processes following each other. This reaction also reduces the equations (7.64) or (7.65) under comparable conditions.

To obtain the overall conversion in relation to time, the rate of disappearance of substance A, n_A must be comparable with the rate of consumption of the solid substance B. Stoichiometry of equation (7.59) results in:

$$n_A = -\frac{\rho_B \, dr_C}{b \, dt} \qquad (7.68)$$

where ρ_B is the molar concentration of the solid substance B and r_C is the radius of the particle of the solid substance at any time. Equations (7.67) and (7.68) give

$$\frac{\mathrm{d}r_C}{\mathrm{d}t} = \frac{\dfrac{bc_{Ab}}{\rho_B}}{\dfrac{1}{k} + \dfrac{1}{k_m}} \tag{7.69}$$

All the parameters, with the exception of k_m in the right hand side of equation (7.69), are independent of r_C. If k_m is regarded as independent of r_C, integration of equation (7.69) should give a straight line. However, in reality k_m changes in relation to r_C, as previously discussed. This relationship can be integrated in order to obtain r_C and, consequently, obtain the conversion as a function of time. A general procedure may be illustrated using the modified equation (7.69) and by integration

$$t = \frac{\rho_B}{bc_{Ab}} \left(\frac{r_0 - r_C}{k} + \int_{r_C}^{r_0} \frac{\mathrm{d}r_C}{k_m(r)_C} \right) \tag{7.70}$$

which describes the relationship between r_C and time. Conversion R depends on r_C through the relationship

$$R = 1 - \left(\frac{r_C}{r_0} \right)^3 \tag{7.71}$$

where r_0 is the initial radius of the solid particle. Equation (7.70) also shows that the time required for obtaining a certain value of r_C (and, hence, a certain degree of conversion) is the sum of the time required to attain the same value of r_C in the absence mass transfer at the time for obtaining the same r_C in the reaction controlled mass transfer. This is an important result applied to any reaction system working in the regime containing several processes with the first order rates assembled in a series.

If a reaction is accompanied by a significant change of enthalpy, both mass transfer and heat transfer must be taken into account. Complete equations including internal and external heat transfer are, however, quite difficult to solve.

Reactions in which a solid product is formed

This type of reaction is frequently encountered in extractive metallurgy. Leaching of minerals from ores is a typical example of such a reaction. This group of reactions is described by the following general equation:

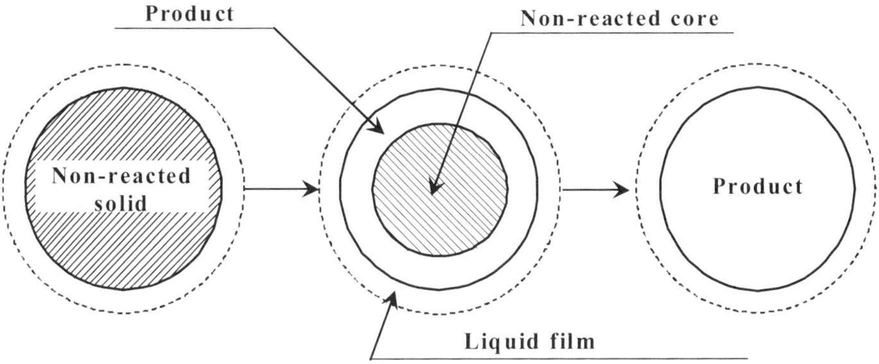

Fig. 7.8. Schematic diagram of a shrinking unreacted core system.

$$A_{(l)} + bB_{(s)} \rightarrow cC_{(l)} + dD_{(s)} \tag{7.72}$$

Figure 7.8 illustrates the course of the reaction in such a system. The overall process may be divided into three steps: external mass transfer, diffusion through the layers of product, and the chemical reaction at the interface between the non-reacting and completely reacted zones.

However, formulation of the equations should include all these steps and indicate the conditions in which one of these steps becomes rate controlling. At the same time, a criterion should be determined for these asymptotic regimes.

In steady conditions, the rate of the chemical reaction at the interface and the rate of the mass transfer process must be the same. When taking into account substance A, the following relationships hold:

for the chemical reaction at the interface

$$-N_A = 4\pi r_c^2 k c_{Ac} \tag{7.73}$$

for diffusion through the layer of the product

$$-N_A = 4\pi r^2 D_e \frac{\mathrm{d}c_A}{\mathrm{d}r} \tag{7.74}$$

for external mass transfer

$$-N_A = 4\pi r_p^2 k_m \left(c_{Ab} - c_{As}\right) \tag{7.75}$$

and it applies that $-N_A$ is the total rate of transfer of substance A into the particle, r_c is the radius of the unreacted core, c_{Ac} is the concentration of A at r_c, D_e is effective diffusivity A in the layer of the product, and r_p is the radius of the particle at any time.

Equation (7.74) assumes that the diffusion of liquid takes place at low concentration of diffusing substances. The solution is obtained by applying the pseudo-steady conditions, which means that movement of r_c is considerably slower in comparison with the time required for the formation of the concentration profile of substance A. This means that if A does not take part in diffusion, r_c does not change and N_A is independent of state. Therefore, equation (7.74) may be integrated for constant N_A together with equation (7.73)–(7.75) as boundary conditions so that the concentration profile is determined as a function of r_c. From the concentration profile obtained by this procedure N_A can be calculated using equations (7.73)–(7.75). Consumption of substance A is related to B by means of the equation

$$N_A = \frac{4\pi r_c^2 \rho_B}{b} \frac{\mathrm{d}r_c}{\mathrm{d}t} \tag{7.76}$$

If the volume of the solid product differs from the volume of the initial solid reactant, the value of r_p will change during the reaction. However, in many liquid–solid phase reactions, this change cannot be ignored.

Solution of the equations (7.73)–(7.75) together with the equation (7.76) gives a linear dependence. For constant r_p the result may be expressed by a means of r_c as follows:

$$\frac{bkc_{Ab}}{\rho_B r_p} t = 1 - \frac{r_c}{r_p} + \frac{kr_p}{6D_e}\left\{1 - 3\left(\frac{r_c}{r_p}\right)^2 + 2\left(\frac{r_c}{r_p}\right)^3 + \frac{2D_e}{k_m r_p}\left[1 - \left(\frac{r_c}{r_p}\right)^2\right]\right\} \tag{7.77}$$

If inert solid substances are mixed with B, ρ_B represents only the number of moles of substance B per unit volume of the entire solid mixture.

The following example describes the relationship for expressing the time dependence of mass transfer for a first order isothermal reaction of non-porous solid substance with a liquid in which the solid substance is in the form of an infinite slab, infinite cylinder, or a sphere:

$$t^* = gF_p(R) + \sigma_s^2\left[pF_p(R) + \frac{2R}{Sh^*}\right] \tag{7.78}$$

where

$$t^* \equiv \frac{bkc_{Ab}}{\rho_B}\left(\frac{A_p}{F_pV_p}\right) \tag{7.79}$$

$$\sigma_s^2 \equiv \frac{k}{2D_e}\left(\frac{V_p}{A_p}\right) \text{(modified Sherwood number)} \tag{7.80}$$

$$Sh^* \equiv \frac{k_m}{D_e}\left(\frac{F_pV_p}{A_p}\right) \tag{7.81}$$

and A_p and V_p are the external area and the volume of the particle, respectively, and F_p is the particle shape factor, which takes the values 1, 2 or 3 for an infinite slab, an infinite cylinder or a sphere. The ratio F_pV_p/A_p is the half thickness of an infinite slab and the radius of an infinite cylinder or a sphere. Other terms of the equation are defined as follows:

$$g_{Fp}(R) \equiv 1-(1-R)^{\frac{1}{F_p}} \tag{7.82}$$

$$p_{Fp}(R) \equiv R^2 \qquad \text{for } F_p = 1$$

$$p_{Fp}(R) \equiv R + (1-R)\ln(1-R) \quad \text{for } F_p = 2 \tag{7.83}$$

$$p_{Fp}(R) \equiv 1 - 3(1-R)^{\frac{2}{3}} + 2(1-R) \quad \text{for } F_p = 3$$

and the following equation is also used

$$R = 1 - \left(\frac{r_c}{r_p}\right)^{Fp} = 1 - \left(\frac{A_pr_c}{F_pV_p}\right)^{Fp} \tag{7.84}$$

Change of the particle size during the reaction

If the particle size changes during a reaction, calculation may be carried out using the following integrated expression:

$$(r_p)_\tau^{Fp} = Z(r_p)_{initial}^{Fp} + (1-Z)_{r_c}^{Fp} \tag{7.85}$$

where Z is the volume of the solid product formed from the unit volume of the reacting solid substance and selected time. If the change of k_m in relation to the particle size is ignored, the solution is thc same as that of equation (7.74) except that the following expression should be used for $p_{Fp}(X)$.

for $F_p = 1$

$$p_{Fp}(R) \equiv ZR^2$$

for $F_p = 2$

$$p_{Fp}(R) \equiv \frac{[Z+(1-Z)(1-R)]\ln[Z+(1-Z)(1-R)]}{Z-1} + (1-R)\ln(1-R)$$

for $F_p = 3$

$$p_{Fp}(R) \equiv 3\left\{\frac{Z-[Z+(1-Z)(1-R)(1-R)]^{\frac{2}{3}}}{Z-1} - (1-R)^{\frac{2}{3}}\right\} \quad \text{(equations 7.86)}$$

If Z approaches unity, the equations (7.86) are reduced to the equation (7.83).

Equation (7.78) shows that the time required to attain a certain degree of conversion is the sum of the times required to obtain the same degree of conversion in the conditions in which the reaction is controlled by three different stages, i.e., chemical reaction, diffusion through the layer of reaction products, and external mass transfer. This is identical with the result obtained for the reactions in which the solid reaction product does not form, described by equation (7.70).

The shrinking unreacted core model is attractive for its simplicity. However, it should be noted that this is valid only for the reaction of a non-porous solid particle occurring at a well-defined sharp reaction interface. Many investigators have applied this model, mainly for its simplicity, to reactions of porous particles where chemical reactions occur in a diffuse zone rathar than at the sharp interface. This approach can be used only for diffusion-controlled reactions with a well developed mathematical apparatus. Generally, the application of the shrinking core model to the reactions of porous solid substances in erroneous analysis of the experimental data, especially in errors in the determination of the reaction rate at different conditions, for example, determination of the activation energy and different physical parameters, for example, the dependence of rate on grain size. Therefore, the reaction of porous particles will be discussed alone.

Reaction of a single porous particle

If the reacting solid substance is initially porous, the liquid reactant diffuses into the solid substance and will also react. The chemical reaction and diffusion take place in parallel in the diffusion zone and not at the sharp interface. The reaction of the porous particle has not been studied so extensively as that of the non-porous particle, only recently studies have appeared concerned with the theoretical fundamentals of this phenomenon. As in the case of the non-porous particle, it is important to understand the role of chemical kinetics and mass transfer. The resistance of matter and heat transfer may greatly affect the apparent activation energy, the apparent reaction order and the dependence of the overall rate on the grain size and the structure of the particle.

Reactions in which no solid reaction product is formed

This type of reaction is found less frequently in hydrometallurgical processes, for example, in leaching of roasted products or some semi-finished products. This type of reaction can be generally described by the scheme:

$$A_{(\text{liquid})} + B_{(\text{solid})} \rightarrow \text{liquid products} \qquad (7.83)$$

In the case of non-porous solid substances, reacting without the formation of a solid reaction product, it was found that the overall rate can be controlled by the chemical reaction or external diffusion. In the case of porous particles, the diffusion of a liquid reagent inside the pores of the solid substance is an additional regime in which the overall reaction depends strongly on the diffusion in porous but is not controlled by this.

At low temperatures when the intrinsic kinetics is slow, the liquid substances may diffuse deep into the solid substance and the reaction takes place everywhere inside the solid substance at the uniform concentration of the liquid reagent. In this case, all the results of kinetic measurements are obtained from intrinsic values.

When increasing the reaction rate, the liquid reagents cannot penetrate deeply into the solid substance without reacting. In these conditions, the reaction takes place in a narrow region of the external surface and reaction consumes the solid substance in the direction from the surface to the centre, Fig. 7.9.

The above situation holds for irreversible reactions. If a reaction takes place mostly in a layer close to the outer surface, the reaction zone may be regarded as flat, neglecting the actual overall

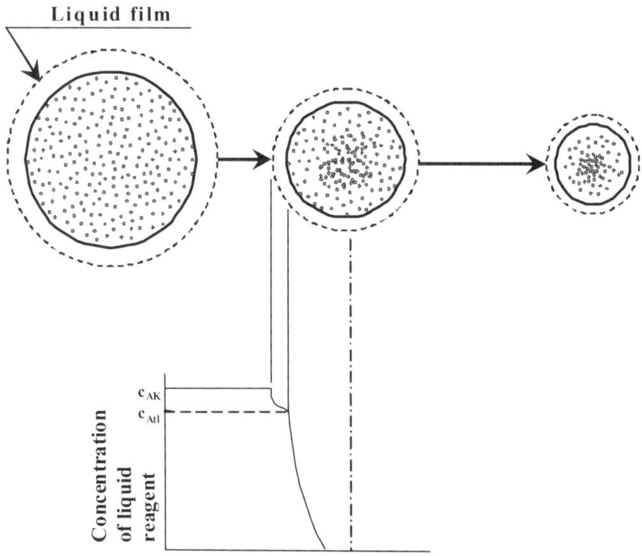

Fig. 7.9. Reaction of the porous particle without forming a solid reaction layer and shrinking in overall size.

geometry of the solid particle. In this case:

$$D_e \frac{d^2 c_A}{dx^2} - kS_v c_A^n = 0 \tag{7.88}$$

where x is the distance normal to the external surface and S_v is the surface area per unit volume, and c_A is the concentration of substance A. The bulk flow is ignored. This holds for equimolar counterflow diffusion or if the concentration of substance A is low.

Many reactions of the liquid–solid phase type may described by the Langmuir–Hinshelwood equation:

$$\text{rate} = \frac{kc_A}{1 + Kc_A} \tag{7.89}$$

If this expression is substituted into equation (7.88), this gives the expression for the amount of the reacted substance:

$$n_A = (2kS_v D_e)^{\frac{1}{2}} \frac{[Kc_A - \ln(1 + Kc_A)]^{\frac{1}{2}}}{Kc_A} \tag{7.90}$$

Reactions in which a solid reaction product is formed

This type of reaction is similar to the reactions of non-porous substances and the reacting solid substance is initially porous. The reaction can again be described by the scheme

$$A_{(l)} + bB_{(s)} \rightarrow cC_{(l)} + \mathrm{d}D_{(s)}$$

In a porous solid substance the reaction takes place mostly in the diffused zone against the reaction at the sharp interface. This results in the gradual conversion of the solid substance in the direction into the centre. Generally, the outer layer will initially react completely and gradually the thickness of the completely reacted layer in the direction into the centre of the porous substance increases. Figure 7.10 shows the reaction of a porous substance on a which a layer of solid reaction product forms.

If the diffusion rate is approximately equal to the rate of the chemical reaction, the concentration of the liquid reactant is the same in the entire volume of the solid substance and the rate of the reaction is the same. However, if the chemical kinetics is considerably faster than the diffusion rate, the reaction takes place only in a narrow region between the non-reacted and completely reacted regions. This situation is identical to the case of diffusion-controlled reactions of the decreasing core of the non-porous particle. Mathematical formulation also includes the internal chemical kinetics and diffusion and determines criteria for the situation in which the rate-controlling stage of the reaction can be determined. This holds for the isothermal system and the

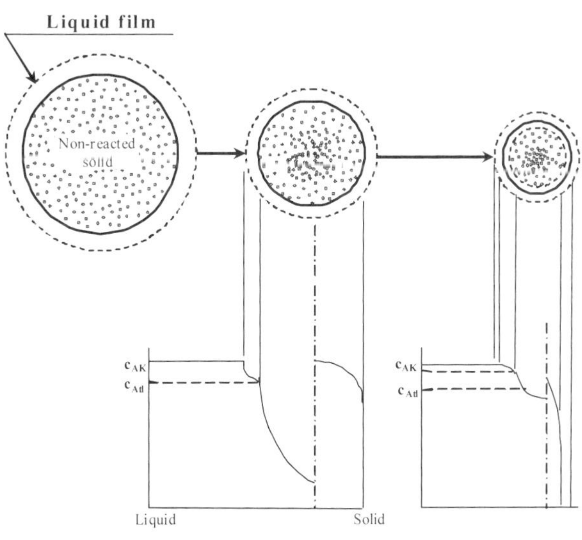

Fig. 7.10. Reaction of the porous particle forming a solid reaction product.

irreversible first order reaction. Possible structural changes have been ignored.

The following exceptions were made when deriving the rate equations [13]:

- The pseudo-steady state approximation is valid for the determination of the concentration profile of the liquid reactant inside the solid porous particle;
- The resistance of external mass transfer is negligible;
- Internal diffusion is either equimolar or occurs at low concentrations of the diffusing species is considered;
- Inside the particle, diffusivity is constant;
- Diffusion through the product layer around the individual grains is fast.

This results in the following equation:

$$X = \frac{\int_0^1 \eta^{F_p - 1}(1 - \xi^{F_g})\mathrm{d}\eta}{\int_0^1 \eta^{F_p - 1}\mathrm{d}\eta} \qquad (7.91)$$

where X is the amount of the reacted substance A, F_p and F_g are the shape factor of the particle and the grain factor, and ξ and η are the dimensionless coefficients of the surface and porosity.

The reaction of the porous solid substance with a liquid reagent includes the chemical reaction and internal diffusion which take place in parallel. Analysis of this system is generally more complicated than in the reaction of the non-porous particle. Far more detailed analysis is essential for analysing the behaviour of the system of many particles which interact during the process [13].

7.2.1.2. Kinetics of leaching processes in multi-particle systems

Previously, the behaviour of individual particles surrounded by an infinite liquid medium was discussed. Various steps or the combination of steps which may control the overall rate of the process were discussed and suitable models for explaining this behaviour were sought. The results will now be applied to the behaviour of multi-particle systems. In practice, these polydispersed systems are encountered in almost all cases. In addition, in the systems in practice the particles often react with each other and/ or with the liquid medium.

In previous studies, the behaviour of multi-particle systems was greatly simplified and the 'average' particle was considered. However, it was shown later that the size distribution of the particles plays a very important role in determining the overall kinetics of the leaching processes and the so-called population equilibrium has been taken into account. Consequently, it is possible to investigate the behaviour of every 'type' of the particle in clustering and the effect on other particles.

Before analysis, it is necessary to take into account several main aspects of a multi-particle system, because of the following:

- the leaching kinetics of the multi-particle system cannot be generalised using the same criteria as the leaching kinetics of single particles;
- the existence of a large number of variables which play a significant role in leaching of multi-particle systems;
- the effect of the distribution of the properties of materials and of particle-liquid interaction on the overall reaction kinetics of a cluster of particles have synergetic tendencies.

As an example, one may consider the oxidation leaching of a chalcopyrite concentrate in a mixed reactor. It is well known that at elevated temperatures and high oxygen pressure the pure chalcopyrite reacts in accordance with the equation:

$$CuFeS_2 + \frac{17}{4}O_2 + H^+ \rightarrow Cu^{2+} + Fe^{3+} + 2SO_4^{2-} + \frac{1}{2}H_2O \qquad (7.92)$$

and the leaching kinetics of the individual particles is controlled by the surface reaction [14]. This and previous discussion show that the kinetic dependence of leaching in the form $1-(1-R)^{1/3}$ in relation to the leaching time for approximately spherical particles of the same size should be linear with the angle of inclination inversely proportional in relation to the particle size. Figure 7.11*a* confirms the validity of this assumption. Figure 7.11*b* shows the same type of dependence but for particles with a wide size range.

It should noted that in this case, although the leaching kinetics of the individual particle is controlled by the rate of the surface reaction, the course of the overall reaction rate, shown in Fig. 7.11, deviates strongly from linearity which shows that the overall kinetics of leaching of the particles with the wide size range is not controlled by the dependence $(1-(1-R)^{1/3})$. Each particle reacts by the characteristic rate in the form

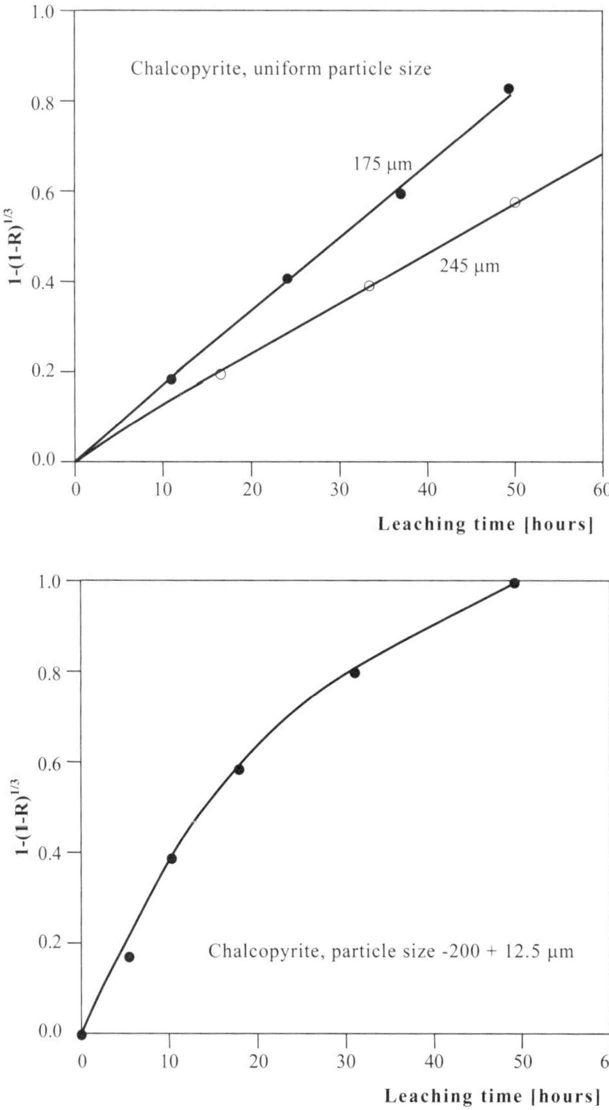

Fig. 7.11. Comparison of the kinetic curves of leaching of chalcopyrite with the uniform particle size and a wide grain size range.

$$1-(1-R)^{\frac{1}{3}} = \left(\frac{K}{d_o}\right)t \tag{7.93}$$

but the overall rate of the cluster of the particles consists of the individual rates of the particles. This function is determined by the size distribution of the particles.

The size distribution of the particles may be expressed by an infinite number of variables characterising the shape and size, and the detailed structure of the distribution depends on the material of the particles and the methods of producing the material. This was the main reason why it was concluded that the distribution of the particles should be characterised using two parameters, the mean size or central distribution tendency μ, and the coefficient of variation or the relative spread of sizes around the mean, *CV*. Figure 7.12 shows the kinetic curves of leaching of the individual grain size groups with the mean size being 50 μm, and the variation factor 0.73, i.e. the mean size varies between 12.5 and 200 μm.

It is evident that the mean grain size distribution has a significant effect on the kinetics. It may expected that as the average grain size of the charge decreases, the overall reaction rate of the entire cluster of the particles will increase.

The effect of the scatter of the size distribution of the particles for the given mean size is shown in Fig. 7.13. The effect is not so dramatic as with a change of the main size but is still significant indicating that the average grain size of the charge is not sufficient for describing comprehensively the leaching characteristics. The curves with a low value of *CV* represent the low initial rate and the relatively high rate at longer leaching times. The curves with a high value of *CV* represent a high initial leaching rate and a low

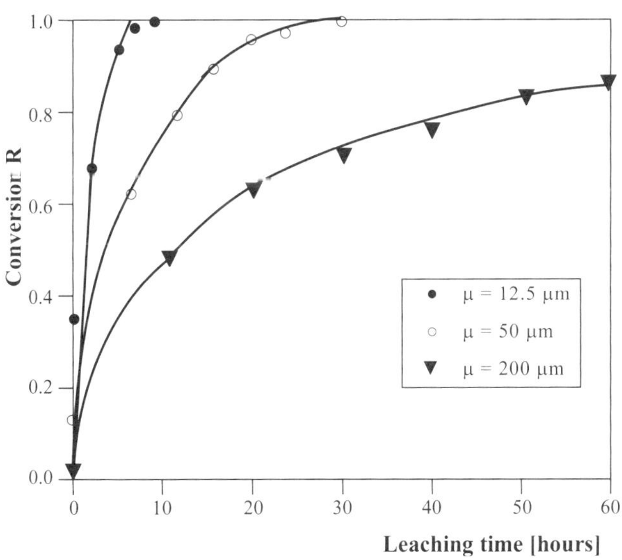

Fig. 7.12. Effect of the change of the mean size of a chalcopyrite charge, $CV = 0.73$.

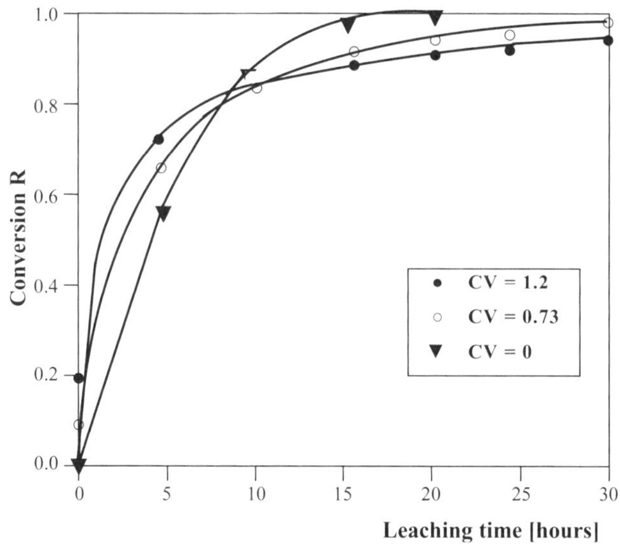

Fig. 7.13. Effect of the change of the distribution coefficient of the charge, μ = 50 μm.

rate at longer times so that the curves in Fig. 7.13 intersect. The leaching rate of the small particles is high and this is reflected in the form of the leaching curves by the rapid increase at the beginning, whereas the leaching rate of the large particles is low and this is reflected in the increase of the leaching curves at longer leaching times.

Figure 7.14 shows the effect of a different type of the grain size distribution of the overall kinetics of the reaction of a cluster of grains. In this case, leaching was carried out on a mixture of 1:1 with approximately the same grain size of chalcopyrite and another copper sulphide in a charging regime. As in the case of chalcopyrite, the leaching kinetics of the second phase is controlled by the surface reaction, but the rate constant of the surface reaction is approximately four times higher than the rate constant of the surface reaction of chalcopyrite. The figure shows the effects of both minerals and also of a mixture of minerals. Again, the mixture does not copy the linear dependence $1-(1-R)^{1/3}$, which controls the individual components of the mixture. This dependence is simply unsuitable for this case.

The principle of the interaction between the particles and the liquid phase is shown in Fig. 7.15. Previous cases related to highly diluted suspensions in which the concentration of the reactants on the surface of the particles did not change greatly during the

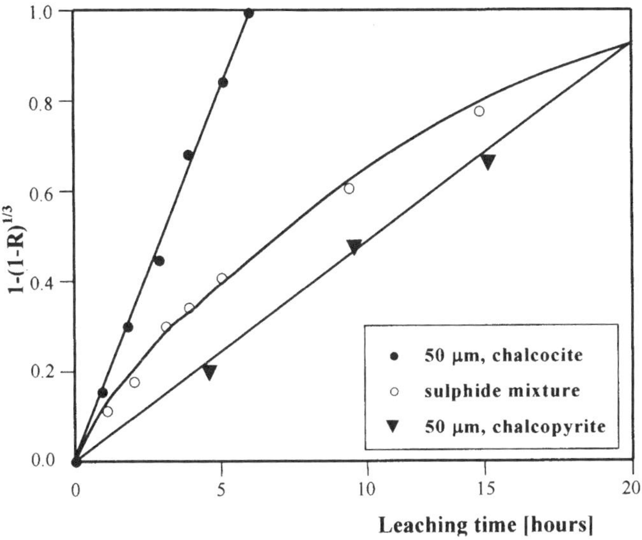

Fig. 7.14. Effect of the composition of the charge on the shape of kinetic leaching curves.

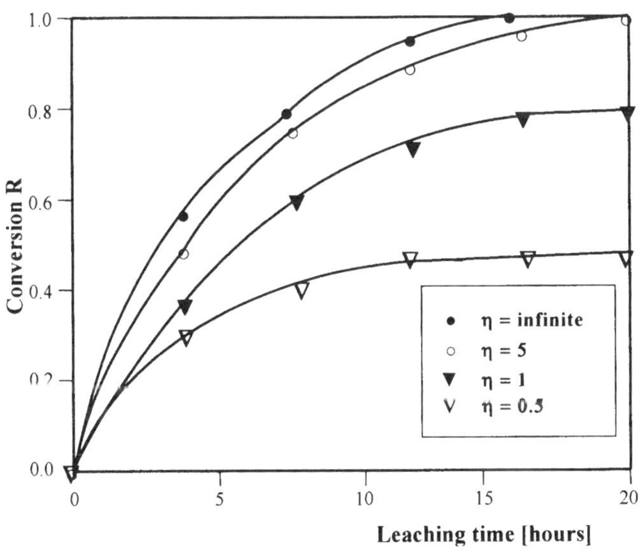

Fig. 7.15. Illustration of the effect of the particle-liquid interaction on chalcopyrite leaching ($\eta \to \infty$ corresponds to diluted suspensions, $\eta \to 0$ to concentrated suspensions).

reaction. However, the consumed amount of the reactant in more concentrated suspensions may be so large that the overall reaction rate decreases. The deceleration of reaction

$$aA_{(s)} + bB_{(l)} \rightarrow cC_{(s)} + dD_{(liquid\ product)}$$

as the result of this reaction may characterised by dimensionless parameter η defined as follows

$$\eta = \frac{a\, Vc_{B_0}}{b\, M_{A_0}} \tag{7.94}$$

where c_{B_0} is the initial concentration of the liquid reactant B, V is the initial volume of the liquid, and M_{A_0} is the initial amount of the mols of the solid substance A. From the physical viewpoint, η is the number of the mols of the substance A which may be changed by the initially present amount B separated by the actual initial amount of the mols of substance A. Figure 7.15 shows the effect of parameter η on the course of the leaching reaction of chalcopyrite particles of the same size. These relationships show that the interactions between the particles and the liquid phase cannot be ignored because of the low values of η especially at high degrees of conversion.

Previously, attention was given to the effect of the grain size distribution and solid particle-liquid phase interaction on the kinetics of the reaction in the multi-particle system represented by acid pressure leaching of chalcopyrite in the presence of oxygen in the batch regime. Similar trends may also be detected in liquid extraction, cementation, precipitation, etc., in batch or also in continuous processes. These effects are so significant that they require general analysis of the leaching kinetics in multi-particle systems.

Determination of leaching variables of a multi-particle system and analysis

Actual leaching systems use a relatively dense suspension formed by the individual particles. Depending on the type of leached material and liquid reagent, the reaction mechanism of these particles differs. After starting the process, the interface is characterised by the formation of either soluble or insoluble reaction products transferred into the surrounding medium or they may remain on the surface. These products are either porous or dense. The density of the resultant products may be similar to or differ from that of the initial material and, depending on this, the apparent form of the reacted grains increases or decreases or may remain constant. A similar situation is also found in the core of the particle. All or only some of the components may react in the particle, etc. There may be a large number of combinations and usually they also change during the process. Consequently, it is not

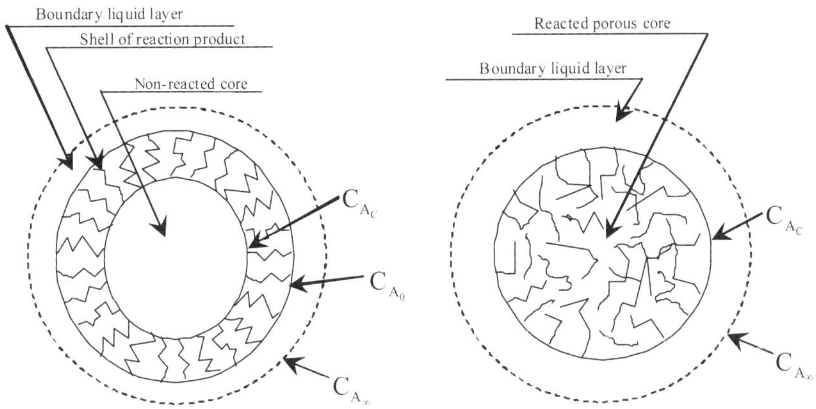

Fig. 7.16. Quasi-spherical models of the reaction of solid particles. The conversion model (the core is covered by the solid reaction product) (left). Extraction model (porous core) (right).

possible to describe the behaviour of such a particle during the process, although these variables are most important for analysis and description of the kinetics of the overall process.

The main view of the possible variations is shown in Fig. 7.16.

In the reaction of conversion of the particle there are three possible cases of the rate-controlling step of the reaction:

- transfer of the reactant mass or product mass through the layer of the liquid reagent adjacent to the surface;
- heterogeneous chemical reaction on the surface of the shrinking unreacted core;
- diffusion through the layer of the reaction product.

There are also three possible cases for extraction:

- mass transfer in the boundary liquid layer;
- heterogeneous reaction on internal surfaces of particles;
- internal diffusion in the pores.

All these six cases of the rate-controlling steps must be considered when deriving the kinetic equations of the reacting particles with four conditions:

- constant flow from the surface of the particle of the constant radius;
- constant flow from the surface of the shrinking core with a change of the radius, determined by the reaction of the material, i.e. mass equilibrium;

- quasi-steady diffusion through the thickening layer of constant chemical potential at the interface with the rate of thickening of the layer controlling the rate of mass transfer through the shell;
- non-steady diffusion (extraction) from a constant radius particle.

These considerations show that the interfacial area may not change in a regular manner. On one side, the actual area through which the reaction takes place may relatively increase because the effect of the internal stresses resulting from internal diffusion leads to the formation of a large number of cracks and cavities on the surface. The leaching solution penetrates into these cavities and since the solution is not renewed, the equalisation of the concentration reduces the leaching rate or leaching may be stopped. Consequently, the area is no longer functional.

On the other hand, a part of or the entire area may be covered by the solid reaction product which may increase or reduce the size of the area. This depends mainly on the density of the product but also on other properties.

The ideal case is the one in which it would be possible, at every measured instant, to determine the actual area of the reaction surface. However, this is not possible. Therefore, in practice we consider the presence of a grain of specific form which the reacted (leached) particle originates during the reaction and substitution of the results into these equations shows whether the reaction follows the proposed course or whether it deviates from it. This procedure therefore uses kinetic modelling equations for determination of the course and mechanism of the heterogeneous reaction.

The mathematical description of the kinetic dependences is based on the kinetic curves of the time dependence of the amount of leached metal. As already mentioned, these reactions are basically controlled either by mass transfer or the rate of the chemical reaction. However, the controlling mechanism is usually not known in advance. A first approximation may also be the shape of the kinetic curves: if the curves are parabolic, this may indicate a reaction controlled by mass transfer, i.e. by diffusion. This reaction is described by Fick's laws which are parabolic. On the other hand, the linear form of the kinetic dependences may indicate the reactions controlled by the chemical reaction since the rate of the chemical reaction is defined by the equation of the straight line. This would be valid, however, only in the cases of ideal experimental set-up, i.e., with all the parameters being constant.

This is possible only in a small number of cases. In addition, the form of the kinetic curves changes as a result of greater 'thickening' i.e., the ratio of the solid and liquid phase for leaching.

A large number of kinetic modelling equations have been derived for the individual types of reactions and mechanisms. The derivation of these equations is based on the behaviour of the reagents in the process. At present, these equations are used widely for describing the leaching of individual minerals in different media. Some of the modelling equations are summarised in Table 7.1.

One of the fundamental equations for a particle shrinking in volume was derived as follows [14]: the main assumption was the spherical form of the particle, although the final equation is applicable to particles of any isometric shape. In principle, the procedure is identical with that described for different geometrical forms of the grains. In suitably selected experimental conditions, the members on the left hand side of the equation, i.e. ratio r_0, density ρ, and concentration c, may be regarded as constant so that the final modelling equation has the form

$$1-(1-R)^{\frac{1}{3}}=kt \tag{7.95}$$

and of course $k=\dfrac{c.k_i}{r_0\rho}$ time^{-1}. The graphic dependence of the left hand side of equation (7.95) in relation to t should be linear with the angle of inclination k, k has the unit of $1/t$.

All particles of the individual minerals in the pulp which have the same initial diameter will react in the same manner in accordance with the derived equation. If the equation (7.95) is used for the pulp containing particles of different diameters, it is necessary to know the mass fractions w_i of the particles with different radius r_i. If the initial average radius of the particles in the mass fraction w_i is defined as r_{i0}, the equation (7.95) can be written in the form

$$1-(1-R_i)^{\frac{1}{3}}=\frac{ck_i}{r_{io}}t \tag{7.96}$$

where R_i is the fraction which has reacted at the mass ratio w_i. The total amount of the reacted solid substance is given by the equation

$$R=\sum_i w_iR_i \tag{7.97}$$

There are a large number of cases in which a mineral particle

Table 7.1. Kinetic modelling equations used in hydrometallurgy, where $f(R)$ is the reaction kinetic model

Model	***f(R)***
random nucleation hindered 1st. order	$-\ln(1-R)$
generalised n-th order	$\frac{1}{n}\left(1-(1-R)^{-(n-1)}\right)$
one – or two dimensional Avrami-Jerofejev	$(-\ln(1-R))^{\frac{1}{2}}$
two – or three dimensional Avrami-Jerofejev	$(-\ln(1-R))^{\frac{1}{3}}$
three- dimensional Avrami-Jerofejev	$(-\ln(1-R))^{\frac{1}{4}}$
beneralised Avrami-Jerofejev	$(-\ln(1-R))^{\frac{1}{n}}$
shrinking area	$2\left(1-(1-R)^{\frac{1}{2}}\right)$
shrinking core	$3\left(1-(1-R)^{\frac{1}{3}}\right)$
Shestak's generalised model	$a^m(1-R)^n(-\ln(1-R))^p$
Prout – Tomkinson's model	$\ln\left(\frac{R}{1-R}\right)$
one dimensional diffusion	R^2
two- dimensional diffusion	$(1-R)\ln(1-R)+R$
three- dimensional diffusion, Jander, Kröger, Ziegler	$\left(1-(1-R)^{\frac{1}{3}}\right)^2$
three- dimensional diffusion, Ginstling-Brounhstein	$\left(1-\left(\frac{2}{3}R\right)-(1-R)^{\frac{2}{3}}\right)$
diffusion through reaction product layer, Carter-Valensi	$\frac{z-(z-1)(1-R)^{\frac{2}{3}}-(1+(z-1)R)^{\frac{2}{3}}}{z-1}$
Zhuravlev	$\left(\left(\frac{1}{1-R}\right)^{\frac{1}{3}}\right)^2$

contains several metals but only one of these metals is leached out during leaching. In another case, several or all metals are leached out from the mineral, but the leaching rate of each metal is different. This may result in the formation of a porous reaction product surrounding each particle of the non-reacted mineral. A special case is the one in which the diameter of such a particle remains equal to the initial radius r_0 of the particle prior to the reaction. In this case, the reaction rate of the decreasing core with the radius r is controlled by the diffusion rate of the reactants to the non-reacted core through the layer of solid reaction products.

If a particle is spherical, the reaction rate may be described by the equation:

$$-\frac{\mathrm{d}n}{\mathrm{d}t}=\frac{4\pi\, r^2}{\sigma}D\frac{\mathrm{d}c}{\mathrm{d}r} \tag{7.98}$$

where n is the number of mols of unreacted mineral in the core, and σ is a stoichiometric factor, the number of moles of the diffusing substance required for releasing one mol of leached metal from the core. Integration of this equation in the range r and r_0, considering the steady state conditions, gives

$$-\frac{\mathrm{d}n}{\mathrm{d}t}=\frac{4\pi\, Dcr_0}{\sigma(r_0-r)} \tag{7.99}$$

and the concentration of the reactant on the reaction interface is low in comparison with c.

The relationship for the total number of moles n in the non-reacted core is given by the equation

$$n=\frac{4\pi\, r_3}{3V} \tag{7.100}$$

where V is the molar volume, equal to M/ρ, and M is the molar weight and ρ is the density of the solid substance.

Combination of the equations (7.99) and (7.100) gives:

$$-\frac{\mathrm{d}r}{\mathrm{d}t}=\frac{VDcr_0}{\sigma\; r(r_0-r)} \tag{7.101}$$

for the speed of movement of the boundary between the core and the reaction product by expressing the unreacted cored radius. Combining the relationships (7.99) with the equation for conversion and its form differentiated in respect of time

$$\frac{dR}{dt} = -3\left(\frac{r^2}{r_0^3}\right)\left(\frac{dr}{dt}\right) \tag{7.102}$$

gives the kinetic equation for the already reacted particles:

$$\frac{dR}{dt} = \frac{3VDc\,(1-R)^{\frac{1}{3}}}{\sigma\; r_0^2\;\; 1-(1-R)^{\frac{1}{3}}} \tag{7.103}$$

Integration for the limit conditions $R = 0$, with $t = 0$, gives

$$1-\frac{2}{3}R-(1-R)^{\frac{2}{3}} = \frac{2VDct}{\sigma\, r_0^2} \tag{7.104}$$

Equation (7.104) may be used efficiently for a large number of leaching processes, for example, leaching of chalcopyrite in ferric sulphate. The graphical representation of the left hand side of the equation in relation to time gives a straight line with the angle of inclination equal to r^{-2}. The elemental sulphur, formed by the reaction:

$$CuFeS_2 + 4Fe^{3+} \rightarrow Cu^{2+} + 5Fe^{2+} + 2S^0$$

forms a layer strongly bonded to the surface of the mineral. The shrinking core model can also be applied to dissolution of Al_2O_3 from bauxite in Bayer's process. Bauxite particles contain a large number of minerals, such as SiO_2, TiO_2, Fe_2O_3, and so on. Some of these react together with Al_2O_3 with NaOH solution, but TiO_2 and Fe_2O_3 remain in the leaching residue in the finely dispersed form.

References

1. Treindl L.: Chemical kinetics [in Slovak], SPN Bratislava, 1978.
2. Roller P.S.: *J.Phys.Chemistry*, 39, 1935, 221–237.
3. Fage A., Towned H.C.H.: *Proc. Royal Society,* London, A135, 1932, 656-667.
4. Bircumshaw L.L., Riddiford A.C.: Transport control in heterogeneous reactions, Quarterly Review, London, 6, 1952, 157–185.
5. Levich V.G.: Physicochemical Hydrodynamics, Prentice Hall, Englewood Cliffs, New Jersey, 1962.
6. Havlík T.: Dissolution of copper sulphides in acid media, Košice, 1982, PhD dissertation.

7. Levich V.G.: *Zh. Fiz. Khimii*, 18, 1944, 335–355.
8. Smith J. M.: Chemical Engineering Kinetics, 2nd.ed., McGraw-Hill, 1970
9. Wadsworth M. E.: Hydrometallurgical processes, Rate Processes of Extractive Metallurgy, Sohn H.Y., Wadsworth M.E. eds, Plenum Press, New York, 1979, 133–186.
10. Havlík T.: Acid oxidation leaching of chalcopyrite and behaviour of sulphur in the process, PhD dissertation, Technical University Košice, April 1996.
11. Havlík T., Škrobian M., Petricko F.: *Acta Metallurgica Slovaca*, 2, 2, 1996, 133–141.
12. Vaughan D. J., Becker U., Wright K.: *Int. J. Miner. Process.*, 51, 1997, 1–14.
13. Sohn H.Y., Wadsworth M.E.: Rate Processes of Extractive Metallurgy, Plenum Press, New York, 1979.
14. Yu P. H., Hansen Ch. K, Wadsworth M. E.: *Met. Trans.*, 4, September 1973, 2137–2144.

Chapter 8

LEACHING IN CHLORIDE MEDIA

Sulphuric acid is used efficiently in many hydrometallurgical processes, with the oxidant being the ferric ion in the form of ferric sulphates. In order to intensify leaching processes, investigations have been and are being carried out into a large number of other media and oxidants and it was soon found that although the ferric ion is an agent leaching metals from ores and concentrates, as a result of using the ferric ions in different forms these are leached at different rates and in different amounts. Using the ferric chloride as the leaching agent in comparison with ferric sulphate with otherwise comparable conditions, the yield is 2–8 times higher for the same leaching time [1].

The chloride medium was already used in the sixteenth century on amalgamation of silver but cuprous chloride was used for direct precipitation of silver sulphide only in 1860 [2]. This date can be regarded as the initial date of direct hydrometallurgy of sulphides. During the following approximately 100 years, chloride metallurgy was used only seldom in industry, especially in extraction of nickel, cobalt and copper and only as a special part of the conventional metallurgical procedure. In addition to using ferric chloride, investigations were also carried out into the possibilities of using cuprous chloride for leaching copper sulphides. One of the main advantages is that the bivalent copper ion is reduced to the monovalent ion so that energy may be saved (theoretically up to 50%) in electrowinning of copper from the solution. This fact was noted and patented already in 1893 by Hoepfner [3]. The fact that pilot plant tests were carried out only after a long period of time was due to the situation which shows that electrolysis from chloride solutions is a far more complicated and sensitive process than electrolysis from sulphate solutions. In addition, in the majority of the last century the dominant process on the world-wide scale was pyrometallurgical production of copper, and refining was carried out using electrolysis with soluble anodes.

8.1. Main aspects of leaching chalcopyrite in a chloride medium

Although the overall reaction of leaching chalcopyrite using $CuCl_2$ is quite simple

$$CuFeS_2 + 3CuCl_2 \rightarrow 4CuCl + FeCl_2 + 2S^0 \qquad (8.1)$$

in reality it is a complicated multistage process controlled by the heterogeneous kinetics of interaction of the solid phase with a leaching medium containing a combination of different ions in different and dynamically changing concentration ratios. In addition, as shown in many experimental studies, the addition of further chlorides results in a change in the activity of the ions present there as a result of the formation of chloride complexes [4–10]. To understand efficiently the behaviour of the system, special attention should be given to certain phenomena such as:

Solubility of chlorides

Solubility of CuCl in cold water is very low, only around 60 mg/l. However, in concentrated chloride solutions the solubility rapidly increases [11]. This fact is very important because the chloride technology of leaching sulphide minerals by $CuCl_2$ would be otherwise not feasible for use in practice. Another important moment is that the solubility of CuCl depends greatly also on the form of presence of the chloride ions. These facts are summarised in Fig. 8.1.

The increase of the NaCl concentration results in an almost linear increase of the solubility of CuCl. On the other hand, in more complex systems, the effect of the presence of other chlorides may differ – for example, the addition of $FeCl_2$ at a constant concentration of NaCl and HCl increases the solubility of CuCl, whereas the addition of $ZnCl_2$ reduces this solubility. Winand [11] relates this to the strength of the resultant complexes. Fe^{2+} forms weak complexes in the chloride medium and, consequently, is a donor of Cl^- ions. On the other hand, the Zn^{2+} ion forms strong complexes and therefore acts as an acceptor of Cl^-.

Similarly, the solubility of $CuCl_2$ is affected to a certain extent by the presence of other chlorides in the solution. However, its solubility in water is considerably higher than that of CuCl. The addition of NaCl in the presence of HCl slightly increases the solubility of $CuCl_2$ but the presence of $FeCl_3$ and/or ZnCl2 affects

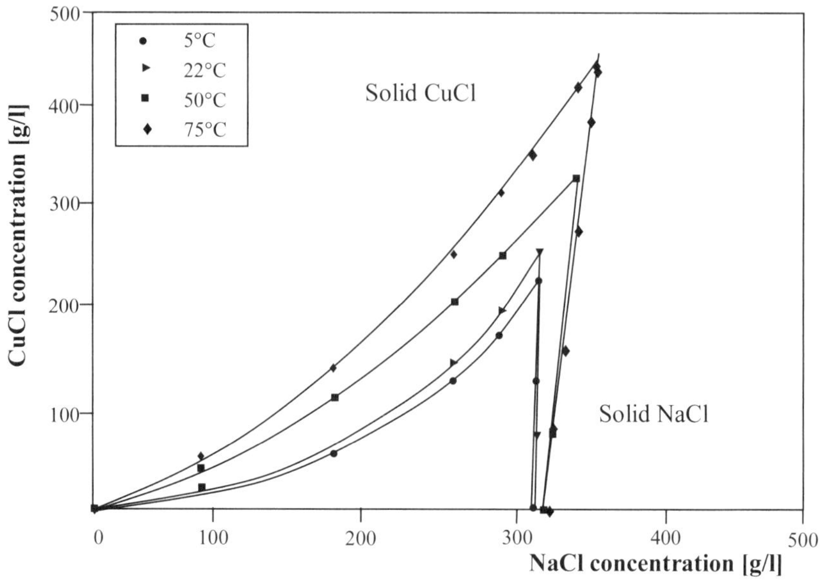

Fig. 8.1 Data for the solubility in CuCl–NaCl–H_2O system at different temperatures in HCl.

its solubility to a greater extent, especially the presence of $FeCl_3$. As indicated by Fig. 8.2, in solutions saturated with both NaCl and $CuCl_2$, the presence of $ZnCl_2$ leads to the precipitation of $CuCl_2 \cdot 2H_2O$, whereas the presence of $FeCl_3$ would cause the precipitation of NaCl.

Another factor affecting the solubility of individual chlorides is temperature. Its effect is stronger in the case of $CuCl_2$, as shown by comparison of Figs. 8.3 and 8.4.

Oxidation potential

From the viewpoint of the leaching process, the oxidation potential of the leaching agent is an important parameter. One of the factors which make the chloride method attractive is that the oxidation potential of the bivalent copper ion in the chloride medium (4 M NaCl + 0.5 M HCl, 30 °C) is almost four times higher than the standard potential of Cu^{2+}/Cu^{+} – 584 in comparison with 153 mV in relation to SHE (standard hydrogen electrode) [11], as shown in Fig. 8.5.

It is also important to note that the oxidation potential of other ions in the chloride medium differs from their standard potential. For example, the reversible potential of the Cu^{2+}/Cu pair is only 84 mV,

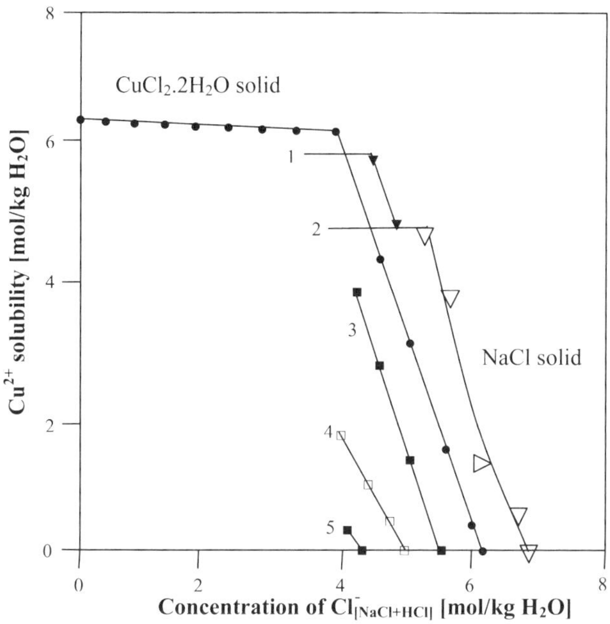

Fig. 8.2 Solubility in complex Cu^{2+} chloride solutions at 50 °C [11]. The system $CuCl_2$–$ZnCl_2$–NaCl–HCl–H_2O: 1) c_{Zn2+} = 0.5 M; 2) c_{Zn2+} = 1.5 M; the system $CuCl_2$–$FeCl_3$–NaCl–HCl–H_2O: 3) c_{Fe3+} = 0.5 M; 4) c_{Fe3+} = 1 M; 5) c_{Fe3+} = 1.5 M.

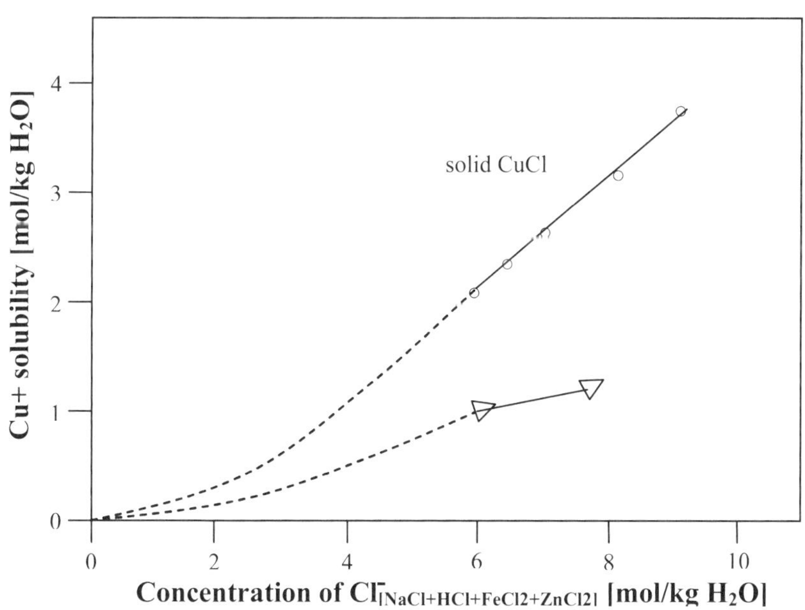

Fig. 8.3 Solubility of Cu^+ in the CuCl–$FeCl_2$–$ZnCl_2$–NaCl–HCl–H_2O system.

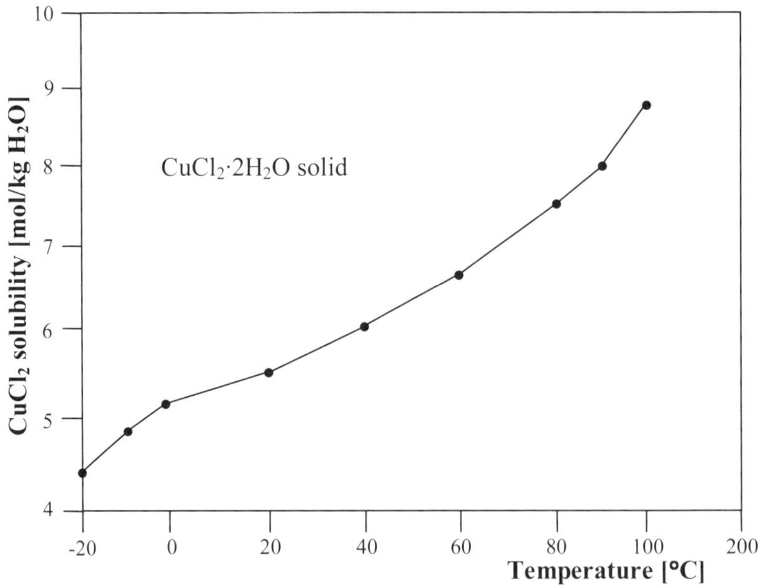

Fig. 8.4 Temperature dependence of the solubility of $CuCl_2$ in water [11].

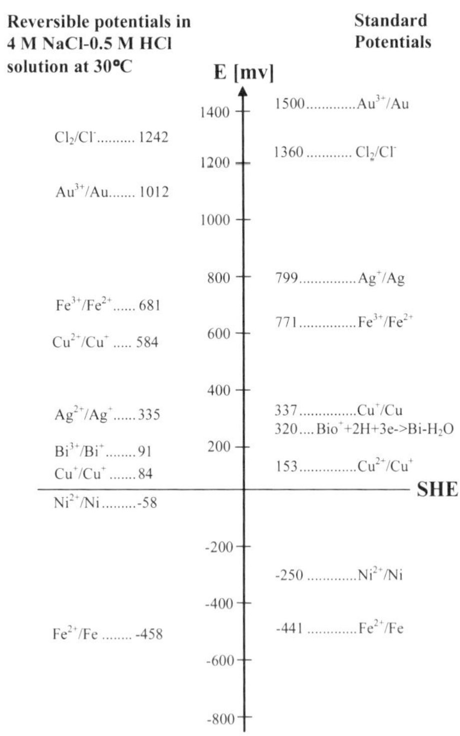

Fig. 8.5 Molar equilibrium potentials *E* measured at 30 °C at the solution for 4 M NaCl–0.5 M HCl compared with a standard potentials at 25 °C [11].

whereas the standard potential of the Cu^{2+}/Cu is 521 mV – this fact has a positive influence in chloride electrolysis of copper. The reversible potential of Fe^{3+}/Fe^{2+} in the chloride medium is 681 mV which is only slightly (by approximately 100 mV) higher than the reversible potential of Cu^{2+}/Cu^{+}. In other words, the oxidation efficiency of the bivalent copper ion in this medium is the same as that of the trivalent iron ion.

On the other hand, the reversible potentials of the ions in the chloride medium are strongly affected by the formation of complexes. However, this means that all essential conditions for the reactions are formed in the leaching process: a relatively dramatic change of the concentration of the individual ions in different leaching stages and large changes of their reversible potentials.

Formation of chlorine complexes

In addition to the fact that the formation of chlorine complexes has a strong effect on the solubility of chlorides of the individual ions, the redox potential of the leaching solution also changes greatly. Winand [11] and Muir [7] summarised information available from different literature sources. For the principle ions present in the solution in leaching of chalcopyrite these data are presented in Table 8.1.

According to Muir [7], the relationship between the potential and the concentration of chloride ions can be expressed as follows:

$$E_{Cu^{2+}/Cu^{+}} = 0.495 + 0.118 \log[Cl^{-}] + 0.0559 \log\left[\frac{Cu^{2+}}{Cu^{+}}\right] \qquad (8.2)$$

This means that the calculated value of the potential for a solution consisting of 4.5 M NaCl and 0.5 M HCl changes from 472 mV (Cu^{2+}:Cu+ = 1:1) to 584 mV (Cu^{2+}:Cu^{+} – 100:1) depending on the ratio of the concentrations of Cu^{2+} and Cu^{+}.

The above mentioned facts – the formation of chlorocomplexes and the variation of the oxidation potentials – may also be described by the potential–pH diagrams. *E*–pH diagrams show differences in the existence of copper ions in the solutions Cu–H_2O, Cu–S–H_2O and Cu–S–Cl_2–H_2O (Figs. 5.11, 5.15).

Thermodynamic studies show that the precipitation of $CuCl_2 \cdot 3Cu(OH)_2$ starts at pH values at which copper in the sulphate solution is still present in the ion form.

Another result is the one which shows that whilst the potential

Table 8.1. Chloride complexes

	Low Cl^- concentration					High Cl^- concentration		
Cu^{2+}	Cu^{2+}	$CuCl^+$			$CuCl_2$		$CuCl_3^-$	$CuCl_4^{2-}$ (or $Cu_2Cl_4^{2-}$)
Cu^+	$CuCl_2^-$			$CuCl_3^{2-}$			$CuCl_4^{3-}$	
Fe^{3+}	Fe^{3+}				$FeCl^{2+}$			$FeCl_2^+$
Fe^{2+}	Fe^{2+}							$FeCl^+$
Zn	Zn^{2+}	$ZnCl^+$			$ZnCl_2$		$ZnCl_3^-$	$ZnCl_4^{2-}$
Pb	$PbCl^+$	$PbCl_2$				$PbCl_3^-$		$PbCl_4^{2-}$
Ni	Ni^{2+}							$NiCl^+$
Co	Co^{2+}							$CoCl^+$
Mn	Mn^{2+}							$MnCl^+$
Cd	Cd^{2+}		$CdCl^+$		$CdCl_2$		$CdCl_3^-$	$CdCl_4^{2-}$
Sb	$SbCl_2^+$	$SbCl_2^+$		$SbCl_3$	$SbCl_4^-$		$SbCl_5^{3-}$	$SbCl_6^{3-}$
Bi	$BiCl_2^+$	$BiCl_2^+$		$BiCl_3$	$BiCl_4^-$		$BiCl_5^{3-}$	$BiCl_6^{3-}$
As					$AsCl_3$			
Ag	$AgCl_2^-$							$AgCl_3^{2-}$
Hg	$HgCl^+$		$HgCl_2$			$HgCl_3^-$		$HgCl_4^{2-}$

of the trivalent iron ion gradually decreases with increasing Cl^- concentration as a result of the formation of ferric chlorine complexes, the potential of the bivalent copper ion increases, Fig. 8.6 [11].

Leaching of copper sulphides

The process of leaching of copper sulphides in a chloride medium takes place by the electrochemical mechanism. In a strong oxidising medium part of sulphur oxidises to sulphate according to the equation:

$$CuFeS_2 + 8H_2O + 16ox \rightarrow Cu^{2+} + Fe^{2+} + 2SO_4^{2-} + 16H^+ + 16ox^- \quad (8.3)$$

However, the majority of chalcopyrite sulphur transfers to the elemental form

$$CuFeS_2 + 4ox \rightarrow Cu^{2+} + Fe^{2+} + 2S^\circ + 4ox^- \quad (8.4)$$

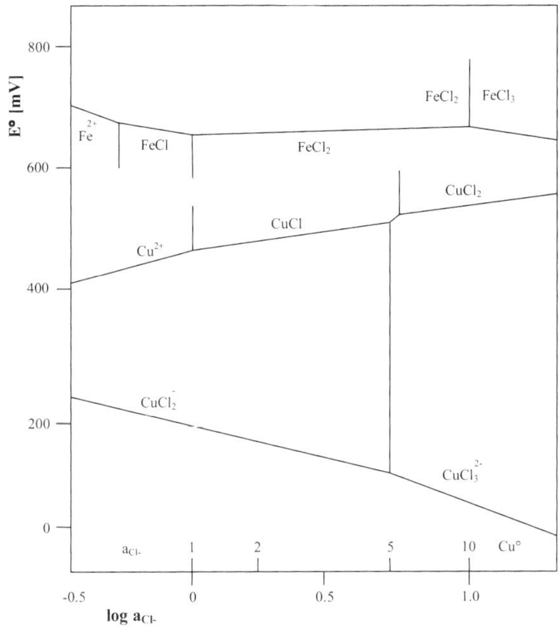

Fig. 8.6 E–a_{Cl^-} diagram for the Fe^{3+}/Fe^{2+}–Cu^{2+}/Cu^{+} system at 25 °C.

In a less oxidation medium CuS may precipitate according to the equation

$$CuFeS_2 + 2ox \rightarrow CuS + Fe^{2+} + S^0 + 2ox^- \tag{8.5}$$

Using the Fe^{3+}/Fe^{2+} plus redox pair in leaching of chalcopyrite, the following anodic reaction takes place:

$$CuFeS_2 \rightarrow Cu^{2+} + Fe^{2+} + 2S^{\circ} + 4e^- \tag{8.6}$$

and cathodic reactions

$$Fe^{3+} + e^- \rightarrow Fe^{2+} \tag{8.7}$$

$$FeCl^{2+} + e^- \rightarrow Fe^{2+} + Cl^- \tag{8.8}$$

$$FeCl_2^+ + e^- \rightarrow Fe^{2+} + 2Cl^- \tag{8.9}$$

$$FeCl_{3(aq)} + e^- \rightarrow Fe^{2+} + 3Cl^- \tag{8.10}$$

The rate-controlling step of these reactions are surface phenomena, and the rate also depends on the overall concentration of the ferric ions. The leaching rate increases with the increase of the concentration of the chloride ions to the value 1M but becomes independent with a further increase of the chloride concentration. At the same time, the leaching rate is independent of the concentration of hydrochloric acid, ferrous chloride and magnesium chloride [12]. Cupric chloride increases the leaching rate, whereas a small amount of the sulphate reduces the rate. If the system contains a large number of sulphate ions, it behaves as in leaching in a sulphate medium.

If the redox pair is formed by Cu^{2+}/Cu^{+}, the main reaction of the process is:

$$CuFeS_2 + 3Cu^{2+} \xrightarrow{NaCl_2, H_2O} 4Cu^{+} + Fe^{2+} + 2S^0 \qquad (8.11)$$

However, this reaction may not take place to the end because it may not take place in the reverse direction, as indicated, but the resultant elemental sulphur may be reduced in accordance with the reaction:

$$2Cu^{+} + S^0 \rightarrow CuS + Cu^{2+} \qquad (8.12)$$

The high value of the Cu^{+}/Cu^{2+} ratio is positively affected by the high concentration of the chloride, high temperature, low pH and short leaching time. If the process requires a solution with only univalent copper, then it is necessary to introduce two-stage leaching or reduction by metallic copper.

If the oxidation conditions in the solution are stronger than those ensured by the redox pair Fe^{3+}/Fe^{2+}, iron may be directly precipitated [13] in the leaching conditions 110 °C, oxygen pressure of 2010 kPa and the concentration 2 M HCl, according to:

$$4CuFeS_2 + 8HCl + 5O_2 \rightarrow 4CuCl_2 + 4FeOOH + 2H_2O + 8S^{\circ} \qquad (8.13)$$

and

$$CuFeS_2 + HCl + O_2 \rightarrow CuCl + FeOOH + 2S^{\circ} \qquad (8.14)$$

The disadvantages which must be taken into account in this case are:

part of the sulphur is oxidised to the sulphate form,
at pH values higher pH = 2.2 $CuFe_2O_4$ may precipitate
CuCl may precipitate in strong acid solutions (higher than 3 M).

The reduction of chalcopyrite by Cu^+ to chalcocite Cu_2S or bornite Cu_5FeS_4 may be realised by the reaction:

$$CuFeS_2 + 4Cu^+ \rightarrow 2Cu_2S + Fe^{2+} + Cu^{2+} \tag{8.15}$$

The rate of the reaction on the fresh surface is high but greatly reduces after the formation of a thin chalcocite or bornite layer. It was found [14] that if the reduction takes place in the presence of metallic copper, the rate of the reaction is high and produces a material which is easier to leach than chalcopyrite.
The probable reaction is

$$CuFeS_2 + Cu^0 + 2Cu^+ \rightarrow 2Cu_2S + Fe^{2+} \tag{8.16}$$

Another method of 'activation' consists of heating chalcopyrite together with elemental sulphur [15] at 475 °C producing simpler sulphides in accordance with the reaction:

$$CuFeS_2 + S^0 \rightarrow CuS + FeS_2 \tag{8.17}$$

However, it has been proven unambiguously that the rate of the process greatly increases if the process takes place simultaneously with dissolution of iron [16].

Electrochemical studies of anodic leaching of sulphides of copper chalcocite and digenite [17–18] show that the first stage is characterised by the preferential leaching of copper forming a compound similar to CuS with a decrease in Cu/S ratio in relation to leaching temperature. The second stage of the reaction is the breakdown of this material into univalent copper and elemental sulphur. The leaching rate increases with the formation of the $CuCl_2^-$ complex at high chloride concentrations. At a low concentration of hydrochloric acid the preferential process is electrochemical oxidation to sulphate. In solutions with a low anodic potential examination showed only Cu^+. Both Cu^+ and Cu^{2+} form at medium potentials, and the main product at higher potentials is Cu^{2+}. At potentials higher than 0.55 V the main product was sulphur but did not show any tendency for oxidation to the soluble form, if a sufficient amount of chloride ions was present.

Regeneration of the leaching agent

The leaching agent may be regenerated after extracting a metal by oxidation of Fe^{2+} to Fe^{3+} or Cu^+ to Cu^{2+}. These reactions are fast and controlled by oxygen diffusion through the liquid–solid phase interface from the side of the solution. The following reaction takes place in the case of copper:

$$4CuCl_4^{3-} + 4H^+ + O_2 \rightarrow 4CuCl_3^- + 2H_2O + 4Cl^- \qquad (8.18)$$

The ions Cu^+ can be oxidised to Cu^{2+} also by anodic electrolysis. This process is fully reversible.

Hydrochloric acid may also be regenerated by spraying roasted chlorides in accordance with the reaction:

$$MeCl_2 + xH_2O \rightarrow MeO + 2HCl + (x-1)H_2O \qquad (8.19)$$

By comparing the method of chloride leaching with conventional sulphate leaching we can define certain potential advantages of the chloride methods:

- easier leaching of complex chloride concentrations;
- transfer of sulphur to elemental form;
- obtaining of concentrated chloride solutions;
- the metal can be extracted from the solution by liquid extraction;
- regeneration of the solution is simple by oxidation with oxygen or chlorine;
- iron can be precipitated from the solution.

However, the main problems associated with chloride processes were defined especially on the basis of pilot plant experience [19]]:

- leaching in a chloride medium is not sufficiently selective and at mixtures or conventional impurities are highly soluble in the leaching medium so that recirculation and cleaning of the leaching agent (bleed stream) should be included in the technological cycle;
- final process produce copper in the form of granules or powder copper with a large specific surface increasing the risk of contamination; in addition, to improve the saleability of the product it is often necessary to remelt or compact the copper;
- high current density is essential in electrolysis – on the one hand, this results in a decrease of capital expenditure but on the other hand increases production costs because the saving of energy in comparison with sulphate electrolyte is not as large as expected;
- solutions of copper chlorides may contain larger fractions of selenium, tellurium and in particular silver because cleaning of the solutions to remove these mixtures is associated with

problems and they subsequently co-precipitate with the produced copper;
- design of electrolytes for electrolysis from a chloride medium is complicated and expensive;
- in contrast to sulphate solutions, chlorine may precipitate on the anode instead of oxygen – because of its toxicity and corrosivity chlorine must be permanently and efficiently removed;
- relatively high vapour tension of chlorides from acid chloride solutions requires a large and efficient extraction system trapping the resultant chlorides followed by their processing;
- in comparison with the sulphate solution, the corrosion strength of chloride solutions is considerably higher and these solutions require special materials for apparatus and support or transport systems, of course this has a detrimental affect on capital expenditure;
- leaching residue from the chloride processes represent a potential environmental risk and require additional processing thus increasing the total production costs.

However, in any case the processes of chloride hydrometallurgy represent at present time the latest trend in the processing of chalcopyrite concentrates. The main representatives of these advanced procedures are the processes Outokumpu HydroCopper and Intec.

References

1. Havlík T.: Acid oxidation leaching of chalcopyrite and behaviour of sulphur in this process, PhD disseration, Technical University, Košice, April 1996.
2. Liddell D. M.: Chlorine Metallurgical Processes, Handbook of Non Ferrous Metallurgy, vol. II, Recovery of Metals, McGraw Hill, New York, 1945.
3. Hoepfner C.: Electrolytic production of metals, US Patent No. 507130, October 24, 1893.
4. Cathro K. J.: Recovery of copper from chalcopyrite by means of a cupric chloride leach, In: International Symposium on Copper Extraction & Recovery, Chapter 40, 1976, 776–793.
5. Wilson J. P., Fisher W. W.: *Journal of Metals*, February 1981, 52–57.
6. Bonan M., Demarthe J. M., Renon H., Baratin F.: *Met. Trans. B*, 12B, June 1981, 269–273.
7. Muir D. M.: Basic Principle of chloride hydrometallurgy, In: Chloride Metallurgy 2002, Vol.II, 32nd Annual Hydrometallurgy Meeting (Ed. Peek E.,

Van Weert G.), 2002, 759–791.

8. Padilla R., Lovera D., Ruiz M. C.: Leaching of chalcopyrite in CuCl–NaCl–O_2 system, In: EPD Congress 1997 (Ed. B. Mishra), 1997, 167–177.
9. Hirato T., Majima H., Awakura Y.: *Met. Trans. B*, 18 B, March 1987, 31–39.
10. Kimura R. T., Haunschild P. A., Liddell K. C.: *Met. Trans. B*, 15B, June 1984, 213–219.
11. Winand R.: *Hydrometallurgy*, 27, 1991, 285–316
12. Palmer B.R., Nebo C.O., Rau M.F., Fuerstenau M.C.: *Met. Trans.* B12, 1981, 595–601.
13. Habashi F., Toor T.: *Met. Trans. B*, 10, 1979, 49–56.
14. Avraamides J., Muir D.M., Parker A.J.: *Hydrometallurgy*, 5, 1980, 325–336.
15. Miškufová A., Kuchár J., Havlík T.: *Acta Metallurgica Slovaca*, 10, 2004, Special Issue 2, (196–205).
16. Subramanian K.N., Kanduth M.: Activation and leaching of chalcopyrite concentrate, CIM Bulletin, 66, 1973, 88–91.
17. Price D.C.: *Met. Trans. B*, 12, 1981, 231–239.
18. Ghali E., Dandapani B., Lewenstam A.: *J. Appl. Electrochemistry*, 12, 1982, 369–376.
19. Hoffmann J. E.: *Journal of Metals*, 8, 1991, 48–49.

Chapter 9

EXTRACTING METALS FROM SOLUTIONS

In leaching, the required metal may be transferred into the solution using a suitable leaching agent, and other elements remain in the insoluble residue and can be extracted by separate processes, especially if some pyrometallurgical operation follows. However, the iron oxides are more or less soluble in all types of conventional acid leaching agents, although this may not be the case in neutral or basic solutions so that the resultant solution with the content of the required metal is almost always contaminated by iron. Of course, other elements may also dissolve and contaminate the leach. Some of these impurities may hydrolyse from the solution as a result of a single increase of pH. On the other hand, there may be a nucleation barrier and kinetically prevent precipitation, so that although the appropriate *E*–pH diagram at elevated temperature predicts the region of stability of a metal in the given conditions, this may not be the case.

For both economical and ecological reasons, the recirculation of the rafinate to the leaching circuit is the standard operation in the production conditions after extracting the metal. However, if the pH of the solution changes as a result of settling of impurities, a large amount of the leaching agents is required to restore leaching capacity. In addition, a large amount of water, required for removing the ion adsorbed on the surface of the precipitate, dilutes the average concentration of the ions in the solution. For these reasons, the majority of technologies are based on a procedure in which only the required metal or metals are extracted from the solution, and the impurities remain in the solution. These are then removed from the solution prior to its recirculation to the leaching circuit using a different procedure. Several methods are used for this.

9.1. Cementation

Cementation is the process of extracting the metals from a solution based on the electrochemical reaction between the cementing metal and the ion of the precipitated metal. Thermodynamic feasibility of cementation is determined from the ratio of the values of the electrode potentials. The electrode potential of the displacing metal must be more negative than that of the displaced metal. The precipitation of the metal is accompanied obviously by a change of its concentration of the solution, and consequently, of its potential. When the equilibrium values are reached, the process stops.

On the basis of the differences in the electrode potentials, we can determine the electrode pairs of the cementing and cemented metal, Table 9.1.

It may be seen that some of the metals may be cemented almost completely, for example, Cu with Zn and Fe, Ni with Zn. However, equilibrium is not established because of kinetic reasons.

Precipitation of metallic copper from the solution by contact with metallic iron has been practised for more than 600 years now and is still justified. The complete cementing reaction may be described as follows

$$Cu^{2+} + Fe^{0} \rightarrow Cu^{0} + Fe^{2+} \tag{9.1}$$

In practice, this means that after immersing a piece of iron into a solution of cuprous ions, the surface of iron is immediately

Table 9.1. Electrode potentials of metal and equilibrium ratio for the pairs of bivalent metals

Metal		**E^0 [V]**		a_{Me1} / a_{Me2}
Me_2	Me_1	Me_2	Me_1	
Zn	Cu	−0.763	+0.34	$1 \cdot 10^{-23}$
Fe	Cu	−0.44	+0.34	$1.3 \cdot 10^{-27}$
Ni	Cu	−0.23	+0.34	$2.0 \cdot 10^{-20}$
Zn	Ni	−0.763	−0.23	$5.0 \cdot 10^{-19}$
Cu	Hg	+0.34	+0.798	$1.6 \cdot 10^{-16}$
Zn	Cd	−0.763	−0.402	$3.2 \cdot 10^{-13}$
Zn	Fe	−0.763	−0.44	$8.0 \cdot 10^{-12}$
Co	Ni	−0.27	−0.23	$4.0 \cdot 10^{-2}$

covered by the precipitated copper layer. The mechanism of the exchange reaction is easy to understand from the viewpoint of electrochemical considerations. The reaction of reduction of Cu^{2+} to metallic copper may be written in the form:

$$Cu^{2+} + e^{-} \rightarrow Cu^{\circ} \tag{9.2}$$

and the reversible electrode potential is then given by the Nernst equation

$$E = E^{0} + \frac{RT}{2F} \ln a_{Cu^{2+}} \tag{9.3}$$

assuming precipitation of pure metallic copper ($a_{Cu} = 1$). The value E^0 of this reaction is +0.34 V for 1 M solution at 25 °C. Similarly, the Nernst equation for the reaction of the iron half cell is

$$Fe^{2+} + 2e^{\circ} \rightarrow Fe^{\circ} \tag{9.4}$$

and has the form

$$E = E^{0} + \frac{RT}{2F} \ln a_{Fe^{2+}} \tag{9.5}$$

and $E^0 = -0.44$ V. The difference between the equations (9.4) and (9.2) gives the reaction of the entire cell, the reaction (9.1), for which the reversible potential at 25 °C has the form

$$E = E^{0}_{Cu} - E^{0}_{Fe} - \frac{RT}{2F} \ln \frac{a_{Fe^{2+}}}{a_{Cu^{2+}}} = 0.34 + 0.44 - 0.0296 \log \frac{a_{Fe^{2+}}}{a_{Cu^{2+}}} \tag{9.6}$$

The reversible potential is positive and, therefore, the change of the standard Gibbs energy $\Delta G = -zFE$ of the overall reaction is negative. This means that even if no voltage is supplied, the electrons are taken from metallic iron for discharging copper ions and iron is also transferred into the solution to maintain electron neutrality. The reaction continues up to reaching the equilibrium state, i.e., until iron is completely dissolved or all copper ions are discharged.

When immersing a cementing metal into a solution containing precipitated metals, and electrochemical reaction takes place resulting in the formation of elementary surfaces covered with the cemented metal – cathodic areas. In order to maintain equilibrium, anodic areas form and a reversible process – ionisation of the atoms of the displacing metal, takes place.

The atoms on the surface of the metal are not equivalent from the energy viewpoint. Energy differences form as a result of the effect of force fields of the adjacent atoms in the solution, structural defects, etc. The cathodic areas form mainly in surface areas in which the electrode potential is higher. The cathodic and anodic areas are mutually conductively connected and, therefore, electrons flow from anodic to cathodic areas where the ions of precipitated metal are discharged, as shown schematically in Fig. 9.1. The outer circuits of such a short circuited cell is an electrolyte and its ohmic resistance depends on the concentration of the ions in the solution.

If the solution contains at the beginning only Cu^{2+} ions, the activity of copper ions decreases and that of iron ions increases from zero until the exchange reaction takes place. At the same time, the value of E for reaction (9.2) decreases and for reaction (9.4) increases, i.e. becomes less negative. If there is an excess of metallic iron, the value of E may be the same for both half cells so that equilibrium is reached. If both potentials of the electrode half cells are equal, the activity coefficient (equation (9.6)) is equal to 4×10^{-27} which indicates that in reality all copper ions will precipitate. However, the rafinate is usually returned to the leaching circuit prior to reaching this stage. The Fe^{2+} ions in the solution may oxidise to the Fe^{3+} form by, for example, bacteria and consequently support the leaching reaction. However, part of iron must be periodically removed to prevent an uncontrollable increase of its content in the solution to which it arises from leached primary products.

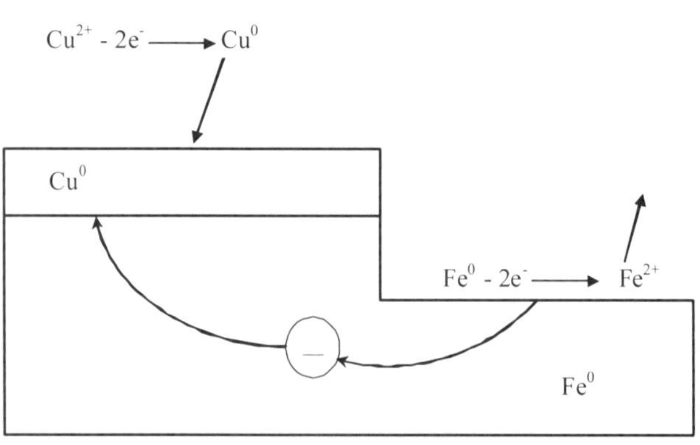

Fig. 9.1. Diagram of the cementation process [1].

Cemented copper is not a pure metal. Every other metal present in this solution whose standard electrode potential is higher than that or iron, for example, Co^{2+}, Ni^{2+}, etc., may also be cemented. On the other hand, if the amount of an element in a solution is small, the difference between the standard potential of the half cell of the element and iron when the equilibrium state is reached is small so that the equilibrium concentration may be higher than the initial amount of the impurity in the solution and, consequently, all the impurities remain in the solution. Cemented copper is also contaminated with small iron particles protected against dissolution by cemented copper on the surface. Therefore, the purity of copper is lower than that obtained in pyrorefining so that further processing is required to obtain the commercial purity of the copper produced by this procedure.

In most cases, the cementing metal is represented by cheap steel scrap. Of course, some mixture elements in the steel will also be cemented. Every metal which is clearly electronegative in relation to the extracted metal, may be used for cementation. For example, zinc is used to precipitate silver or gold from cyanide or bromide solutions:

$$Zn + 2Au(CN)_2^- \rightarrow Zn(CN)_2 + 2Au \qquad (9.7)$$

and the excess zinc is removed by oxidation in remelting noble metals which do not oxidise. Cementation is not selective and almost all other metallic elements present in the solution are cemented also by zinc. This behaviour may be used for purification of solutions of strongly electronegative elements. For example, the addition of finely ground Zn into the $ZnSo_4$ solution results in cementation precipitation of impurities of the type Co, Cd, Cu, Ni, Sb and Th and a pure solution is obtained from which zinc is extracted by a different method. Metallic zinc particles are periodically renewed and large quantities of more 'exotic' metals may be extracted from the cementate.

9.2. Cementation of amalgams

A large number of metals are capable of dissolving in mercury forming a solution – amalgam. Some of the metals are characterised by relatively high solubility in mercury, others by low or are inert, Table 9.2.

In hydrometallurgical practice, cementation is carried out mostly on a zinc or sodium amalgam. The advantages of cementation on

Table 9.2. Solubility of metals in mercury at 25 °C

Metal	at. %	Metal	at. %
In	70.3	Bi	1.6
Tl	43.7	Ca	1.48
Cd	10.06	Au	0.133
Ga	3.6	Ag	0.079
Zn	6.4	Al	0.015
Na	5.3	Cu	0.0079
Pb	1.93	Co	$3.4 \cdot 10^{-6}$
Sn	1.21	Fe	$1.8 \cdot 10^{-6}$
K	2.3	Ni	$7.6 \cdot 10^{-6}$

amalgams are based on the following:

- the overvoltage of hydrogen generation on mercury and amalgams is high. Therefore, metals which cannot be cemented because of the generation of hydrogen in the absence of mercury, can be cemented on the amalgam;
- the precipitation potentials of metals on mercury are displaced to more electronegative values in comparison with the precipitation potentials on the cathode made from the same metal. This enables selective cementation of some of the metals in the presence of others with positive electrode potentials;
- mixing of the amalgam results in constant renewal of its surface, in contrast to classic cementation on the solid surface of the cementing metal, when the latter decreases with progressive cementation.

The electrode potential of the amalgam–metal system is expressed by the equation:

$$E = E_1^0 + \frac{RT}{zF} \ln \frac{a_{\mathrm{Me}}^z}{a_{\mathrm{Me(Hg)}}} \qquad (9.8)$$

where $a_{\mathrm{Me(Hg)}}$ is the activity of a metal in amalgam, and E_1^0 is the normal potential of the amalgam electrode.

The amalgam potential is then a function of the ratio of the concentration of the metal ion in the solution to the concentration of the metal in the amalgam.

Mercury may react with a metal with a formation of an intermetallic compound solution in mercury. The potential of the diluted amalgam in relation to the diluted aqueous solution of the salt may be the same as the concentrated potential taking into account the appropriate concentrated aqueous solution. Therefore, the amalgam potentials are usually determined for systems in which the concentration of the metal and in water is the same. These are referred to as half-wave potentials and are determined, for example, polarography.

Metals may be divided into three groups:

- alkaline metals with a high affinity for mercury. Half-wave potential of these metals are displaced to the positive values in relation to the normal potentials;
- the metals which dissolve in mercury but do not form compounds with it or form only weak compounds (Zn, Cd, ln, Bi). There normal potentials slightly differ from the potentials of the amalgams of the given metal;
- metals which do not react with mercury and its solubility in mercury is low (Cu, Ni, Co, Fe, Cr, Mn). The half-wave potentials of these metals are displaced in the direction of negative values in relation to the normal potentials.

Since the density of mercury is high, any metals can be concentrated in mercury in a small volume of the amalgam. The metals are extracted from all amalgams by acid alkali or by anodic dissolution at a specific potential.

Cementation on the amalgams may be used to produce thallium, indium and cadmium from the solution. It may also be used to separate samarium and europium ions by sodium amalgam in separation of these metals from other rare-earth metals [1].

9.3. Reduction with gaseous hydrogen

The standard electrode potentials of the half cells of the type Me-Me^+ relate to the standard hydrogen electrode defined by the relationship

$$2H^+ + 2e^- \rightarrow H_2 \tag{9.9}$$

which was assigned conventionally the zero value at all temperatures for 1 M solution of H^+ ions ($a_{H+} = 1$) at pH = 0 and $p_{H2} = 0.1$ MPa. Therefore, this means that every metal with a positive electrode potential may be extracted from the solution by gaseous hydrogen. Even the metals with a slightly negative electrode potential may be extracted using hydrogen. The Nernst equation for a hydrogen electrode has the form

$$E = E^0 + \frac{RT}{2F}\ln a^2_{H^+} - \frac{RT}{2F} \tag{9.10}$$

and at 25 °C

$$E = -0.0591\,\text{pH} - 0.0296\log \text{pH}_2 \tag{9.11}$$

This equation defines the position of the lower broken line in *E*–pH diagrams presented in the section on thermodynamics. The equilibrium potential of the hydrogen half cell decreases, i.e. *E* becomes more negative, with increasing gas pressure and decreasing activity of the hydrogen ion or with increasing pH of the solution. The equilibrium potentials of the metal–ion half cells also decrease with a decrease of the activity of the metallic ion. Figure 9.2*a* shows the change of the potential with the pH value of the hydrogen electrode and with the activity of the ion related to an infinite diluted standard state for metallic electrodes at 25 °C.

The metal ions may be extracted from the solution by hydrogen in the conditions described by the upper part of the diagram, above the hydrogen lines, but it is clear that the concentration of the ions of other metals in the solution increases with increasing electronegativity of the metal, if equilibrium is established by gaseous hydrogen. The increase of the gas pressure to 10 MPa has only a small effect on the concentration of residual ions. The increase of temperature to 200 °C, Fig. 9.2*b*, has a stronger effect but the reduction of strongly electronegative elements, such as Al and Zn, is not possible even at 200 °C and a hydrogen pressure of 10 MPa.

Another complication arises when pH of the acid solution is increased a state may be reached in which oxides or hydroxides start to precipitate from the solution. Critical pH values are indicated of the equilibrium lines of every metal in Fig. 9.2*a*. It is clear that Co and Ni are electronegative elements which may precipitate from acid solutions using hydrogen, but the amount of these metals, obtained from the solution, is not significant. This

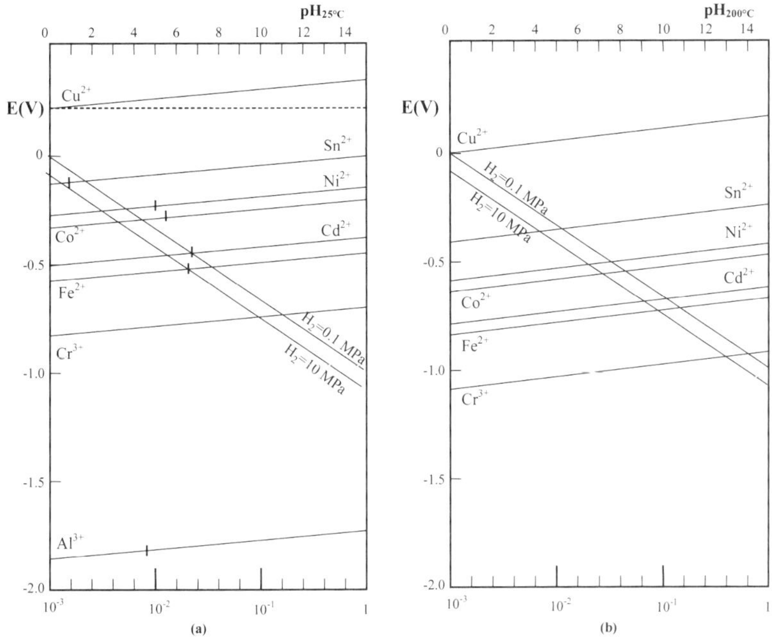

Fig. 9.2. Electrode potential as a function of pH for hydrogen and the function of activity of ions of different metals in a sulphate solution for 25 and 200 °C [2].

problem may be eliminated by leaching in alkaline solutions when it is possible to obtain Co and Ni highly efficiently by reduction with hydrogen from ammonia carbonate solutions. If both these metals are present in the solution, the majority of nickel may be obtained by reduction with hydrogen prior to the start of precipitation of metallic cobalt.

As a result of the high dissociation energy of the hydrogen molecule, the rate of reduction by hydrogen at room temperature and pressure is very low. Therefore, reduction is usually realised at higher temperatures and pressure in an autoclave and the system is mixed in order to increase the contact between the liquid and gaseous phases. A powder metal is also added, the so-called inoculant, to overcome the nucleation barrier in precipitation. This results in a pure metal suitable for, for example, powder metallurgy applications.

Gaseous hydrogen may also be used for precipitating less electronegative impurities from solutions containing more electronegative metals, but the relatively high cost of the gas and the risk of explosion, especially if hydrogen is under pressure, places this method between less extensively used procedures.

9.4. Liquid extraction [3]

The ion exchange in the liquid phase by liquid extraction is an effective method of both cleaning and concentration of ion species in the solution which is to be processed electrolytically. In this method, an organic phase, immiscible in water is added to the aqueous solution of the leaching agent. The phase should be relatively cheap so that petroleum or benzene is used. This organic component contains a chemical compound which forms a compound with the required metallic ion, dissolved in the leaching agent. During extraction of metals, the reagent should selectively separate the required metal ion and the separated metal ion should be easily extractable from the organic phase. This phase should be regenerated and used again in extraction. The reagent and the complex, formed with the metallic ion, should be highly soluble in the organic component and insoluble in the water phase. There is a relatively wide range of extragents available for this purpose. They are divided on the basis of the type of ion for which they are selective in the presence of other impurities, on the basis of the pH of the solution and the type of acid, or alkali, used for leaching the starting material.

The typical structure of the hydroxine reagent LIX, used in extraction of ions from the acid solutions, is shown in Fig. 9.3.

The exchange mechanisms may be described by the scheme:

$$Me^{2+}_{(aq)} + 2RH_{(org)} = R_2Me_{(org)} + 2H^{+}_{(aq)} \quad (9.12)$$

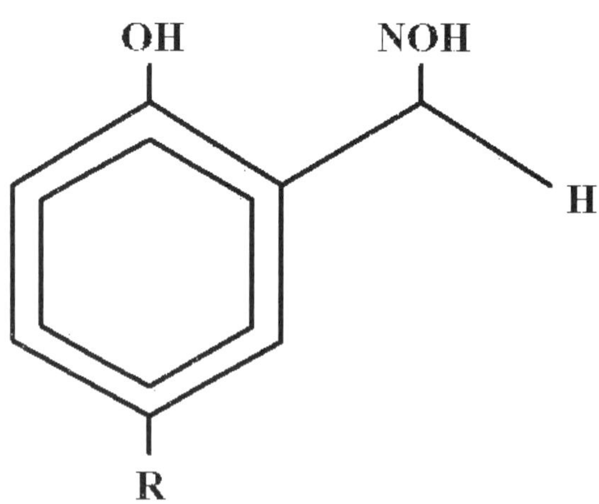

Fig. 9.3. Typical structure of the hydroxine reagent used for liquid extraction of ions from solid solutions $R = C_9H_{19}$ or $C_{12}H_{25}$.

Equation (9.12) shows that the H^+ ions exchange with metallic ions, and the equilibrium state is established by the concentration of the hydrogen ions. The acidity of the aqueous solution during introduction of the organic phase increases during extraction. It is therefore necessary to control the limiting acidity, or the number of the ions in the leaching solution to ensure that pH of the solution is controlled. Every extractant has its maximum efficiency at a specific pH value. For example, the reagent aldoxine shown in Fig. 9.4 with the exchangeable ion H^+, operates better at pH values higher than 2.0. Of course, more acid solutions can also be processed, for example, using the radicals of type CH_3 which replace the exchangeable hydrogen to the form of the keton reagent.

The supplied organic phase with a content of metallic ions is then separated from the aqueous solution and the metallic ions are produced by mixing-in the reagent with the acid solution with a lower pH value which increases a_H, and the reaction (9.12) will run in the opposite direction.

The conditions of introducing and stripping the extractant are shown in Fig. 9.4.

Copper can be extracted completely from the leaching solution naphthalene acid at pH of approximately 4.2, and Cd, Co, Ni and Zn remain in the solution. Copper may be stripped from the organic phase at a pH value of approximately 2.5. Ions of some impurities are also extracted when pH increases to values higher than 5.0, but no copper is extracted when pH drops below 4.0. The Fe^{3+} ions from the solution are also extracted by the naphthalene acid but

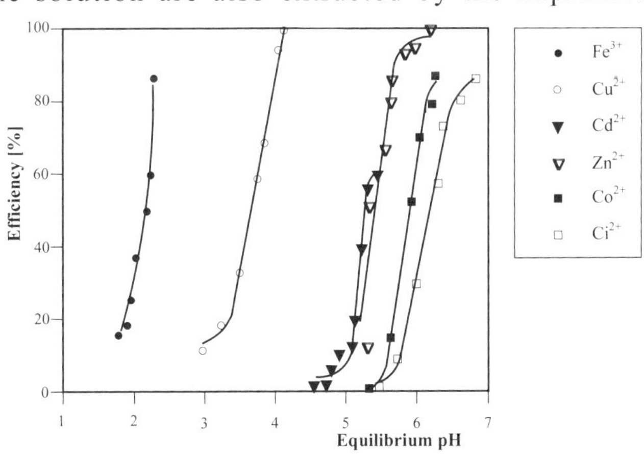

Fig. 9.4. Dependence of the efficiency of extraction of ions in naphthalene acid on pH [4].

remain in the organic phase, if stripping is carried out pH = 2.5. The gradual increase of the amount of Fe^{3+} ions during recycling of the organic phase could reduce the amount of copper which would be extracted in the unit volume and, therefore, iron and similar elements should be removed from the leaching solution prior to the solution making contact with the organic phase.

In some cases, it is difficult to determine the extraction curves of the given metal from the group of the extraction curves of the impurities, as shown in Fig. 9.5.

If it is required to remove the impurity Im completely from the organic phase, the value of pH should not exceed the value indicated by the broken line. Even in this case, less than half of the metal Me can be extracted, even if pH does not decrease during the reaction. The amount of extracted metal increases as a result of transfer of the leaching solution, which is in equilibrium with the organic phase, to one or several additional containers in which it is mixed with the fresh extragent. If the extraction curves of two or three precious metals are located close to each other, it may be financially efficient to extract all these elements simultaneously and then separate them by stripping at different pH values or in different acids. These techniques are used in separating rare metals, such

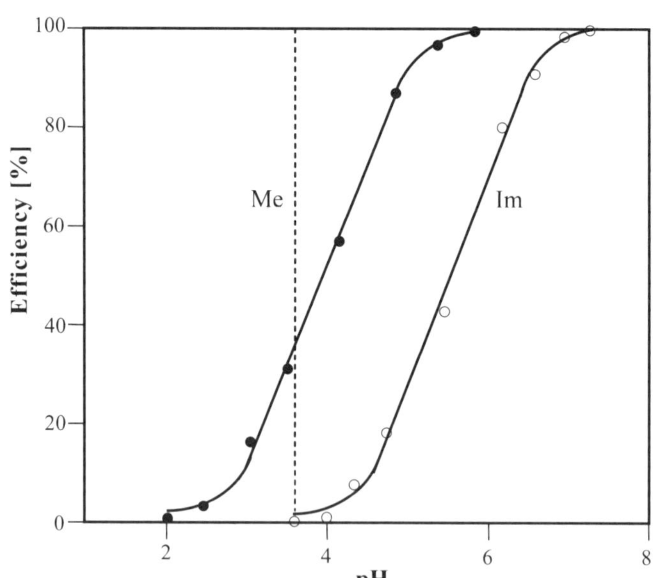

Fig. 9.5. Schematic representation of the effect of displacement of the elements along the pH axis on the purity of metal extracted from the solution by liquid extraction. Me and Im indicate the curves for the metal and the impurity in the same solution.

as Zr and Hf, which are difficult to separate by other methods in the aqueous solution.

There are also reagents which enable selective extraction of one element from the solution in the presence of other valuable elements; in the normal conditions, all these valuable elements are extracted simultaneously using different extraction methods. For example, cobalt is slightly more electronegative than nickel and the majority of its compounds is more stable than the appropriate nickel compounds. On the other hand, cobalt may be extracted from a sulphate leaching solution by substances based on phosphoric acid, for example, Cyanex 272, which leaves the nickel in the aqueous phase [5]. Copper, manganese and zinc would also be extracted by this reagent and, therefore, they must be removed from the solution prior to applying the reagent. Nickel ions are extracted by different reagents, after removing cobalt from the solution.

The reaction described by equation (9.12) considers the exchange of H^+ ions in acid solutions, but the exchange mechanism may also be applied to alkali solutions where, for example, the exchangeable ion is the aminogroup:

$$Me(NH_3)^{2+}_{4(aq)} + 2RH_{(org)} + 2NH^+_{4(aq)} + 2NH_{3(aq)} \qquad (9.13)$$

In other extragents, the exchange mechanism is ensured by an organic pair, for example, in reagents CLD 50 for chloride solutions:

$$Me^{2+}_{4(aq)} + 2Cl^-_{(aq)} + 2R_{(org)} + R_2MeCl_{2(org)} \qquad (9.14)$$

In most cases, liquid extraction is based on adding an organic phase into the solution of a water phase, with continuous intensive stirring. Two or three stirrers are used and they are connected in series to increase the efficiency of extraction of metal ions. The liquid flows into a gravitational cementation container where the organic phase is separated by floating to the surface into two mutually separated liquid phases. The chemical agents used are expensive and it is therefore necessary to separate efficiently both phases with the minimum losses and, on the other hand, the residual amounts of the organic phase, which remain in the leaching solution, could have a detrimental effect on leaching after returning the solution to the leaching cycle.

The volumes of the leaching agent and the organic phase at introduction may be approximately identical. However, if stripping is carried out using a solution of a strong acid or hydroxide, the volume of the extract is lower in comparison with the volume of the leaching agent and the concentration of the metallic ions in the solution appropriately increases.

For a suitably selected extractant and the ratio of the pH values and the volumes of the aqueous and organic phases, the achieved enrichment of the concentration of the ions of the acquired metal could be up to approximately 50:1. At the same time, the ratio of the same metal to the ions of the appropriate impurity in the stripping solution is higher than 5000:1. Usually, these solutions are sufficiently pure and concentrated to be used for direct electrolytic extraction of the metal. However, in some cases it may even be possible to extract the metal directly by precipitation from the organic phase or stripping solution. In the case of copper it is for example difficult to extract copper by acid stripping from Kelox 100 reagent in petroleum, but the metal is easily obtained by the reduction with hydrogen from the organic phase at higher temperatures and pressures in the autoclave [6].

At present, the extent of application of the liquid extraction method, especially in the production of copper and uranium, continuously and rapidly increases. Liquid extraction – electrolysis (SX-EW) is one of the most widely used in hydrometallurgical production of copper.

9.5. Ion exchange

Some natural minerals such as, for example, alumosilicate clays and natural or synthetic zeolites, cellulose, coal, etc., are characterised by a residual electron charge. If these materials are immersed in a solution, the ions of some substances with a suitable charge and size may be trapped on this material. The trapped ions may be subsequently released – eluated, by rinsing with a liquid. These substances are referred to as ion exchangers. They are basically a gel dispersion system in which the dispersion medium is represented by a suitable low-molecular solvent (water in most cases), and the dispersion fraction by the three-dimensional skeleton of the exchanger. From other types of gels they differ by the presence of active groups, also ionogenus, functional or exchange. Active groups are bonded directly to the skeleton, or by a means of another group, such as $-SO_3H$, $-COOH$, $PO(OH)_2$, $-NH_2$, $-NR_2$, etc.

In most cases, metals are extracted using polymeric synthetic resins with cross bonding, for example, polystyrene and acids for bonding functional groups of the type Cl, NO_3, CO_3OH and COOH. The first two operate as exchangeable anions and others as H^+ cations. The bonded ion is positively charged for cation extraction:

$$R^-X^+ + Me^+ = R^-Me^+ + X^+ \quad (9.15)$$

and negatively charged for removing anions

$$R^+X^- + Me^- = R^+Me^- + X^- \quad (9.16)$$

where R is resin and X is the bonded ion. For example, the ions of a uranium complex are produced by the reaction

$$UO_2(SO_4)_3^{4-} + 4RNO_3 = R_4UO_2(SO_4)_3 + 3NO_3^- \quad (9.17)$$

In production, resin is usually in the form of highly porous granules to ensure a large surface area per unit volume. They are placed in vertical columns through which the solution flows, and the granules are gradually saturated with the required metal ions and the solution is depleted. The mother solution is then exchanged for the solution containing ions which eluate the metallic ions from the resin. The ion exchange kinetics increases with a decrease of the granule diameter but this also increases the resistance to the flow of the liquid and the pressure loss. It has been established that the granule diameter of approximately 1 mm is optimum taking these factors into account. Faster exchange and eluation may be achieved by applying resins in the form of fibres which are used at present as cloth filters or infinite bands [7].

Ion exchange is far less efficient in selective extraction of ions of individual metals in comparison with liquid extraction. Generally, it is easier to extract large ions with a large charge than small ions with a small charge. However, a certain degree of selectivity may be achieved by changing the strength of the functional groups. For example, copper may be extracted using an exchanger with a weak cation, but zinc is removed from the solution only by an exchanger with a strong cation. Selectivity may also be achieved by subsequent stripping of the elements using solutions with increasing ionic force.

Usually, liquid extraction is cheaper than ion exchange for solutions with the content of extracted ions greater than 1 g/l, whereas ion exchange is preferred in extracting elements such as gold or uranium from solutions containing only approximately 0.1 g/l of metal ions. Ion exchange is also highly useful in removing heavy metals from effluence prior to transfer to the heaps, since large volumes of the liquid may be cleaned to remove very low residual concentrations prior to the resin reaching the saturated state.

9.6. Precipitation of sparingly soluble compounds

The main quantitative characteristic of the solubility of sparingly soluble compounds is the solubility product. A sparingly soluble compound precipitates up to reaching equilibrium between the produced salt and its ions in the solution. The chemical potential in the equilibrium state is equal to the sum of the chemical potentials of the ions in the solution:

$$\mu^0_{M_mA_n} = m\left(\mu^0_M + RT\ln a_M\right) + n\left(\mu^0_A + RT\ln a_A\right) \tag{9.18}$$

where $\mu^0_{M_mA_n}$ and μ^0_M, μ^0_A are the standard chemical potentials of the solid salt and ions in the solution, and a_M and a_A are ion activities.

The activity of the solid phase is unit and equation (9.18) shows

$$a^m_M a^n_A = e^{\frac{\mu^\circ_{MA_n} - m\mu^\circ_m - n\mu^\circ_A}{RT}} \tag{9.19}$$

The right hand side is constant for the given temperature so that

$$a^m_M a^n_A = L \tag{9.20}$$

or

$$[M]^m[A]^n \gamma^m_M \gamma^n_A = L \tag{9.21}$$

where [M] and [A] are the concentrations of the ions and γ_M and γ_A are activity coefficients.

In diluted solutions containing a sparingly soluble salt, it is valid that $\gamma_M = \gamma_A = 1$ and

$$[M]^m\ [A]^n = L \tag{9.22}$$

If $[M]^m\ [A]^n < L$, the solid salt will dissolve up to reaching the saturation state, and if $[M]^m\ [A]^n > L$, the salt will precipitate up to reaching the equilibrium state.

For sparingly soluble salts, the solubility product is expressed by very small numbers and it is therefore convenient to use logarithmic values, so that pL = –log L. Then

$$pL = -\log L = -m \log a_M - n \log a_A \qquad (9.23)$$

Solubilities M_m and A_n are denoted by S, and the concentration of the ions [M] and [A] will be equal to *mS* and *nS*. Then

$$S^{m+n} = \frac{L}{m^m n^n} \quad \text{or} \quad S = \sqrt[m+n]{\frac{L}{m^m n^n}} \qquad (9.24)$$

Equation (9.24) determines the solubility of salts. The solubility of sparingly soluble salts is affected especially by:

– *Effect of identical ions*

The effect of the residual amount of one of the ions on the solubility of the salt is indicated directly by the solubility product. If the residual concentration of the ion is considerably higher than the concentration given by the solubility of the salt, then the former is negligible. In this case:

$$S = \sqrt[n]{\frac{L}{c_M^m n^n}} \qquad (9.25)$$

where c_M is the residual concentration of the ion M^+.

– *Effect of the concentration of hydrogen ions*

This concentration influences the solubility of salts formed from weak acids or alkali. For a salt of composition MA_n, where A^- is the anion of the principal monobasic acid:

$$MA_n \leftrightarrow M^{n+} + nA^-$$

and

$$L = [m^{n+}][A^-]^n$$

Since this is a very weak acid, it is necessary to consider its dissociation constant

$$HA = H^+ = A^-, \quad K_a = \frac{[H^+][A^-]}{[HA]}$$

If the total concentration A^- in the solution is equal to c_A, then

$$c_A = [A^-] + [HA]$$

If the dissociated fraction of the ions A^- is denoted by α_1, then

$$A\text{–}([A^-]) = \alpha_1 c_A$$

from which it follows that

$$L = [M^{n+}][A^-]^n = [M^{n+}]\alpha_1^n c^n_{A^-} \tag{9.26}$$

Expressing this by a means of K_a and concentration $[H^+]$ gives

$$\alpha_1 = \frac{[A^-]}{[A^-]+[HA]}; \qquad [A^-] = \frac{K_a[HA]}{[H^+]}$$

$$\alpha_1 = \frac{K_a[HA]}{[H^+]\left(\frac{K_a[HA]}{[H^+]}+[HA]\right)} = \frac{K_a}{K_a^+[H_+]} \tag{9.27}$$

Substitution α_1 in equation (9.26) gives

$$L = [M^{n+}]\left(\frac{K_a}{K_a^+[H^+]}\right)^n c^n_{A^-} \tag{9.28}$$

This relationship may be used to calculate the solubility of the salt of a weak monobasic acid at the given concentration of the hydrogen ions.

– ***Effect of hydrolysis of the cation***

The cations of the metals are extensively hydrolysed in an aqueous solution and this affects the solubility of the precipitated compounds. In a simple case, the cation hydrolyses

$$M^{n+} + H_2O \leftrightarrow MOH^{n-1} + H^+; \qquad K_1 = \frac{[MOH^{n-1}][H^+]}{[M^{n+}]}$$

$$MOH^{n-1} + H_2O \leftrightarrow M(OH)_2^{n-2} + H^+; \qquad K_2 = \frac{[M(OH)_2^{n-2}][H^+]}{[MOH^{n+}]}$$

The total concentration of the element M is c_M and

$$c_M = [M^{n+}] + [MOH^{n-1}] + [M(OH)_2^{n-2}] \tag{9.29}$$

The solubility product for type MA_n compound is

$$L = [A^-]^n c_M \alpha$$

where α is the fraction M which is in the ionic state M^{n+}

$$\alpha = \frac{[M^{n+}]}{[M^{n+}] + [MOH^{n-1}] + [M(OH)_2^{n-2}]} \tag{9.30}$$

Expressing concentration by a means of $[H^+]$ and dissociation constants

$$\alpha = \frac{[H^+]^n}{[H^+] + K_1[H^+]^{n-1} + K_1K_2[H^+]^{n-2} + \ldots K_n} \tag{9.31}$$

where K_1 and K_2 are the constants of progressive hydrolysis.

However, in many cases, hydrolysis is far more complicated than previously thought. The hydrolysis of Fe^{3+} (this is a very frequent case in hydrometallurgy) may obtain a number of species, in addition to $Fe(OH)^{2+}$, $Fe(OH)^+_2$ and $Fe_2(OH)^{4+}_2$, forms as well. The hydrolysis of Al^{3+} is even more complicated. In addition to the ions $Al(OH)^{2+}$ and $Al(OH)^+_2$ the solution also contains polymeric anions $Al_2(OH)^{4+}_4$, $Al_3(OH)^{3+}_6$ and larger complexes, for example $Al_{13}(OH)^{7+}_{32}$ [8].

– ***Effect of the formation of complexes***

In the presence of a stable complex it is possible that the metal ion is not precipitated in the solid phase. This depends on the mutual ratio of the values of the solubility product and the constants of the formation of the complex . At the defect of the complexing anion *X*, equilibrium is achieved in the system during the formation of the complexes MX and MX_2

$$MA_{tuhe} \leftrightarrow M + A; \qquad L = [M][A]$$

$$M + X \leftrightarrow MX; \qquad \beta_1 = \frac{[MX]}{[M][X]}$$

$$MX + X \leftrightarrow MX_2; \qquad \beta_2 = \frac{[MX_2]}{[MX][X]} = \frac{[MX_2]}{\beta_1[M][X]^2}$$

The fraction α of the ions M not bonded in the complex is:
The total concentration M is

$$c_M = [M] + [MX] + [MX_2] = [M] + b_1[M][X] + b_1 b_2[M][X] \quad (9.32)$$

or

$$\alpha = \frac{[M]}{c_M} = \frac{1}{1 + \beta_1[X] + \beta_1\beta_2[X]^2} \quad (9.33)$$

This gives

$$L = [M][A] = \alpha\, c_M[A] \quad (9.34)$$

In the absence of the common ion, the solubility of the salt is given by:

$$S = c_M = [A]; \qquad L = \frac{\alpha S^2}{\sqrt{\frac{L}{\alpha}}} \quad (9.35)$$

9.7. Crystallisation

A compound crystallises from the solution only if its concentration is higher than the solubility at the given temperature, i.e., the solution must be supersaturated. There are two types of crystallisation: isothermal and isohydric. Isothermal crystallisation occurs at constant temperature and the supersaturation of the solution is reached by evaporation or by another method of removing the solution. Isohydric crystallisation is realised by cooling the solution, and the amount of the solvent does not change. Supersaturation of the solution occurs as a result of a decrease of the solubility of the salt due to decreasing temperature.

The equilibrium crystallisation conditions are expressed by the solubility diagram. The effect of pressure on equilibrium between the liquid and the solid phase in this case is very small and is therefore disregarded. The phase rule in crystallisation has the form:

$$S = K + 1 - F$$

where K is the number of components, S is the number of degrees of freedom, F is the number of phases.

Two methods of presenting the solubility diagrams have been accepted for the salt-water two-component system: by a means of the solubility polytherm and using the composition-temperature solubility diagram.

Temperature is an independent variable in the polytherm, and the dependent variable is the solubility of the salt. If the curves are smooth, no polymorphous transformation take place in the system, and crystal hydrates do not form. The solubility usually increases with increasing temperature or, in a small number of cases, it may decrease. If a chemical reaction takes place between the salt and water with the formation of crystalline hydrates, the solubility polytherm shows an inflection point. This is observed in, for example, the solubility polytherm of ferrous sulphate, Fig. 9.6.

In the composition–temperature solubility diagram, composition is the independent variable and temperature the dependent variable. The salt-water systems are characterised by eutectic diagrams or by the formation of a chemical compound which dissolves and also breaks down, or dissolves without dissociation. A suitable example is the solubility diagram of the $MgCl_2$–H_2O system, shown in Figure 9.7. This diagram is used efficiently in processing sea water for production of $MgCl_2$ for electrolytic production of magnesium.

A supersaturated solution must be prepared for precipitating crystals of the given substance from the solution. The supersaturated solution may be produced by slow cooling close to saturation of the solution. This method is used for substances whose solubility increases with increasing temperature. The solvent is evaporated at atmospheric or reduced pressure, or a chemical reaction leading to the formation of a new compound, whose concentration is higher than its solubility in the given solution, is used. Salt displacement may also be considered, i.e., a substance which reduces the solubility of the compound present in the solution, is added to this solution. For example, this is a compound with a common ion and the dissolved substance or an addition of alcohol

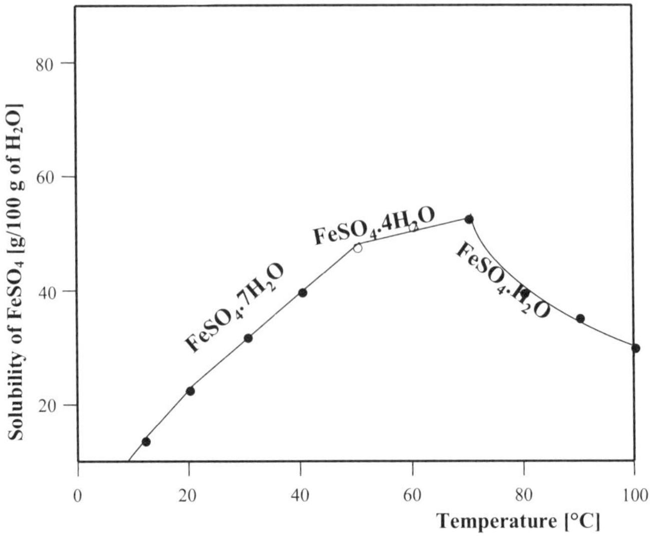

Fig. 9.6. Solubility polytherm of ferrous sulphate in water.

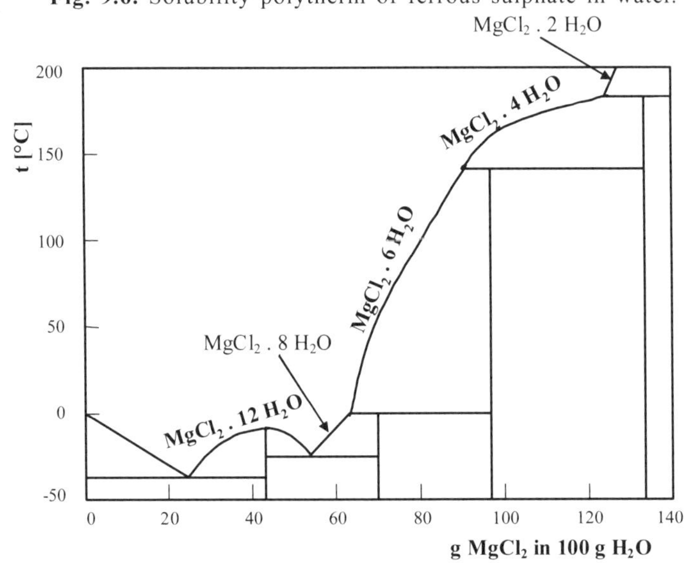

Fig. 9.7. Solubility diagram in the $MgCl_2$–H_2O system.

which bonds water and causes salt precipitation.

9.8. Electrolysis and electrolytic refining

Metals may be extracted from aqueous solutions by electrolysis (electrowinning) or they may be electrolytically refined to remove impurities (electrorefining). The principles of these processes are based on considerations of the galvanic cell and have been described in detail in the appropriate literature, for example [8-11]. Suitable examples are metallic copper and zinc electrodes, both immersed in an electrolyte formed by the aqueous solution of sulphuric acid. Zinc and copper are soluble in this acid. The reactions of half cells may be expressed as follows

$$Cu^{2+} + 2e^- = Cu^0 \qquad E^0 = +0.34\ V \qquad (9.36)$$

$$Zn^{2+} + 2e^- = Zn^0 \qquad E^0 = -0.76V \qquad (9.37)$$

If both metals are connected by an external electrical circuit, the electrons are transferred from zinc on to copper because zinc is more electronegative in comparison with copper. The transfer of electrons from zinc causes the latter to dissolve and zinc leaves the electrode in the form of Zn^{2+} ions and is transferred into the solution. Therefore, the reaction (9.37) takes place in the opposite direction. The electrons, arriving at the copper electrode, are discharged by transformation of the Cu^{2+} ions to Cu^0 atoms which settle on the copper electrode so that the reaction (9.36) also takes place in the direct direction. The reaction of the entire cell is

$$Zn + CuSO_4 \rightarrow Cu + ZnSO_4 \qquad (9.38)$$

This shows that the zinc electrode receives a negative charge and becomes an anode, whereas copper is a positively charged cathode. It has therefore been assumed that the current flows in the direction opposite to that of the electrons, i.e., from the cathode to the anode.

If an electrolyte contains ions or both metals at unit activity and the electrodes are connected through a high external resistance causing that the reactions are slower and reversible, the maximum voltage which can be generated in the cell is given by the sum of the electrode potentials of two half cells:

$$E = E^0_{Cu} - E^0_{Zn} = +0.34 - 0.76 = +1.10\ V \qquad (9.39)$$

If a voltage potential slightly higher than 1.10 V is introduced between the electrodes, then the copper electrode is more negative than the zinc one and theoretically the reactions can be reversible, i.e., copper is dissolved on the negatively charged pole (anode), and zinc precipitates on the positively charged cathode. However, in practice, the reaction taking place in the reverse direction requires a higher potential difference and, consequently, other undesirable factors could enter the process.

The sequence of the naturally occurring direct and induced reverse reaction corresponds to the charging and discharging cycle of the electrical battery. The discharge stage is in fact the main stage of obtaining metal by cementation, as already mentioned. Basically, the charging stage is similar to reactions which take place during electrolytic precipitation. If different metals are replaced by the electrodes produced from the same metal, one from the pure metal and the other one from the metal contaminated with impurities, the standard electrode potential of the cell is almost zero, depending on the deviation of the activity of the metal of the contaminated electrode. Adding a very low direct voltage potential in order to shift the potential of the contaminated electrode to more negative values should, according to theory, result in the dissolution of the metal of the contaminated electrode and in displacement to the pure metal electrode where it will precipitate. However, in practice there must be again a high voltage between the two electrodes to ensure the reaction. The elements that are insoluble in the electrolyte remain in the solid state as sludge, whereas more electronegative elements remain in the solution. This is the principle of electrorefining.

9.8.1. Electrowinning of metals

Almost all metals can be produced from their ores by electrolytic precipitation. However, in reality, the amount of metals produced by this procedure is very small in comparison with pyrometallurgical methods but on the other hand the total amount of the individual metals produced by the electrolytic method differs. This is caused by the overall economic parameters of the process and by the requirements imposed on the produced metal. In most cases, non-ferrous metals are produced by electrolytic precipitation because the impurity of the product must be very high.

Electrolysis from aqueous solutions

The electrolytically produced metals are extracted in most cases from a sulphuric acid solution. Metallic copper, as a typical example, may be produced from a sulphate leaching solution by electrolytic precipitation of the metal on the inert electrode prepared from lead sheets and anodes of contaminated copper. This process cannot be described by the reactions (9.36) and (9.37) because according to these reactions the anode should be consumed and the electrolyte should be greatly enriched with the ions of the anode metal at copper precipitation on the cathode. It is therefore necessary to use anodes from electrochemically inert metal which does not dissolve in the electrolyte. Lead is suitable for this because it does not dissolve in sulphuric acid, although it has a higher electronegativity than copper.

If the anode does not dissolve, the electrons, used for precipitation of copper, must be obtained from another anodic reaction. They may be replaced by ionisation of water

$$H_2O \rightarrow 2H^+ + \frac{1}{2}O_2 + 2e^- \qquad E^0 = 1.23 \quad V \qquad (9.40)$$

and this reaction is also accompanied by the release of gaseous oxygen on the anode. H^+ ions are bonded with $SO^{2-}{}_4$ anions which diffuse through the electrolyte towards the anode and they regenerate the acid so that the solution may be recycled for use in leaching or liquid extraction after obtaining the given metal from it. However, on the other hand, the presence of H^+ ions in the solution enables an additional cathodic reaction

$$2H^+ + 2e^- = H_2^- \qquad E^0 = 0 \text{ V} \qquad (9.41)$$

with the release of gaseous hydrogen on the cathode. Since the reaction of release of hydrogen acts against precipitation of the metal taking into account the electrons that are available, the amount of precipitated metal by the transfer of the given amount of electrical energy decreases with increasing amount of released hydrogen. If copper electrolytically precipitates, the standard electrode potential is more positive for the copper half cell than for the release of gaseous hydrogen, and copper precipitates preferentially from $CuSO_4$ solution. The amount of generated hydrogen may increase greatly only in electrowinning of more electronegative elements lie zinc, chromium, etc.

Both these reactions include the change of the concentration of the H^+ ion, and this shows that the electrode potentials depend on the pH of the solution. If the activity of the water is unit and p_{O2} in equation (9.40) is equal to 0.1 MPa, the equilibrium potential is defined as a function of pH by the Nernst equation

$$E = 1.23 - \frac{RT}{2F}\ln a_{H_+}^2 = 1.23 - 0.0591\text{pH} \qquad (9.42)$$

and, similarly, for the equation (9.41)

$$E = 0 - 0.0591 \text{ pH} \tag{9.43}$$

The relationships (9.42) and (9.43) describe the position of the dotted lines in the potential–pH diagram described in the part concerned with thermodynamic studies. In addition to these lines, Fig. 9.8 also shows the standard electrode potentials of some metals. However, the positions of these lines are displaced with a change of the partial pressure of hydrogen or the change of the activity of the components taking part in this reaction.

Table 9.3 gives the values of the standard electrode potentials of some metals at 25 °C.

The equilibrium between the metal and its ions in the solution does not depend on the pH of the solution which means that the Me–Me^{2+} interface is represented by a horizontal line. This shows that the equilibrium voltage which must be exceeded in order to precipitate the metal on the cathode and release gaseous oxygen on the anode, decreases with increasing pH of the solution. The possibility of the formation of gaseous hydrogen on the cathode in precipitation of electronegative elements also decreases with

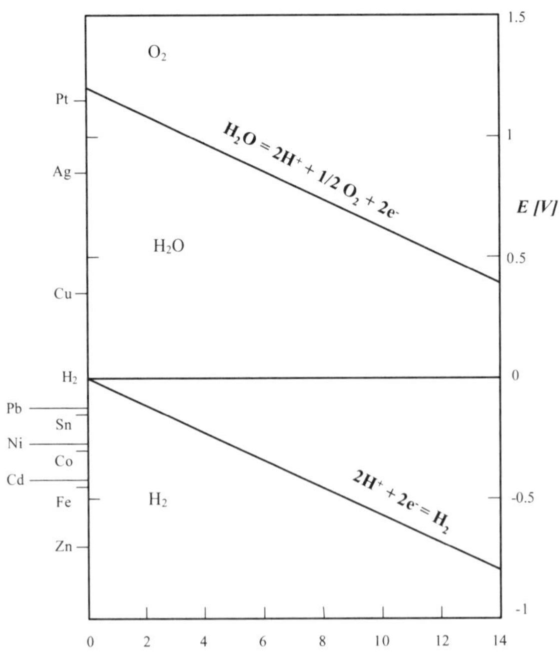

Fig. 9.8. Area of stability of water as a function of potential and pH of the solution.

Table 9.3. Standard electrode potentials of metals at 25 °C

Couple	Potential [mV]	Couple	Potential [mV]
Li/Li^+	–3.04	Co/Co^{2+}	–0.28
K/K^+	–2.92	Ni/Ni^{2+}	– 0.23
Ca/Ca^{2+}	–2.87	Sn/Sn^{2+}	–0.14
Na/Na^+	–2.71	Pb/Pb^{2+}	–0.13
Mg/Mg^{2+}	–2.37	**$H_2/2H^+$**	**±0.00**
Mn/Mn^{2+}	–1.18	Cu/Cu^+	+0.34
$2H_2O/H_2 + 2OH^-$	–0.83	$2Hg/Hg_2^{2+}$	+0.79
Zn/Zn^{2+}	–0.76	Ag/Ag^+	+0.80
Cr/Cr^{3+}	–0.74	Hg/Hg^{2+}	+0.85
Fe/Fe^{2+}	–0.56	Pt/Pt^{2+}	+1.20
Fe/Fe^{3+}	–0.44	$Cl_2/2Cl^-$	+1.36
Cd/Cd^{2+}	–0.40	Au/Au^+	+1.50
Ti/Ti^{2+}	–0.34	$F_2/2F^-$	+2.87

increasing pH, which is again indicated by the *E*–pH diagrams. However, these diagrams also show that there is a limit up to which pH can increase for every considered metal, in addition to the fact that the risk of formation of the oxide, hydroxide or some other salt of the given metal increases. In addition to this, water is a very poor conductor at pH = 7.0 and the resistance of the acid electrolytes increases with increasing pH. The pH of the solution is not constant but gradually decreases with the release of H^+ ions in accordance with the equation (9.40), together with the precipitation reaction.

The concentration and, therefore, activity of the metal ions in the solution decreases with increasing amount of the metal precipitated on the cathode. This increases the equilibrium voltage which must be exceeded for the reaction to take place, but this effect is very weak in comparison with the effect of the change of pH on the hydrogen release reaction. For example, the standard electrode potential of pure copper in equilibrium with 1 M solution of Cu^{2+} is +0.34 V.

If the activity of copper ions decreases to 0.1 as a result of precipitation of the metal, then according to the Nernst equation the electrode potential changes by

$$E = E^0 + \frac{RT}{2\mathrm{F}} \ln 0.1 = +0.34 - 0.0591 = +0.28 \text{ V} \qquad (9.44)$$

The equilibrium potential decreases further with gradual decrease of the concentration of the ions in the solution but in practice the recirculation of the electrolyte starts prior to the moment at which all the ions precipitate on the cathode.

The voltage required for the precipitation reaction differs from the equilibrium potential corresponding to the Nernst equation. The actual voltage must be higher in order to overcome various internal resistances of the process. In most cases, these are ohmic resistances at inputs of electric voltage and electrical contacts of the cathode and the anode. These resistances may be minimised by efficient arrangement and operation of the electrolysers and by controlling the conductivity of the electrolyte.

Another loss of the potential inside the electrolysis baths is caused by polarisation. If the ions are discharged by the passage of current, the electrolyser ceases to behave in the thermodynamically reversible manner.

Consequently, the electrode potential of the reactions at the half cells is not equal to the values determined by reversible conditions. The change of the electrode potential is caused by polarisation and the magnitude of the change is represented by the symbol η which is referred to as the overvoltage. Two main components, concentrational and activation polarisation, should be considered here.

The first expression is regarded simply as a potential decrease as a result of the change of the ion concentration in the region between the anode and cathode. The electrolyte circulates between the electrodes in order to reduce the concentration gradients and cause the ions to separate inside the volume of the solution by mass transfer. However, precipitation of the metal on the cathode reduces the ion concentration at the liquid–solid phase interface leading to a rapid change of the composition in the layer in the vicinity of the interface, similar to the concentration gradient through the boundary diffusion layer in the liquid-liquid phase reactions. Transfer takes place by diffusion through a relatively static layer. Consequently, the concentration of the metal ions may decrease to a very low level above the equilibrium value in the volume of the solution in order to overcome the concentration polarisation.

Activation polarisation forms as a result of the effect of the required activation energy of the atom in the solid anode on its transfer in the ionic form to the solution and its re-precipitation on the cathode.

One of the conditions for partial dissolution of the ion in the immediate vicinity of the electrode surface is the formation of an activated complex in transfer of the atom to the ionic state. If the electrode is in equilibrium with the electrolyte, the free energies of the metal atoms in the solid and ionic states in the solution are identical. The activation energy of transfer is the same in both directions (i.e. from the liquid to solid phase and from the solid phase to liquid) so that the dissolution rates and precipitation rates on the electrodes are identical. However, the electrode potential becomes polarised, with the overvoltage η, whilst the equilibrium is being disrupted and the atoms are displaced preferentially to (cathodes) or from (anodes) the electrode. The value of η depends on several factors, including the composition of the electrodes and the electrolyte and the reactions taking place on the electrode.

Activation polarisation on the anode increases ΔG and also the energy of the anode metal by the value $zF\eta$ and also $\Delta G = -zFE$. The energy of the activated complex also increases by the value $azF\eta$, where a is the factor of symmetry defined by the position of the maximum in the reaction co-ordinate, Fig. 9.9. The reversed conditions, with a different value of η, relate to the cathode.

The current in the electrode as a result of this polarisation may be described by the equation [12]

$$I = I_a e^{\left[\frac{(1-a)zF\eta}{RT} - \frac{-azF\eta}{RT}\right]} \tag{9.45}$$

where I_a is the difference between the supplied and outgoing current of the electrode, or the change of the current density. This is the so called Butler–Volmer relationship. Since all the expressions in the equations are constants with a given electrode process at constant temperature and composition, the equation is used in the following form:

$$\eta = b \log I_0 \pm b \log I \tag{9.46}$$

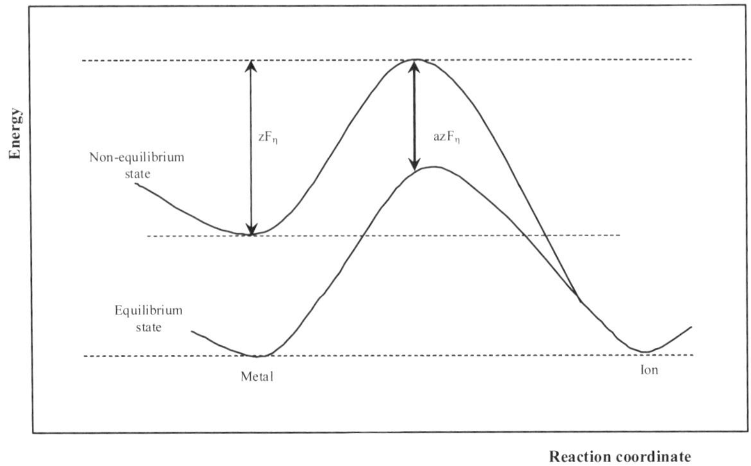

Fig. 9.9. Effect of activation polarisation on the activation energy of electrolytic reactions.

where $b = \dfrac{2.303RT}{(1-a)zF}$

which is known as the Tafel equation. Constant b, which determines the angle of inclination η against log I, is the Tafel slope of the process. The plus sign in equation (9.46) indicates anodic overvoltage, whereas the minus sign indicates activation polarisation on the cathode.

One of the stages of the anodic reaction is the release of oxygen in the gaseous form using an inert anode. The nucleation barrier must be overcome prior to the formation of gaseous bubbles and the gases then not released into the atmosphere when the voltage increases above the equilibrium value. The excess energy, required for release of oxygen is anodic overvoltage which is in fact anodic polarisation for the release of gaseous oxygen which is usually considerably greater than the energy required for dissolution of precipitation and forms an important part of the overall requirements on voltage. Many methods are used to reduce the overvoltage, including coating of anodes with platinum or other noble metals that catalyse the generation of gas (this is very expensive), or additions of sulphite compounds which oxidise to the sulphate with release of electrons:

$$H_2SO_3 + H_2O \rightarrow SO_4^{2-} + 4H^+ + 2e^- \qquad E^0 = 0.17\,\text{V} \qquad (9.47)$$

or using anodes with diffusion of hydrogen to compensate for the dissociation of water by the hydrogen oxidation reaction.

The Nernst equation shows that the voltage required theoretically for the precipitation of copper from 1 M solution is (+1.23–0.34) = +0.89. However, in reality, this voltage contains the Nernst component, the anodic and cathodic overvoltage and voltage required for overcoming ohmic resistance and electrolyte resistance:

$$V = E + \eta_{(anode)} + \eta_{(cathode)} + IR \tag{9.48}$$

This shows that the electrolytic precipitation of copper requires a voltage of 2.0–2.5 V. It may be seen that the efficiency of utilisation of energy is low.

Figure 9.8 shows that the standard electrode potentials of Ag, Cu and Pt are positioned above the hydrogen line and these metals may be precipitated from the solution without releasing gaseous hydrogen. Silver and platinum, together with gold and other electropositive noble metals, are insoluble or only slightly soluble in sulphate solutions. Traces of silver in the solution can also be easily precipitated by low concentrations of chloride ions in the electrolyte. These metals originate from ores and do not precipitate with other metals during electrolytic precipitation.

Other common metals are more electronegative than the standard hydrogen electrode. The released gaseous hydrogen consumes preferentially free electrons in comparison with the precipitated metal in the equilibrium conditions of electrolytic precipitation of these metals from aqueous solutions. Fortunately, in practice the equilibrium is not reached because cathodic release of gaseous hydrogen also requires an overvoltage for nucleation of hydrogen. This overvoltage increases with increasing temperature of the solution and with increasing current density. The voltage required for the release of hydrogen also changes greatly depending on the precipitated metal. Lead, tin and zinc cathodes are very weak catalysts for the reaction of the gas and zinc with the voltage of the half cell E^0 = –0.76 V can be extracted electrolytically from the solution without the formation of any significant amount of hydrogen. However, other metals, such as Co, Fe, Ni and Sb are less electronegative than zinc and more difficult to extract because they have higher values of polarisation activation and operate as better catalysts of gas formation.

In electrolytic precipitation of electronegative elements the extent of release of hydrogen may decrease with the increase of pH and, consequently, the acidity of the solution is maintained on the lowest level at which the metal ions remain stable in the solution. In precipitation of nickel from its sulphate solution, at least 15% of the supplied energy is lost in the generation of gaseous hydrogen at the lowest practical value of pH. Manganese with $E^0 = -1.1$ V is the most electronegative element which may be produced by electrolysis from aqueous solutions, although with low efficiency, using solutions with pH of approximately 9.9 which is the lowest values at which $Mn(OH)_2$ does not precipitate from the solution. The elements with high electronegativity, for example, aluminium, alkali elements and alkali rare earth metals, cannot be produced by electrolysis from aqueous solutions, not even highly basic aqueous solutions.

As a result of the formation of a rough cathode surface, the overvoltage of hydrogen decreases for every precipitated metal because this leads to the formation of cracks and crevices in which bubbles can easily form. To prevent the formation of bubbles, very small quantities (often in ppm) of organic substances, such as glue, and thiourea are added to the electrolyte. These substances settle preferentially on growing projections, protruding from the cathode and, therefore, inhibit further growth of the cathode in these areas. Consequently, there is sufficient time for the growth of the cathode area in other regions so that the resultant surface is smoother. Excessive amounts of these additions inhibit settling on the entire surface and, consequently, the voltage for the reaction would have to be increased. This shows that the optimum concentration of these substances in the electrolyte must be strictly controlled. Another possibility is the electrolytic polishing of the cathode surface by periodic switching of the voltage for several seconds. This results in the formation of a smooth surface even in electrolytic metallising using alternating partially rectified current.

Copper is more electropositive in relation to the usual impurities dissolved in the sulphate solution in the given pH range, it is evident that the impurities will not precipitate from the solution together with copper on the cathode, if the voltage determined for the precipitation of copper is used. This produces electrolytically the pure metal with the impurity content expressed in ppm. In the other extreme case, all electronegative impurities precipitate before zinc, as they are present in the $ZnSO_4$ electrolyte, so that the solution must be cleaned prior to electrolysis to produce pure zinc. It is very important to remove all the impurities because they may catalyse the reaction of the formation of gaseous hydrogen and, consequently, reduce cathodic overvoltage. The only impurity which may be tolerated in the zinc solution is manganese, but in this case must be ensured that the oxygen overvoltage is not sufficient for the precipitation of MnO_2 on the anode. It is not necessary to remove all the traces of impurities whose electronic activity is similar to that of the precipitated metal and they do not affect the hydrogen overvoltage, because the steady potential of precipitation of the impurity decreases with a decrease of its concentration in the solution.

According to the first Faraday law, the charge required for the precipitation of one gram equivalent of the element, dissolved in the electrolyte at 100% efficiency, is 1 Faraday, i.e. 96.487 C. One gram equivalent is equal to the atomic mass of the element divided by its valency at which it is present in the solution. For bivalent copper ions with the atomic mass of 63.57 one gram equivalent is equal to 31.79.

The second Faraday law defines the maximum amount W of the element which may be electrolytically precipitated during time t in seconds

$$\frac{dW}{M} = dm = \frac{1}{nF} dt \tag{9.49}$$

where M is the molecular mass, m is the number of mols, I is current and n is the valency of the ion. This means that if the efficiency of the electrolyser is 100%, the transfer of 1 A in 1 hour generates $60 \cdot 60/n \cdot 96487$ mols. The theoretical amounts which may be precipitated by this amount of energy are:

Metal	Co	Cu	Fe	Ni	Pb	Zn
Grams	1.100	1.184	0.695	1.095	3.864	1.219

This is also shows that 1 gram equivalent of metal is precipitated by a current of 1 A in 96487 seconds, or 26 hours and 48 minutes. Of course, the rate of precipitation increases with increase of the current density, i.e. A m^{-2} of the electrode.

However, the amount of precipitated metal in practice is smaller than the theoretical amount because part of the energy is used for undesirable reactions, for example, generation of gaseous hydrogen. The main undesirable loss of energy is caused by the presence of iron ions in the solution. Iron is more electronegative than copper, lead, tin and nickel and remains partly or completely in the solution during electrolysis of these metals. However, iron can be oxidised to Fe^{3+} on the anode and reduced to Fe^{2+} on the cathode, and the electrons are consumed in the process and the amount of precipitated metal decreases. For this reason, the maximum rational amount of Fe^{2+} ions in the solution in copper electrolysis is approximately 0.06 M/l Fe^{2+}.

Current efficiency is usually defined as the precipitated amount of metal multiplied by 100 and divided by the amount which should precipitate theoretically according to Faraday's law. In electrolysis from aqueous solutions, the current efficiency is usually between 80 and 95% but may decrease to 70%, if the iron content of the solution is high and may be even lower if gaseous hydrogen also forms during precipitation of the metal.

Electrolysis of metals from aqueous solutions in practice

A typical electrolyser contains approximately 50 anodes and cathodes, each with the area of approximately 1 m^2, arranged in a series. The decrease of voltage in the electrolyser is then equal to the decrease of voltage in the individual anode-cathode pairs. The electrolysers are connected in parallel and the buzz bar which supplies power to the cathodes in an electrolyser is also connected to the anodes in the adjacent electrolyser, so that the overall decrease of voltage in the system is equal to the direct voltage supplied by the generator. To produce nickel or zinc, this voltage is approximately 3.5 V and for copper it is around 2.5 V. The electrolyte flows from vat to vat under gravity.

In most cases, the electrolyte is heated to 50 °C to reduce its resistance. Current density is usually in the range 100–500 Am^{-2}. According to Faraday's law, the precipitated amount should increase linearly with the increase of current density, but the voltage required for overcoming activation and concentration polarisation in the electrolyte also increases with the increase of the rate of precipitation and this increases the overall consumption of energy for every unit of the metal precipitated from the solution. The increase of current density also increases the roughness of the cathode surface thus increasing the risk of generation of gaseous hydrogen and other losses.

The anodes are produced usually from lead alloys, but their inertness in the strong acid solutions, for example, for liquid extraction of copper, does not have to be very high. The pH of the solution must be increased by dilution in such cases. Surface oxidised sheet of titanium or titanium alloys or sheets coated with oxides of noble metals, or with titanium carbide for solutions with low pH, are used quite often. TiC is a good conductor and both Ti and TiC are passivated in anodic polarisation. The cathode sheets are produced from thin sheets of precipitated metal or from other cold-rolled metal to ensure its resistance to shape changes. Consequently, the copper can be precipitated on titanium sheets electrolytically coated with copper. Aluminium sheets are used in electrolysis of zinc. The cathodes are replaced by fresh starting sheets when their thickness is several centimetres and the gap between the anode and the cathode is too small for efficient circulation of the electrolyte.

The gap between the anode and cathode is usually 20–50 mm and decreases during precipitation of the metal on the cathode. To ensure the required potential gradient in the electrolyte, the initial distance is as small as possible in order to permit the increase of the thickness of the cathode by precipitation of the metal without any risk of short circuiting because of irregularities or distortion of the surface, and this should permit efficient circulation of the electrolyte between the anode and cathode areas. However, the gap must increase if it is necessary to separate every anodic cathode by a porous membrane or diaphragm in order to prevent the transfer of solid substances or impurities to the surface of the cathode. The diaphragm is also used in electrolysis of nickel to separate the acid anolyte solution from the less acid catholyte to reduce the risk of generation of gaseous hydrogen.

Sulphate solutions are used mostly for electrolysis from aqueous solutions but some elements, such as Co, Ni and Zn can also be produced easily from chloride solutions. Copper is more difficult to obtain by this procedure because copper forms chloride anions in the solution. It is also quite difficult to prevent the growth of dendrites on the cathode and the release of gaseous chlorine on the anode.

9.8.2. Electrolytic refining

The vat for electrolytic refining is similar to that used for electrolytic precipitation but in this case the inert anodes are replaced by plates of impure metal (several centimetres in thickness) which is to be refined. The anodes are replaced after they have dissolved to a large degree and have become too thin so that they may break up. The cathodes are produced from the starting sheet of the metal to be refined or sheets of stable material, coated with the refined metal. In normal conditions, the

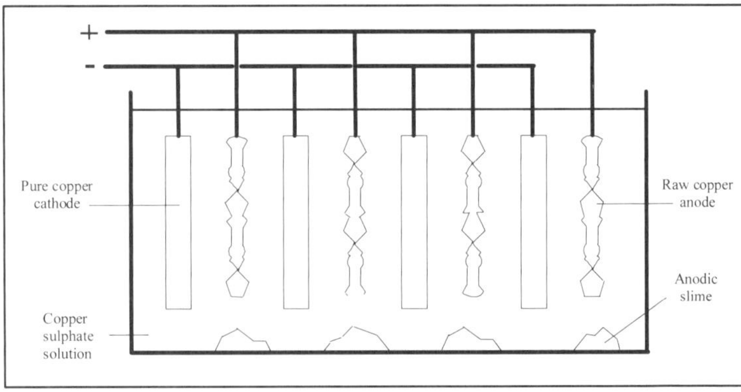

Fig. 9.10. Scheme of copper electrorefining.

Fig. 9.11. A copper electrorefining plant.

electrolyte is an acid solution in which the anode metal may dissolve but which in the ideal case does not dissolve the impurities present in the material.

Figure 9.10 shows schematically copper electrorefining and Fig. 9.11 shows a plant for electrolytic refining of copper.

The reversible potential of the vat is equal to zero if the anode and the cathode are produced from the same metal. The anode is not pure, and with the exception of nickel anodes, the content of the impurities seldom exceeds 1% and the content of impurities is often considerably lower so that the activity of the main metal on the anode is very close to the unit activity and the reversible potential is very low. As in electrowinning of metals, a high potential is required for overcoming other resistances in the circuit. Since the electrons arriving on the anode are consumed by the formation of metal ions, they do not take any part in the generation of oxygen as in electrowinning of metals. In this case, the largest proportion of the potential is usually represented by the resistance of the electrolyte. The solutions are usually heated to 50–70 °C to increase conductivity. The voltage for refining of copper is the range 0.2-0.25 V and the voltage for refining relatively contaminated nickel is 1.5–2.0 V.

Electrorefining is suitable mainly for removing the impurities such as bismuth, which has a highly detrimental effect on the properties of the metal and cannot be removed from metals such as copper or lead by pyrometallurgical procedures, and these impurities are also insoluble in the electrolyte. It is also suitable for removing impurities from remelted metals which are reduced by carbon to the same or greater extent. Gold, silver and other noble metals can be extracted from the ore by pyrometallurgical procedures. The gold content of copper anodes is usually 8–73 ppm, the silver content 90–7000 ppm. These valuable components are either insoluble in the electrolyte or if they are soluble then they can be easily precipitated by cementation because they are more electropositive than all other conventional metals. The insoluble noble metals concentrate in the anode sludge on the bottom of the electrolyser together with the compounds of the elements such as Se, Te and they are produced by different procedures. Consequently, the electrorefining process becomes economically more efficient.

The metals are refined mostly in electrolytes containing sulphuric acid but other solutions may also be used. Lead is insoluble in sulphuric acid so it is refined in the solutions of hydrofluoric acid, H_2SiF_6 where it is present as $PbSiF_6$.

In addition to silver and gold, the copper anodes may also contain nickel, lead and oxygen in concentrations of around 0.4%. Complex oxides may also be found and they may contain As, Bi, Ni, Pb and Sb, including selenides and tellurites. Some of these impurities dissolve partially or completely in sulphuric acid, but the majority of the compounds are transferred to the anode sludge in different quantities. Small amounts of As, Ni and Sb are transferred to the cathode only if the oxygen content of the anode is very low. Since soluble impurities are present only in very small amounts and are more electronegative than copper, they remain in the electrolyte solution. The solution is circulated in electrorefining vats and then in the external circuit in which the impurities are removed to prevent the increase of their concentration, i.e., activity during which they could precipitate on the cathode together with copper.

In electrorefining of nickel from sulphate solution the anode may contain 2% or more of copper, together with a variable amount of cobalt and iron as the main impurities. The standard electrode potential of these elements is higher or approximately the same, Table 9.3, so that they can precipitate together with nickel on the cathode. This is prevented by using a pervious membrane or diaphragms between the anode and the cathode. The elements dissolve from the anodes in the anolyte and the solution flows into the external circuit. In the circuit, copper is cemented on metallic nickel or precipitated as a sulphide by adding sulphur. Cobalt and iron are precipitated by oxidation using chlorine or air. The purified and filtrated solution is returned to the catholytic section from where nickel is electrolytically precipitate on the cathode. The hydrostatic pressure in the catholyte should be higher than that in the anolyte to ensure the flow of the electrolyte through the diaphragm from the catholyte to the anolyte and prevent the arrival of impurities on the cathode. The presence of the diaphragm and, therefore, the larger gap between the anode and the cathode greatly increase the requirements on voltage for electrolysis of nickel. The resistance of the electrolyte is also higher than in refining of copper because the pH value of the solution must be higher, around 4, to prevent the formation of gaseous hydrogen.

The diaphragm electrolyser is also used for the production of pure nickel from anodes cast from Ni_3S_2 to matte. The anodic reaction is:

$$Ni_3S_2 \rightarrow 3Ni^{2+} + 2S + 6e^- \tag{9.50}$$

for which $E^0 = -0.35$ V and the voltage should be appropriate increased. The solution is again purified in the external circuit but the amount of the sludge produced in the anolyte greatly increases as a result of precipitation of sulphur so that the anodes are placed in cloth bags to trap the sludge. This may also be a process suitable for the production of pure copper directly from the matte so that the formation of SO_2 in pyrometallurgical converting may be prevented. Unfortunately, the quality of the surface of Cu_2S is not suitable for this operation and the reactions taking place would be efficient only if the voltage were considerably higher.

The sludge formed from anode impurities usually builds up at the bottom of the vat but may also coat the anode surface. Porous sediments may be separated when dissolving the main metal. Non-porous sediments of the compounds such as Cu_2O or Cu_2S may passivate the anode surface and interrupt the process. Therefore, the production conditions must be efficiently controlled to avoid the formation of an adhesive layer. Similarly, the skeleton of sulphur residue, which remains on the surface during electrolysis of the anodes produced from the nickel matte, prevents the diffusion of ions to and from the dissolved surface so that the voltage must be again increased to ensure further dissolution of the anode.

References

1. Kmetová D.: Theory of pyro-, hydro- and electrometallurgy [in Slovak], ES VŠT Košice, 1980.
2. Richardson F.D.: Extraction and Refining, Institute Metallurgist Review Course, 1970.
3. Ritcey G.M., Ashbrook A.W.: Solvent Extraction: Principles and Applications to Process Metallurgy, Elsevier, Amsterdam, 1984.
4. Fletcher A.W.: *Metals and Materials*, 3, 9, 1969.
5. Rickelton W.A., Flett D.S., West D.A.: Solvent Extraction Ion Exchange, 2, 6, 24, 1984.
6. Stubina M.M., Distin P.A.: *Trans. Inst. Min., Metall.*, 96, C27, 1983.
7. Kotze M. H.: *Journal of Metals*, May 46, 1992.
8. Bard A. J., Faulkner L. R.: Electrochemical Methods: Fundamentals and Applications (2nd ed), John Wiley and Son Ltd., December 2000, pp. 833.
9. Brett Ch. M. A, Brett A. M. O.: Electrochemistry: Principles, Methods and Applications, Oxford Science Publ., January 1993, pp. 427.
10. Hamnett A., Vielstich W., Hamann C.H.: Electrochemistry, John Wiley and Son Ltd., January 1998, pp. 423.
11. Oldham K.B., Myland J.C.: Fundamentals of Electrochemical Science, Academic Press, Inc., San Diego, 1994, pp.459.
12. Bockris J. O. M., Reddy A. K. N.: Modern Electrochemistry, Plenum Press, New York, 1976.

CHAPTER 10

EFFECT OF THE ELECTRONIC STRUCTURE ON LEACHING OF SULPHIDE SEMICONDUCTORS

It has been mentioned several times that oxidation leaching may be regarded as an electrochemical reaction including the cathodic reduction of the oxidant and the anodic oxidation of the sulphide. Many sulphides are leached electrochemically and this behaviour may be described by the model of the mixed potential which originally describes corrosion processes but its validity has also been expanded to the leaching kinetics of sulphides.

When considering the model of the mixed potential in leaching kinetics, it is assumed that the potential at the interface on the side of the solid (the region of the spatial charge) remains constant so that the potential difference will be realised only through the Helmholtz laycr. This behaviour is typical of metals and may be concluded that the coefficient of charge transfer of a single electron in the ideal case is 0.5. However, many sulphide minerals show semiconductor behaviour and the semiconductor model assumes that the majority of the potential gradient is located in the region of the spatial charge in the ideal case with the charge transfer coefficient of 1. This means that the results will differ if the leaching kinetics is described by the model of the mixed potential or by the chemical model.

Garrett and Brattain [1] were amongst the first researchers who investigated in detail the semiconductor-solution interface and laid foundations of semiconductor electrochemistry. Three semiconductor–solution interfaces were proposed:

- idealised model [2];
- recombination models [3, 4];
- the model of the connected Fermi levels [5].

Band semiconductor theory

The band structure of semiconductors is characterised by the forbidden band between the filled valency band and empty conductivity bands. In a semiconductor there are two types of charge carriers: the electrons in the conductivity band and holes in the valency band. The probability of formation of some quantum state of the band occupied by the electron or a hole is controlled by the Fermi–Dirac distribution law.

The band model of the surface of the semiconductor differs from the model of the surface of the metal because of the excess density of the surface charge of the solid referred to as the spatial charge layer. The density of the charge and the surface increases as a result of completion of the spatial structure resulting in the formation of Tamm or Shockley states on the surface. Excess charge density in the region of the spatial charge results in the so called bending of the band at the interface. This means that the bands in the volume are thinner than those on the surface if the potential difference is positive, and, conversely, if the potential difference is negative. The 'single atom' band model of a type *n* semiconductor as shown in Fig. 10.1. The concentration of charge carriers on the surface is an exponential function of the potential difference between the surface and the volume of the semiconductor.

The majority of the sulphides may exist as non-stoichiometric compounds and, consequently, affect the electronic properties of the solid. Compounds with an excess of metal, such as, for example, chalcopyrite, show *n*-type semiconductivity whereas compounds with an excess of anions show *p*-type semiconductivity.

Model of the energy level of ions in the solution

The energy description of the ions in a solution is far more complicated than in solids or on their surface. The energy levels of the ions in the solution are separated as a result of thermal fluctuations of dipoles which surround the ion during solvation. Electrostatic polarisation theory indicates that the separation of the energy levels is of the Gauss parabola type. The probability of transfer of an electron between the electrode and the ion in a solution depends on the energy level of the ion at the moment of transfer, indicating the importance of fluctuations [7]. The energy required for reaching the activated state of the ion is the difference of the energies between the oxidant E_{ox} and E^* and the probability

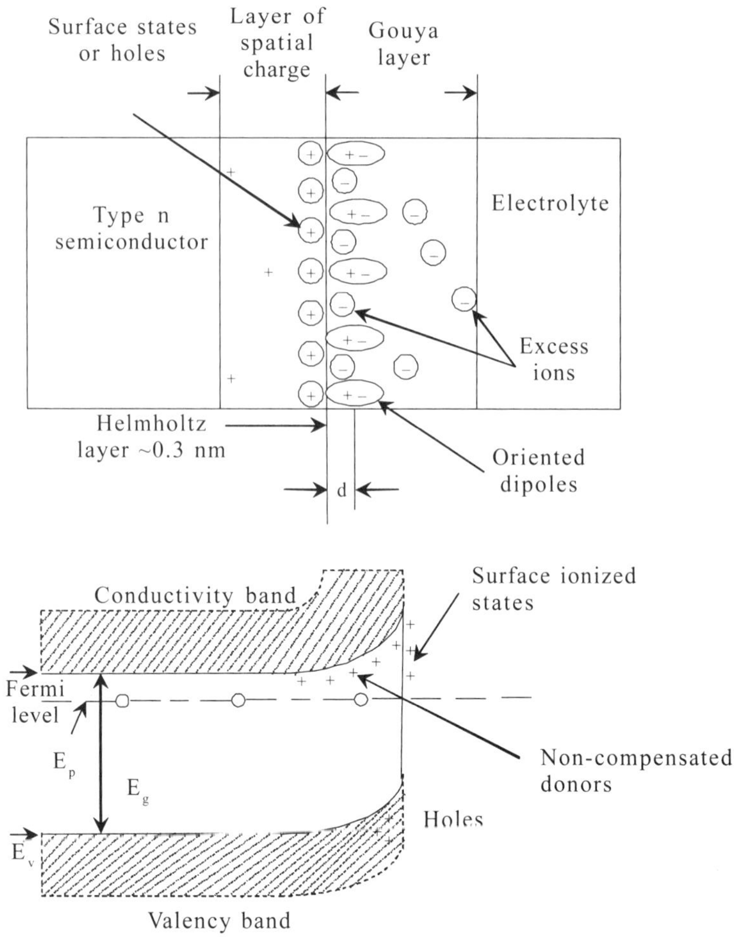

Fig. 10.1 Diagram of the distribution of the charge and energy on the interface between the type and semiconductor and the solution [6].

W_{ox}, and the energy level will be E^* according to the relationship

$$W_{ox} = (4\pi\lambda kT)^{-\frac{1}{2}} \exp\left[-\frac{(E_{ox} - E^*)^2}{4\lambda kT} \right] \qquad (10.1)$$

where λ is the energy including the energy of contributions from expansion of the bands inside the co-ordination sphere and the rotation of the molecules of the solvent in the external sphere, k is the Boltzman constant and T is temperature. These energy fluctuations are not small and the standard deviation $(\lambda kT)^{1/2}$ may amount to several tens of electron volts. The values of λ for a large number of oxidants and reduction agents were published in [8].

The energy levels of the ions in the solution may be determined from the standard redox potential of the redox pair E_{redox}, equal to 1/2 $(E_{red}+E_{ox})$ and are directly proportional to the redox potentials presented in tables in handbooks [9].

Semiconductor–solution interface

The semiconductor–solution interface consists of a double layer in which the charge of the solid is balanced in the volume of the solution in the vicinity of the surface. Charges are concentrated in the region of the spatial charge, in the Helmholtz area and the Gouy layer and may be described quite accurately by approximations as condensers connected in series. A typical interface is shown in Fig. 10.1.

The semiconductor and solution are in equilibrium when the electrochemical potential of the electron in the solid is equal to the solution potential, i.e. $E_F = E^0_{redox}$. When a solid is immersed in the redox solution of an electrolyte the existing charges are redistributed to the interfacial region so that the excess density of the charge in the solution is modified by the same density of the charge in the solid (the layer of the spatial charge) to equilibrium [2, 10].

For the majority of semiconductors it was found that the potential depends greatly on pH as the shift of 0.59 V per unit pH as a result of the adsorption of OH^- or H^+. Similar shifts of pH were recorded on sulphide electrons as a result of the effect of HS^- and H^+ [11].

Charge transfer between the semiconductor and the solution

Electron transfer may take place only between the states with the same energy level, i.e. without transfer of energy by radiation or in a coating. Using the expressions of the fluctuating energy model, the ion must show a very high probability of fluctuation to the energy level of the band in which electron transfer takes place. For electron transfer to take place, E^0_{redox} must be close to the conductivity or valency interface. Comparison of the interface of the valency band E_{vs}, and conductivity band E_{cs} with E^0_{redox} provides a qualitative image as to whether electron transfer is possible or not. The redox pair may cause exchange by only one of the bands. If the redox potential reaches the value $E^0_{redox} > 0.5$ V, the processes take place in the valency band, and if $E^0_{redox} < 0$ V, they take place in the conductivity band.

The rate of electron transfer from the semiconductor to the solution is proportional to the number of occupied positions in the semiconductor $D_{sc}(E)f(E-E_{sc})$ and the number of unoccupied positions in the redox electrolyte, $D_{ox}(E)$, at the same energy E. The value $D_{sc}(E)$ is density of positions in the semiconductor, $f(E-E_{sc})$ is Fermi's function describing the number of occupied positions, and $D_{ox}(E)$ is the density of the positions in the oxidant in the solution. The total current from the semiconductor to the electrolyte is the integral of this probability in the range of the total energy level:

$$j^{-} = q\int_{-\infty}^{\infty} \nu^{-}(E)D_{sc}(E)f(E-E_{sc})D_{ox}(E)\mathrm{d}E \tag{10.2}$$

where $\nu^{-}(E)$ is the transfer coefficient, $D_{ox}(E) = C_{ox}W_{ox}$, where C_{ox} is the concentration of the oxidant in the solution.

The current–voltage dependence of the semiconductor-solution ideal system is often described as Schottky's barrier [12]. Substitution of (10.1) into (10.2) integration gives

$$j_v = j_{v,0}\left(\frac{qV_s}{kT}\right) \quad 1 \tag{10.3}$$

where j_v is the current density of the valency band, and $j_{v,0}$ is the change of current density. This equation is valid for type *p* semiconductors.

The characteristic feature of Schottky's barrier is that only the cathodic process is affected by the potential in the conductivity band and in the valency band only the anodic process depends on voltage. The anodic process in type *n* semiconductors is limited by the diffusion of carriers (holes) on the surface and represents limiting current. This shows that the rate of anodic dissolution of type *n* minerals is lower than that of type *p* minerals.

The voltage–current curves for the reaction of the single electron on the surface of an ideal semiconductor, equation (10.3), assume the charge transfer coefficient of 1 and the Tafel constant of 0.059 V. However, if the number of surface positions exceed 10^{13} positions per cm^2, the Fermi level in the semiconductor is constant and the electron potential gradient forms through the

Helmholtz layer. In this case, the Fermi layer is disrupted, the transfer coefficient is equal to 0.5 and the Tafel constant is 0.118 V decade^{-1} which represents the behaviour of charge transfer in the metal-electrolyte system. Similar Tafel constants are determined in degeneration of the semiconductors. However, on the whole it may be said that even pure element semiconductors show deviations from the ideal behaviour.

Anodic corrosion of semiconductors

The bonds between adjacent atoms in the crystal exist as a result of the interaction of binding orbitals of adjacent atoms. The binding electronic state with the valency band of the solid and the presence of holes in this band indicate that one of the binding electrons has been removed and the bond weakened. The holes may be replaced by internal electrons as a result of electric current or in the case of corrosion and leaching they may be brought by the oxidant into the valency sphere.

In semiconductors with the forbidden band larger than 1 eV, anodic dissolution takes place almost completely by the supply of holes to the valency sphere [13]. Table 10.1 shows the contribution of holes to the anodic breakdown reaction.

Oxidation breakdown of M^+S^- sulphide semiconductors take place in the following stages:

$$\left(M^{\delta+}S^{\delta-}\right)_s = X \rightarrow \left(MX\right)^+ + \bullet S_s + e^- \tag{10.4a}$$

$$\left(M^{\delta+}S^{\delta-}\right)_s = X + h^+ \rightarrow \left(MX\right)^+ + \bullet S_s \tag{10.4b}$$

$$\bullet S_s + Y \rightarrow \left(SY\right)^+ + e^- \tag{10.4c}$$

$$\bullet S_s + Y + h^+ \rightarrow \left(SY\right)^+ \tag{10.4d}$$

$$\bullet S_s + \bullet S_s \rightarrow \left(S - S\right)_s \tag{10.4e}$$

$$\bullet S_n + \bullet S_s \rightarrow S_{n+1} \tag{10.4f}$$

where h is the hole, X and Y represent the ion substances in the solution. It may be assumed that the electropositive component, M^+, i.e., the element associated with the conductivity band, is transferred into the solution whereas in many cases the more electronegative element undergoes a recombination reaction by which this element is transferred to the zero valency state.

Table 10.1. Contribution of holes to anodic dissolution of semiconductors [13]

	E_g [eV]	Product of oxidation breakdown	Hole contribution [%]
Ge	0.66	$Ge^{4}+$	55–70
Si	1.09	Si^{2+} (Si^{4+})	~100
GaAs	1.35	Ga^{3+}, As^{3+}	~99.8
CdTe	1.5	Cd^{2+},Te^{4+}	?
CdSe	1.74	Cd^{2+},Se	~96
CuO	1.95	Cu^{2+}, 0.5 O_2	?
CdO	2.2	Cd^{2+}, 0.5 O_2	?
GaP	2.35	Ga^{3+}, P or P^{3+}	~100
CdS	2.4	Cd^{2+}, S	~100
ZnO	3.3	Zn^{2+}, 0.5 O_2	~100

The general requirement on steady-state corrosion is that the anodic and cathodic current should be equal $j^+ = j^-$. The surface electrical field of the metal has the reversed effect on these two processes and ensures identical current conditions. Corrosion may be described by the superposition of anodic and cathodic currents. In semiconductors, this equality means that

$$j_c^+ + j_v^+ = j_c^- + j_v^- \tag{10.5}$$

and the current in every band will have the reversed sign

$$j_c^+ + j_v^- = j_c^- + j_v^+ \tag{10.6}$$

However, to understand fully the leaching processes it is necessary to consider the crystal structure and its behaviour, the type of semiconductor, and the characteristics of the semiconductor-solution interface.

10.1. Leaching kinetics and electrochemistry of sulphides

Koch [14] and Hiskey and Wadsworth [15] investigated the

electrochemistry of metal sulphides and Wadsworth [16] and Dutrizac and MacDonald [17] published a review of the leaching of metallic sulphides. Burkin [18] published data on the occurrence of phase transformations in the solid state in the leaching of sulphides, stressing the formation of the solid interfacial product at the leaching interface preventing further reaction, and Peters [19] discussed experimental procedures in the investigations of electrochemical phenomena in leaching of sulphides. Shuey [20] published a review of the electronic properties of all minerals and Vaughan and Craig [21] discussed the mineral chemistry of sulphides.

The metal sulphides are characterised by a wide range of structural and electronic properties, as shown in Table 10.2.

10.1.1. Leaching of chalcopyrite

Anodic leaching of chalcopyrite is governed by the equations

$$CuFeS_2 \rightarrow Cu^{2+} + Fe^{3+} 2S^\circ + 5e^- \qquad (10.7)$$

and

$$CuFeS_2 + 8H_2O \rightarrow Cu^{2+} + Fe^{3+} 2SO_4^{2-} + 16H^+ + 17e^- \qquad (10.8)$$

The majority of sulphide sulphur is transferred into the elemental form, the remainder (3–15%) to the sulphate form. The results obtained by several authors show that the oxidation products are

Table 10.2. Properties of some metallic sulphides [15]

Mineral	Formula	Resistance [Ω m]	Semiconductor type	Residual potential [V]
pyrite	FeS_2	$0.1\text{–}3\cdot10^{-2}$	*n, p*	0.63
chalcopyrite	$CuFeS_2$	$0.2\text{–}9\cdot10^{-3}$	*n*	0.53
chalcocite	Cu_2S	$10^{-2}\text{–}10^{-5}$	*p*	0.44
covellite	CuS	$10^{-4}\text{–}10^{-6}$	metallic *p*	0.42
galenite	PbS	10^{-5}	*n, p*	0.28
sphalerite	ZnS	$10^{5}\text{–}10^{7}$	-	–0.24

resistant to further leaching. Sulphur [22], semiconductors of the type *p* or *n* such as CuS and polysulphides with a shortage of metals [23, 24] inhibit the transfer of electrons or ions to the interface

Chalcopyrite is a good conductor with a resistance of approximately 10^{-3} m [20]. It is a type *n* semiconductor taking into account the surplus of metal with the forbidden band of approximately 0.6 eV. The model of ion bonding of chalcopyrite is regarded as $Cu^{+}Fe^{3+}(S^{2-})_2$. Of course, in this case there is no direct covalent bond between the anions, but there is high covalency between the cations and the anions [20]. Unfortunately, direct calculations of the energy band in chalcopyrite is difficult because of its antiferromagnetisms [25]. However, studies have been published of other compounds of a simpler type indicating that the lower part of the conductivity band of chalcopyrite is mostly Fe3d, and the upper part of the valency band is Cu3*d* and S3*p*. The appropriate 'single-electron' band diagram is shown in Fig. 10.2.

Parker, et al. [26] carried out extensive electrochemical studies of chalcopyrite leaching in a chloride medium in the temperature range 70–90°C and the potential range 0.2-0.6 eV. They found that the rate of reduction of the oxides can be arranged in the following sequence:

$$Br_2, CuCl_2, Fe(CN)_6^{3-} > FeCl_3 > Fe_2(SO_4)_3, Cl_2$$

The location of the bands and redox potentials of these oxidants is shown in Fig. 10.2. In order to 'keep' the electron or add a hole, the energy level of the oxidant must be sufficient to overcome the edge of the conductivity or valency band. The diagram shows that the rate of reduction of $CuCl_2$ and I_3^- is high and is determined by the energy level of these ions overlapping the edge of the conductivity band; the rate of reduction of Br_2 is also high because the energy level of bromine extends through the edge of the valency band. On the other hand, the levels of the ferric chloride and sulphate are located inside the forbidden band and their reduction rate is low.

Parker [26] also published the data on the anodic saturation currents typical of type *n* semiconductors in which the number of holes is the limiting factor for migration and oxidation of Fe^{2+} in the solution. Similarly, Beckstead, et al. [27] published results confirming that the leaching rate increases with an increase of the amount of Fe^{3+} and then reaches a steady value indicating saturated currents.

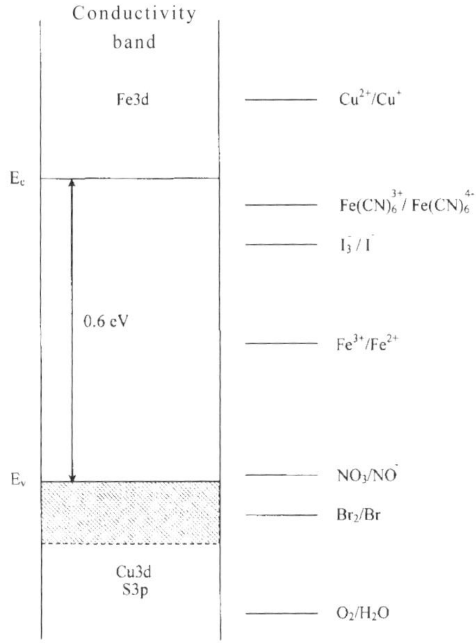

Fig. 10.2. Diagram of the energetic level of chalcopyrite indicating the contribution Cu3*d* to the valency band and the contribution of Fe3*d* to the conductivity band.

The presence of the saturated currents has not been confirmed for I_3^- and $CuCl_2^-$, but the energy levels of these ions exceed the edge of the conductivity band. Consequently, it may be concluded that the interactions I_3^-/I^-, $CuCl_2/CuCl_2^-$ and $Fe(CN)_6^{3-}/Fe(CN)_6^{4-}$ are the processes taking place in the conductivity band, whereas Fe^{3+}/Fe^{2+} and Br_2/Br^- take place in the valency band. Zevgolis and Cooke [28] published the results indicating the existence of the anodic limiting current as a result of migration of holes in type *n* chalcopyrite.

The addition of $CuCl_2$ to the $FeCl_3$ solutions increases the current density of the chalcopyrite electrode. Table 10.3 gives the data showing that the anodic dissolution of chalcopyrite takes place with the contribution of holes in the valency band and of electrons in the conductivity band. This is the case of many semiconductors, for example Ge, with the size of the forbidden band smaller than 1 eV [29].

Formation of a layer on the surface of chalcopyrite in leaching

The positive potential on the chalcopyrite electrode induces anodic current leading to electrode oxidation. If the electrode is connected

Table 10.3. Comparison of corrosion currents of chalcopyrite [26]

Ox	Red	E_{corr} [V]	j_{corr} [kA/m²]
Fe^{3+}	Fe^{2+}	0.39	55
Fe^{3+}/Cu^{2+}	Fe^{2+}	0.48	110
Cu^{2+}	Cu^{+}	0.28	2.5

to an open circuit, and the application of the potential results in reactivation and high current which again decrease with time. This was interpreted [26] as the formation of a polysulphide layer with a shortage of metal on the leached surface which reduces the rate of transfer of Cu^{+} and Fe^{2+} ions into the solution. It is assumed that this layer degradates by the reactivation of the electrode producing elemental sulphur.

Munoz et al. [30] concluded that the kinetics of leaching of chalcopyrite in sulphate solutions is controlled by diffusion through the layer of elemental sulphur coating the leached surface. The high value of the activation energy was attributed to the rate-controlling transfer of electrons through the sulphur layer and the kinetics were described by Wagner's theory. However, on the other hand, this sulphur layer is porous and also contains a large amount of impurities. The removal of sulphur [31] from the leached surface slightly increases the leaching rate but not as expected. This shows that it is not clear whether the formation of elemental sulphur on the surface is an inhibiting factor.

McMillan et al. [32] published the values of anodic currents which decrease with time. They used a series of potential pulses in the range 0.2–0.6 V in relation to the standard electrode and observed a blue layer product which was regarded as a copper-enriched sulphide phase. This was interpreted as an example of the formation of a protective layer through which the rate-limiting ions must diffuse. The authors also compared the electrochemistry of the specimens of semiconductors of type *n* and *p*. However, their conclusion according to which the type of semiconductor has no effect on the behaviour of the electrode appears to be slightly premature. The main reason for this is that they used natural crystals and their type *p* specimens contained quite large quantities of pyrite inclusions which probably induced possible galvanic reactions.

The formation of the layer on the surface increases the value of the Tafel constant above 0.118 V decade^{-1} [33]. McMillan *et al.* [32] obtained a value of 0.2 V decade^{-1} as a result of the formation of a clearly visible blue CuS layer or non-stoichiometric sulphides of Cu_xS type. However, if such a layer forms on the electrode surface, the current should decrease until a stable layer of copper sulphide appears.

The layer of copper sulphide or some other layer does not reduce the leaching rate in chloride solutions until the thickness of the layer reaches a steady value. This results in linear kinetics and a strong dependence on the ferric chloride concentration. The layer formed in sulphate solutions apparently inhibits leaching and the kinetics is parabolic. The differences may be given by the capacity of chloride ions to be adsorbed on the surface; sulphate ions do not have this capacity. The formation of chloride species is also important. This may result in the formation of surface states inside the forbidden band of the layer which may support the passage of current and, consequently, catalyse electrochemical dissolution of the layer.

It is assumed that the formation of the sulphate as a final product of chalcopyrite sulphur takes place by the mechanism in which the added holes, as the contribution of Cu3d to the valency band, are capable of oxidising water and OH radicals oxidise directly the sulphur of the valency band to the sulphate. The rate of oxidation of elemental sulphur is very low and is not regarded as the factor forming the sulphate.

Galvanic interactions during leaching

If two minerals with different potentials are in contact, the mineral with the more positive potential is reduced and that with the more negative potential oxidised. There potential difference is the driving force of galvanic corrosion.

Although there are large number of experimental studies, the majority of the results have been interpreted only semi-quantitatively. The potential differences of galvanic pairs of natural minerals forming the sequence: galenite–pyrite > chalcocite > pyrite > chalcocite–chalcopyrite > galenite–chalcopyrite > chalcopyrite > pyrite, are known.

Hepel [34] considered the effect of a strong bond between two semiconductor minerals of the type *p* and *n* on the galvanic corrosion current. The polarisation of the *p–n* mineral junction depends obviously on the potential of each mineral. The potential of each mineral is determined by measuring the Fermi level or the electrochemical potential of the electrons in the solid taking into account the effect of the half-cell. The potentials of the open circuit of the substances of the type *p* and *n* of the same mineral differ in the differences in the Fermi level and the maximum is shown by the energy of the forbidden band is maximum. This is observed in pyrite and galenite. If the potential of a mineral of type *p* is higher than that of mineral of type *n* the potential difference at the junction is small and high currents may flow. If the type *p* mineral has more negative potential than type *n* mineral, the potential difference at the junction is high and may be sufficiently high to prevent any current at the junction. Hepel [34] derived a general relationship for determining the electrochemical parameters of the system and the coefficient of anodic acceleration. The experimental results published by Hepel confirm the directional properties of the galenite–galenite, galenite–chalcopyrite and chalcocite–pyrite electrodes, although they did not show ideal Schottky's barriers. The formation of surface states and disruption of the Fermi level was also detected. Despite the presence of a large number of impurities in the natural minerals, it is also necessary to consider the ohmic and semiconductor properties of the minerals forming the solid bond, if galvanic interactions are considered.

Figure 10.3 shows the galvanic interaction between chalcopyrite and metallic copper [35].

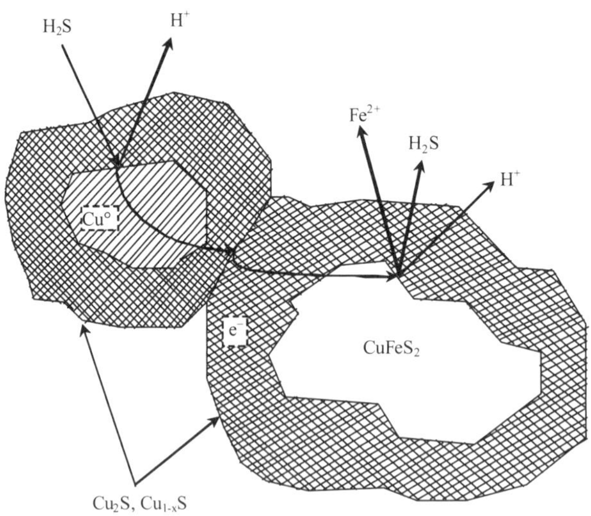

Fig.10.3. Example of galvanic interaction between chalcopyrite and opper.

References

1. Garrett C.G.B., Brattain W.H.: *Physical Review*, 99, 1955, 376-391.
2. Gerischer H.: Semiconductor electrochemistry. In: Physical Chemistry – an Advanved Treatise, vol. IXA, Eyring H., Henderson D., Jost W eds., Academic Press, New York, 1970.
3. Frank S.N., Bard A.J.: *Journal of American Chemical Society*, 97, 1975, 7427-7432.
4. Peter L.M., Li J., Peat R.: *Electroanalytical Chemistry*, 165, 1984, 29-36.
5. Bard A.J.: *Journal of American Chemical Society*, 102, 1980, 3671-3685.
6. Dewald J.F.: Surface Chemistry of Metals and Semiconductors, John Wiley and Sons, New York, 1980.
7. Marcus R.A.: *Annual Rev. Physical Chemistry*, 15, 1964, 155–161.
8. Frese K.W.: *Journal of Physical Chemistry,* 85, 1981, 3911–3920.
9. Morrison S.R.: *Journal of Vacuum Sci. Technology*, 15, 1978, 1417–1427.
10. Gerischer H.: Advances in Electrochem. and Electrochem. Engineering, vol.1, Interscience, 1960.
11. Ginley D.S., Butler M.A.: *Journal of Electrochemical Society,* 125, 1978, 1968–1973.
12. Morrison S.R.: Electrochemistry of Semicondutors and Oxidized Metal Electrodes, Plenum Press, New York, 1980.
13. Gerischer H., Mindt W.: *Electrochimica Acta*, 13, 1968, 1329–133.
14. Koch D.F.A.: Modern Aspects of Electrochemistry, vol .10, Plenum Press, New York, 1975.
15. Hiskey J.B., Wadsworth M.E.: Process and fundamental consideration in selected hydrometallurgical systems, SME-AIME, New York, 1981, 303-313.
16. Wadsworth M.E.: Electrochemical reactions in hydrometallurgy, Metallurgical Treatises, AIME, Elliot ed., 1981, 1-22.
17. Dutrizac J.E., MacDonald R.J.C.: *Minerals Sci. Engng.*, 6, 2, April 1974, 59-101.
18. Burkin A.R.: Solid-state transformations during leaching, *Min. Sci. Engng.* 1, 1, 1969, 4–15.
19. Peters E.: Trends in Electrochemistry, Bockris J.O.M. ed., Plenum Press, New York, 1977.
20. Shuey R.T.: Semiconducting Ore Minerals, Elsevier, Amsterdam, 1975.
21. Vaughan D.J., Craig J.R.: Mineral Chemistry of Metal Sulfides, Cambridge University Press, Cambridge, 1978.
22. Munoz P.B., Miller J.D., Wadsworth M.E.: *Met. Trans. B*, 10B, June 1979, 149–158.
23. Biegler T., Swift D.A.: *Electrochimica Acta*, 24, 1979, 415–420.
24. Linge H.G.: *Hydrometallurgy*, 2, 1976, 51–64.
25. Crundwell F.K.: *Hydrometallurgy,* 21, 1988, 155–190.
26. Parker A.J., Paul R.L., Power G.P.: *Journal of the Electroanalytical Chemistry,* 118, 1981, 305–316.

27. Beckstead L.W., Munoz P.B., Sepulveda J.L., Herbst J.A., Miller J.D.: Acid ferric sulfate leaching of attritor-ground chalcopyrite concentrates, Extractive Metallurgy of Copper, vol.II., Yannopoulos & Agarwal eds., 1976, 611-632.
28. Zevgolis E.N., Cooke S.R.B.: Electrochemical properties of the semoiconductor mineral chalcopyrite, In: XI. Int. Min. Proc. Congress, Cagliary, Paper 16, 1975.
29. Khan S.U.: Aspects of Modern Electrochemistry, vol. 15, White R. ed., Plenum Press, New York, 1983.
30. Munoz P.B., Miller J.D., Wadsworth M.E.: *Metallurgical Transactions B*,10B, June 1979, 149–158.
31. Havlík T.: Acid oxidation leaching of chalcopyrite and behaviour of sulphur in this process [in Slovak], PhD dissertation, Faculty of Metallurgy, Technical University, Košice, April 1996, p.291.
32. McMillan R.S., MacKinnon D.J., Dutrizac J.E.: *Journal of Applied Electrochemistry*, 12, 1982, 743-749.
33. Vijh A.K.: Electrochemistry of Metals and Semiconductors, Marcel Dekker, New York, 1973.
34. Hepel T.: In: Int. Symp. on Electrochem. in Mineral and Metal Processing, Richardson P.E. ed., Electrochem. Society, Pennington, 1984.
35. Peters E.: The Physical chemistry of hydrometallurgy, Proc. Int. Symposium on Hydrometallurgy, Evans and Shoemaker eds., AIMMPE, 1973, 205-228.

CHAPTER 11

EXPERIMENTAL METHODS OF INVESTIGATING HYDROMETALLURGICAL PROCESSES

The direct result of examination of the kinetics of a heterogeneous reaction of the solid–liquid phase type is the kinetic dependence of the conversion on time in the defined conditions which often cannot be described by a simple function. Generally, the main part of the kinetic relationship may be described by the expression of the type:

$$R = kt^n \tag{11.1}$$

where R is conversion, t is the reaction time, n is the order of the reaction, and k is the rate constant. The value of n may vary in a wide range from 0.125 up to 22.8 [1]. Of course, in this case, the coefficient n cannot be regarded as identical with the order of the reaction because the concentration of the reagents in the reaction mixture is not an unambiguous function of time and changes in a discrete manner in the volume of the reaction mixture. The mathematical processing of the kinetic data may be realised by two procedures:

- purely formally in order to find the equation describing most accurately of the experiments, and the variables of the equations do not have any specific physical meaning;
- by means of kinetic relationships based on the defined models of the reaction of the volumes of the solid substances, and the variables of these equations correspond to the nature of the processes taking place.

The formal kinetic equation is based on the application of the law of the effect of the mass with may be presented in the form:

$$\frac{d(1-R)}{dt} = \frac{dR}{dt} = k(1-R) \tag{11.2}$$

where n is the exponent, the value reaches formally identical with the reaction order.

Determination of the rate constant and the reaction order

The values of the parameters n and k are determined by the integral and differential methods. The breakdown of the variables in the equation (11.2) and integration in the range from 0 to t gives the following relationship:

$$\frac{1}{1-n}\left[1-(1-R)^{1-m}\right]=kt \tag{11.3}$$

Using the experimental values of R and t in equation (11.3), it is possible to calculate the value of n at which k is a constant.

The general rate equation

$$v=-\frac{\mathrm{d}m}{\mathrm{d}t}=kA.c$$

where A is the reaction surface, c is the concentration of the reagent and k is the rate constant, may be written in the new form:

$$(1-R)^{1-n}=1-(1-n)kt \tag{11.4}$$

and the function on the left-hand side can be expanded into a binomial series

$$(1-R)^{1-n}=1-(1-n)R-\frac{n}{2}(1-n)R^2 \tag{11.5}$$

Comparison of the relationships (11.4) and (11.5) gives

$$R=kt-\frac{n}{2}R^2 \tag{11.6}$$

Since at low values of R the right-hand part of equation (11.6) is approximated by the first term in $R^2 = kRt$, the equation (11.6) has the following form:

$$\frac{t}{R}=\frac{1}{k}-\frac{n}{2}t \tag{11.7}$$

Equation (11.7) makes it possible to obtain rapidly, although approximately, the values of n and k.

Another possibility of determining the values of k and n is the differential method, based on the relationship for the reaction rate in the logarithmic form:

$$\ln\frac{dR}{dt} = \ln k + n\ \ln(1-R) \tag{11.8}$$

The possibilities of the differential method are restricted by the low accuracy of determination of the rate of transformation dR/dt.

Formally, the values of n and k can also be determined graphically; the rate constant is determined as the tangent at the given point of the kinetic curve, Fig. 11.1.

The reaction order may be determined using several procedures:

a – dependence of the logarithm of the reaction rate on the concentration logarithm gives a straight line whose tangent gives the reaction order. The scheme of the graphical solution of equation (11.8) is shown in Fig. 11.2;

b – in the initial stages of the reaction, it may be assumed that $c = c_0$. For the initial phases of the reaction, the general kinetic equation has the form:

$$v = kc_{\circ}^{n} \tag{11.9}$$

After two measurements with different initial concentrations c_{o1} and c_{o2}:

$$v_1 = kc_{\circ 1}^{n} \qquad\qquad v_2 = kc_{\circ 2}^{n} \tag{11.10}$$

After comparison and taking the logarithm:

$$n = \frac{\ln v_1 - \ln v_2}{\ln c_{o1} - \ln c_{o2}} \tag{11.11}$$

c – the dependence of the concentration on time for the reactions of the first, second and third order is expressed graphically and it is determined which of the relationships satisfies the equation of the straight line;

d – the dependence of the half-time of breakdown on the initial concentration is determined:

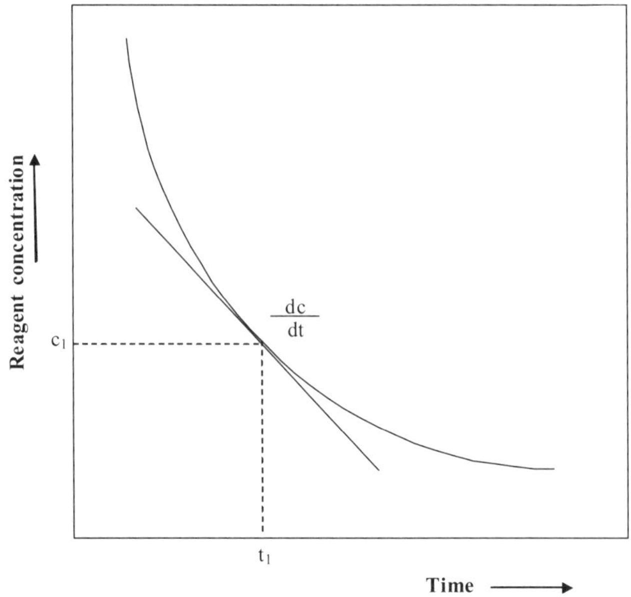

Fig. 11.1. The rate of the chemical reaction at time t1 equal to dC/dt.

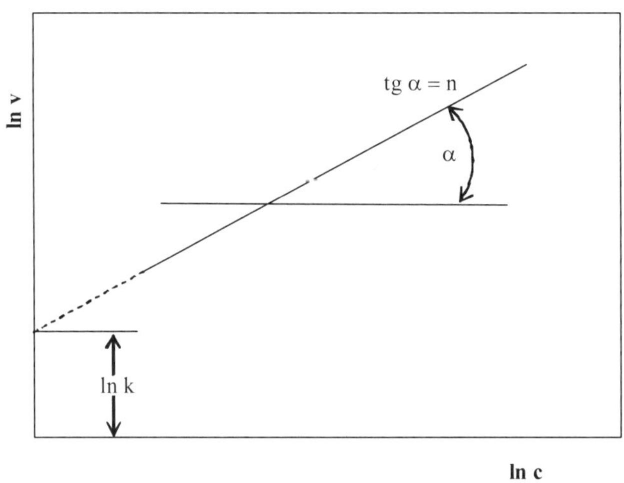

Fig. 11.2. Graphical determination of the reaction order.

$$\tau = \frac{\ln 2}{k} = \frac{0.693}{k}$$ for the first order reactions

$$\tau = \frac{1}{kc_{\circ}}$$ for the second order reactions

$$\tau = \frac{3}{2c_o^2 k} \qquad \text{for the third order reactions}$$

According to these equations, the half-time τ of the first order reaction is independent of the initial concentration, and in the case of the third order reaction, the half-time is indirectly proportional to the square of the initial concentration. Therefore, if the initial concentration changes, for example, five times, in the third order reaction it changes by a factor of 25.

If the rate is proportional to different powers of the concentrations of several substances, it is necessary to find the dependence of the rate on the concentration of each substance and the reaction rate is given by the equation:

$$v = k\, c_A^a c_B^b c_C^c \qquad (11.12)$$

The overall order of the reaction is $(a+b+c)$. The partial orders are determined experimentally by examining the reaction rate in such a manner that all the components, with the exception of the investigated component, are present in a large excess. If the excess of the substances A and B is large, the concentration of the substances may be regarded as constant so that the rate is proportional to only c_A^a. The order is determined by one of the previously mentioned methods and the value of exponent a is also determined using this procedure. The entire procedure is repeated by determination of the kinetics at the excess of the substances A and C, and exponent b is determined together with exponent c at the excess of the substances A and B. This is the so-called elimination method for the determination of the reaction order.

Determination of activation energy

The Arrhenius equation shows that, in a general case, the values of the rate constant k and the activation energy E are not known, whereas the value R and the specific temperature at which the reaction takes place, are unavailable. Since the logarithmic form of the Arrhenius equation is a straight line, it is sufficient to determine theoretically the values of the rate constant at two temperatures in the form:

$$\ln k_1 = \ln A - \left(\frac{E\,1}{R\,T_1}\right) \qquad (11.13)$$

and

$$\ln k_2 = \ln A - \left(\frac{E\,1}{R\,T_2} \right) \tag{11.14}$$

The difference of the logarithms is their ratio and since they are constants, are the result is again a constant $\ln k_1 - \ln k_2 = \ln \frac{k_1}{k_2} = \ln k$ and, the same time, the values of the logarithms of frequency factor A are negated.

Comparison of the relationships (11.13) and (11.14) gives:

$$\ln k = -\frac{E}{R}\left(\frac{1}{T_1} - \frac{1}{T_2} \right) \tag{11.15}$$

The graphical representation of this function in Fig. 11.3 makes it possible to obtain the activation energy from the angle of inclination of the straight line.

However, in practice the two values of the rate constants of a guarantee that the determined value of the activation energy is accurate, because of two reasons:

- to determine the optimum course of the straight line it is necessary to optimise this relationship by one of the available methods, for example, the method of least squares. As the number of available points increases, the accuracy of the form of the straight line also increases;
- as mentioned later, the activation energy also represents the main mechanisms of the reaction, i.e., whether the reaction is controlled by diffusion (external or internal), or whether it is controlled by the kinetics of the chemical reaction and, in both cases, in which temperature range. If the reaction mechanism changes, this is reflected in the form of the Arrhenius function by an inflection point at the temperature at which the reaction mechanism also changes.

For these reasons, it is always necessary to carry out experiments in order to determine the rate constant of the reaction in relation to temperature in the largest possible number and, in the ideal case, they should be repeated several times to ensure reproducibility.

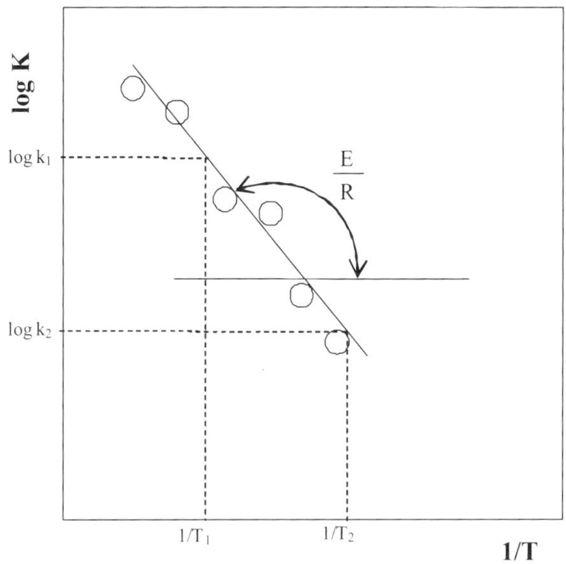

Fig. 11.3. Graphical determination of the constant of the Arrhenius equation.

11.1. Experimental methods of examination of leaching processes

The formulation of a proposal of leaching experiments should be of course based on the general knowledge of the investigated system, i.e., understanding of the expected chemical reaction, or the reactions and their conditions. Both can be estimated on the bases of thermodynamic investigations and also the already published studies. The result of the experimental examination of leaching should be the determination of the parameters affecting the kinetics of the leaching process and determination of the leaching mechanism. In experimental practice, the leaching kinetics are studied by the method of investigating a constant leached surface (the rotating disk, cylinder or ring method), and also leaching of the volume samples. In the first case, the constant area of the interface results in a low reaction rate and, consequently, an extremely long experiment time. In addition, these parameters do not involve other factors present in actual processes. Therefore, the examination of leaching of powder samples with a defined grain size is a more popular method. These examples are similar to materials used in production but it is impossible to investigate continuously the changing interfacial area in this case. In these cases, it is necessary to use the model kinetic equations for the determination of the course and mechanism of a heterogeneous reaction. Since the rate of the heterogeneous reaction is a function significantly influenced by individual variables:

$$v = f(T, \tau, c_i, A, y, \omega, p_i, K{:}p, \ldots)$$

where T is temperature, τ is time, c_i is the concentration of reagents, A is the interfacial area, y is the thickness of the diffusion layer, ω is the rate of mixing of the pulp, p_i is the pressure of gas components, $K{:}P$ is the ratio of the liquid to the solid phase in the pulp, etc, the experiment is setup in such a manner that after minimising the number of variables to the required number by heuristic methods and determination of their minimum and maximum values of the individual variables gradually change whilst others remain constant. The results can be presented in the graphical form and used to determine all the required values, such as the reaction mechanism, the effect of temperature, the effect of concentration of reagents, etc. The integral part of the investigations is phenomenological investigations, the morphology of leached particles, examination of the structure of the particles, etc.

In order to eliminate of the effect of other variables, they must be present in a sufficient surplus amount, and the effect of a single variable is investigated. This is a very important fact which is often underestimated because otherwise the effect of another variable may also be present. For example, the leaching rate at room temperature is usually low so that when using the low leaching concentration, this has no effect. The increase of temperature usually increases the leaching rate several times and the initial concentration of the leaching agent, sufficient for leaching at room temperature, becomes a limiting factor, instead of the leaching temperature. The result will then be confusing.

Also, large surpluses of the reagents may not be suitable. For example, large surpluses of the leaching agent in comparison with the amount of the sample (for example, litres in comparison with the grams) to appear as suitable for maintaining of the constant concentration difference, but it could happen that of the analytical be determined amount of the investigated leached metal in the solution will be very low in relation to dilution and, a systematic analytical error would be caused. Usually, it is therefore necessary to determine the minimum required amount of the reagent by calculation from the course of the expected chemical reactions, and the investigated variables are varied around this value. Another important fact is the requirement of reality, i.e. nobody will use concentrated sulphuric acid for leaching sulphides in production. The heuristic approach and experience are of extreme importance in the formulation of experiments [2].

For example, the leaching reaction in the investigations of the leaching of chalcopyrite using ferric chloride is:

$$CuFeS_2 + 4FeCl_3 \rightarrow CuCl_2 + 5FeCl_2 + 2S^\circ \qquad (11.16)$$

At a mass of the specimen of 5 g, 4 moles of $FeCl_3$ are required for the reaction of 1 mole of $CuFeS_2$, which means that at least 17.54 g $FeCl_3$ is required for 5 g $CuFeS_2$. When using, for example, 500 ml of the leaching solution with this amount, at least the 0.2 M solution of $FeCl_3$ should be used, as indicated by elementary calculations. However, this does not mean that when using the solution, the leaching kinetics will be optimum. Experimental investigations determine the optimal values of all the parameters.

Generally, it would be necessary to construct a matrix of experiments which would consider all the investigated effects and then it would be necessary to change gradually all the variables, whilst the other variables would be maintained constant. However, this is unacceptable both from the time viewpoint and financial considerations. There are sophistical approaches in which the significant effects on the processes may be determined using and acceptable number of experiments. These are the methods of factorial experiment design [3] which are highly suitable in cases in which it is necessary to investigate high large number of variables, for example, in a series of production electrolysers. The largest disadvantage of these methods from the viewpoint of fundamental research is, however, that these methods do not make it possible to obtain time relationships nor graphical interpretation. The time relationships are then used to determine the main kinetic relationships, for example, the activation energy is determined from the graphical dependence.

Therefore, the experimental plan is determined on the bases of the requirements for investigations. In leaching, the main requirement is usually the determination of the kinetic equation, the effect of temperature, reagent concentration, the effect of the slowest reaction stage and, consequently, the determination of the overall reaction mechanism. It is not known in advance whether the process is controlled by some type of diffusion or by the rate of the chemical reaction. The procedure may be used in, for example, the following steps:

11.1.1. Leaching equipment

One of the most important stages of research is the development and construction of suitable equipment. Equipment must be prepared taking into account the conditions of the investigations, and must be capable of maintaining all essential conditions for the course of the experiment in order to maximise the elimination of all systematic and random experimental errors. The diagram of basic leaching equipment is shown in Fig. 11.4.

Figure 11.5 shows laboratory equipment designed for the investigation of leaching processes with automatic collection and evaluation of temperature data, pH and the redox potential. The computer interface enables further services, depending on the type of sensor used in the investigations [4]. Equipment also enables to simultaneous experiments, as indicated by the close-up in Fig. 11.6.

This type of equipment enables continuous control of the process

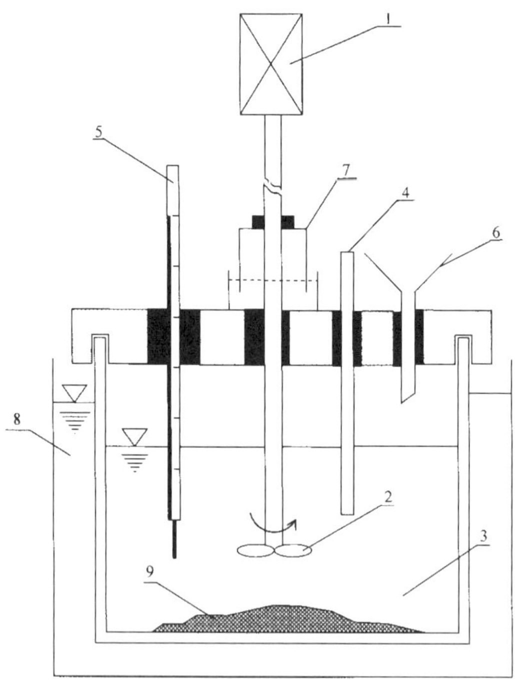

Fig. 11.4. Diagram of equipment for leaching in the solid–liquid phase system. 1) the stirrer engine with constant adjustable revolutions, 2) stirrer; 3) leaching pulp; 4) sampling of the liquid; 5) thermometer; 6) the orifice for introducing the sample and the leaching agent; 7) siphon seal; 8) thermostat; 9) solid sample.

Fig. 11.5. PClab equipment for investigating leaching.

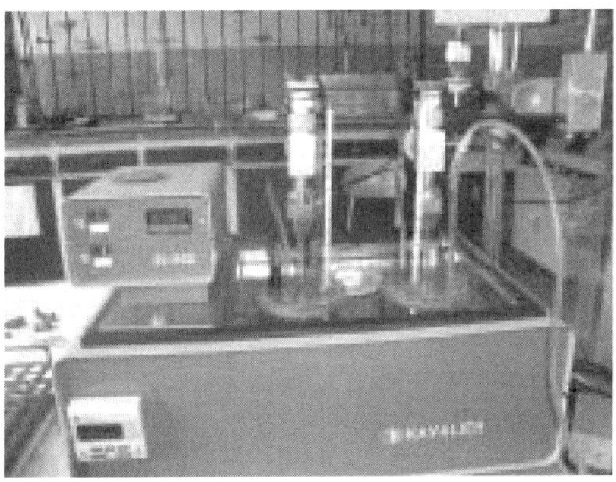

Fig. 11.6. Detailed view of PClab equipment.

and taking samples in any time period. At the same time, it is characterised by flexible introduction and an application of other sensors, for example, electrodes for measuring pH, or the redox potential. At the same time, other instruments or a personal computer, can be connected.

In the experiments using a gas atmosphere, it is necessary to supply the atmosphere to the leaching reactor. If the gases used in the experiments are corrosive, or dangerous (O_3, Cl_2, SO_2, etc), the reactor must be efficiently insulated and enable safe sampling. Figure 11.7 shows the scheme of the apparatus for leaching using a gas atmosphere, containing ozone. This equipment, used initially for the investigations by the rotating disk method [5], may be used efficiently for investigating the leaching of sample specimens by simple exchange of the tip of the rotating disk for a blade stirrer.

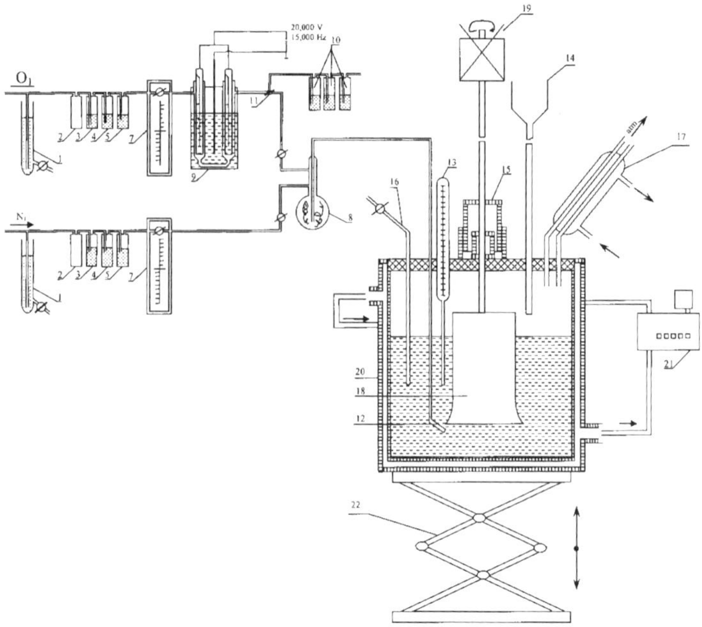

Fig. 11.7. Equipment for leaching using a gas atmosphere, containing ozone, by the rotating disk method. 1) manostat; 2) the empty, safety scrubber; 3) silica gel; 4) concentrated sulphuric acid; 5) phosphoric oxide; 6) pyrogallol; 7) the capillary flow rate meter; 8) gas mixer; 9) ozoniser; 10) gas analyser; 11) valve; 12) glass frit; 13) thermometer; 14) burette; 15) siphon seal; 16) sample areas; 17) reflux condenser; 18) rotating disk; 19) the drive with adjustable revolutions; 20) reactor; 21) thermostat; 22) pantographic table.

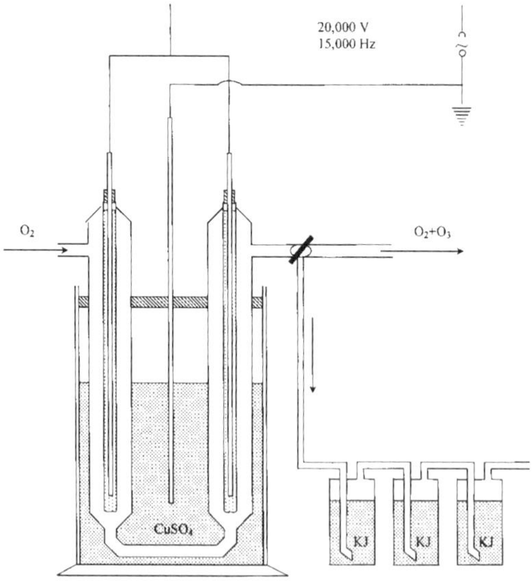

Fig. 11.8. Siemens ozoniser, detailed view of Fig. 11.7.

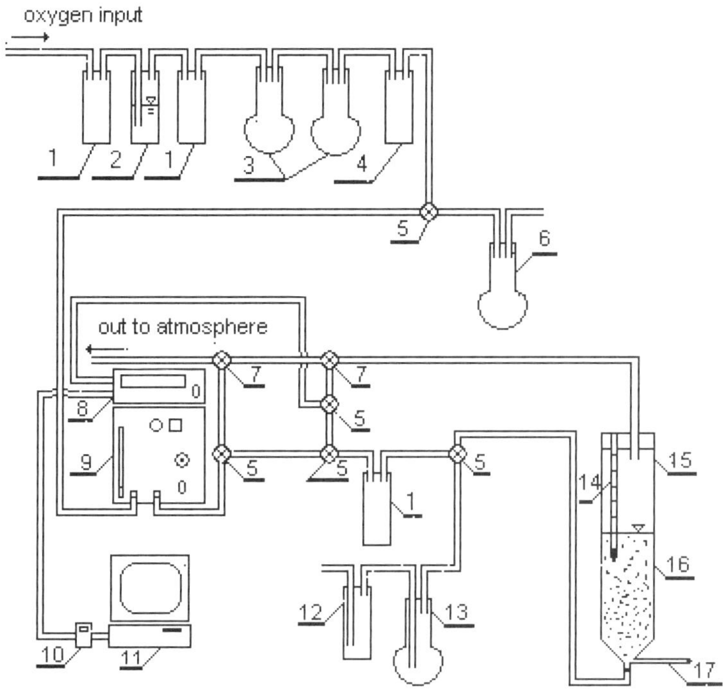

Fig. 11.9. Diagram of the apparatus for leaching in the whirling layer using an ozone-containing gaseous atmosphere. 1) empty safety scrubbers; 2) concentrated sulphuric acid; 3, 4) silica; 5, 7) gas distribution valve; 6) pressure reduction; 8) ozone analyser; 9) ozoniser; 10) AD converter; 11) continuous data collection; 12, 13) KJ solution for inspection titration analysis of ozone; 14) thermometer; 15) reactor; 16) leaching pulp; 17) sampling.

In this case, the ozone was generated using a Siemens laboratory ozone generator, based on silent discharge in dry oxygen, detailed view in Fig. 11.8.

Figure 11.9 shows the diagram and Fig. 11.10 the view of modified equipment for leaching using an ozone-containing gas atmosphere. In this equipment, the reactor in which mixing was carried out with a mechanical stirrer, is replaced by a column in which leaching was carried out in a whirling layer and the solubility of ozone in the liquid was investigated on the basis of hydrostatic pressure [6]. The self-supporting' ozoniser, Fig. 11.8, was replaced by a commercial ozoniser with an ozone analyser working online.

High-energy fields may be applied using equipment shown in Fig. 11.11 and 11.12.

The principle of equipment in which leaching is investigated in the extreme conditions of temperature and pressure – high-pressure leaching – is identical with equipment for leaching in the normal

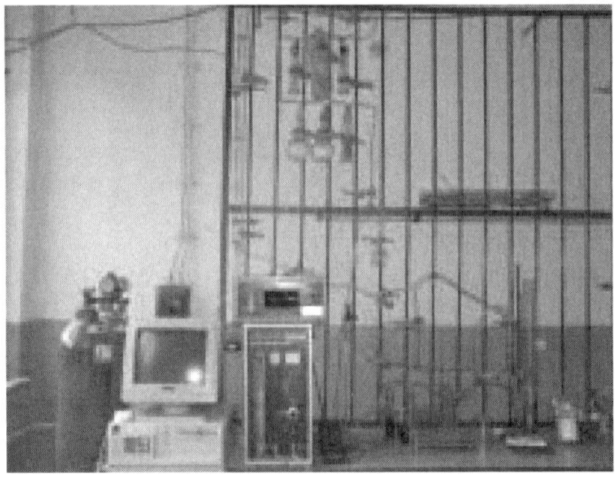

Fig. 11.10. Equipment for leaching in the whirling layer using an ozone-containing atmosphere.

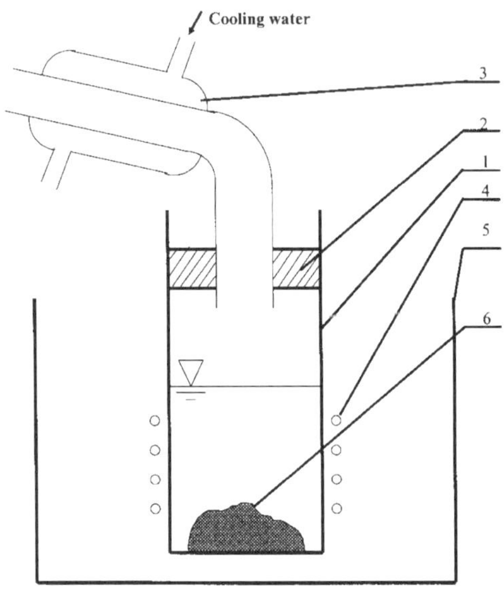

Fig. 11.11. Diagram of the apparatus for leaching in a high-a frequency field [7]. 1) the reactor; 2) the seal of the reactor; 3) reflux condenser; 4) high-frequency heating; 5) HF generator; 6) charge.

conditions of temperature and pressure, Fig. 11.3. Since the conditions are extreme, investigations are carried out using professional systems, autoclaves, which are capable of withstanding these conditions. However, qualified personal is essential for operating these systems.

Of course, these systems are always universal to a certain

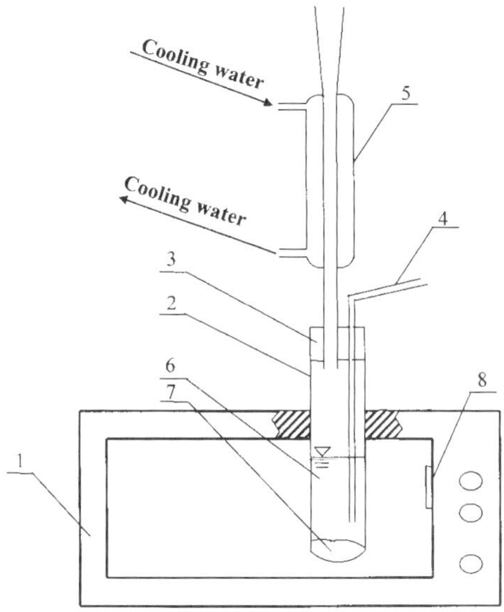

Fig. 11.12. Diagram of the apparatus for leaching in a microwave field [8, 9]. 1) microwave equipment; 2) the reactor; 3) the seal of the reactor; 4) sampling; 5) the reflux condenser; 6) the leaching pulp; 7) the charge; 8) the microwave emitter.

Fig. 11.13. The view of the autoclave for investigating leaching [10].

degree so that they can be used for various leaching experiments, for example, leaching of industrial waste [11].

Figure 11.13 shows the autoclave which enables leaching up to 350 °C and 18 MPa.

11.1.2. Experimental procedure

Effect of mixing

Mixing of the pulp does not have any significant effect on the leaching rate, provided the leached particles are in the suspended state. The problem is that the critical value of mixing depends on many factors, and also on the type and nature of the leached particles. On the other hand, the excessively high mixing rate may also cause problems and it is therefore necessary to determine the critical value of the mixing rate, i.e., the number of revolution of the mixer per unit time.

The dissolution rate in diffusion-controlled dissolution is described by the equation of convective diffusion [12]:

$$\frac{\mathrm{d}c_{\text{solution}}}{\mathrm{d}t} = k_1\,\omega^{\frac{1}{2}}\left(c_A - c_{\text{solution}}\right) \tag{11.17}$$

where ω is the angular speed of the stirrer. The value k_1 increases as a result of a decrease of the concentration gradient, and when the reaction is no longer inhibited by external diffusion, k_1 becomes constant. The so-called critical speed of revolution is determined by plotting the graph of the dependence of the rate of dissolution on the angular speed of rotation of the disc.

The experiments, formulated in this manner, were carried out at constant temperature and concentration conditions are shown in Fig. 11.14. The figure indicates that the yield increases up to approximately 250 min^{-1} and at more than 600 min^{-1} the yield again decreases. This decrease is caused by trapping of the liquid medium by the rotating surface which again increases the length of diffusion of the reagents. The determined critical revolutions are estimated to be approximately 250 min^{-1}, but because of the possible variant of inhibiting the reaction by diffusion through the Nernst layer, in experiments in all possible extreme conditions, the value N = 300 min^{-1} was always used. These results then ensure the equivalent hydrodynamic conditions in industrial reactors.

Process kinetics

Figure 11.15 shows the typical course of leaching of a chalcopyrite concentrate in a chloride medium in the temperature range 20–95 °C at the atmospheric pressure [13].

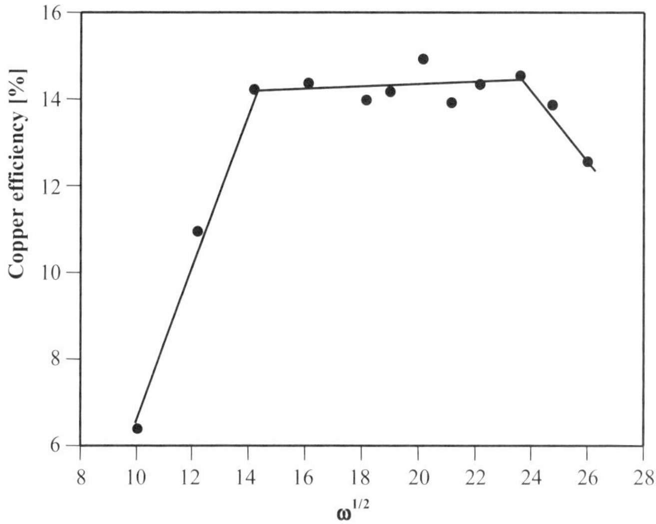

Fig. 11.14. Determination of the critical revolution of the mixer.

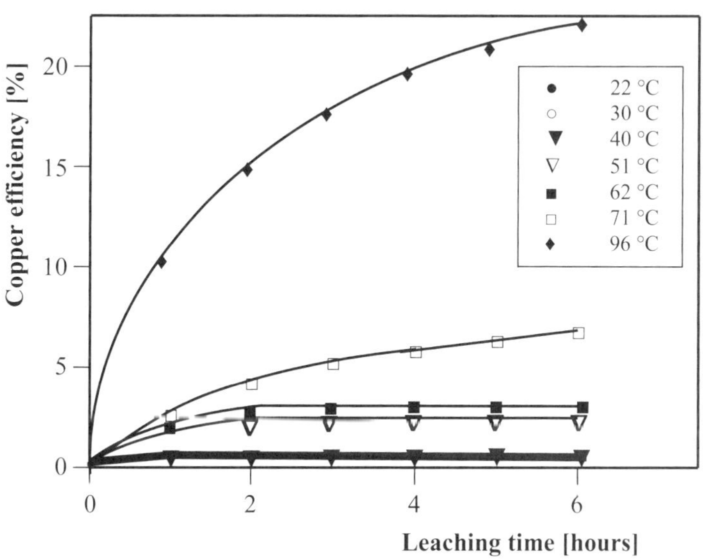

Fig. 11.15 Time dependence of leaching of Cu from chalcopyrite (–110+60 μm) at different temperatures in 0.5 M $FeCl_3$ + 0.5 M HCl.

Table 11.1 shows the yield of copper for the kinetic curves shown in Fig. 11.5 for the leaching conditions in the solution of 0.5 M $FeCl_3$ + 0.5 M HCl for the chalcopyrite fraction –110+60 μm.

The individual curves are used to determine the rate constants as a tangent to the curves of the given point using a suitable mathematical procedure. Of course, this must also have a physical

Table 11.1. The values of yield for the kinetic curves in Fig. 11.15

Time [h]	Yield of copper in the solution [%]						
	22 °C	30 °C	40 °C	51 °C	62 °C	71 °C	95 °C
1	0.29	0.37	0.43	1.49	1.74	2.14	11.00
2	0.33	0.41	0.52	1.61	2.1	3.61	14.85
3	0.35	0.45	0.58	1.91	2.43	4.55	17.56
4	0.37	0.45	0.61	1.94	2.69	5.36	19.65
5	0.39	0.5	0.67	2.08	2.73	5.89	21.00
6	0.41	0.54	0.69	2.23	2.85	6.34	21.56

meaning: in leaching of powder samples using a suitable modelling kinetic leaching equation, as discussed in the previous chapters. Table 11.3 contains the rate constants, determined in these experiments.

Effect of the grain size

The investigations of the effect of the grain size are fundamental and it is well-known [14, 15] that the leaching rate increases with decreasing grain size. Of course, this is given by the increase of the interfacial contact area through which chalcopyrite is leached. In addition to the determined standard grain size –110+60 μm, investigations also carried out of natural concentrate with a grain size of –314+200 μm, i.e., the grain size resulting from flotation [16]. The reason for this is that the grain size is not modified prior to leaching because of a possible increase of production costs so that the results obtained in this manner would be more similar to the considerations of leaching in practice.

Comparison of the typical leaching curves of chalcopyrite with different grain size is shown in Fig. 11.16 and Table 11.2 and indicates a number of facts.

Despite the objective result according to which the apparent leaching rate decreases with decreasing interfacial area, Fig. 11.16 shows the reverse. Examination of the initial sample by electron microscopy shows that although the experiments were carried out using the chalcopyrite fraction –315+200 μm, this fraction does not represent the actual particle size of the sample. Figure 11.17 shows

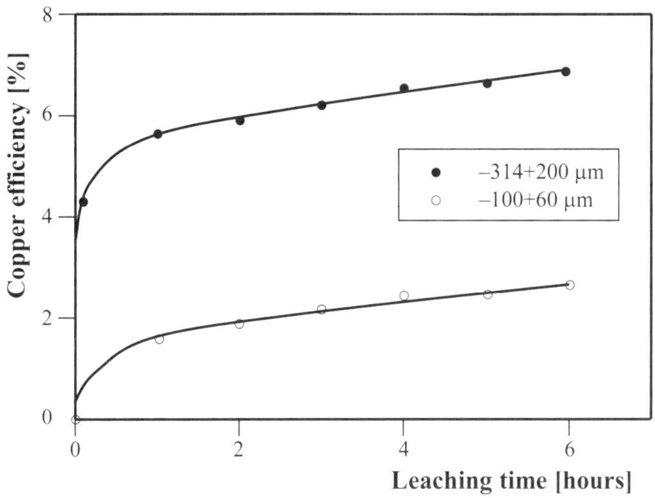

Fig. 11.16. Kinetic curves of leaching of chalcopyrite in ferric chloride at 60°C for the grain size of –314+200 μm and –110+60 μm.

Fig. 11.17. Initial chalcopyrite sample, –314+20 μm.

the presence of clusters of small particles which bond with each other physically and apparently form large particles. After introducing into the solutions, the particles are dispersed and take part individually in leaching. The small particles are leached quite rapidly and only larger pieces are leached subsequently. As a result of this decrease of the contact interfacial area, the overall leaching rate also greatly decreases.

Comparison of the difference of the copper yield into the solution for the individual fractions gave the value of approximately 4%. In the curve of the fraction –314 +200 μm this value is obtained almost immediately after the start of leaching (within 1 min).

In addition to the previously mentioned fact of the presence of

Table 11.2. Numerical values of the kinetic curves of leaching of chalcopyrite of the grain size –314+200 μm and –110+60 μm at 60°C

Leaching time [h]	Yield [%] –314+200 mm	Yield [%] –110+60 mm	Δ [%]
0	4.39	–	–
1	5.58	1.74	3.84
2	6.11	2.11	4.01
3	6.43	2.43	4.00
4	6.78	2.69	4.09
5	6.89	2.73	4.16
6	7.02	2.85	4.17

submicroscopic particles, leaching is also affected by the presence of easily soluble particles on the surface of the starting material prior to leaching. Simple experiments with rinsing of the initial concentrate with the distilled water after 10 min showed repeatedly that approximately 3.99% Cu was transferred into the solution, which is in excellent agreement with the differences shown in Table 11.2. These easily soluble fractions are formed on the surface of chalcopyrite probably as a result of longer storage or due to the effect of milling with subsequent weathering. This is not probably the singular result; it has been confirmed exactly in [19–21] in leaching of tetrahedrite and probably complicated the evaluation of results of several authors, judging by the form of the published leaching curves [15, 17, 18].

Effect of acid concentration

The presence of acid in the leaching medium in acid oxidation leaching has a single role: prevent hydrolytic precipitation of iron from the solution. A certain minimum amount is sufficient for this and a further increase of the acid concentration in the leaching medium causes problems, especially in production and ecology.

In order to determine the effects of the acidity of the solution on the leaching kinetics, leaching experiments were carried out at 70°C using ferric chloride with a concentration of 0.5 M, and the

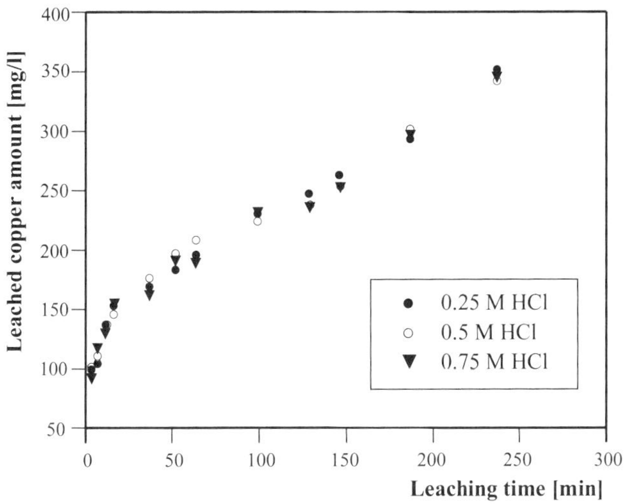

Fig. 11.18. Effect of acidity on the leaching kinetics of chalcopyrite.

concentration of hydrochloric acid was varied in the range 0.25, 0.5 and 0.75 M. Lower concentrations were not used because of the possibility of hydrolysis of iron. The results presented in Fig. 11.18 show that the change of the acidity of the leaching solution in the range 0.25–0.75 M HCl has no significant effect on the leaching kinetics.

Effect of the concentration of ferric ions

The dependence of the leaching rate of chalcopyrite on the concentration of the ferric ions is shown in Fig. 11.19. The

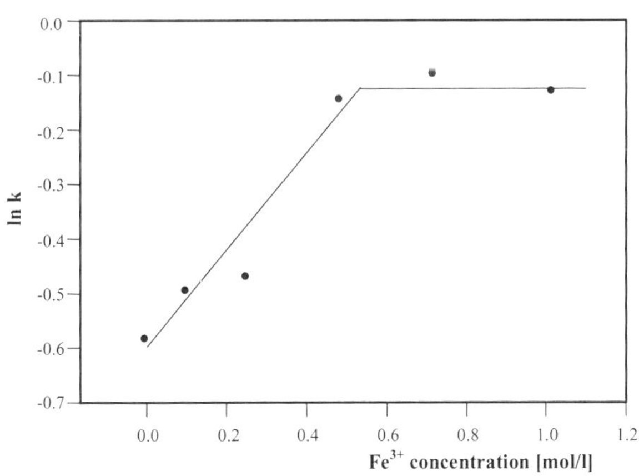

Fig. 11.19. Effect of the concentration of ferric ion on the leaching kinetics of chalcopyrite.

insufficient amount of the ferric ions in the leaching medium obviously reduces the leaching rate as a result of external diffusion to the leaching interface. As already mentioned, the limiting concentration for this case was estimated at approximately 0.2 M $Fe3^{+}$, but as shown by Fig. 11.9, only the concentration of ferric ions higher than 0.5 M is sufficient for ensuring the sufficient number of the ferric ions required for leaching not inhibited by external diffusion.

Effect of temperature

In general, oxidation leaching of chalcopyrite in the acid medium is characterised by the increase of the leaching rate with increasing temperature, as indicated in Fig. 11.15. The quantitative representation of the effect of temperature is possible using the Arrhenius relationship, shown in Fig. 11.20 for the concentrations of 1.0, 0.75 and 0.5 M $Fe3^{+}$. The values of the rate constants, essential for calculations, are summarised in Table 11.3.

The determined value of the activation energy indicates that, in this case, the process takes place to some extent in the kinetic regime, although close to the boundary determining the interface between the kinetic and diffusion regime. In this case, the temperature is sufficiently high for a fast chemical reaction resulting also in the formation of elemental sulphur on the leached surface, as indicated by the measurements of X-ray diffraction phase

Table 11.3. Values of the rate constant *k* for leaching in a chloride medium

	Rate constant k		
Temperature [°C]	**0.5 M**	**0.75 M**	**1 M**
22	$6.8982\cdot 11^{-4}$ (0.999)	$7.345\cdot 11^{-4}$ (0.991)	$8.457\cdot 11^{-4}$ (0.994)
30	$9.5627\cdot 11^{-4}$ (0.977)	$1.009\cdot 11^{-3}$ (0.988)	$1.298\cdot 11^{-3}$ (0.992)
40.5	$1.5344\cdot 11^{-4}$ (0.973)	$1.790\cdot 11^{-3}$ (0.987)	$1.969\cdot 11^{-3}$ (0.986)
51	$2.6970\cdot 11^{-4}$ (0.999)	$2.931\cdot 11^{-3}$ (0.989)	$3.435\cdot 11^{-3}$ (0.993)
62.5	$5.2356\cdot 11^{-4}$ (0.995)	$5.871\cdot 11^{-3}$ (0.993)	$6.861\cdot 11^{-3}$ (0.989)
71.5	$1.1365\cdot 11^{-4}$ (0.979)	$1.411\cdot 11^{-2}$ (0.988)	$1.742\cdot 11^{-2}$ (0.979)
95	$1.7113\cdot 11^{-4}$ (0.961)	$2.882\cdot 11^{-2}$ (0.976)	$2.583\cdot 11^{-2}$ (0.985)

analysis and examination of morphology. The rate-controlling step in this case is probably the formation of elemental sulphur on the reaction surface. However, on the other hand, as indicated by scanning electron micrographs, elemental sulphur does not form a completely compact layer on the leached surface which efficiently protects the supply and removal of the reactants to/or the leached surface.

The formed sulphur is also permanently 'attacked' in movement of the pulp and by milling. It appears that the overall process contains another possible partial process which has not so far been considered. External diffusion of the reactants is eliminated and this obviously results in the regime to which the chemical reaction of the formation of elemental sulphur S_8 (or of another products) and also the diffusion of the reactants through these products, contribute. However, it is difficult to determine the role of these two components in the overall rate, taking into account the constantly changing active area through which the leaching process takes place, on the one hand, as a result of the continuous decrease of the area after leaching and, on the other hand, as a result of coating with sulphur and removal of sulphur as a result of peeling. It is likely that some of the measured normal values of the concentration of copper in the solution are a manifestation of this behaviour.

In the investigated concentration range of the leaching medium 0.5 M–1 M $FeCl_3$, there were no significant deviations in the values of activation energy, as indicated by Fig. 11.20, which means that the mechanism of the process in this range is the same.

However, this may not always be the case. The reaction mechanism may change from diffusion-controlled to chemically-controlled, and vice versa. This change may take place as a result of the change of the concentration of the reagents in the solution, pressure or temperature. In chemical or electrochemical reactions, the rate of the chemical reaction at of the low concentration of the reagents in the volume of the solution is low and, consequently, may determine the overall rate of the process. The increase of the reagent concentration also increases the rate of the chemical reaction and the slowest stage changes to the diffusion of the dissolved ions from the interface into the solution through the boundary layer. In graphical representation of the kinetic curves, this is also reflected in their non-linear form. In a relatively large number of cases, the regime of the heterogeneous reactions at low temperatures is different from that at high or higher temperatures.

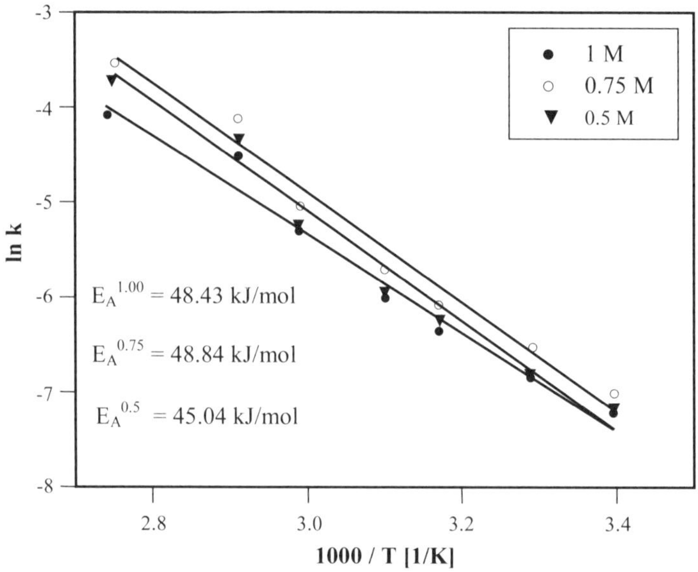

Fig. 11.20. Arrhenius plot of the effect of temperature on the leaching of chalcopyrite in a chloride medium.

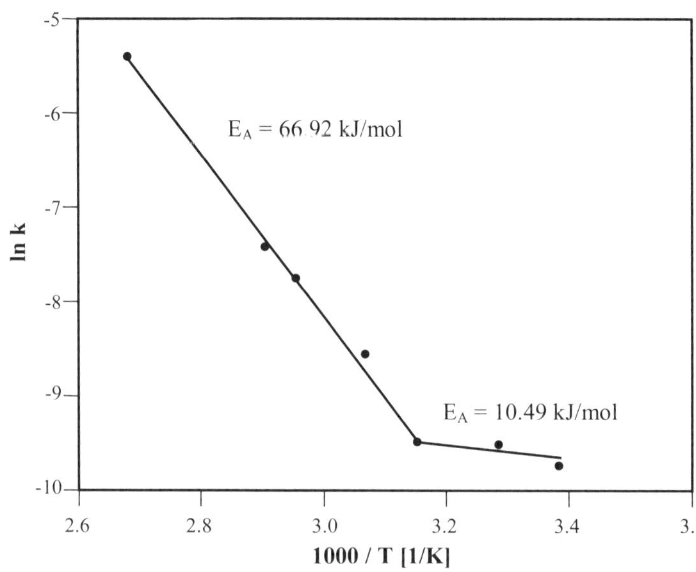

Fig. 11.21. Arrhenius plot of the effect of temperature on leaching of chalcopyrite in a sulphate medium.

plot of the effect of temperature on the leaching of chalcopyrite in ferric sulphate [13, 29] for the concentration of 0.5 M Fe^{3+}.

In the investigated temperature range, 20–90°C, the Arrhenius plot shows a break at 40°C, indicating a change of the reaction mechanism. In the range at lower temperatures (20–40°C), the activation energy was E_A = 11.96 kJ/mole [CI = 0.847]. The apparent activation energy at higher temperatures was E_A = 68.75 kJ/mole (CI = 0.998) (CI is the correlation index).

The change of the mechanism of chalcopyrite leaching in ferric sulphate depending on temperature is a completely new observation. On the other hand, it is not in any contradiction with the published results because no results have been published on leaching in such a wide temperature range. It is generally known that low temperature greatly reduces the rate of the leaching process and this is obviously the main reason why the authors working in this area have considered the temperature range interesting for commercial application.

The activation energy determined in the temperature range 20–40°C indicates that diffusion is the rate-controlling stage. No formation of elemental sulphur, coating the leached surface, was detected in this temperature range. This means that there is no diffusion of reactants through the layer. Diffusion is usually very fast [22]. On the other hand, the chemical reaction is very slow at these low temperatures and, at the same time, the viscosity of the leaching medium increases. Therefore, it appears that the rate-controlling stage in this reaction is the external diffusion of the reactants from the volume of the leaching agent to/or the leached surface, as a result of the higher viscosity of the medium at lower temperatures [23, 24]. This was also confirmed by the experimental investigations including the filtration of sulphate and chloride media at which the rate of filtration of the sulphate solution was considerably lower than that of the chloride solution.

The apparent activation energy E_A = 69 kJ/mole, determining the temperature range 20–99°C, are in excellent agreement with the published results were the average value 69.2 kJ/mole, as indicated by Table 11.4. The determined activation energy is valid for the reaction taking place in the kinetic regime.

This example of the examination of leaching is not exhausting, and it cannot be. It is always important to know which type of reaction should be studied, which investigations are possible, and the type of equipment available. The formulation of the experiments will differ if both the liquid and solid phase will contribute to the reaction, and also the fact when only non-oxidation reactions will take place, and also differ if electrochemical reactions out to be studied, etc. Analysis of the results may also be purely

Table 11.4. Values of apparent activation energy E_A in the measured temperature range for the leaching medium $Fe_2(SO_4)_3$

t [°C]	E_A [kJ/mol]	Reference
27-85	84	27
30-95	38-63	26
32-50	75	28
50-94	71	17, 25
60-90	84	22
40-99	67	13
average	*68.75*	

phenomenological on the basis of chemical elemental analysis of dissolved amounts, or chemical analysis of solid leaching residue which may also be studied by phase analysis (X-ray diffraction phase analysis, etc), morphological analysis (scanning electron microscopy), elemental microanalysis (EDAX, LINK).

Other methods include electrochemical methods for measurement of the changes of the electrochemical potential in situ in the process (cyclic voltametry, chronopotentiometry, etc). Each method has its advantages and enables the process to be examined from a specific angle, but all these methods are available only in a small number of cases and the time and financial viewpoint is also very important. Examples of the published results of leaching of sulphide will be discussed in the following sections, in which different investigations methods are used.

11.2. Changes of pH in relation to temperature

As already mentioned, the course of the hydrometallurgical processes is usually more favourable at higher temperatures. Of course, due to this, it is necessary to measure the parameters of variables of the process. With the exception of temperature and, possibly, the redox potential, it is the actual value of pH. These measurements were taken in most cases using a glass electrode in the temperature range up to 110°C. However, it should be stressed that the measurement of pH at elevated temperatures results in a number of problems. In particular, the measuring system must be calibrated to ensure accurate measurements. This is carried out using a series of seven buffering solutions, calibrated between zero and 95°C [30, 31].

Measurements were taken using a measuring hydrogen electrode and an Ag–AgCl electrode as the reference electrode immersed in a potassium chloride solution. Three different concentrations for each buffer were used. The considered cell was:

$$\mathrm{Ph(H_2)} \mid \text{buffer, KCl (m)} \mid \ \text{Ag–AgCl, Pt}$$

where Pt was replaced by Pd or In, if necessary. These experiments yielded sufficiently accurate values of $E^0_{\mathrm{Ag,AgCl}}$ for the temperature range 25–275 °C, resulting in the EMF value of the hydrogen electrode. The cell does not have a liquid junction and, consequently, the reactions of the half cell are:

$$\mathrm{H + e = \frac{1}{2} H_2} \quad \text{and } \mathrm{AgCl + e = Ag + Cl^-}$$

with the potentials of the half cell

$$E_{\mathrm{H^+,H_2}} = E^0_{\mathrm{H,H_2}} - \frac{RT}{F} \ln\left(\frac{\{\mathrm{H_2}\}^{\frac{1}{2}}}{\{\mathrm{H^+}\}}\right)$$

$$E_{\mathrm{Ag,AgCl}} = E^0_{\mathrm{Ag,AgCl}} - \frac{RT}{F} \ln\left(\frac{\{\mathrm{Ag}\}\{\mathrm{Cl^-}\}}{\{\mathrm{AgCl}\}}\right)$$

or the potential of the entire cell is:

$$E = E_{\mathrm{Ag,AgCl}} - E_{\mathrm{H^+,H_2}} = E^0_{\mathrm{Ag,AgCl}} - \frac{2.303RT}{F} \log\left(\frac{\{\mathrm{Cl^-}\}\{\mathrm{H^+}\}}{\{\mathrm{H_2}\}^{\frac{1}{2}}}\right)$$

The value $\{\mathrm{Cl^-}\}$ in the buffer solution is determined by the molality of added KCl and it means that $\{\mathrm{Cl^-}\} = m_{\mathrm{KCl}}\gamma_{\mathrm{Cl^-}}$, so that

$$p\left(\{\mathrm{H}\}\gamma_{\mathrm{Cl^-}}\right) = -\log\left(\{\mathrm{H^+}\}\gamma_{\mathrm{Cl^-}}\right) = \frac{E - E^0_{\mathrm{Ag,AgCl}}}{\frac{2.303RT}{F}} - \log m_{\mathrm{KCl}} + \frac{1}{2}\log\{\mathrm{H_2}\}$$

In fact, the measured value was $p(\{HCl\}\gamma_{Cl-})$ of a series of chloride concentrations in every buffer and the values, obtained at individual temperatures, were extrapolated to zero concentration giving the value $p(\{HCl\}\gamma_{Cl-})$ which is thermodynamically completely defined. The value of pH of the buffer solution in the absence of the chloride is:

$$p\{H\} = p\left(\{H\}\gamma_{Cl^-}\right)^0 + \log\gamma^0_{Cl^-}$$

and it is assumed that for $\gamma^0_{Cl^-}$:

$$\log\gamma^{\circ}_{Cl^-} = -\frac{AI^{\frac{1}{2}}}{1+1.5I^{\frac{1}{2}}}$$

where A is the Debye–Hückel limiting constant and depends on temperature. This assumption is only the non-thermodynamic part of the definition of $p\{pH\}$ and makes the value γ_{Cl-} equal to the mean activity coefficient of NaCl in the solution up to 0.5 M. The ionic force of the buffer solution I in the absence of KCl is calculated from the data $p(\{HCl\}\gamma_{Cl-})$ and other appropriate equilibrium constants of the buffer salts.

Seven standard buffer solutions are presented in Table 11.5.

Table 11.5. Standard buffer solutions and the pH values of the solutions at 25°C

Substance	Concentration [M]	pH at 25 °C
KH_3 oxalate·$2H_2O$	0.05	1.679
KH tartrate	saturated at 25 °C, 0.0341	3.557
KH phtalate	0.05	4.008
KH_2PO_4 Na_2HPO_4	0.025 0.025	6.865
KH_2PO_4 Na_2HPO_4	0.008695 0.03043	7.413
$Na_2B_4O_7$·$11H_2O$	0.01	9.180
$Ca(OH)_2$	saturated at 25 °C, 0.0203	12.454

The first two solutions are thermally unstable and dissociate at relatively low temperatures. The phthalate solution is thermally highly stable and has excellent buffer properties. The phosphate and borate buffers are also quite suitable for application at higher temperatures but their pH values are too high for the Ag–AgCl electrodes. The solubility of calcium hydroxide $Ca(OH)_2$ decreases with increasing temperature. The determined and published values of pH for the measured temperatures are presented in Table 11.6.

Table 11.6. Values of pH at elevated temperatures for some buffer solutions [31]

Temperature [°C]	Oxalate		Tartrate		Phthalate	
	pH	$-\log \gamma^{o}_{Cl}$	pH	$-\log \gamma^{o}_{Cl}$	pH	$-\log \gamma^{o}_{Cl}$
95	1.779	0.114				
110	1.792	0.115	3.676	0.093	4.254	0.113
115	1.804	0.116	3.690	0.094	4.284	0.114
111	1.817	0.117	3.705	0.095	4.313	0.115
115	1.830	0.118	3.721	0.096	4.340	0.117
120	1.843	0.120	3.738	0.097	4.366	0.118
125			3.753	0.099	4.391	0.111
130			3.768	0.110	4.416	0.111
135			3.781	0.111	4.441	0.113
140			3.791	0.113	4.465	0.115
145			3.800	0.114	4.490	0.116
150			3.804	0.116	4.514	0.118
155			3.806	0.117	4.537	0.120
160			3.802	0.119	4.560	0.122
165					4.582	0.123
170					4.602	0.125
175					4.619	0.127
180					4.633	0.130
185					4.645	0.131
190					4.651	0.134
195					4.654	0.136
200					4.651	0.138

The graphical for of the temperature dependence of the pH values of the buffers is shown in Fig. 11.22.

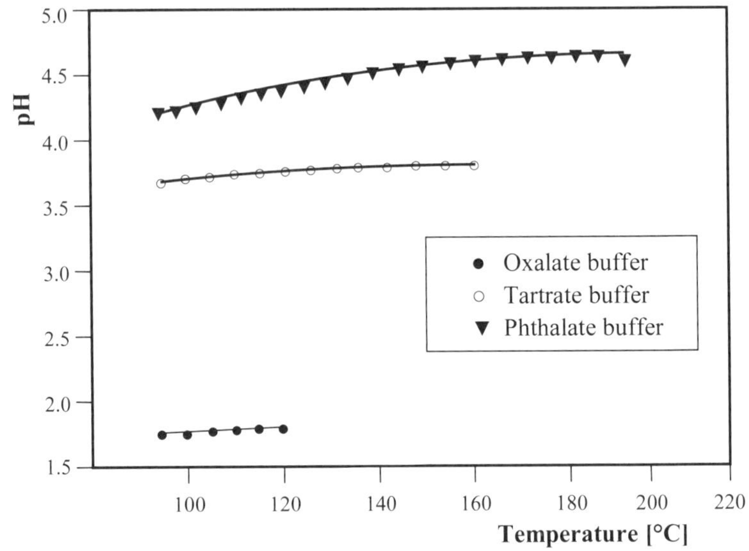

Fig. 11.21. Temperature dependence of buffer pH.

References

1. Tretjakov Yu.D.: Solid-phase reactions [in Russian], Khimiya, Moscow, 1978, 164–192.
2. Havlík T., Flórián K., Havlík M.: Fundamentals of experimental procedures [in Slovak], Štrofek Publishing House, Košice, 1998, 238.
3. Erfurth H., Just G.: Modelling and optimisation of chemical processes [in Czech], SNTL, Prague, 1979, 200.
4. Ukašík M., Havlík T.: *PCLab* – Laboratory measurement system developed at Department of Nonferrous Metals and Waste Treatmeny, Faculty of Metallurgy, Technical University, Košice.
5. Havlík T.: Study of dissolution of copper sulphides in an acid medium, Dissertation, Faculty of Metallurgy, Technical University, Košice, 1982.
6. Ukašík M., Havlík T.: *Hydrometallurgy*, 77, 2005, 139–147.
7. Havlík T., Šulek K., Briančin J., Kammel R.: *Metall*, 52, 7–8, 1998, 1–4.
8. Havlík T., Popovičová M.: *Acta Metallurgica Slovaca*, 6, 2, 2000, 171–177.
9. Havlík T., Popovičová M., Ukašík M: *Metall*, 55, 11, 2001, 332–335.
10. Havlik T., Friedrich B., Stopic S. : *Erzmetall*, 57, 2, 2004, 83–90.
11. Miškufová A., Krištofová S.: Processing neutralisation sludge from electroplating [in Slovak], Proc. seminar Waste Recycling VI, Košice, 2002, 165–170.
12. Leviè V.: *Zh. Fiz. Khimii*, XXII, 6, 1948, 702–729.
13. Havlík T.: Acid oxidation leaching of chalcopyrite and behaviour of sulphur in this process, PhD dissertation, Technical University, Košice, April 1996.
14. Dutrizac J.E.: *Met. Trans.* B., 12B., June 1981, 371–378.
15. Majima H., Awakura Y., Hirato T., Tanaka T.: *Can. Metall. Quarterly*, 24, 4, 1985, 283–291.
16. Havlík T., Škrobian M., Baláž P., Kammel R.: *Int. Journal of Mineral Processing*, 43, 1995, 61–72.
17. Dutrizac J.E., MacDonald R.J.C.: *Met. Trans.*, 2, 1971, 2311–2312.
18. Braithwaite J.W., Wadsworth M.E.: Oxidation of chalcopyrite under simulated conditions of deep solution mining, Extract. Metall. of Copper, vol.II., Yannopoulos & Agarwal eds., TMS-AIME, 1976, 752–775.
19. Havlík T., Škrobian M.: *Trans. Tech. Univ. Košice*, 2, 1992, 139–144.
20. Havlík T., Škrobian M., Dudáš D.: *Hutnícke listy*, XLVI, 11–12, 1991, 76–80.
21. Havlík T., Škrobian M., Baláž P.: *Erzmetall*, 47, 2, 1994, 112–119.
22. Munoz P.B., Miller J.D., Wadsworth M.E.: *Met. Trans.* B., 11B, June 1979, 149–158.
23. Daucík K, Lisá M., Kordík D., Daučík P.: Chemical laboratory tables [in Slovak], Alfa, Bratislava, 1984.
24. Březina F., Mollin J., Pastorek R., Šindelár Z.: Chemical tables of inorganic compounds [in Czech], SNTL Prague, 1986.
25. Dutrizac J.E., MacDonald R.J.C., Ingraham T.R.: *Trans. Met. Soc. AIME,* 245, May 1969, 955–959.
26. Dutrizac J.E.: *Met. Trans. B.*, 9B, September 1978, 431–439.

27. Baur J.P., Gibbs H.L., Wadsworth M.E.: *Met. Soc. AIME*, 72-B-96, 1972.
28. Lowe D.F.: The kinetics of the dissolution reaction of copper and copper-iron sulphide minerals using ferric sulphate solutions, Ph.D Thesis, University of Arizona, 1970.
29. Havlík T., Škrobian M., Petričko F.: *Acta Metallurgica Slovaca*, 2, 2, 1996, 133–141.
30. Bates R.G.: *J. Res. Natl. Bur. Std.*, A66, 1962, 179–184.
31. Manning G.D.: *J. Chem. Soc. Faraday Trans.*, 1, 74, 2434–2451.

Chapter 12

LEACHING OF COPPER SULPHIDES

Copper is found in the nature in more than 160 minerals. Of course, copper producers strive to process the starting materials with the maximum efficiency. In addition to processing carried out to produce a suitable concentrate, another important factor is the mineralogical and chemical composition of the starting material and also its metal content. From this viewpoint, the ideal starting material is the copper suphide Cu_xS, especially chalcocite, Cu_2S, with almost 80 wt.% of copper. However, there are almost no natural deposits of chalcocite and covellite ores and, therefore, sulphides of the type $Cu_xFe_yS_z$, mostly chalcopyrite $CuFeS_2$, are processed in most cases, because they are found in sufficient quantities in the nature. Minerals of the type $Cu_xMe_yS_z$ are also interesting (here Me is a metal other than iron). These minerals include tetrahedrite $Cu_{12}Sb_4S_{13}$, famatinite Cu_3SbS_4, chalcostibite $CuSbS_2$, lautite CuAsS, enargite Cu_3AsS_4, sinerite $Cu_6As_4S_9$, tenantite $Cu_{12}As_4S_{13}$, claprotite Cu_3BiS_3, stanite Cu_2FeSnS_4, sromeyerite CuAgAs, and others. Of course, prior to starting production, it is necessary to test all aspects of leaching on the laboratory and pilot-plant scale.

12.1. Copper sulphides of the $Cu_xFe_yS_z$ type

Since chalcopyrite $CuFeS_2$ is the most widely found copper mineral, the possibilities of metallurgical processing of this mineral are the subject of special attention. One of the most important effects exerted on the process is the effect of leaching. Therefore, work is being carried out constantly to examine the kinetics of leaching in various media in order to find the optimum conditions for the effective hydrometallurgical process of processing chalcopyrite. Many aspects of leaching of this mineral have not been completely explained and the results are characterised by a relatively large scatter and are interpreted in different ways.

The solubility of chalcopyrite in water is very low: at 120 °C, solubility is $2.3\cdot10^{-6}$ mol/l [1]. Stanczyk and Rampacek [2] investigated surface reactions of chalcopyrite, although the mineral was not leached. Microscopic examination of chalcopyrite particles, suspended in distilled water and heated in an autoclave for 2 h at 200 °C, showed that the particles are coated with a thin surface layer of copper sulphide and ferrous sulphide. The reaction was described by the equation:

$$CuFeS_2 \rightarrow CuS + FeS \tag{12.1}$$

The reaction of leaching chalcopyrite in an acid medium without air was described by Ichikuni [3] who found that only iron is transferred into the solution. The rate of reaction was very low, even at temperatures of 100 °C using concentrated hydrochloric acid [4]

$$CuFeS_2 + 2H^+ \rightarrow CuS + Fe^{2+} + H_2S \tag{12.2}$$

In a reduction medium, for example, in the presence of SO_2, the following reaction takes place:

$$SO_2 + 2H_2S \rightarrow 3S^0 + 2H_2O \tag{12.3}$$

$$4CuS + 4H^+ + SO_2 \rightarrow 4Cu^+ + 5S^0 + 2H_2O \tag{12.4}$$

Meyers et al. [5] published data on the dissociation of chalcopyrite in an aqueous solution of hydrochloric acid with SO_2 at 180 °C and a pressure of 1220 kPa. Elemental sulphur was released in accordance with the reaction:

$$4CuFeS_2 + 3SO_2 + 12HCl \rightarrow 4CuCl + 4FeCl_2 + 12S^0 + 6H_2O \tag{12.5}$$

It is much easier to leach chalcopyrite in the presence of an oxidation agent. Many oxidation agents have been investigated. Of these, only oxygen does not require regeneration and return to the process. On the other hand, the application of oxygen is efficient only at high temperatures and pressures which is also economically inefficient.

Chalcopyrite is leached in a neutral aqueous medium with the formation of sulphates

$$CuFeS_2 + 4O_2 \rightarrow CuSO_4 + FeSO_4 \tag{12.6}$$

This reaction is slow at room temperature, but the reaction rate increases with increasing temperature. In these conditions, iron is hydrolysed in accordance with the equations:

$$Fe^{2+} + 2H_2O \rightarrow Fe(OH)_2 + 2H^+ \quad (12.7)$$

$$4Fe(OH)_2 + 2H_2O + O_2 \rightarrow 4Fe(OH)_3 \quad (12.8)$$

However, it is possible that oxidation precedes hydrolysis in accordance with the following reactions:

$$4Fe^{2+} + 4H^+ + O_2 \rightarrow 4Fe^{3+} + 2H_2O \quad (12.9)$$

$$Fe^{3+} + 3H_2O \rightarrow Fe(OH)_3 + 3H^+ \quad (12.10)$$

The oxidation of the iron ion is catalysed by the copper ion as follows:

$$Cu^{2+} + Fe^{2+} \rightarrow Cu^+ + Fe^{3+} \quad (12.11)$$

Depending on temperature, the iron hydroxide is rapidly transformed to oxyhydroxide or oxide according to with the equations:

$$Fe(OH)_3 \rightarrow FeOOH + H_2O \quad (12.12)$$

$$2FeOOH \rightarrow Fe_2O_3 + H_2O \quad (12.13)$$

According to Stanczyk and Rampacek [5], after leaching a 5% chalcopyrite suspension in an aqueous medium at 200 °C for 30 min, 16.8 g/l of copper, 0.23 g/l of ferrous ions, 0.86 g/l of ferric ions and approximately 5 g/l of sulphuric acid were transferred into the solution. The residue contained mainly hydrolytically precipitated iron on which copper was absorbed.

The copper sulphides are not only in the acid but also in basic media. In ammoniacal media, copper forms complex ions, referred to as aminocomplexes. In this case, iron precipitates in the form of hydroxide and sulphides oxidised to higher positive valencies or even to sulphates. In an ammoniacal medium, oxidation of sulphur takes place in steps with the formation of several intermediate compounds, such as thiosulphates, polythionates, dithionates, hydrogen sulphites and sulphoxylates.

The overall reaction of pressure leaching of chalcopyrite according to Evans et al. [6] is:

$$2CuFeS_2 + 17/2O_2 + 12NH_3 + (2+n)H_2O \rightarrow$$
$$\rightarrow 2Cu(NH_3)_4SO_4 + 2(NH_4)_2SO_4 + Fe_2O_3 \cdot nH_2O \qquad (12.14)$$

at a temperature of 85 °C, a pressure of 710 kPa, and 4.5–5.0 moles of free ammonia to the total amount of metal. After nine hours, 96% of copper and 80% of zinc was leached. Iron precipitated as the hydrated ferrous oxide and also as ferric arsenate and antimonate.

This process was not introduced into industry because of problems with sulphur which must be bonded in ammonium sulphate.

Kuhn et al. [7] carried out investigations to find improvements for the possible application of this process in the dispersion of oxygen into the sludge. Ammonium sulphate was dissociated with lime to ammonia which was recycled back to the process and to calcium sulphate.

The reaction of chalcopyrite with sodium hydroxide at 60–120 °C at an oxygen pressure of 500–1500 kPa is described as follows:

$$2CuFeS_2 + 12NaOH + 17/2O_2 \rightarrow$$
$$\rightarrow 2Na_2Cu(OH)_4 + Fe_2O_3 + 4Na_2SO_4 + 2H_2O \qquad (12.15)$$

According to Haskett et al. [8], the reaction takes place within 0.5–3 h in the presence of a sufficient excess of sodium hydroxide. The latter maintains unstable cuprate ions in the solution and prevents the hydrolysis of copper hydroxide. However, a large problem in this process is the regeneration of sodium hydroxide.

The initial studies by Trail and McClelland [9] were concerned with the examination of leaching of a chalcopyrite concentrate in a concentrated solution of ferrous chloride at 95 °C. The yield of copper was approximately 90%, but the yield of iron was only 60–70%, which was probably caused by its hydrolysis and subsequent precipitation during the experiment. Only 5% of elemental sulphur was produced.

Pike et al. [10] investigated the leaching of a chalcopyrite concentrate on the pilot plant scale. Of the chloride oxides and ferrous sulphate investigated, the chloride proved to be more efficient than the sulphate. In this study, the yield of copper was between 85 and 90 °C using solutions with 4–10% of ferric chloride.

Brown and Sullivan [11] and Sullivan [12] investigated the leaching of a chalcopyrite concentrate of ferric chloride and ferric sulphate. The concentrate was finely ground because chalcopyrite is a mineral characterised by high resistance to leaching and this method increased the interfacial area. Using a 1% solution of ferric sulphate for 57 days at 35 °C, the yield of copper was 33% but the leaching rate was considerably increased by increasing temperature. In 14 days at 50 °C, 44% Cu was leached and leaching at the boiling point for 5 h produced 38% Cu.

The leaching reactions proposed by the above authors are as follows:

$$CuFeS_2 + 4Fe^{3+} \rightarrow Cu^{2+} + 5Fe^{2+} + 2S^0 \qquad (12.16)$$

$$CuFeS_2 + 4Fe^{3+} + 2H_2O + 3O_2 \rightarrow Cu^{2+} + 5Fe^{2+} + 2H_2SO_4 \qquad (12.17)$$

In the reaction (12.16), approximately 75% of chalcopyrite was dissolved and the residue dissolved in accordance with the reaction (12.17). Small amounts of sulphate formed in the presence of dissolved oxygen. However, these authors also found that the ferric chloride is a better leaching agent than the ferric sulphate.

Klec and Liopo [13] investigated the reaction of a chalcopyrite ore in an acidified solution of ferric chloride. They proposed the following leaching reaction:

$$CuFeS_2 + 4FeCl_3 \rightarrow CuCl_2 + 5FeCl_2 + 2S^0 \qquad (12.18)$$

Subsequently, sulphur was oxidised in accordance with the reaction:

$$8S^0 + 6H_2O \rightarrow H_2S_3O_6 + 5H_2S \qquad (12.19)$$

Yermilov, Tkachenko and Tseft [14] investigated the leaching kinetics of chalcopyrite in the solutions of ferric chloride in the temperature range 60–106 °C. The reaction rate increased with increasing temperature and the apparent activation energy was approximately 50 kJ/mole. The leaching rate depended directly on the concentration of ferric ions in the solution. The sulphur, formed during the reaction, did not interfere with the leaching kinetics.

Ichikuni [15] confirmed that the leaching of chalcopyrite takes place in accordance with the equation (12.18). Later, he concluded [16] that small deviations in the ratio of leached copper and iron, detected in the initial stages of leaching in ferric chloride, were the consequence of the preferential leaching of iron from the structure of chalcopyrite. He also found that the ferric ions attacked the mixture of chalcopyrite and pyrite with a low degree of selectivity. He concluded that the galvanic effect resulting from the presence of pyrite does not operate in leaching of chalcopyrite.

Hiskey and Wadsworth [17] investigated the galvanic reaction between chalcopyrite and copper. This reaction results in the rapid transformation of chalcopyrite to chalcocite in accordance with the reaction:

$$CuFeS_2 + Cu^0 + 2H^+ \rightarrow Cu_2S + Fe^{2+} + H_2S \tag{12.20}$$

The rate-controlling parameter was the current in relation to the effective surface of the anode and the cathode of the galvanic cell. The experimental activation energy was E_A = 48 kJ/mole. An electrochemical model of leaching has been proposed for the overall mechanism of the reaction.

Dutrizac, MacDonald and Ingraham [18] investigated the leaching of synthetic chalcopyrite in the solution of ferric sulphate by the rotating disk method. The leaching reaction was identical with the relationship (12.18). In the investigated temperature range of 50–94 °C, leaching follows the parabolic kinetics and the apparent activation energy was 71 kJ/mole. At concentrations lower than 0.01 M Fe^{3+}, the rate-controlling step of the process was the diffusion of ferric sulphate through the layer of the solid reaction product on the leached product – elemental sulphur. At higher values, the rate was no longer dependent on the concentration of the ferric ion in the solution, nor on the concentration of sulphuric acid in the solution in the range 0.001–1.0 M H_2SO_4. The addition of the ferrous sulphate to the solution resulted in a large decrease of the leaching rate. On the other hand, the addition of the ferric chloride to the solution of the ferric sulphate greatly increased the leaching rate of chalcopyrite [19]. The addition of $FeCl_2$, even in large quantities, to the solution of ferric chloride reduced only slightly the leaching rate at 85 °C [20]. Jones and Peters [21] showed, in similar leaching experiments, that the addition of 0.1 M of $FeSO_4$ to 0.1 M $Fe_2(SO_4)_3$ solution reduced the amount of dissolved copper from 20 to 4% in the same leaching conditions. On the other hand, Beckstead et al. [22] obtained approximately identical leaching relationships after adding 20 or 100 g/l of $FeSO_4$ to the leaching medium $Fe_3(SO_4)_2$ at 93 °C.

Haver and Wong [23] proposed an integrated hydrometallurgical process of treatment of chalcopyrite concentrates using the acidified solution of ferric chloride. The process takes place in the temperature range 30–106 °C, and the yield of copper at higher temperatures is 99%. Leaching is controlled by the parabolic kinetics as a consequence of mass transfer through the layer of elemental sulphur, formed as a reaction product at the interface. It was found that at the ratio of $FeCl_3$ to $CuFeS_2$ of 1:2.7, the entire amount of dissolved copper was indeed in the monovalent form. Consequently, the leaching reaction should be described by the relationship:

$$CuFeS_2 + 3FeCl_3 \rightarrow CuCl + 4FeCl_2 + 2S^0 \qquad (12.21)$$

Approximately 70% of the total amount of sulphur was transferred into the elemental form from chalcopyrite, and the residue was oxidised, probably to the sulphate.

Baur, Gibbs and Wadsworth [24] investigated the leaching of natural powder chalcopyrite in the temperature range 27–85 °C in ferric ions in the concentration range 0–0.69 M Fe^{3+} at pH of 1.25. The results show that the chalcopyrite is leached in accordance with the parabolic kinetics and this was attributed to the effect of the diffusion of reaction products through the layer of elemental sulphur formed as a reduction product at the interface. The relatively low leaching rate was also attributed to this fact. The calculated activation energy was 84+21 kJ/mole, which is in agreement with the rate-controlling step of the mechanism of diffusion in the solid-state stipulated previously. The oxidation rate did not depend on the concentration of the ferric ions above 0.01 M Fe^{3+}.

Similar parabolic kinetics was observed by Dutrizac, MacDonald and Ingraham [18] for sulphate media and by Haver and Wong [23] for chloride media.

Dutrizac [25] compared the leaching of chalcopyrite in the solutions of ferric chloride and sulphate in comparable conditions with the same sample of chalcopyrite in the temperature range 45–100 °C. He confirmed that the leaching rate in the chloride medium is higher than in the sulphate medium and determined the values of the apparent activation energy for leaching in the chloride as E_A = 42 kJ/mole and for leaching in the sulphate as E_A = 75 kJ/mole. The presence of acid in the solution has no effect on the leaching rate; the role of the acid is to prevent hydrolytic precipitation of iron from the solution. It was also confirmed that the rate is proportional to the surface area of chalcopyrite and this applies to both investigated media.

Munoz, Miller and Wadsworth [26] investigated the leaching of chalcopyrite powder and confirmed that the rate-controlling step is the mass transport through the layer of elemental sulphur formed on the leached surface. Applying Wagner's theory of oxidation, they proposed that the rate-controlling step may be the transfer of electrons through the elemental sulphur layer. The reaction rate, predicted on the basis of the physical–chemical properties of the systems such as the conductivity of elemental sulphur and the change of the free energy of the reaction, is in agreement with the experimentally determined rate. The activation energy was E_A = 83.7 kJ/mole, and this value was approximately the same as the activation energy of transfer of the mass of electrons through the elemental sulphur layer, E_A = 96.3 kJ/mole.

Warren, Wadsworth and El-Raghy investigated the electrochemical oxidation of chalcopyrite in various acidified solutions [27, 28] using the method of potentiodynamic polarisation. The anodic polarisation of chalcopyrite was sensitive to the pH value at higher potentials, but showed no sensitivity at low potentials in the sulphate solution. In addition to this, measurement of the dissociation current at a constant potential copied the Sato–Cohen logarithmic model for the formation of the thin layer. On the basis of equilibrium measurements of current and mass, they proposed a mechanism of the formation of intermediate sulphide phases, formed in the sequence $CuFeS_2 \rightarrow S_1 \rightarrow S_2$. At higher potentials, the increase of current is compatible with the dissociation of water into the form of chemisorbed oxygen releasing copper and forming sulphate ions.

By examining the electrochemical reaction of chalcopyrite in sulphuric acid, Warren et al. [29] detected the formation of intermediate products of chalcocite Cu_2S and djurleite $Cu_{1.96}S$ as the products of the leaching reaction as follows:

$$(2-x)\mathrm{CuFeS_2} + (6-4x)\mathrm{H^+} + (2-2x)\mathrm{e^-} \rightarrow$$
$$\rightarrow \mathrm{Cu_{2-x}S} + (2-x)\mathrm{Fe^{2+}} + (3-2x)\mathrm{H_2S} \quad (12.22)$$

$$\mathrm{CuFeS_2} + (3-2x)\mathrm{Cu^{2+}} + (4-4x)\mathrm{e^-} \rightarrow 2\mathrm{Cu_{2-x}S} + \mathrm{Fe^{2+}} \quad (12.23)$$

The first reaction applies in the absence of cupric ions and the second one to the presence of Cu^{2+} ions in the solution.

Further aspects of the electrochemical leaching of chalcopyrite have also been investigated by Wadsworth [30] and Ammou-Chokroum et al. [31].

Comparison studies of the electrochemical and chemical leaching of chalcopyrite in the acidified solution of ferric chloride were published by Hirato et al. [32]. The leaching rate of chalcopyrite is affected quite strongly by the concentration of $FeCl_3$, but is independent of the concentration of $FeCl_2$. An electrochemical mechanism was proposed for the leaching process, including the oxidation of chalcopyrite and reduction of the $FeCl_2^+$ ion as the main carrier of Fe^{3+}. In the investigated temperature range 50–85 °C, the activation energy for chemical leaching was E_A = 69 kJ/mole, and for electrochemical leaching it was E_A = 59.5 kJ/mole.

Hirato et al. [33] also leached natural chalcopyrite in an acidified ferric sulphate solution and investigated the morphology and electrochemical properties of chalcopyrite. Initially, chalcopyrite was leached by parabolic kinetics and later only by linear kinetics. In the initial stage, a layer of elemental sulphur formed on the leached surface but after some time the layer peeled off. In the stage of linear leaching the formation of elemental sulphur on the surface was no longer observed. The activation energy was in the range 77–88 kJ/mole which indicates that the process is chemically controlled.

The same authors [34] investigated, in similar conditions, the leaching of chalcopyrite in $CuCl_2$ by electrochemical and chemical methods, with special attention paid to the morphology of leached particles. The formation of elemental sulphur on the leached surface with the morphology similar to that in leaching using $FeCl_3$ was detected. The leaching kinetics was linear, with a break representing the acceleration of the reaction as a result of the increase of the interfacial surface. The leaching rate was proportional to the square root of the concentration of $CuCl_2$ and inversely proportional to the square root of the concentration of CuCl. The electrochemical mechanism of leaching, including the oxidation of chalcopyrite and $CuCl_2^-$ and the reduction of $CuCl^+$, was proposed. The activation energy determined for the investigated temperature range 60–85 °C was E_A = 81.5 kJ/mole for chemical leaching. For electrochemical leaching, with the addition of NaCl the activation energy was E_A = 59.5 kJ/mole.

Braithwaite and Wadsworth [35] simulated the leaching of chalcopyrite *in situ* in order to obtain information on the leaching mechanism, the formation of end products, the effect of pH, the effect of temperature, the partial pressure of oxygen, the size of the particles and the optimum composition of the leaching solution in order to group the results in three areas: economic, environmental and the increase of the extent of utilisation of the available natural resources. The experiments were carried out on natural chalcopyrite, leached in diluted sulphuric acid in an autoclave in the temperature range 30–150 °C, at a partial pressure of oxygen of 0.3–12 MPa, in the pH range 0.86–5.9.

The possibility of oxidation leaching of chalcopyrite in the given conditions was confirmed, with the reaction products being ferric, cuprous and ferrous ions, Fe_2O_3, elemental sulphur and sulphate ions. The concentration of ferrous and ferric ions in the solution initially rapidly increased but subsequently decreased to very low values (below 50 ppm). Iron precipitated from the solution in the form of haematite, and the content of the ferric ions in the solution was negligible at pH higher than 2. Sulphides were partially oxidised to elemental sulphur and partially to sulphate ions, depending on the variables. The experimental value of activation energy was E_A = 93 kJ/mole for temperatures above 120 °C and E_A = 37 kJ/mole for lower temperatures, indicating a change of the reaction mechanism. The mechanism of the process was the model of mixed kinetics, containing the diffusion of reactants and the reactions on the leached surface.

Yu et al. [36] investigated leaching of chalcopyrite in the temperature range 125–175 °C and the partial pressure of oxygen of 0.5–3 MPa in sulphuric acid. The reaction products were the soluble ions of copper and iron and elemental sulphur. The linear kinetics of leaching under the effect of constant adsorption of oxygen and the chemical reaction of the surface was detected. The activation energy was E_A = 138 kJ/mole.

This review of the published results indicates the relatively large scatter in the results and, consequently, differences in interpretation. In order to unify the results, Dutrizac et al. [37] investigated the leaching of chalcopyrite from various deposits in ferric sulphate and chloride. Eleven samples of chalcopyrite from various deposits showed approximately similar leaching rates and also similar mechanisms, with the release of more than 95% of sulphur in elemental form.

Linge et al. [38] also investigated the leaching of chalcopyrite in ferric ions from various Australian deposits and found that the reactivity depends on the composition of chalcopyrite and impurities and mixtures. No galvanic interaction was found between the individual minerals present in the sample. In reality, a certain amount of minor substances is always present in actual concentrates and this may have a significant effect on the leaching kinetics. Therefore, Dutrizac and MacDonald [39] investigated these processes on synthetic specimens prepared by mixing defined amounts of the individual 'impurities' with chalcopyrite. The examples prepared in this manner were leached in the solution of ferric sulphate by the rotating disk method. The results show that the effect of inert impurities has no other consequences, but the presence of cubanite $CuFe_2S_3$, bornite Cu_5FeS_4, pyrite FeS_2, molybdenite MoS_2 and stibnite Sb_2S_3 increases the leaching rate and, on the other hand, the presence of galenite PbS and sphalerite ZnS reduces this rate. These results confirm the mechanism of galvanic corrosion, except the behaviour of molybdenite in the system.

The copper–iron–sulphur system contains a large number of different minerals with a very wide non-stoichiometry range [40]. If the natural form of α-chalcopyrite is heated in inert conditions, part of sulphur is lost with the formation of the compound of β-chalcopyrite $CuFeS_{1.83}$. Both these types of chalcopyrite were leached in an acidified solution of ferric sulphate by Ferreira and Burkin [41]. The concentration of the leaching agent, the grain size and temperature were varied. They found that whilst the leaching rate of natural chalcopyrite is very low, the leaching rate of β-chalcopyrite is considerably higher and both iron and copper are transferred into the solution although at different rates, as indicated by the content of copper and iron in partially leached specimens and by the results of measurements by x-ray diffraction powder diffractometry.

Comparison of the leaching results of chalcopyrite in different media indicates the effect of the environment on the leaching kinetics. Even when using the same oxidation agent, the ion Fe^{3+} in different forms, the leaching kinetics differed. If the ferric ion is in the sulphate form, the rate of leaching of chalcopyrite is 5–10 times lower than when using the chloride medium, which clearly indicates the significant effect of the chloride ions on the leaching process. Palmer et al. [42] investigated the leaching of chalcopyrite in ferric chloride and sodium chloride, acidified with hydrochloric acid. The process took place in the linear kinetics regime, indicating the mechanism of controlling the reaction on the leached surface. The reaction rate depended on the overall concentration of the ferric ions. In addition to this, the reaction rate increased with increasing concentration of the chloride ions up to the value 1 mol/l and became independent at higher concentrations. These results are in agreement with the proposed electrochemical mechanism of the rate-controlling step. The anodic reaction includes the oxidation of chalcopyrite or copper and ferrous ions and also elemental sulphur:

$$CuFeS_2 \rightarrow Cu^{2+} + Fe^{2+} + 2S^0 + 4e^- \quad (12.24)$$

However, the cathode is characterised by the occurrence of four parallel reactions, representing the reduction of the non-complexed ferric ion

$$Fe^{3+} + e^- \rightarrow Fe^{2+} \quad (12.25)$$

and the reduction of the first, second and third chlorine complex of trivalent iron:

$$FeCl^{2+} + e^- \rightarrow Fe^{2+} + Cl^- \quad (12.26)$$

$$FeCl_2^+ + e^- \rightarrow Fe^{2+} + 2Cl^- \quad (12.27)$$

$$FeCl_{3(aq)} + e^- \rightarrow Fe^{2+} + 3Cl^- \quad (12.28)$$

O'Malley and Liddell [43, 44] also confirmed that the presence of chloride ions accelerates the kinetics of the leaching reaction of chalcopyrite. The final ratio Cu^+/Cu and the resultant degree of leaching of chalcopyrite depend on the actual value of the concentration of the chlorides in the solution. The entire amount of the ferric ions was used for the reaction and the copper ions Cu^{2+}, formed in the process, also take part in the reaction, but only to the extent determined by the concentration of the chloride ions. In the solutions with medium to high concentration (~greater than 3 M) approximately 90% Cu^{2+} was reduced to Cu^+. At a lower chloride concentration, the extent of the reaction with the copper ion was proportionately lower.

The specific role of NaCl as a carrier of chloride ions was investigated by Dutrizac and MacDonald [19]. They confirmed that the presence of NaCl increases the rate of extraction of copper from chalcopyrite at leaching temperatures higher than 50 °C, although there were also exceptions from this rule when testing several natural chalcopyrites. However, this was not explained.

The study of chalcopyrite leaching by potentiometric titration in ferric nitrate were published by Linge [45] for constant pH and the oxidation potential in the temperature range 25–40 °C. Parabolic kinetics was detected in accordance with some previous investigations, but in this case examination shows the occurrence of an initial reaction in which the amount of dissolved iron was greater than that of copper. This resulted in a high rate of formation of the new surface chalcopyrite-like phase, depleted in the metal. The subsequent leaching reaction with the activation energy E_A = 50.6 kJ/mole was far more sensitive to the change of temperature and its rate was considerably lower than the diffusion rate of the oxidation agent (E_A = 8.4 kJ/mole) and was interpreted as diffusion-controlled in the crystal structure of chalcopyrite.

Bardt [46] and later Bjorling [47–51] and Habashi [52] investigated oxidation by air in of sulphuric acid with a small addition of nitric acid. The role of nitric acid was the leaching of sulphides and the production of elemental sulphur. Nitric acid acted as a catalyst, and gaseous nitrogen oxides were released during leaching and in the presence of oxygen and water they were regenerated back to HNO_3, according to the equations:

$$3MeS + 2HNO_3 + 6H^+ \rightarrow 3Me^{2+} + 3S^0 + 2NO + 4H_2O \quad (12.29)$$

$$4NO + 3O_2 + 2H_2O \rightarrow 4HNO_3 \quad (12.30)$$

In the case of chalcopyrite, the course of the reaction may be described as follows [53]:

$$3CuFeS_2 + 20HNO_3 \rightarrow$$

$$\rightarrow 3Cu(NO_3)_2 + 3Fe(NO_3)_2 + 6S^\circ + 8NO + 10H_2O + 3O_2 \quad (12.31)$$

The reaction is highly exothermic. A partial process of the reaction (12.31) is the oxidation of elemental sulphur

$$2HNO_3 + S \rightarrow H_2SO_4 + 2NO \quad (12.32)$$

When using a concentrated acid, the course of the reaction could not be controlled.

Comparison of the leaching of a chalcopyrite ore in the leaching media H_2SO_4, HCl, $FeCl_3$ and HNO_3 was described by Sen et al. [54]. They confirmed that the rate of the process is the lowest in sulphuric acid, and the diffusion is the rate-controlling step of the reaction. However, the kinetic characteristics were not determined, because the reaction rate was very low.

Leaching in HNO_3, HCl and $FeCl_3$ for 6 h was controlled by the chemical reaction with the determined values of the activation energy E_A = 69 kJ/mole (HNO_3), E_A = 67 kJ/mol (HCl) and EA – 47 kJ/mol ($FeCl_3$).

The company Kennecott Copper in Utah [55) built a pilot plant for the processing of a chalcopyrite concentrate in the HNO_3-H_2SO_4 mixture and found that the processes visible but not economical because of the formation of nitrogen oxides N_2O which cannot be regenerated to nitric acid resulting in losses and also damage to the environment. Similar studies but on a smaller scale were carried out by Brenneck [56, 57].

However, the slow leaching reaction of chalcopyrite in the medium of ferric ions is greatly accelerated by adding silver ions having the function of a catalyst of the reaction. Pawlek [58] investigated two samples of finely ground chalcopyrite subjected to pressure leaching at 120 °C and 0.2 MPa of O_2. In the absence of the silver catalyst 25% of the coarser and 51% of the finer chalcopyrite reacted for 30 min. In the presence of the silver catalyst, 30 or 95% of chalcopyrite reacted of the same period of time.

McElroy and Duncan [59] investigated bacterial leaching of ground chalcopyrite using Tiobacillus ferooxidans (TBF) in the presence and absence of silver nitrate, respectively. Without catalyst, the yield of copper was 20–60%. When adding 0.05% of the nitrate, the yield increased to 60–100%.

Snell and Sze [60] proposed leaching with acidified ferric sulphate containing silver ions in the range 0.004–0.04% of the mass of the concentrate. The yield of copper in the optimum conditions was 95% after three hours of leaching at 90–120 °C.

The mechanism of acceleration of the leaching reaction of chalcopyrite using the silver ions has been described by Miller and Portillo [61] and Price and Warren [62]. According to these authors, the reaction rate is controlled by the electrochemical reaction:

$$CuFeS_2 + 4Ag^+ \rightarrow 2Ag_2S + Cu^{2+} + Fe^{2+} \qquad (12.33)$$

$$2Fe^{3+} + Ag_2S \rightarrow 2Ag^+ + S^0 + 2Fe^{2+} \qquad (12.34)$$

The formation of Ag^+ by the reaction (12.34) and its subsequent reaction (12.33) increase the rate of extraction of copper in the presence of silver. The electrochemical reaction (12.34) takes place on the surface of the crystallites of silver sulphide throughout the entire thin layer of Ag_2S. The crystallites act as short circuits of microelectrodes, resulting in the discharge of Fe^{3+} and release of Ag^+. In these conditions, the kinetics should depend on the concentration of Ag^+, Fe^{3+} and Fe^{2+} and should be independent of the initial particle size, i.e. the rate depends to a greater extent on the internal area of the crystallites than on the external area of particles. Elemental sulphur covers the crystals of Ag_2S forming a porous layer. The resultant silver ions play the role of a carrier, and the thin layer of the silver sulphide, required to prevent the formation of a compact sulphide layer on the chalcopyrite surface, is constantly renewed.

Other possibilities of intensifying the leaching of chalcopyrite were investigated by Murr and Hiskey [63] and later by Antojevic et al. [64]. In these studies, the oxidation agent was potassium dichromate $K_2Cr_2O_7$, acidified with sulphuric acid. Investigations were carried out into the effect of mixing of the pulp, temperature, concentration of Cr^{6+} and H_2SO_4 and the particle size on the leaching rate of chalcopyrite. Activation energy was E_A = 48–54 kJ/mole, indicating that the rate is controlled by a chemical or electrochemical reaction. It was established that the mixing rate and the concentration of chromium ions have no significant effect on the leaching rate of chalcopyrite. In addition to this, it should be taken into account that hexavalent chromium is a strongly carcinogenous substance.

The possibilities of intensifying leaching by high-intensity milling have been studied since the beginning of the 70s of the previous century. Gerlach et al. [65] were among the first investigators to show the beneficial effect of vibration milling on the rate of subsequent leaching. A similar effect was also demonstrated by Beckstead et al. by milling in an attritor [22]. However, the views on the reasons for the intensification effect of milling differ. Gerlach et al. [65] attributed the increase of the kinetics to the increase of the size of the surface and energy buildup in the structure of the milled material, Beckstead et al. [22] assumed the controlling effect of the increase of the size of the surface in milling. Tkacova and Balaz [66] have confirmed that the reaction rate of oxidation leaching of chalcopyrite increases with an increase of the size of the specific surface and with a decrease of the content of the crystalline phase. However, they did not say which of these factors has a controlling effect.

A detailed and systematic review of the mechanochemical activation of several sulphides and comparison of the processes of leaching of activated and non-activated sulphides has been published by Balaz et al. [67, 131].

Cobble et al. [68] combined the operation of activation milling in an attritor and acid oxidation leaching of chalcopyrite in a single process. Milling in the attritor at 90 °C in acidified ferric sulphate resulted in the yield of copper of 99% after 6 hours. The authors propose a flow diagram of the hydrometallurgical production of copper combined with liquid extraction and electrolysis. Similar results have also been obtained by Havlik and Kammel [69].

Glaum and Walli [70] have described the oxidation of sulphides of nonferrous metals in an aqueous medium with the formation of elemental sulphur as the reaction product into which a liquid organic agent, insoluble in water, was added in order to dissolve the produced elemental sulphur. Butinelli et al. [71, 72] investigated the leaching of a complex copper–zinc concentrate in ferric and copper chloride with the addition of an organic solvent for sulphur. The main kinetic characteristics and the values of activation energy were determined but similar studies have not been carried out on pure chalcopyrite.

Ozolins and Rushikina [73] found that when applying direct electrical fields or ultrasound the solubility of chalcopyrite in an acidified solution of ferric sulphate at 50 °C increases. The efficiency of the direct electrical field was higher than that of ultrasound. However, the application of ultrasound is undoubtedly an interesting intensification element in hydrometallurgical processes, as indicated by the published results [74–77].

Similarly, the application of other high-energy fields has a positive effect on leaching. Florek et al. [134] leached a chalcopyrite concentrate using an acid solution of ferric sulphate and also in a microwave field. Comparison of the results of leaching using conventional heating and leaching in a microwave field at 900 W and a frequency of 2.45 GHz showed that leaching is more efficient in the microwave field.

Antonucci and Correa [120] leached a chalcopyrite concentrate in concentrated sulphuric acid in a microwave field. The leaching products were elemental sulphur and cupric ions. Like the previously mentioned authors, they confirmed that the efficiency of leaching in the microwave field is higher than in conventional heating.

A relatively slow reaction of leaching of chalcopyrite resulted in an effort to accelerate this process by different methods. Heating of copper and iron sulphides at temperatures of approximately 500 °C with the access of air resulted in sulphation of the sulphides [78]. Ferric sulphide dissociates into ferric oxide in the temperature range 670–710 °C by the reaction:

$$2CuFeS_2 + 15/2O_2 \rightarrow 2CuSO_4 + Fe_2O_3 + 2SO_2 \qquad (12.35)$$

At the temperature of roasting processes of approximately 680 °C, copper or zinc sulphates are formed preferentially.

Warren et al. [79] investigated pressure leaching of pre-roasted chalcopyrite concentrates at 825 °C without air, with part of sulphur removed by thermal dissociation. This resulted in the formation of a material which was far easier to leach than the initial chalcopyrite. The heating of chalcopyrite in the presence of sulphur at 475 °C resulted in a phase transformation and in the formation of new phases, covellite CuS and pyrite FeS_2 [80].

This principle was used in several pilot plant and production systems, for example, DOWA, CSIRO, U.S. Smelting [80] or TREADWELL [81]. However, it should be considered whether the more favourable conditions in the leaching process can outweigh the considerable difficulties in roasting and whether this could satisfy increasingly stringent environmental legislative.

Bartlett [82] used therefore a different method of enrichment of chalcopyrite, i.e. hydrothermal change of chalcopyrite to digenite directly by oxygen in a continuous process in a copper sulphate medium. The process took place in an autoclave at 200 °C with the duration of less than 1 hour according to the reaction:

$$3CuFeS_2 + 6CuSO_4 + 4H_2O \rightarrow 5Cu_{1.8}S + 3FeSO_4 + 4H_2SO_4 \quad (12.36)$$

The entire process was investigated on the pilot plant scale, with the process realised into autoclaves placed next to each other. In the first autoclave, chalcopyrite was processed by pressure leaching in order to produce copper sulphate

$$CuFeS_2 + 4O_2 \rightarrow CuSO_4 + FeSO_4 \qquad (12.37)$$

and $CuSO_4$ reacted with the added concentrate without oxygen in the second autoclave in accordance with the reaction (12.36). The produced digenite was then processed far more easily and faster by conventional hydrometallurgical procedures.

The process of enrichment of chalcopyrite also includes other reduction agents, for example SO_2 [5, 83].

Investigations of the bacterial leaching of a relatively pure chalcopyrite, chalcopyrite concentrate, or ores containing chalcopyrite have been carried out in [84, 85]. In leaching chalcopyrite in an acidified solution in the presence of bacteria *Thiobacillus ferooxidans* (TBF), the leaching solutions contains after some time a certain fraction of ferric ions. The leaching of chalcopyrite using bacteria probably takes place by two mechanisms: direct interaction of the bacteria with the sulphide in the structure of chalcopyrite, and indirectly by leaching using ferric ions. However, it is not easy to determine the proportion and significance of these two mechanisms. Sutton and Corrick [86] believe that chalcopyrite is 'attacked' only by the indirect leaching mechanism. The mechanism which they propose includes the initial oxidation of chalcopyrite by air by reaction (12.37). The ferric sulphate, produced by the bacteria, was rapidly oxidised by air in the presence of bacteria in accordance with the reaction:

$$4FeSO_4 + H_2SO_4 + O_2 \xrightarrow{\text{bacteria}} 2Fe_2(SO_4) + 3H_2O \qquad (12.38)$$

The ferric sulphate, produced in this manner, leached chalcopyrite forming increasing amount of ferrous sulphate which again took part in the entire cycle:

$$CuFeS_2 + 4Fe^{3+} \rightarrow Cu^{2+} + 5Fe^{2+} + 2S^0 \qquad (12.16)$$

The sulphur, produced by indirect leaching of chalcopyrite, is oxidised to the acid which prevents the hydrolysis of iron:

$$2S^0 + 3O_2 + 2H_2O \xrightarrow{\text{bacteria}} 2H_2SO_4 \qquad (12.39)$$

The reaction (12.37) is slow and, for this reason, the initial leaching rate is relatively low, but gradually increases up to a certain limit.

A similar mechanism was proposed by Malouf and Prater [87] to explain the leaching of chalcopyrite during heap leaching. Comparison of the leaching experiments in the presence of TBF and in sterile conditions showed the relatively high leaching rate in the presence of bacteria, but only after a long period of time (hundreds of days).

In certain conditions, it is possible to achieve the direct oxidation of chalcopyrite by some types of bacteria which produce copper and iron in the soluble form, although the accurate mechanism of this process is not yet completely clear. Additions of certain compounds accelerate the process. For example, the addition of saccharose [88] or surface-active substances [89] may increase the leaching rate of certain chalcopyrite ores. Similarly, the presence of some other minerals, like pyrite, increases the oxidation rate of chalcopyrite, probably by the effect of galvanic corrosion. On the other hand, the presence of molybdenite suppresses the process [90]. This is associated with the residual electrochemical potentials of these minerals. According to Wyckoff [91], the leaching of chalcopyrite ores and concentrates from different deposits showed certain differences.

Generally, it may be assumed that if the basic essential conditions for the leaching process are fulfilled from the theoretical viewpoint, i.e. the sufficiently high redox potential in the solution, then the effect of the type of leaching medium is not very important. However, from the practical viewpoint, the type of leaching medium is very important, even so that it underlies practical operations. Both viewpoints, economic and ecological, are at present dominant in the evaluation of suitability of a specific leaching medium for a specific sulphide. Therefore, various leaching media have been used in the experiments with leaching of chalcopyrite and the results have contributed significantly to understanding the entire process of leaching of sulphides.

One of the most important parameters, the effect of temperature on the leaching of chalcopyrite, is not defined unambiguously. This is given by differences in the leaching conditions, in particular, by defects in the structure of the leached material. This results in the scatter of the determined activation energy by different authors, as indicated by Table 12.1.

The results of examination of the possibilities of leaching of chalcopyrite show a considerable scatter from the viewpoint of the yield, the kinetic dependence of the effect of the individual variables of the process, and also the interpretation of the mechanism of the process. In addition to different approaches used by the individual authors, the strongest effect on the situation is exerted by the nature of the materials used in the experiments. It has been shown unambiguously that the content of impurities, even in the minimum quantities, has a very strong effect on the leaching characteristics. Similarly, the nature and chemical composition of these impurities strongly influenced the leaching characteristics. The process is also greatly influenced by the internal structure and defectiveness of the mineral. These are the main reasons for the large scatter of the results, of course, in addition to the experimental procedures and media, which form part of the system, and it is also the reason for continuing studies of the leaching of chalcopyrite from different deposits.

Another impetus for the research in the area of leaching of

Table 12.1. The results of investigations of the leaching kinetics of chalcopyrite in different media by different authors

Material	Medium	E_A [kJ/mol]	t [°C]	Rate-controlling step	Ref.
$CuFeS_2$	$FeCl_3$	50	60-106	not given	14
$CuFeS_2$ natur.	$FeCl_3$	High	35-100	not given	12
$CuFeS_2$ synt	$FeCl_3$	46±4	50-100	mass transfer	20
$CuFeS_2$ natur.	$FeCl_3$	63±8	40-100	mass transfer	25
$CuFeS_2$ natur.	$FeCl_3$	High	30-106	parabolic kinetics, mass transfer	23
$CuFeS_2$ natur.	$FeCl_3$	42±4	30-100	mass transfer	20
$CuFeS_2$ natur.	$FeCl_3$	69	50-85	electrochemical mechanism	32
$CuFeS_2$ natur.	$FeCl_3$	62	75-96		42
$CuFeS_2$ natur.	$FeCl_3$	47	35-85	chemical reaction	54
$CuFeS_2$ natur.	$FeCl_3$	55	3.5-95	chemical reaction of S° creation	98,99
$CuFeS_2$ natur.	$FeCl_3 + CCl_4$	31.2	45-80	diffusion	100
$CuFeS_2$ synt.	$FeCl_3$ electrochem.	38±4	25-75	mass transfer	31
$CuFeS_2$ natur.	$FeCl_3$ electrochem	59.5	50-85	electrochemical mechanism	32
$CuFeS_2$ natur.	$Fe_2(SO_4)_3$	High	35-100	not given	12
$CuFeS_2$ synt.	$Fe_2(SO_4)_3$	71	50-94	parabolic kinetics mass transfer	18
$CuFeS_2$ synt.	$Fe_2(SO_4)_3$	38-63	30-95	mass transfer	20
$CuFeS_2$ natur.	$Fe_2(SO_4)_3$	75	32-50	linear kinetics chemisorption on surface	101
$CuFeS_2$ natur.	$Fe_2(SO_4)_3$	84	27-85	parabolic kinetics mass transfer	24
$CuFeS_2$ natur.	$Fe_2(SO_4)_3$	84	60-90	transfer of electrons	26
$CuFeS_2$ natur.	$Fe_2(SO_4)_3$	77-88	50-90	chemical reaction	33
$CuFeS_2$ natur.	$Fe_2(SO_4)_3$	12	20-40	external difusion	97
$CuFeS_2$ natur.	$Fe_2(SO_4)_3$	69	40-90	chemical reaction of S° creation	97
$CuFeS_2$ natur.	$CuCl_2$ chemical	81.5	60-85	electrochemical mechanism	34
$CuFeS_2$ natur.	$CuCl_2$+NaCl electrochem	59.5	60-85	electrochemical mechanism	34
$CuFeS_2$ natur.	H_2SO_4+ pressure	93	> 128	mixed kinetics model diffusion+surface reaction	35
$CuFeS_2$ natur.	H_2SO_4+ pressure	37	< 128	mixed kinetics model diffusion+surface reaction	35
$CuFeS_2$ natur.	H_2SO_4	138	125-175	adsorption of O_2 chemical reaction	36
$CuFeS_2$ natur.	$H_2SO_4+O_3$	60	25-50	chemical reaction	102
$CuFeS_2$ natur.	$Fe(NO_3)_3$	58.6	25-40	parabolic kinetics	38
$CuFeS_2$ natur.	HNO_3	69	35-75	chemical reaction	54
$CuFeS_2$ natur.	HCl	67	35-85	chemical reaction	54
$CuFeS_2$ natur.	$K_2Cr_2O_7$+ H_2SO_4	48-54	30-80	electrochemical mechanism	64

chalcopyrite is the effort for intensification of the process, as already mentioned previously. Of course, it is necessary to achieve the highest possible leaching efficiency, in the shortest possible time, in the industrially accepted medium and for the method used. Work continues in the search for a suitable leaching medium and in the application of the efficient leaching method. The former includes the application of oxidation agents with a high redox potential, for example H_2O_2 [102, 121], O_3 [108], chromates [64], permanganate [122–125]; etc., work is also carried out to use again $CuCl_2$ as the oxidation agent [126, 127], the application of strong reduction agents such as metallic iron [128, 129], metallic copper [17], SO_2 [83], hydrogen [130], and so on.

The directions of search for the optimum leaching method include the work in the area of pressure leaching [36], bacterial leaching [85–91], mechanochemical leaching [131, 132], thermal processing of concentrates [79–81], the application of high-energy fields such as ultrasound [73–77], high-frequency [123] and microwave fields [114–120]. However, there are still problems with the comparison of the results for the previously mentioned reasons of the uniqueness of every chalcopyrite mineral. Therefore, Havlik et al. [97–100, 106–121, 123–129] carried out long-term research into the leaching of chalcopyrite using, in every experiment, a sample of the same chalcopyrite at comparable leaching parameters and different methods. The results may be summarised as follows.

From the industrial practice viewpoint, the two most important oxidation agents are ferric chloride and ferric sulphate. That is why the first experiments were carried out using these oxidation agents. The natural chalcopyrite concentrate imported from Cuba was used for all these experiments, the grain size –100+60 μm, which is approximately the flotation grain size. The chemical composition of the sample was as follows: Cu = 32.3+0.6%, Fe = 28.4+0.5%, S = 31.9+2%, oxides 5.61%. The results of X-ray diffraction phase analysis show that the sample contained chalcopyrite $CuFeS_2$ and a small amount of quartz SiO_2.

12.1.1. Leaching of chalcopyrite by ferric sulphate [97]

The leaching medium was a solution with the concentration 0.5 M $Fe_2(SO_2)_3$ in a 0.5 M solution of H_2SO_4.

Figure 12.1 shows the course of leaching of the chalcopyrite concentrate in the temperature range 20–99 °C at atmospheric pressure in the sulphate medium.

In the group of the tested kinetic models, the measured values were described most accurately by the shrinking core model. This means that, in accordance with the published results, it is probably the process controlled by the chemical reaction at the interface.

The effect of temperature is expressed by the Arrhenius relationship, graphically shown in Fig. 12.2 for the concentration of 0.5 M Fe^{3+}.

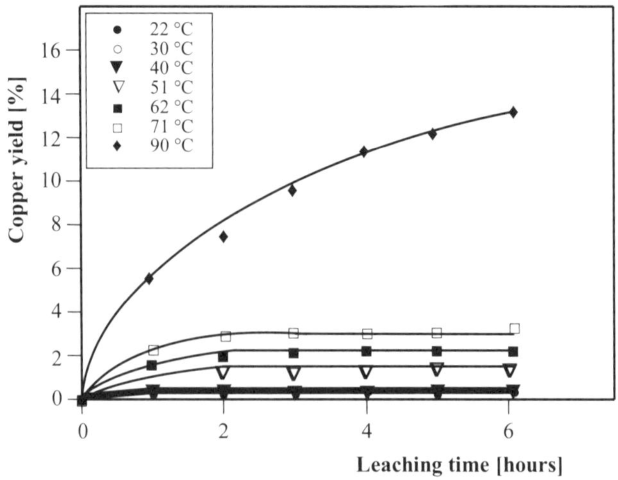

Fig. 12.1. Kinetic curves of leaching of chalcopyrite in a sulphate medium (0.5 M $Fe_2(SO4)_3$ +0.5 M H_2SO_4).

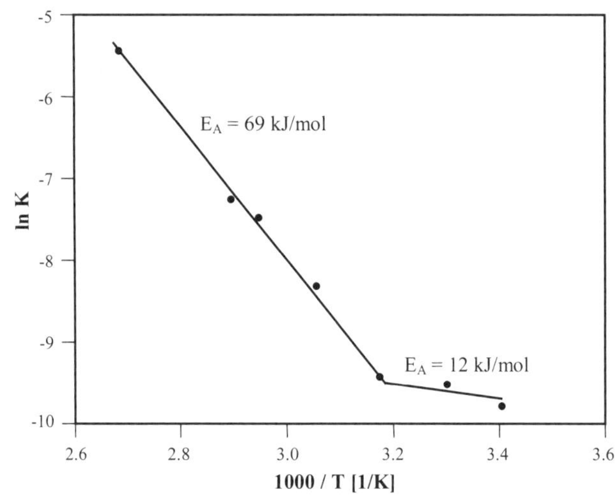

Fig. 12.2. The Arrhenius plot for chalcopyrite leaching in a sulphate medium.

In the investigated temperature range 20–99 °C, the Arrhenius relationship shows a break at 40 °C, indicating a change of the reaction mechanism. The apparent activation energy determined in the range of lower temperatures (20–40 °C) was E_A = 12 kJ/mole (more accurately 12.96 kJ/mole (IK = 0.945)). At higher temperatures, the apparent activation energy was E_A = 69 kJ/mole (68.75 kJ/mole (CI = 0.998)).

The very fact of the course of leaching of chalcopyrite in the sulphate medium with a change of the mechanism in relation to temperature is a completely new result. On the other hand, this does not contradict any of the previously published results because no results have been published of leaching in such a wide temperature range. It is generally well-known that low temperature reduces greatly the leaching rate and this is obviously the main reason why all the authors have investigated the temperature range interesting for commercial applications.

The activation energy determined in the temperature range 20–40 °C indicates that diffusion is the rate-controlling step. The formation of a layer of elemental sulphur, covering the leached surface, was not observed in this range so that the diffusion of reactants through the layer cannot be considered. Diffusion within the crystal structure is extensive [26]. On the other hand, the rate of the chemical reaction in this region was very low because of low temperatures but, the same time, the viscosity of the leaching medium increases. Therefore, it appears that the rate-controlling factor in this range is the external diffusion of the reactants of the bulk of the leaching agent to/from the leached surface due to higher viscosity of the media at lower temperatures. This has also been confirmed by experiments carried out with the filtration of sulphate media in which the rate of filtration of the sulphate solution was considerably lower than that of, for example, the chloride solution.

The apparent activation energy E_A = 69 kJ/mole, determined in the temperature range 40–99 °C, is in good agreement with the published results, in which the average value was 70.28 kJ/mole, as indicated by Table 12.1. The determined value of the activation energy specifies the reaction taking place in the kinetic regime as also indicated by the nature of the leaching relationships.

Examination by X-ray diffraction phase analysis and scanning electron microscopy showed the formation of elemental sulphur of the surface of partially leached chalcopyrite. The behaviour of sulphur during leaching will be discussed elsewhere. In general, the particles showed only a slightly attacked surface and the process was controlled by linear kinetics. The smaller amount of elemental sulphur, produced by the leaching reaction, is at least partially responsible for this course because it does not cover completely the leached surface and, at the centre, since the degree of conversion is low, the interfacial area changes only slightly.

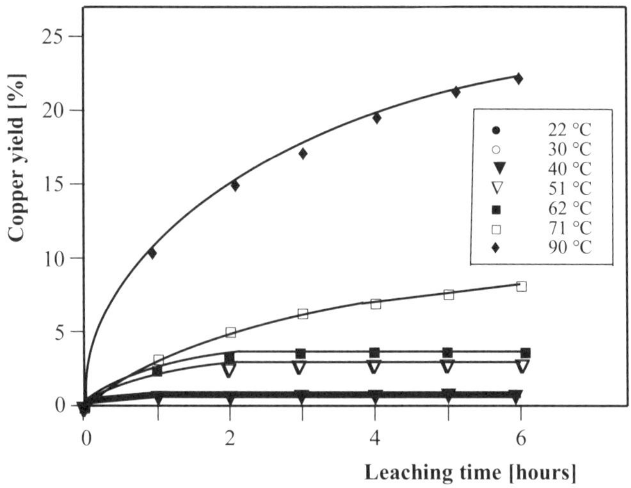

Fig. 12.3. Chalcopyrite leaching kinetic curves in the chloride medium 0.5 M $FeCl_3$+ 0.5 M HCl.

12.1.2. The leaching of chalcopyrite by ferric chloride [98, 99]

Experiments were carried out in the solutions of 0.1, 0.25, 0.5, 0.75 and 1.0 M $FeCl_3$ in M HCl as a leaching medium.

Figure 12.3 shows the leaching kinetic curves in the temperature range 3.5–80 °C at atmospheric pressure.

The product of chalcopyrite leaching by ferric ions is, in addition to the soluble metallic ions, also elemental sulphur covering more or less the leached surface thus preventing the diffusion of reagents. The presence of the acid in the leaching medium in acid oxidation leaching has the only role: to prevent hydrolysis of iron. Only a small minimum amount of the acid is required, and any further increase of the concentration of the acid in the leaching medium causes problems, mainly production and ecological. In order to determine the effect of the acidity of the solution of the leaching kinetics, leaching experiments were carried out at 70 °C, using ferric chloride with the concentration of 0.5 M and the concentration of hydrochloric acid was varied in the range 0.25, 0.5 and 0.75 M. Lower concentrations were not used because of the possibility of hydrolysis of iron. The results confirm that the change of the acidity of the leaching solution in the range 0.25–0.75 M HCl has no significant effect on the leaching kinetics and, therefore, all the experiments were carried out with the acid concentration of 0.5 mol/l.

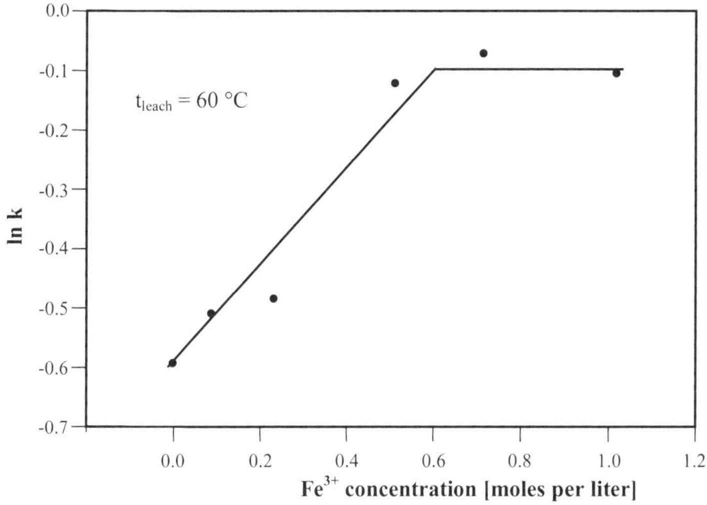

Fig. 12.4. Dependence of the leaching rate of chalcopyrite on the concentration of the ferric ion.

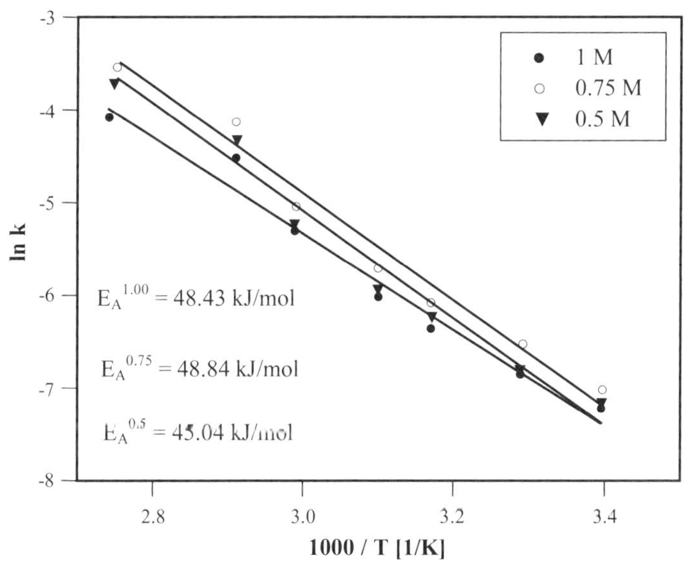

Fig. 12.5. Arrhenius plot for chalcopyrite leaching in ferric chloride.

The dependence of the leaching rate of chalcopyrite on the concentration of the ferric ion is shown in Fig. 12.4. Of course, the insufficient amount of the ferric ion in the leaching solution reduces the leaching kinetics as a result of external diffusion to the leaching interface. Figure 12.4 shows that the concentration of ferric ions higher than 0.5 M is sufficient for ensuring the presence of a sufficient number of the ferric ions required for leaching reactions not inhibited by external diffusion.

The quantitative expression of the effect of temperature for this case is indicated by the Arrhenius plot shown in Fig. 12.5 for the investigated concentrations of 1.0 of, 0.75 and 0.5 M Fe^{3+}.

The apparent activation energy E_A = 47 kJ/mole, determined in the temperature range 25–95 °C, is in agreement with the published results. Dutrizac [25], who obtained the values of 38–63 kJ/mole with a scatter, even concludes that the most probable value is 42–46 kJ/mole. The determined activation energy indicates that a reaction taking place in the mixed regime. In this case, temperature is sufficient for a fast chemical reaction leading to the formation of elemental sulphur on the leached surface, as confirmed by X-ray diffraction phase analysis and examination of morphology.

12.1.3. Leaching of chalcopyrite by ferric chloride with the addition of carbon tetrachloride [100]

Elemental sulphur is one of the products of leaching of chalcopyrite by the ferric ion. The sulphur partially covers the leached surface thus reducing the leaching rate. One of the methods of intensifying leaching is the addition of other agents to the leaching system to

Table 12.2. Solubility of sulphur in carbon tetrachloride [105]

t [°C]	Solubility [g S/100 g of solvent]
15	0.641
12	0.748
25	0.831
35	1.155
45	1.564
54	2.008
60	2.150
90	8.150
100	18.00
120	100.00

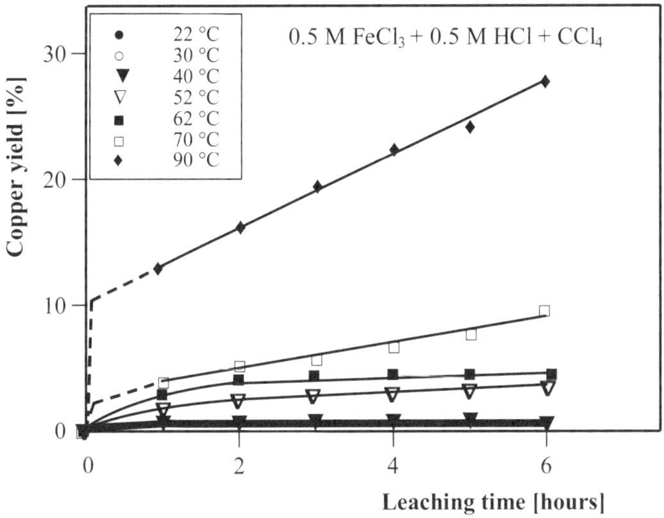

Fig. 12.6. Kinetic curves of leaching of chalcopyrite at different temperatures in $FeCl_3$+ CCl_4.

remove the produced elemental sulphur [70, 71, 103, 104].

In [100] the leaching medium was a solution of 1.0 M $FeCl_3$ in 0.2 M HCl. The calculated amount of CCl_4 was added in advance to the solution so that it was capable of dissolving the entire amount of sulphur present in the sample at 60 °C, as indicated in Table 12.2 [105].

Figure 12.6 shows the kinetic curves of leaching of chalcopyrite in $FeCl_3$+CCl_4 in the temperature range 40–90 °C at the atmospheric pressure.

Comparison with the leaching curves of chalcopyrite in $FeCl_3$ without CCl_4 [99], Fig. 12.3, shows that by adding the tetrachloride the curves become linear, especially at higher temperatures, clearly indicating the prevention of diffusion through the layer of elemental sulphur. This is also confirmed by the values of the yield, shown in Table 12.3, indicating almost doubling of the yield when adding CCl_4 to the leaching medium.

The use of such agent has other advantages – because of high vapour tension it is east to evaporate (it is not flammable and does not explode) and a result of re-condensation can be returned the process. After evaporation, elemental sulphur is obtained without any further dressing.

Table 12.3. The yield of copper after four hours

Temperature [°C]	Cu yield [%]	
	without CCl_4	with CCl_4
40	4.42	9.42
60	7.49	–
65	–	19.12
80	16.15	23.15
90	–	30.26

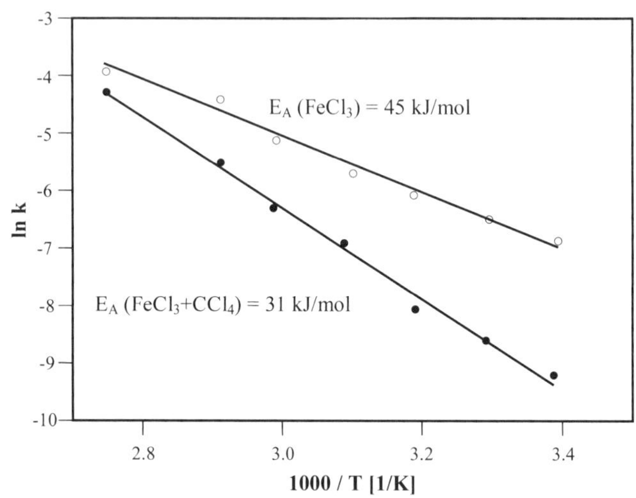

Fig. 12.7. The Arrhenius plots for chalcopyrite leaching in 0.5 M $FeCl_3$ and 0.5 M $FeCl_3$+ CCl_4.

The effect of temperature is represented by the Arrhenius plots, shown in Fig. 12.7. The calculated value of activation energy E_A = 31.2 kJ/mole is lower than when using the ferric chloride and indicates a mixed process mechanism. The diffusion component is probably formed by the external diffusion of CCl_4.

12.1.4. Leaching of chalcopyrite in sulphuric acid using ozone as the oxidation agent [106–121]

For the transfer of copper ions into the solution in acid leaching of chalcopyrite $CuFeS_2$ in sulphuric acid, the redox potential of the acid medium must be theoretically higher than +0.4 V. These conditions may be ensured using ozone as an oxidation agent. The principle of application of ozone as the oxidation agent in oxidation leaching of ores and concentrates of nonferrous metals is based on the high value of the oxidation potential of ozone, +2.07 V, which oxidises (with the exception of platinum, gold and iridium) all metals and their compounds in the solution. This was used as a basis in the leaching experiments of chalcopyrite [107–120, 122] which confirmed the high efficiency of ozone as the oxidation agent.

Ozone was prepared in a commercial ozoniser of Fischer® Model 502 (Switzerland) whose operating principle is based on quiet discharge by the effect of high-voltage with a high frequency in oxygen, using a Siemens ozoniser. The amount of ozone in the working atmosphere, produced by this equipment, is in the range 20–60 g/m^3 of O_3 in 1 m^3 of the atmosphere and depends on the flow rate of oxygen through the ozoniser. The leaching medium was represented by the aqueous solutions of sulphuric acid, concentration 0–1.0 M.

The leaching of chalcopyrite using ozone took place according to the reaction

$$3CuFeS_2 + 8O_3 \rightarrow 3CuSO_4 + 3FeSO_4 \qquad (12.40)$$

Experiments with acid oxidation leaching of chalcopyrite in sulphuric acid using ozone as the oxidation agent have been expanded by the examination of conversion of iron in the solution. When using the ferric ion as the oxidation agent, either in the form of sulphate or chloride, this was not possible because of the minimum change of the concentration of iron during the process in the leaching medium in comparison with the initial concentration, given by the composition of the medium.

Figure 12.8 shows the kinetic curves of the yield of copper, obtained by leaching a chalcopyrite concentrate in a 0.5 M solution of H_2SO_4 in the temperature range 3.5–75 °C using ozone as the oxidation agent in the amount of approximately 2.5 volume % in the oxygen atmosphere with continuous bubbling through the working solution during the experiment. Similarly, Fig. 12.9 shows the kinetic curves of the yield of iron from chalcopyrite.

The curves are exponential and since no continuous layer of elemental sulphur or any other reaction product forms during the process on the leached surface, they are described most efficiently by the shrinking core model without the formation of a solid reaction product.

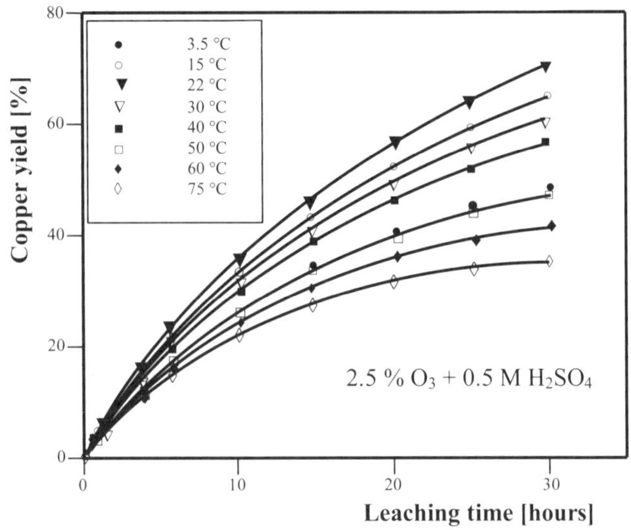

Fig. 12.8. Dependence of the yield of copper on the leaching time of chalcopyrite using ozone.

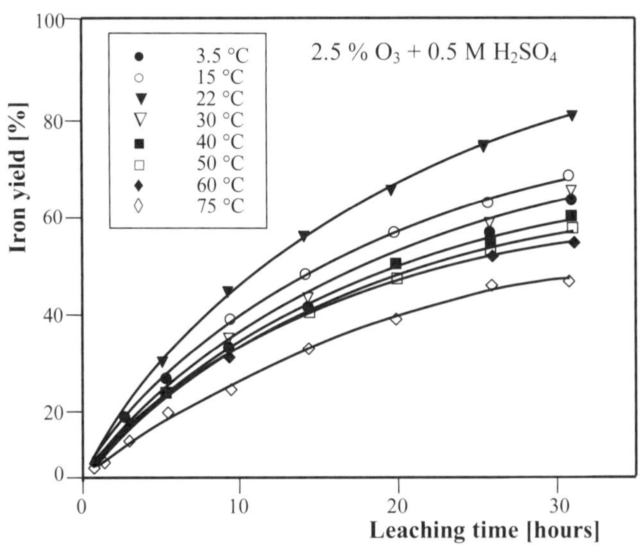

Fig. 12.9. Dependence of the yield of iron on the leaching time of chalcopyrite using ozone.

The oxidation leaching of sulphides in the acid medium is characterised by the increase of the leaching rate with increasing temperature. However, Figs. 12.8 and 12.9, showing the kinetic curves of chalcopyrite leaching with ozone indicate however, the

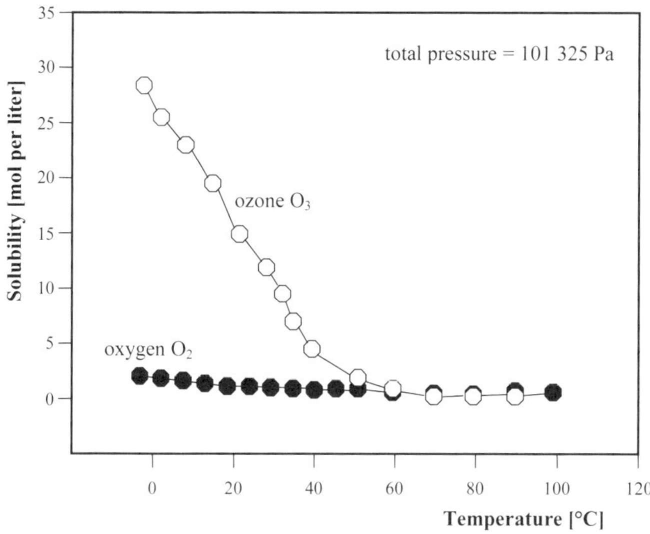

Fig. 12.10. The temperature dependence of solubility of oxygen and ozone in water.

complete opposite. This is given by a rapid decrease of the solubility of ozone under the effect of temperature, as shown by Fig. 12.10.

Although the data for the solubility of ozone have been obtained only for temperatures of up to 50 °C, the experiments show that the effect of ozone is also strong at higher temperatures, for example at experiment temperatures of 90 °C the amount of dissolved copper was approximately 5 times greater when using ozone in comparison with the one when using oxygen. According to [122], the solubility of ozone at 60 °C is 0 and it therefore appears that the oxidation capacity of ozone is also strong at higher temperatures. Gazo et al. [10] described the existence of activated states of oxygen which have, like ozone, oxidation properties, but it is very difficult to estimate the contribution of the activated form of oxygen and of ozone to the reaction.

Two competing phenomena take place during the process: the increase of the dissolution rate under the effect of temperature, and a decrease of the rate under the effect of the decrease of the amount of ozone in the solution. The curves at 3.5 and 15 °C should be the manifestation of these tendencies. The form of these curves slightly differs from the other curves and the absolute amount of dissolved copper or iron in unit time is lower than at 22 °C, but higher than at other temperatures.

Since the highest yield of the metals has been obtained at room

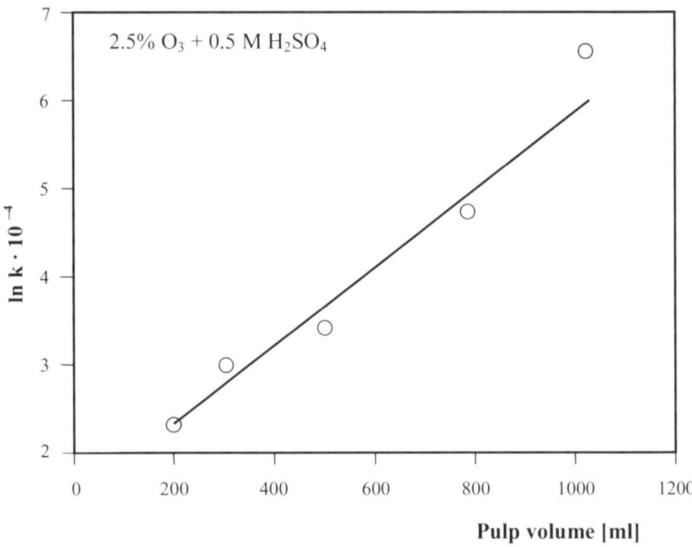

Fig. 12.11. Dependence of the leaching rate on the amount of the leaching solution.

temperature, the effect of the amount of dissolved ozone in the leaching medium is evident. Therefore, investigations were also carried out into the effect of the contact time of gaseous ozone with the liquid and its capacity to dissolve in copper. Figure 12.11 shows the course copper leaching from chalcopyrite in different volumes of the solutions of sulphuric acid. It may be seen that the amount of leached copper from chalcopyrite increases with increase of the volume of the leaching agent, i.e. with increase of the height of the level of the column of the leaching solution. The initial assumption according to which the leaching kinetics depends on the amount of ozone in the solution has been confirmed.

The dependence of the leaching rate on the amount of the solution, Fig. 12.11, shows that the solution is not saturated even when using 1000 ml of the solution which is the maximum height from the design viewpoint of the equipment, because the leaching rate increases in direct proportion. This means that to increase the efficiency of leaching it is necessary, in these conditions, to continue increasing the height of the liquid column.

The amount of ozone in the working atmosphere has a strong effect on the amount of metal leached into the solution. Figure 12.12 shows the kinetic curves of leaching of copper in relation to the amount of ozone in the working atmosphere.

Figure 12.13 is the dependence of the leaching rate of copper

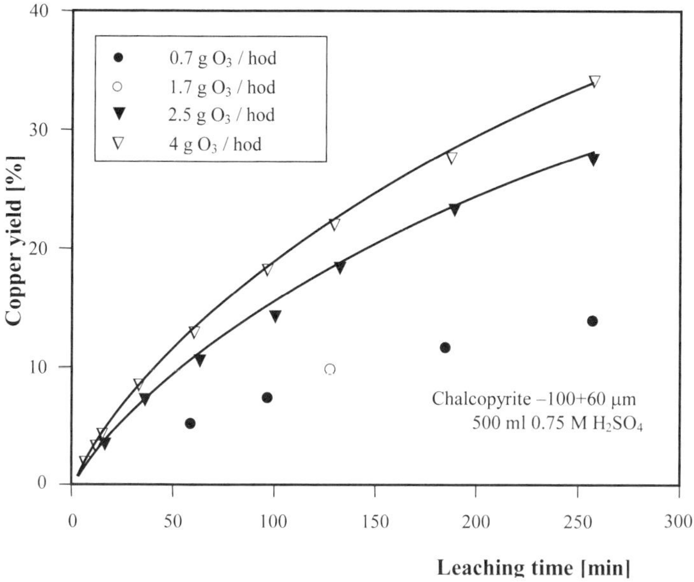

Fig. 12.12. Kinetic curves of copper leached depending on the amount of ozone in the working atmosphere.

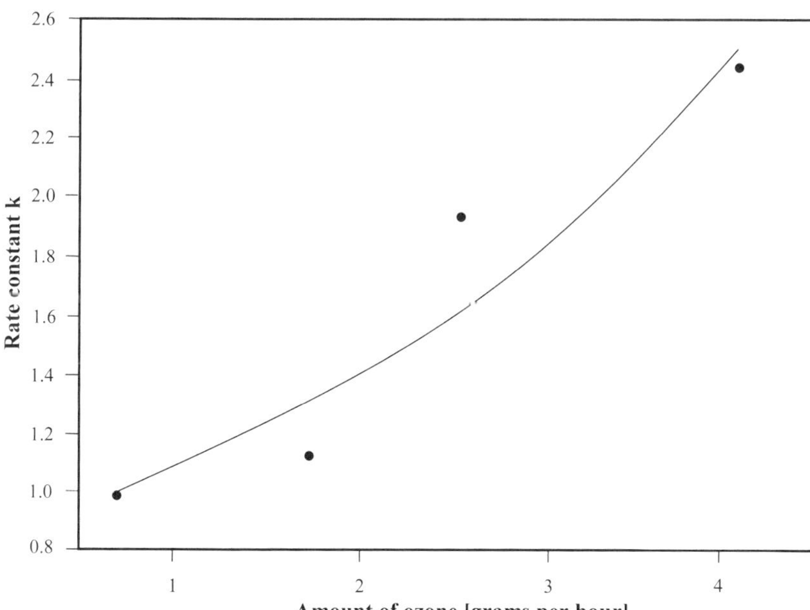

Fig. 12.13. Dependence of the amount of leached copper on the amount of ozone in the working atmosphere.

from chalcopyrite on the amount of ozone in the working atmosphere. In may be seen that the increase of the amount of ozone in the working atmosphere increases almost proportionately to the leaching rate.

These results show the controlling effect of ozone on the leaching rate of chalcopyrite. To determine the exact effect of all the factors influencing leaching, it is necessary to know the exact amount of ozone in the solution in the given conditions. However, it has not as yet been possible to solve the problem satisfactorily because of technical reasons. Consequently, it has not been possible to determine the activation energy nor other kinetic characteristics. Despite these facts, it may be seen that the leaching rate is the highest at a temperature of 22 °C. Taking into account the temperature dependence of the solubility of ozone, shown in Fig. 12.10, it may be assumed that although the amount of the oxidation agent in the solution at low temperatures is sufficiently large, a controlling role is played by the temperature and viscosity of the leaching medium.

The rate-controlling step could be the external diffusion of ozone to the leached surface and the removal of the reaction products through the layer of the viscous medium and/or slow reaction under the effect of low temperature. At higher temperatures, the shortage of the oxidation agent in the leaching medium is already evident and the diffusion of the agent to the leached surface from the volume of the leaching solution obviously limits the overall reaction rate. The amount of the leached metal increases with increase in the volume of the leaching solution, i.e., with increase of the height of the liquid column, and this results in an increase of the contact time of the gas bubble, containing ozone, with the pulp.

The highest yield was obtained at a leaching temperature of approximately 20 °C. This is a positive result – in practice, the cost of heating, holding at a temperature and cooling of the large volumes of the diluted solution would be greatly reduced. In addition, this oxidation agent does not contaminate the solution with foreign anions and ensures complete transfer of sulphur to the soluble form.

An important advantage of the use of ozone as the oxidation agent in acid oxidation leaching is also the availability of high performance commercial ozone generators (Fisher, Brown Boveri) used, for example, for ozonation of drinking water. Therefore, it appears that it may also be used as an efficient and cheap oxidation agent in the hydrometallurgy of copper and may improve the economic parameters of this production process.

12.1.5. Leaching of chalcopyrite in the high-frequency field [123]

One of the promising methods of intensification of the leaching of chalcopyrite is the application of high-frequency induction heating of the leaching pulp. This is a completely original contribution to this problem. No mention of this method has been made in the available literature. The concept is basically simple and has already been applied in metallurgical practice, because high-frequency induction heating is used quite widely in the melting of metals. The condition for efficient functioning of this system is that the heated material must be electrically conducting.

Induction heat forms always when electromagnetic waves impact the electrically conducting material [133]. Some of the waves are reflected and returned, the others penetrate into the material and induce eddy currents in the material. The energy of the electromagnetic waves is used for releasing free electrons so that they can move in the material. The electrons move in the direction of intensity of the electrical field and acquire high velocities over short distances and, consequently, a relatively high kinetic energy. When the electrons collide with the atoms of the conductor, the electrons transfer their kinetic energy to the atoms that increasing the intensity of vibrations of the atoms, i.e., the temperature of the material increases. The electromagnetic waves are damped in the wall and their energy is converted to thermal energy.

This is the basis of the concept of application of high-frequency heating in hydrometallurgy. Since leaching solutions are electrolytes, i.e., conduct electrical current, they can be used in this application.

Natural chalcopyrite with a grain size of −100+60 μm was leached in an aqueous solution of ferric chloride with the concentration in the range 0–1 M $FeCl_3$ acidified by the addition of HCl with different concentrations in the range 0–1 M. The leaching time was 3 h. Because of the principle of the methods, it was not possible to change or set the temperature which in every experiment was stabilised at the boiling point of the solution, i.e., +104 °C. Mixing was generated by intensive boiling.

The kinetic curves of leaching of copper in relation to the concentration of Fe^{3+} ions are shown in Fig. 12.14.

The following results were obtained: in the low concentration

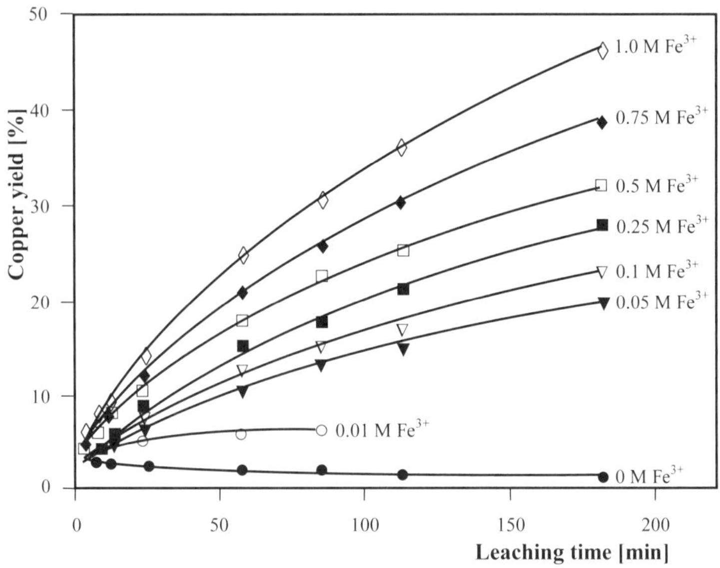

Fig. 12.14. Kinetic leaching curves of copper leached from chalcopyrite in relation to the concentration of the ferric ion.

range, approaching zero {0–0.05 M $FeCl_3$} the extent of leaching is low but when the concentration is increased above 0.1 M $FeCl_3$ the rate of the process greatly increases. It is important to note that the leaching rate was not stabilised even at the selected limiting value of 1 M $FeCl_3$, which is in contradiction with the results obtained in classic leaching experiments in the normal conditions [98]. In this case, 0.2–0.5 M $FeCl_3$ is usually sufficient to obtain the limiting concentration, and with a further increase of concentration the leaching rate was no longer affected. In this case, a positive intensification effect may be exerted either by efficient mixing or by the high-frequency field.

Figure 12.15 shows the dependence of the logarithm of the rate constant on the concentration of ferric ions. It shows a sharp break at approximately 0.1 M $FeCl_3$. The dependence clearly shows a further increase of the leaching rate as a result of increase of the concentration. However, this is not rational because of practical reasons, for example, because of the unacceptably aggressive environment.

The acidity of the medium has also a significant role in the investigated processes. Figure 12.16 shows the kinetic curves of leaching of copper from chalcopyrite using 0.5 M $FeCl_3$ in relation

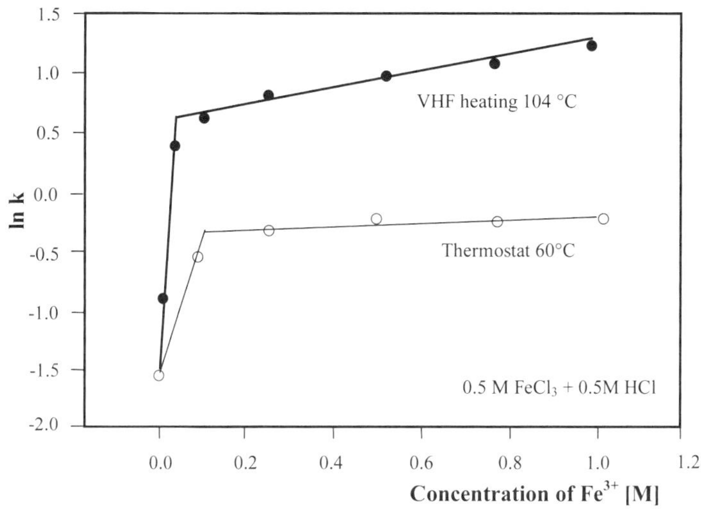

Fig. 12.15. Dependence of the leaching rate on the concentration of ferric ions.

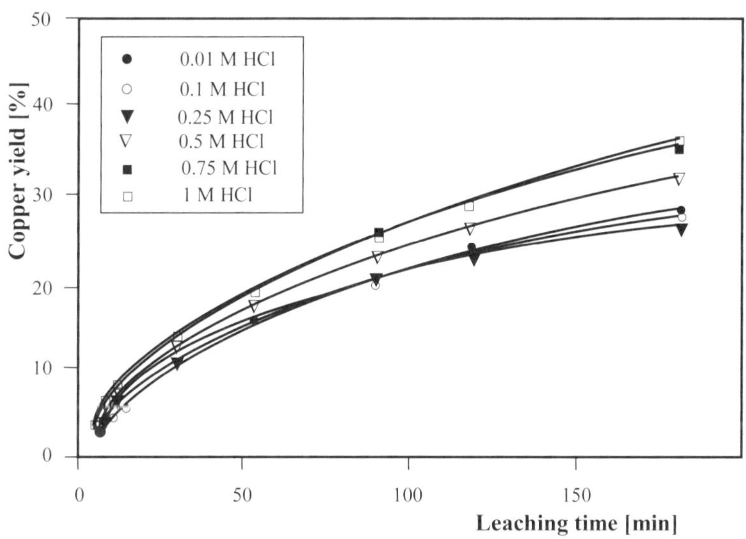

Fig. 12.16. Kinetic curves of leaching of copper from chalcopyrite in relation to the acidity of the medium.

to the acidity of the medium. In this case, it was again showed that of the increase of acidity (up to 1.0 M HCl) has a positive effect on the leaching kinetics which was not evident in leaching in the conventional conditions because according to the generally accepted view, of the acidity of the medium is maintained to prevent

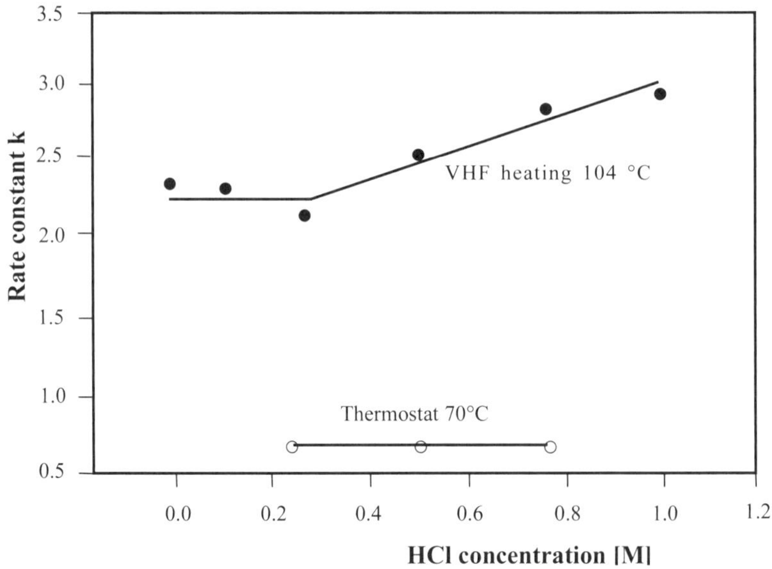

Fig. 12.17. Dependence of the leaching rate on the concentration of HCl.

hydrolytic precipitation of iron from the solution and above a certain limit of the increase of the concentration is no longer efficient.

In this case, Fig. 12.17, it may be seen that the increase of the acidity of the medium results, as in the previous case, in a small increase of the leaching rate.

The resultant values of the yield of copper from chalcopyrite are comparable and higher in comparison with the method of conventional heating in the experiments in open reaction vessels. The heating rate of the leaching medium is considerably higher than in the classic reactors and the time required to reach the working temperature is expressed by tens of seconds, whereas in conventional heating in comparable volumes it is tens of minutes. It is also important to note that in high-frequency heating, the yield is positively affected by both the increase of the concentration of ferric ions and of the acidity of the medium. However, this is not so in the conventional experiments.

12.1.6. Leaching of chalcopyrite in the microwave field [124–129]

The principle of application of the microwave field in leaching is the same as in the previous case, but a different type of high-energy field is used. Microwave energy is non-ionised electromagnetic radiation in the frequency range 300 MHz–300 GHz. Microwaves reflects on the surfaces of metallic materials and do not interact with insulating materials. Industrially manufactured equipment uses frequencies of 915 MHz and 2450 MHz.

Microwave heating differs from the classic heating methods by the fact that heat forms by rotation of polar particles, i.e., molecules and ions in the high-frequency electromagnetic field. They try to orient themselves in the direction of radiation, but the electrical field changes its direction very rapidly in accordance with the frequency – at a frequency of 2450 MHz the direction of the electrical field changes $2.45 \cdot 10^9$ times per second. The rotation of the molecules results in heating them and, in addition to this, the molecules do not manage to follow this change of the direction of the electrical field and, consequently, absorb radiation energy.

Leaching experiments were carried out using solutions with the concentration 0–1.0 M $FeCl_3$ in 0.5 M HCl. Leaching time was 120 min. Mixing was ensured by intensive boiling and the temperature was stabilised at 104 °C, which is the boiling point of the solution.

The increase of the output of the microwave emitter results in an increase of the leaching efficiency, as shown by Fig. 12.18.

The dependence of leaching efficiency on the concentration of

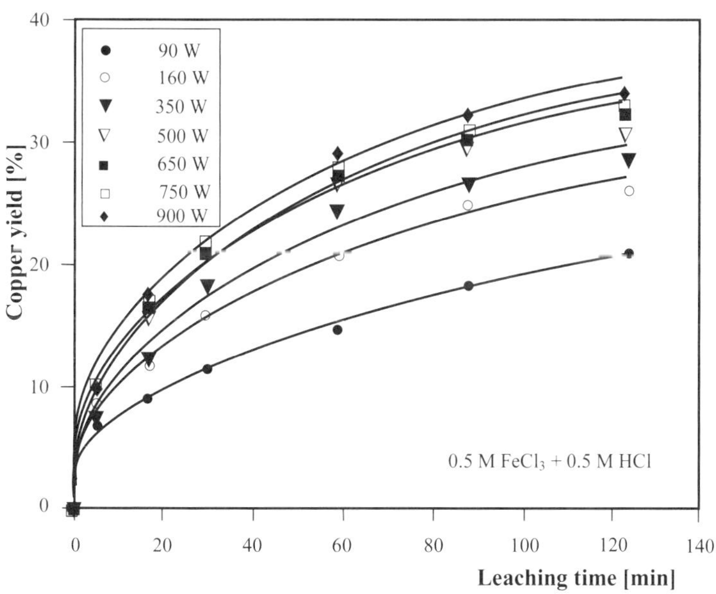

Fig. 12.18. Dependence of the amount of leached copper from chalcopyrite on the performance of the emitter.

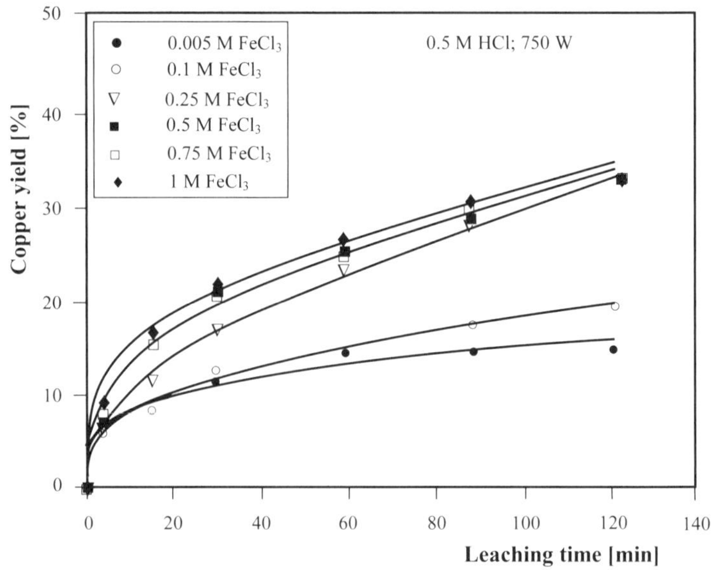

Fig. 12.19. The dependence of the amount of leached copper from chalcopyrite on the concentration of $FeCl_3$.

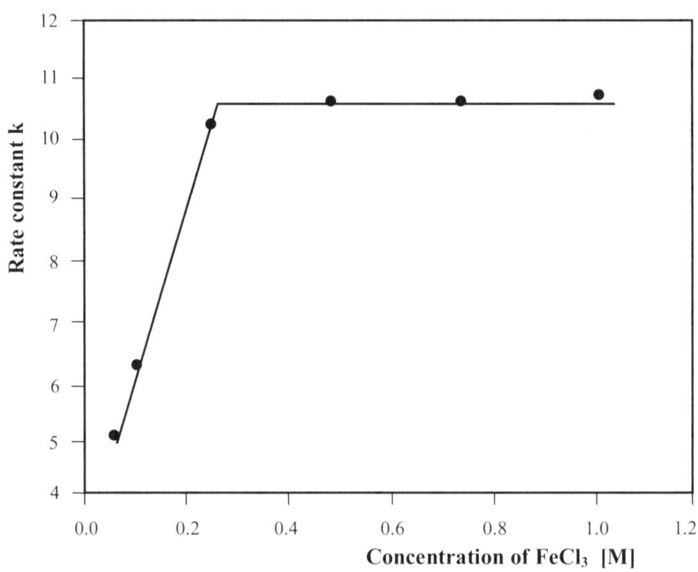

Fig. 12.20. Dependence of the leaching rate of the concentration of $FeCl_3$.

ferric ions in the solution is shown in Fig. 12.19. The increase of the concentration of the ferric ions in the solution also increases the efficiency of leaching copper from chalcopyrite. At lower concentrations, the leaching rate is high and subsequently

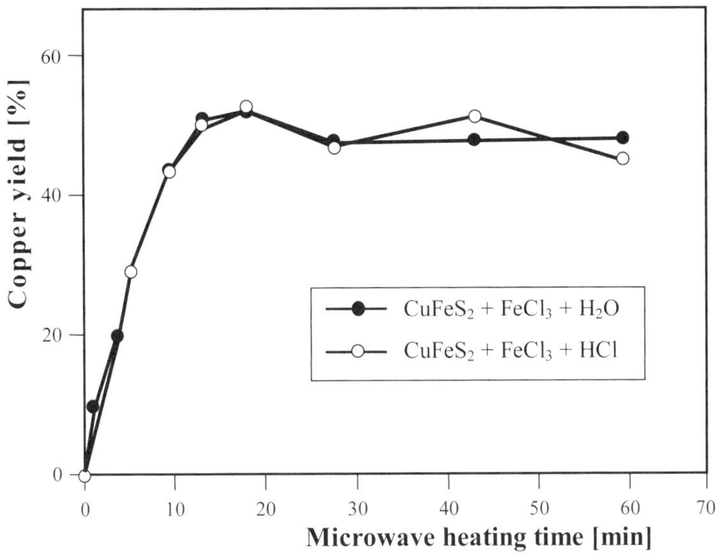

Fig. 12.21. Kinetic leaching curves of copper from chalcopyrite in water after microwave heating of a mixture of chalcopyrite and $FeCl_3$ moistened with water or HCl.

decreases, Fig. 12.20.

Other possibilities of using the microwave field in the leaching of chalcopyrite have been examined. Instead of using the conventional acidified aqueous solution of ferric ions into which chalcopyrite was added and the pulp was placed in the microwave field, the following procedure was used: a dry mixture of chalcopyrite and ferric chloride was prepared in the equimolar ratio and the mixture was subjected to microwave radiation. Solid-state reactions resulted in the transfer of copper to easily soluble solutions which were transferred into the solution as a result of short-term leaching in cold water. The leaching efficiency was considerably greater than in the previous cases.

Figure 12.21 shows the kinetic curves of leaching of copper from chalcopyrite in relation to time, with the mixture moistened with water or 0.5 M HCl. The amount of leached copper is almost the samc in both cases. The maximum amount of leached copper was already reached after 15 min of microwave heating.

The dependence of the amount of leached copper on the duration of microwave heating and on the subsequent leaching time is shown in Fig. 12.22. The optimum duration of the effect of microwave energy is 15 min. The entire amount of leached copper was transferred almost instantaneously into the solution and no further

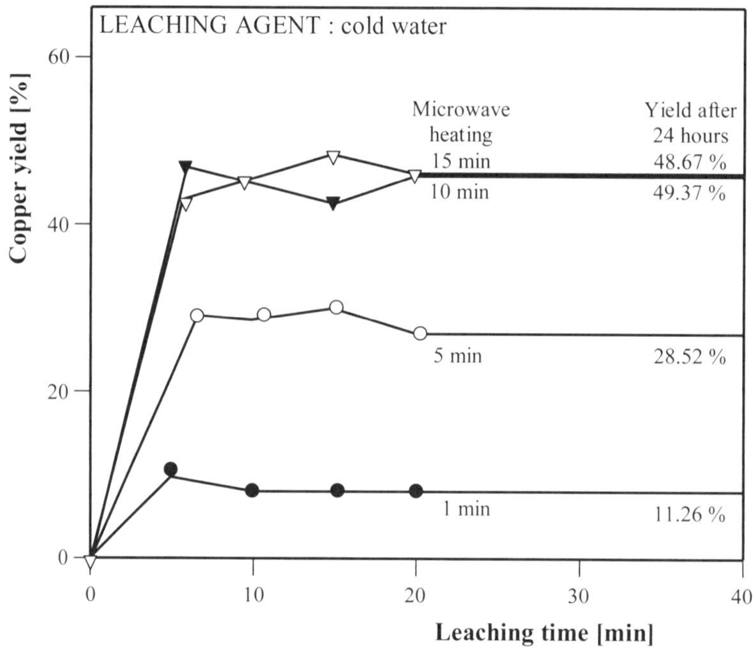

Fig. 12.22. Kinetic curves of leaching of copper from chalcopyrite water in relation to the duration of microwave heating of the mixture of chalcopyrite and $FeCl_3$.

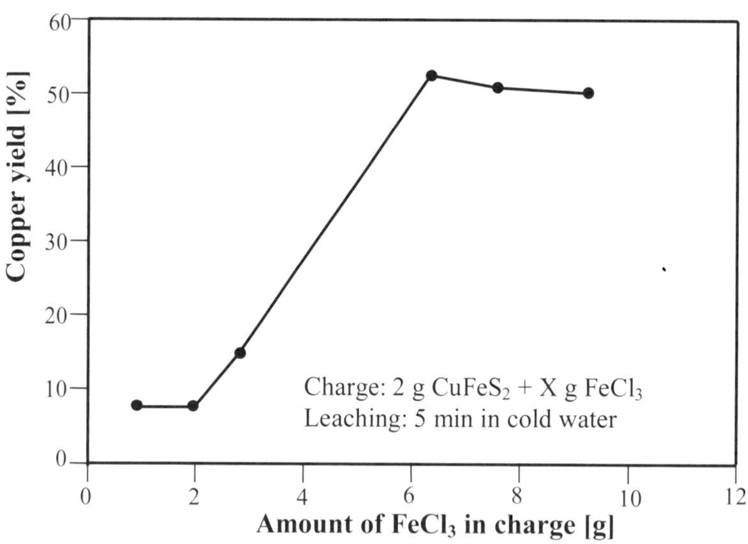

Fig. 12.23. The effect of the amount of ferric chloride in the charge on the amount of leached copper.

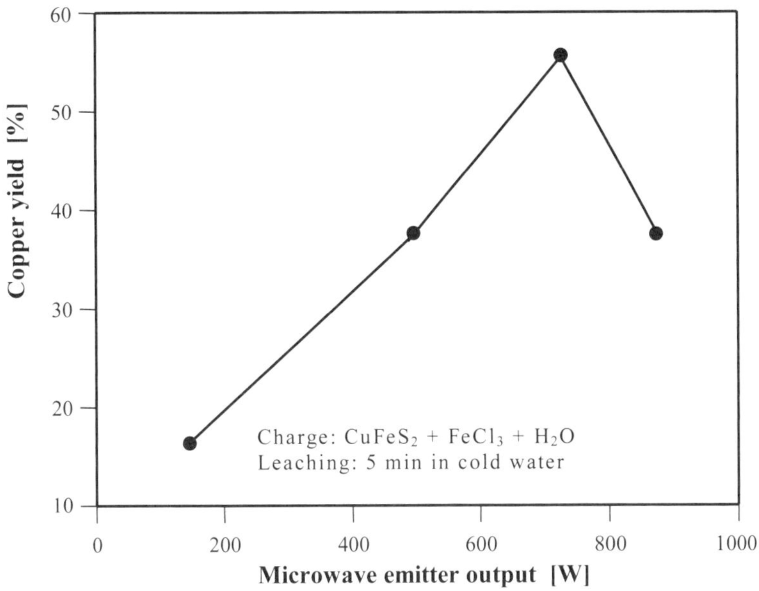

Fig. 12.24. Effect of the performance of the microwave emitter on the amount of leached copper.

copper was leached even after 24 hours. This shows that 5 min of leaching in cold water is fully sufficient for the transfer of copper into the solution.

Another important factor is the amount of $FeCl_3$ in the charge. Figure 12.23 shows the effect of the amount of ferric chloride in the charge on the amount of leached copper. The optimum amount of ferric chloride in the charge is equal to the stoichiometric amount.

Figure 12.24 shows the results of experiments carried out to investigate the effect of the intensity of microwave irradiation on the charge on the amount of leached copper. The results show that with the increase of the performance of the emitter (MW) the yield of copper increases, but only up to 750 W. With a further increase of performance above 900 W, the yield of copper decreases. This may be also caused by the fact that in heating the rate of heating of the mixture was too high resulting in partial dissociation of ferric chloride.

Comparison of this method of microwave leaching with other methods, using the 'classic' pulp for leaching, shows:

- the process is simple, leaching takes place under the normal conditions of temperature and pressure;
- very short reaction time and leaching time;
- cheap and neutral leaching agent – water;
- high yield of copper;
- ecological advantages – no acid is used in the process;
- availability of commercial microwave equipment.

12.1.7. Leaching of chalcopyrite in the presence of deep nodules as an oxidation agent [124, 125]

Deep sea polymetallic manganese nodules contain a high proportion of manganese oxide which are a generally strong oxidation agent. Therefore, in addition to other applications, investigations were also carried out to determine the use of these nodules as a possible oxidation agent in leaching of sulphides [122, 123] in a solution of HCl with different concentrations at different temperatures. The chalcopyrite, which is not leached in the HCl solution, was leached efficiently in the presence of deep sea nodules in the solutions of 3–4 M HCl in accordance with the general leaching reaction:

$$5MnO_2 + 2CuFeS_2 + 20HCl \rightarrow 5MnCl_2 + 2CuCl_2 + 2FeCl_3 + 10H_2O + 4S^0 \quad (12.41)$$

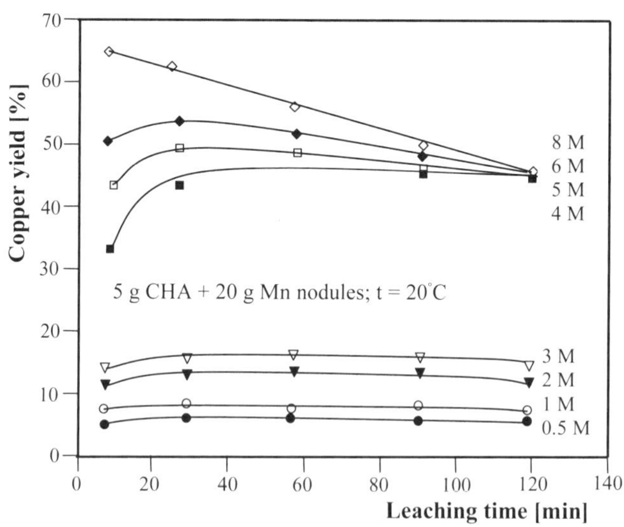

Fig. 12.25. The leaching of chalcopyrite in the presence of deep sea Mn nodules in HCl at 20 °C.

Subsequent experiments were carried out with deep sea nodules with 32% Mn and solutions of HCl in the range 0.5–8 M HCl at temperatures of 20 and 70 °C, and the main factors influencing the process were investigated.

The effect of HCl concentration

Figure 12.25 shows the amount of leached copper using different HCl concentrations. The optimum HCl concentration was 4 M.

The effect of temperature

Figure 12.26 shows the kinetic curves of leaching of copper at temperatures of 20, 50, 60 and 70 °C and the HCl concentration of 4 M, as determined from the previous dependence. The optimum leaching temperature appears to be 50 °C.

Of course, the amount of the oxidation agent in the charge controls the efficiency of the process. Figure 12.27 shows that the ratio of the amount of the nodules to chalcopyrite of 4:1 is optimum for efficient leaching.

Several important facts follow from the experimental results:

- chalcopyrite is sufficiently leached in the presence of deep sea manganese nodules in HCl;

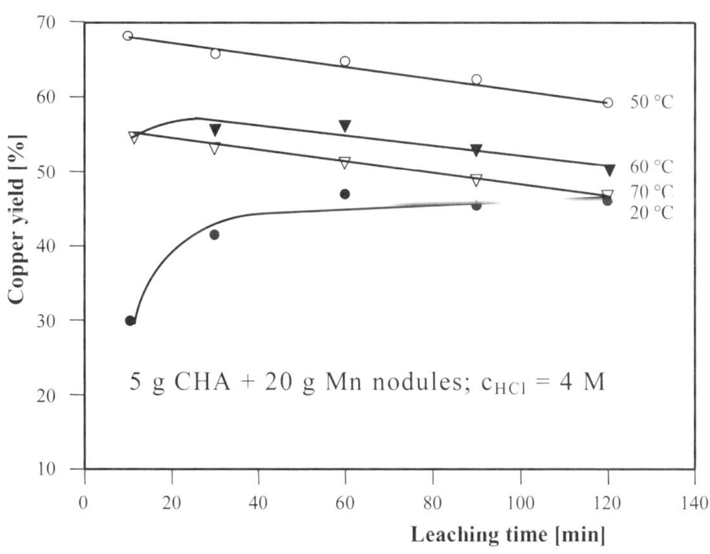

Fig. 12.26. Leaching of chalcopyrite in the presence of the deep sea Mn nodules in HCl at different temperatures in 4 M HCl.

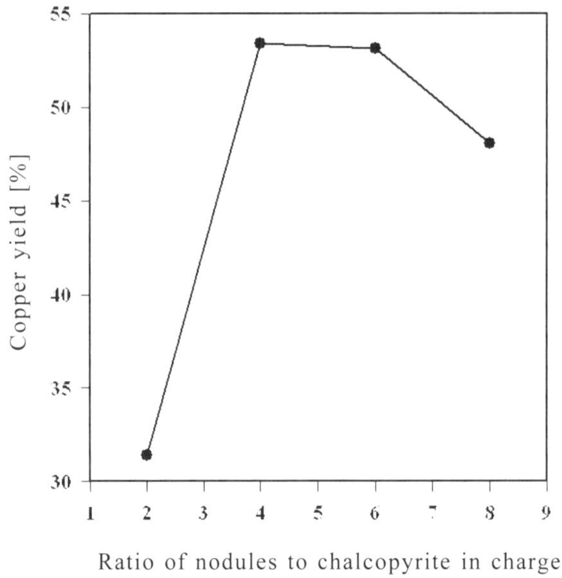

Fig. 12.27. The leaching of chalcopyrite in the presence of different amounts of deep sea Mn nodules at 20 °C.

- the amount of leached copper is the highest of all tested method;
- the optimum concentration of HCl for leaching is 4 M;
- the optimum ratio between the nodules and chalcopyrite in the charge is 4:1.

All the experiments described here were carried out using the same chalcopyrite concentrate and the experimental conditions which may be compared to a certain extent. The author knows of no such detailed investigation carried out. The results obtained in this manner could then be reliably compared and it would be possible to estimate the efficiency of a specific method. This is also important owing to the fact that the methods, used in the pilot and full plant, are based on the application of ferric sulphate or chloride and have the lowest efficiency of all the investigated methods.

Some of the characteristics obtained in the experiments are summarised in Table 12.4.

Comparison of the leaching results obtained for the same chalcopyrite by different methods, described in Table 12.4, in the conditions in which the given method was characterised by the highest efficiency of leaching copper, is shown in Fig. 12.28.

Table 12.4. Results of leaching kinetics of chalcopyrite and its mechanism

Material	Leaching medium	E_A [kJ/mol]	t [°C]	Cu yield *(after 2hours)*	Mechanism	Leaching feature
	$Fe_2(SO_4)_3$ + H_2SO_4	12	20-40	0.36 40 °C	mass transfer	elemental sulphur creation
	$Fe_2(SO_4)_3$ + H_2SO_4	69	40-95	7.99 95 °C	chemical reaction	elemental sulphur creation
	$FeCl_3$ + HCl	55	3.5-95	14.85 95 °C	chemical reaction	elemental sulphur creation
$CuFeS_2$ *Cuba*	$FeCl_3$ + HCl+ CCl_4	31.2	45 -80	16.57 80 °C	mass transfer, sulphur dissolution in CCl_4	no elemental sulphur on interface
	H_2SO_4+O_3		22	23.4 22 °C	chemical reaction	no elemental sulphur on interface
	$FeCl_3$ VF ohrev		104	27.4 104 °C	parabolic kinetics, mass transfer	solid state reaction
	$FeCl_3$ microwave heating		104	33.2 104 °C	parabolic kinetics, mass transfer	900 W
	$CuFeS_2$ + $FeCl_3$ paste; microwave heating			55.2 20 °C *	dissolving of soluble Cu compunds	* leaching in cold water, 5 min

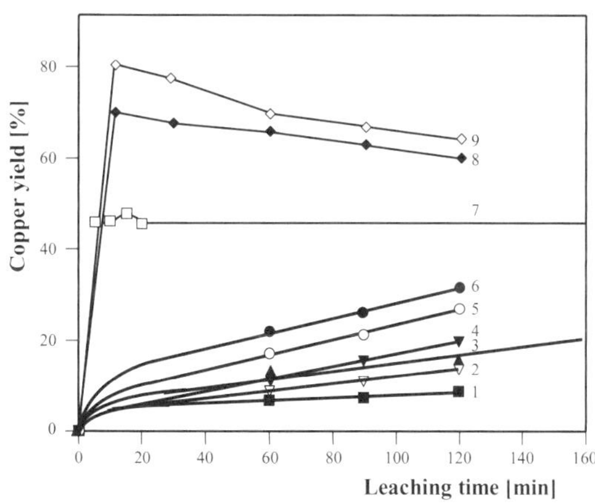

Fig. 12.28. Comparison of the kinetic curves of leaching copper from chalcopyrite by different methods: 1) $Fe_2(SO_4)_3$+ H_2SO_4; 99 °C, 2) $FeCl_3$ + HCl, 95 °C; 3) $FeCl_3$ +CCl_4 + HCl, 90 °C; 4) O_3+H_2SO_4, 22 °C; 5) high-frequency leaching, $FeCl_3$+ HCl, 104 °C; 6) microwave leaching, 900 W, $FeCl_3$ + HCl, 104 °C; 7) microwave heating of the dry mixture $CuFeS_2$+$FeCl_3$, leaching in cold water; 8) non-milled Mn nodules, HCl, 50 °C; 9) milled manganese nodules, HCl, 50 °C.

12.2. Copper sulphides of the Cu_xS type

Sullivan [135] published the first detailed view of the dissolution of chalcocite in an acid oxidation medium using ferric sulphate. He found that chalcocite dissolves in two stages according to the following reactions:

$$Cu_2S + Fe_2(SO_4)_3 \rightarrow CuS + CuSO_4 + 2FeSO_4 \quad (12.42)$$

$$CuS + Fe_2(SO_4)_3 \rightarrow CuSO_4 + 2FeSO_4 + S^0 \quad (12.43)$$

Stage I is considerably faster than stage II, and the former is almost completely independent of temperature which indicates that the reaction rate is controlled by mass transport. The second stage already depends greatly on temperature and on the surface area of the dissolved particles. The experiments have confirmed that if the amount of copper transferred into the solution is less than 50%, elemental sulphur does not form and the solid phase CuS forms after transfer of 50% of copper into the solution [136].

Two distinctive stages of the reaction were also found by Warren [137] in pressure leaching of chalcocite in sulphuric acid. The activation energy of the first stage was 27.6 kJ/mole, and that of the second stage only 7.5 kJ/mole. This value was attributed to the diffusion of reactants through the sulphur layer covering the surface of the leached particles, formed by the reaction (12.43). The activation energy of the second stage, determined by Sullivan, was approximately 36 kJ/mole. Stanczyk and Rampacek [138] also observe two stages of the solution of chalcocite in the process of pressure leaching in sulphuric acid. At temperatures above 120 °C, the resultant elemental sulphur melts and coats the leached particles, which greatly reduces the leaching rate.

Fisher and Roman [139] did not detect stage II in the examination of the leaching kinetics of chalcocite in sulphuric acid with oxygen as the oxidation agent. The final product of the solution was covellite CuS, and the experimental temperature was up to 65 °C. The measured activation energy was 27 kJ/mole. Since Dahms et al. [140] obtained sulphur as the reaction product when using the same media at temperatures of up to 120 °C, it is evident that stage II is slow at temperatures up to 65 °C.

It was also found that the addition of halogenous anions to the leaching agent (Cl^- or Br^-] greatly increases the efficiency of the process of dissolution of chalcocite. Subsequently, Roman and Brenner [141] proposed a process in which the sulphuric acid, containing chloride ions, may be a useful leaching agent of copper sulphides and describes the advantages of this medium.

Thomas, Ingraham and Macdonald [142] proved, however, that this process is more complicated. They investigated the kinetics of leaching of chalcocite in the solution of ferric ions by the rotating disc method and confirmed the two reaction steps, proposed by Sullivan. However, they found that the reaction (12.42) contains in fact several intermediate stages characterised by the formation of djurleite $Cu_{1.93}S$, digenite $Cu_{1.8}S$ and a new phase $Cu_{1.4-1.05}S$, which was later described by Moh [143] and named the 'blue' covellite. Normal covellite was not found in dissolution. These sulphides were detected by X-ray diffraction phase analysis. The activation energy of the first stage was approximately 20–25 kJ/mole. The rate-controlling step in the first stage was the diffusion of the ferric ion through the boundary layer, surrounding the surface of the mineral. During the first step of the reaction, the time dependence of the amount of leached copper was linear. The rate of the second leaching stage, represented by the equation (12.43) at 25 °C, was considerably lower, but at 50 °C both the rates were already identical. The activation energy of the second part of the reaction is higher which corresponds to the chemically controlled reaction rate.

Tkachenko and Tseft [144] investigated the leaching of high-purity natural chalcocite in acidified solutions of ferric chloride in the temperature range 60–105 °C. In this temperature range, the dissolution rate was high and was controlled by mass transport in the solutions.

Mulak [145] confirmed the dissolution of chalcocite through the formation of the transitional product CuS. At temperatures below 60 °C, the dissolution rate was controlled by mass transport to the leached surface with the activation energy of 6.3 kJ/mole. At temperatures higher than 60 °C, the rate of the second stage of dissolution with the activation energy of 22 kJ/mole was controlling. The first stage of dissolution was not sensitive to the change of the acid concentration but depended on the concentration of the ferric ion in the solution in the range 0.005–0.05 M. Similar results were also obtained by Kopylov and Orlov [146].

King, Burkin and Ferreira [147] investigated the kinetics of dissolution of synthetic chalcocite by the acidified solution of ferric chloride. They confirmed that the first, fast stage of the solution produces CuS and is followed by the second stage in which the rate of the dissolution of this intermediate product is considerably lower. The activation energy of the first stage of the dissolution reaction was 3.43 kJ/mole and depended greatly on the concentration of $FeCl_3$ in the range 10^{-4}–10^{-2} M. The activation energy of the second stage of dissolution was 101–122 kJ/mole and this was interpreted as the chemically controlled reaction of oxidation of the sulphide ion. Elemental sulphur formed only in the second part of the reaction.

The transformation of chalcocite to covellite was investigated by X-ray diffractometry and electron microscopy. The results show that the new sulphides formed change continuously and the sulphides Cu_2S, $Cu_{1.96}S$, $Cu_{1.8}S$ and CuS were found, and the existence of the region of non-stoichiometric copper sulphides in the range $Cu_{1.5}S$ to CuS, i.e. covellite with an excess of copper, was confirmed. This study showed that the extraction of copper from the surface of the mineral to the surrounding solution resulted in the formation of a diffusion gradient penetrating deeply into the solid phase. The rate-controlling step of dissolution was diffusion of copper inside the crystal structure through the phase $Cu_{2-x}S$.

The existence of the non-stoichiometric phases between chalcocite and covellite was also confirmed by the results obtained by Burkin [148].

Kmetova et al. [149] investigated the oxidation leaching of chalcocite in nitric acid in the temperature range 25–85 °C at different partial pressures of oxygen in order to find a method of efficient leaching (from the energy viewpoint) at room temperature. They found that at the HNO_3 concentration of up to 1 mol/dm^3 and at a temperature of 25 °C the rate of the process is relatively low with the activation energy of 19.5 kJ/mole which corresponds to the first stage of the process [150]. The rate-controlling step of the reaction was probably the diffusion of the oxidant to the leached surface.

Peters and Loewen [151] investigated the mechanism of oxidation leaching of chalcocite in perchlorate acid at a higher pressure of oxygen in the temperature range 105–140 °C. They found that the first stage of dissolution of chalcocite is completed within 30 min and requires very fast diffusion of copper ions inside the crystal structure of chalcocite. The second stage is slower, the resultant pores and cracks indicate the electrochemical nature of dissolution. The activation energy of the second stage of the solution was 47.7 kJ/mole which was interpreted as the control of the heterogeneous reaction on the surface of the mineral.

Dahms et al. [140] investigated pressure leaching of chalcocite in an acid medium and explained the mechanism of the process. The overall reaction consists of two stages, with the first stage taking place without oxidation of sulphur by the reactions

$$Cu_2S + \frac{1}{2}O_2 \rightarrow CuS + CuO \tag{12.44}$$

$$CuO + 2H^+ \rightarrow Cu^{2+} + H_2O \tag{12.45}$$

which means that the presence of the acid is essential for the reaction. The second stage contains oxidation leaching of covellite. The kinetic equation of the amount of dissolved copper was determined in the form:

$$\frac{d\left[Cu^{2+}\right]}{dt} = 7.35 \cdot 10^3 \, p_{O_2}^{\frac{1}{3}} A^{-\frac{20770}{RT}} \tag{12.46}$$

Examination of the effect of the partial pressure of oxygen suggested the conclusion that the adsorption of oxygen on the surface plays a very important role in the entire mechanism. Consequently, there is a relationship between the reaction rate and the surface:

$$\frac{dx}{dt} = k' A \tag{12.47}$$

and the surface area A is described by the Langmuir adorption isotherm in the form:

$$A = W p_{O_2}^{\frac{1}{3}} \tag{12.48}$$

The entire process is conditioned by the number of defects in the crystal structure on which the process of oxygen adorption takes place.

Investigations also carried out into the possibilities of electrochemical dissolution of chalcocite [151], especially in regard of the application of methods for processing copper matte. Intermediate products, digenite, blue and natural covellite, formed immediately after the start of the process of anodic dissolution.

Mackinnon [152] also investigated anodic dissolution of chalcocite. He used the method of leaching in a flow reactor and the results of leaching for different leaching agents and experimental conditions show that anodic dissolution of chalcocite takes place by means of the formation of intermediate sulphides, with the blue covellite $Cu_{1.1}S$ at the end of this series of sulphides.

When using sulphuric acid, bubbled with oxygen, the yield of copper was approximately 50% which indicates the occurrence of the stage I of the overall dissolution reaction of chalcocite. When using the addition of KBr in the leaching solution, approximately 95% of chalcocite was dissolved.

The company Broken Hill Associated Smelters Pty Ltd (BHAS) in Port Pirie, southern Australia, developed a process for processing secondary products from refining of copper–copper–lead sulphide matte using sulphuric acid bubbled with oxygen containing sulphate and chloride ions. Leaching was carried out at a temperature of 85 °C with oxygen as the oxidant. Copper was dissolved whilst lead remained in the solid residue as lead sulphate together with elemental sulphur. The copper from the solution was immediately transferred into the sulphate electrolyte by liquid extraction. From the electrolyte it was electrolytically precipitated. Lead sulphate was recycled in a lead-making shaft furnace [153, 154].

Havlik [155, 156] investigated the leaching of synthetic chalcocite by the rotating disk method in the medium of sulphuric acid using a gas atmosphere containing ozone at the room temperature and pressure. Since the high oxidation potential of the ozone resulted in the oxidation of sulphur in the soluble form, the sulphur did not cover completely the reaction surface so that detailed examination of the mechanism of the leaching reaction was possible. The results confirmed that the process takes place in two stages, the faster stage, controlled by diffusion, and the slower stage, controlled by the kinetics of the chemical reaction. With time and with the increase of the deficit of copper in the crystal structure, the two stages are equalised as regards the rate. The results also confirmed the formation of non-stoichiometric compounds of the type Cu_2S on the leached surface [157] by the methods of examination of the fine structure of the mineral.

Thomas, Ingraham and Macdonald [142] investigated the leaching of synthetic digenite in acid solutions of ferrous sulphate in the temperature range 5–80 °C. They found that the course of dissolution is similar to of chalcocite, with a similar two-stage reaction mechanism. The first stage is described by the relationship:

$$5Cu_{1.8}S + 4Fe_2(SO_4)_3 \rightarrow 4CuSO_4 + 5CuS + 8FeSO_4 \qquad (12.49)$$

The activation energy of the first stage was 21–25 kJ/mol. The second stage of dissolution is basically similar to stage II of reaction (12.43), is slow at 25 °C, with a high activation energy indicating the control of the reaction rate by the formation of elemental sulphur and covellite. The rates of the two dissolution stages were equalised at temperatures above 60 °C.

The expedience of processing a rich copper concentrate has been described by Bartlett [158]. According to his results, the flotation chalcopyrite concentrate may be hydrothermally converted to digenite directly by oxygen in the continuous process in the autoclave at temperatures higher than 200 °C by the reaction:

$$1.8CuFeS_2 + 0.8H_2O + 4.8O_2 \rightarrow Cu_{1.8}S + 1.8FeSO_4 + 0.8H_2SO_4 \quad (12.50)$$

The process was also tested on the pilot plant scale at the United States Gold Plants [159].

Sullivan [160] found that covellite is leached in the solution of ferric sulphate by the relationship:

$$CuS + Fe_2(SO_4)_3 \rightarrow CuSO_4 + 2FeSO_4 + S^0 \quad (12.51)$$

He also found that the leaching rate is considerably lower than that in the leaching of chalcocite and is lower than the dissolution of the product formed by the first stage of the reaction of dissolution of chalcocite. This shows that this product is not identical with natural covellite, although its chemical formula is CuS. King [161] assumed, however, on the basis of diffraction measurements and electron microscopic examination, that these minerals are identical and explains the differences in the reaction rate by large porosity and, consequently, a large reaction surface of CuS, formed as the reaction product of leaching of chalcocite. The activation energy determined by King was 105 kJ/mole.

Thomas and Ingraham [162] investigated the leaching of synthetic covellite in an acidified solution of ferric sulphate in the temperature range 25–80 °C by the rotating disk method. Although the measured reaction rate was considerably higher than that published by Sullivan, the authors believed that the synthetic CuS dissolved in accordance with the stage II of the equation (12.43). Two rate-controlling steps were found in the investigated temperature range. At temperatures below 60 °C, the rate was controlled by the chemical reaction on the dissolved surface with the activation energy of 136 kJ/mole. At higher temperatures, the rate was controlled by the transport of the solution with the activation energy of 33.5 kJ/mol. This dissolution mechanism of covellite has not been confirmed by other authors.

Baur, Gibbs and Wadsworth [163] leached pieces of covellite and found that this mineral initially reacts very rapidly but after a short period of time the leaching rate decreases. This was explained by the formation of a layer of elemental sulphur on the leached surface.

The results obtained by Peters and Loewen [150] for pressure leaching of natural covellite in perchlorate acid show that the leaching rate of synthetic covellite, produced as the product of dissolution of covellite, is higher than that of natural covellite. Investigations carried out at a temperature of 125 °C show that the reaction surface is covered with a layer of liquid elemental sulphur which greatly reduces the rate of the process of dissolution of covellite.

Dutrizac and MacDonald [164] studied the leaching of pure synthetic CuS by the rotating disk method and also of high-purity natural covellite in acidified solutions of ferric sulphate in the temperature range 15–95 °C. In both materials, the rate of dissolution was relatively low and increased in the initial stages of dissolution, but at the end the rates were linear. The activation energy determined in the initial leaching stages was 75 kJ/mole. During leaching, cracks formed on the leached surface as a result of mass transport and increased the active surface thus increasing the rate in the initial stage. Like Thomas and Ingraham [162], these authors observed the formation of a small amount of sulphates, in addition to elemental sulphur.

The overall reaction of leaching of covellite is described by the reaction:

$$CuS + (2+8x)Fe^{3+} + 4xH_2O \rightarrow Cu^{2+} + (2+8x)Fe^{2+} + (1-x)S^0 + xSO_4 \tag{12.52}$$

The values of x varied in a range which enabled dissolution of 0–10% of sulphur to the sulphate form. The leaching rate of copper from synthetic CuS decreased rapidly with the increase of the concentration of ferrous sulphate in the original leaching medium. At a concentration of the ferric ions in the solution higher than 0.05M, the leaching rate was already independent of the concentration of Fe^{3+}. Natural covellite was leached at the rate which was identical with the leaching rate of synthetic covellite and the effect of temperature on the two materials was also approximately identical. The rate-controlling step of the leaching reaction of both materials was the chemical reaction on the surface of CuS. Although the presence of the pores and dimples in the sample indicated the microgalvanic dissolution process, no proof was presented that the rate of transfer of the electrons is the rate-controlling step for the leaching of CuS.

Mulak [165] investigated the leaching of synthetic CuS in acidified solutions of ferric sulphate in the temperature range 30–90 °C. Leaching was linear and the rate was independent of mixing. The activation energy of the process was approximately 84 kJ/mole and this value, as well as the linear course of dissolution, showed that the leaching rate is controlled by the slow chemical reaction on the surface of CuS.

Lowe [168] investigated the leaching of covellite in the acid solution of ferric ions in the temperature range 40–70 °C. He found the linear course of the reaction rate with the activation energy of 59 kJ/mole. The leaching rate was independent of the acid concentration but slightly increased with the increase of the concentration of the Fe^{3+} ions. The rate-controlling step of the reaction was the chemisorption process on the surface of covellite. These results confirmed Sullivan's conclusions [160].

Walsh and Rimstidt [167] studied the kinetics of reaction of leaching of covellite and blue covellite in ferric ions media. The activation energies, 51.8 kJ/mole in the case of blue covellite and 58.3 kJ/mole for natural covellite, are in the range of the published data. However, in any case they show that the reaction is not simple and the rate of reaction of blue covellite is higher than that of the stoichiometric compound. At the same time, the authors concluded that in most cases the natural form of covellite is blue-nonstoichiometric covellite.

The kinetics of leaching of covellite in the solution of sulphuric acid with a variable addition of NaCl, bubbled with oxygen, was also investigated by Cheng and Lawson [168]. The reaction rate was determined by the model of the shrinking core and the activation energy was 77 kJ/mole in the temperature range 75–95 °C. After six hours at 90 °C approximately 85% of the total amount of copper was leached, approximately 10% of sulphide sulphur oxidised to sulphate and the remainder precipitated as elemental sulphur.

The leaching rate of covellite was very low without the addition of chloride ions, probably under the effect of the formation of a passivating compact layer of elemental sulphur on the leached surface with a crypto-crystalline amorphous structure. The effect of chloride ions is based on the support of the formation of large sulphur crystals enabling penetration of the reactants through the layer of elemental sulphur on the leached surface.

Examination of the anodic leaching of covellite [169] showed that the current density should be maintained below 1 mA/cm^2, and the anodic potential remains low and leaching takes place in accordance with the reaction:

$$CuS \rightarrow Cu^{2+} + S^0 + 2e^- \tag{12.53}$$

The products of the anodic reaction were elemental sulphur with small particles of non-reacted CuS. At current densities higher than 1 mA/cm^2 the anodic potential increases to high values resulting in the rapid generation of oxygen on the surface of the electrode. This is undesirable for practical application.

Kametani [170] found that to maintain the electrochemical reaction with the CuS anode, it is necessary to add ferrous sulphate to the electrolyte. Consequently, the anodic reaction is as follows:

$$Fe^{2+} \rightarrow Fe^{3+} + e^+ \tag{12.54}$$

and

$$2Fe^{3+} + CuS \rightarrow Cu^{2+} + 2Fe^{2+} + S^\circ \tag{12.55}$$

Mackinnon [171] continued his studies of anodic leaching of chalcocite in a fluid reactor [152] using synthetic covellite. Leaching of covellite in a sulphuric acid electrolyte is accompanied by the generation of a large amount of oxygen on the platinum electrode, with the exception of the case of using the unrealistically low current density. In mixed electrolyte, such as H_2SO_4–NaCl or H_2SO_4–KBr, anodic leaching takes place at acceptable current densities by the mechanism of mutual charge exchange in the Cl^-–$1/2Cl_2$ or Br^-–$1/2Br_2$ pairs.

Comparison of the results of chemical and electrochemical leaching of covellite in an acidified chloride solution of ferric ions was published by Hirato et al. [172]. On the basis of the results, the author recommended the electrochemical mechanism of leaching, including oxidation of CuS and reduction of $FeCl_2^+$. The activation energy of chemical leaching was 46.7 kJ/mole and in electrochemical leaching 50.4 kJ/mole.

This review indicates the current state of knowledge in the area of leaching of copper sulphides of the chalcocite–covellite type, especially from the viewpoint of theoretical understanding of this process and its mechanism. From the practical viewpoint, the factor of the leaching medium is again important. Therefore, it is essential to mention that various media have been tested for experiments with leaching of sulphide minerals and the results have contributed significantly to understanding the entire leaching process. The most important processes include, in addition to the already mentioned ones, the application of EDTA [173], perchlorate acid [74], pressure leaching [175], fluosilicate acid [176], and others.

In comparison with chalcopyrite, it may be said that the leaching mechanism of chalcocite consists of two stages, and the first stage takes place by a relatively complicated mechanism, as discussed later. On the other hand, the effect of temperature on the leaching of copper sulphides of the type Cu_xS has not been unambiguously defined. This is given by the differences in the leaching conditions, especially by the defects of the structure of the leached material.

Table 12.5. Results of leaching of chalcocite and covellite obtained by various authors

Mineral	Material	Leaching medium	t [°C]	E_A [kJ/mol]	Ref.
Cu_2S	synthetic	$Fe^{3+} + H_2SO_4$	25-60	20-25	142
→	synthetic	$Fe^{3+} + HCl$	22-80	3.5	147
CuS	natural	$O_2 + HClO_4$	105-140	7.53	150
	synthetic	$O_2 + HNO_3$	25	19.5	149
	natural	$O_2 + H_2SO_4$	>100	27.6	177
		$O_2 + H_2SO_4$		27	137
	natural	$Fe^{3+} + HCl$	35-95	36	135
	synthetic	$Fe^{3+} + H_2SO_4$	30-90	1.5	145
	synthetic	$Fe^{3+} + HCl$	20-80	0.8	161
	natural	$Fe^{3+} + H_2SO_4$	28-70	6.7	166
CuS	natural	$O_2 + H_2SO_4$	65	27	139
→	natural	$Fe^{3+} + H_2SO_4$	25-60	92	162
Cu^{2+}	intermediate product of Cu_2S leaching	$Fe^{3+} + HCl$	20-120	104	161
	natural	$Fe^{3+} + H_2SO_4$	40-90	59	166
	synthetic	$Fe^{3+} + H_2SO_4$	30-90	84	164
		$O_2 + H_2SO_4$		7.5	136
	natural	$Fe^{3+} + H_2SO_4$		58.3	178
	synthetic	$O_2 + H_2SO_4/HCl$	75-95	77	168
	synthetic	$Fe^{3+} + HCl$	20-90	46.7	172
	synthetic	$Fe^{3+} + HCl$ (electrochemically)	20-90	50.4	172
	synthetic	$Fe^{3+} + H_2SO_4$	20-90	20	165
	natural	$Fe^{3+} + H_2SO_4$	15-95	18	164

This results in the scatter of the determined activation energies obtained by different authors, Table 12.5.

12.3. Copper sulphides of the type $Cu_xMe_yS_z$

A representative of this type of sulphides is tetrahedrite which has not been been extensively processed in the past but is found as the raw material in Slovakia and, consequently, the special attention should be paid to it because it is not only a potential raw material for the production of copper and antimony but it also interesting because of its content of mercury and silver.

In most cases, the tetrahedrite is described by the formula $Cu_{12}Sb_4S_{13}$, but in the nature it is present in the composition which may be described by the formula $(Cu, Ag)_{10}$ (Cu, Zn, Fe, Cr, Hg, $Cu)_2$ $(Sb, Bi, As)_4S_{13}$ [179]. No optimum technology is as yet available for pyrometallurgical processing because it is not possible to separate efficiently copper from antimony, and also there are problems with environmental protection.

There have been a relatively small number of studies concerned with the leaching of tetrahedrite in an aqueous medium, although as early as in 1914 Nishihara [180] published the results of studies carried out to examine the leaching of a tetrahedrite ore. He found that the rate of dissolution of this mineral in acid solutions of ferric sulphate was relatively high.

Brown and Sullivan [181] investigated the leaching of tetrahedrite also in an acid solution of ferric sulphate. Highly contaminated tetrahedrite was used. This material was leached in a 0.5% solution of sulphuric acid, containing a 1–5% solution of ferric sulphate. 52–95% of copper was leached at a temperature of 35 °C after 30 days. On the other hand, tetrahedrite could not be leached in the H_2SO_4 without ferric ions.

Dutrizac and Morrison [182] leached synthetic tetrahedrite with and without iron by the method of rotating disc in an acid solution of ferric ions. The process was very slow, even at high temperatures of 65–95 °C. Leaching was described by linear kinetics with the activation energy of 64–83 kJ/mole, in relation to the iron content of tetrahedrite. The process was controlled by chemical or electrochemical dissolution. The rate of the process depended strongly on the concentration of $FeCl_3$ and was relatively independent of the acidity of the leaching agent and also increased with the increase of the total concentration of the chloride ions. The product of the leaching reaction were the ions of Fe^{2+} and

Cu^{2+}, together with elemental sulphur. Antimony was oxidised to Sb^{5+} which subsequently precipitated as tripuhyite $FeSbO_4$.

Correia et al. [183] investigated the kinetics of leaching of tetrahedrite in $FeCl_3$–NaCl–HCl solutions in the temperature range 60–104 °C and the concentration of ferric ions of 0.001–1 M at atmospheric pressure. The activation energy was 65+6 kJ/mole with the order of the reaction in relation to the concentration of the ferric ions in the range 0.35+0.03. The rate–controlling step was the chemical reaction on the leached surface. No sulphate formed during leaching.

Havlik and Skrobian [184] leached natural powder tetrahedrite in an acid solution of ferric chloride at the normal conditions of temperature and pressure. The process was also slow, with the parabolic kinetics, in the temperature range 40–70 °C. The activation energy of 65 kJ/mole indicated the process is controlled by the chemical reaction, probably by the formation of elemental sulphur on the leached surface. 25% of copper was extracted into the solution during three hours of leaching. The behaviour of antimony was not investigated.

Havlik et al. [185, 186] later leached natural tetrahedrite in acid solutions of ferric sulphate at the normal conditions of temperature and pressure. The determined value of the activation energy (42–49 kJ/mole) indicated the process is taking place in the mixed regime and is not controlled by the chemical reaction kinetics, as in leaching in ferric chloride. It was also found that the surface of tetrahedrite is eroded by the effect of external conditions and contains approximately 10% of easily soluble compounds containing copper, probably in the sulphate form.

In addition to copper, the behaviour of antimony and sulphur during leaching was investigated. Copper is transferred to the dissolved form as bivalent. Antimony is partially transferred into the solution in the soluble form $Sb_2(SO_4)_3$ in contrast to the results obtained by Dutrizac and Morrison [182] who described the reaction of antimony with the ferric ions with the formation of tripuhyite $FeSbO_4$ which gradually precipitated from the solution.

Since the leaching of tetrahedrite in the acid medium is a relatively slow process when using standard leaching agents, several authors have investigated pressure leaching.

Gerlach et al. [187] investigated pressure leaching of several sulphides, including tetrahedrite, in order to explain the relationship of certain physical properties and the change of the structure with the leaching capacity of copper and antimony, with the formation of slightly soluble antimony sulphates. During one hour, 12% of metal was leached from the tetrahedrite. The apparent activation energy was 23.45 kJ/mole. The differences in the formation of soluble and insoluble antimony compounds were explained by mechanochemical effects in processing of tetrahedrite concentrate as well as by the limited solubility of antimony in the given conditions.

Caldon [188] patented the process of pressure leaching of tetrahedrite concentrates using the solutions of H_2SO_4 and HNO_3 at an oxygen pressure of 0.1–0.3 MPa.

Scheiner et al. [189] processed a tetrahedrite concentrate after alkaline leaching in sodium hydroxide and sodium sulphide in order to remove antimony by pressure leaching at 0.3 MPa in the solution of $FeCl_3$ at 100–125 °C, to obtain copper and, in the third stage in the cyanide solution, to obtain silver.

Bahr and Priesemann [190] also investigated the pressure leaching of a copper concentrate with 27% of copper presented chalcopyrite, tetrahedrite, tenantite and covellite, and 2.7% of silver, present mainly in tetrahedrite. This concentrate was leached in the solution $CaCl_2$+HCl at 100–190 °C and a partial pressure of oxygen of 4 MPa. Copper and silver were leached after 90 min, and arsenic and iron precipitated from the solution as oxides.

Neiva Correia et al. [191] used a chloride medium for the pressure leaching of tetrahedrite. They used different solutions, cupric chloride with ammonium chloride, cupric chloride and sodium chloride and ferric chloride with sodium chloride for leaching. The best results were obtained with ammonium chloride were almost the entire amount of tetrahedrite was leached after three hours at 130 °C at the oxygen pressure of 0.3 MPa. When using ferric chloride and cupric chloride at the oxygen pressure of 0.3 MPa and a temperature of 130 °C, the yield of copper was approximately 60%. When using ferric chloride and the oxygen pressure of 0.6 MPa, the copper yield was up to 40%.

Other authors investigated the bacterial leaching of tetrahedrite. Brynner et al. [192] in the experiments with bacterial leaching achieved leaching of only 3 mg of copper from 1 g sample of tetrahedrite after leaching for 42 days.

Frenay [193] described bioleaching of a tetrahedrite concentrate with 13–20% Cu, 8% Sb, and 750 g/t Ag. The yield of copper was up to 90% and the leaching residue contained antimony and silver which were subsequently leached, initially with sodium sulphide with the leaching efficiency of antimony 190% and then in cyanide with the leaching efficiency of 85% of silver.

Similarly, Brunner et al. [194] and Kusnierova [195,196] published the results of bioleaching of tetrahedrite using *Thiobacillus ferooxidans*. Kusnierova describes the formation of new minerals, such as oxides, sulphates, and secondary sulphides.

In the majority of the published data on leaching of tetrahedrite, the authors have used alkaline leaching in Na_2S and NaOH as the leaching agent [197–200]. Balaz et al. [201–203] carried out alkaline leaching in selective removal of antimony, mercury and arsenic from a concentrate with the tetrahedrite content of approximately 41%. The process included mechanochemical processing of the concentrate using an attritor and a Na_2S–NaOH solution, followed by leaching of arsenic and mercury. Similarly, Balaz et al. [204, 205] published the results of leaching of mechanically activated tetrahedrite in thiourea to extract silver. The procedure was included in a proposal for a technology termed MELT [131, 206].

The alkaline leaching of tetrahedrite in the Na_2S may be described by the following simplified relationships:

$$2Cu_3SbS_3 + Na_2S \rightarrow 3Cu_2S + 2NaSbS_2 \tag{12.56}$$

$$NaSbS_3 + Na_2S \rightarrow Na_3SbS_3 \tag{12.57}$$

The soluble Na_3SbS_3 containing trivalent antimony is oxidised to the pentavalent form from the solution in accordance with the reaction:

$$(x-1)Na_3SbS_3 + Na_2S_x \rightarrow (x-1)Na_3SbS_4 + Na_2S \tag{12.58}$$

The mercury present in tetrahedrite is leached by sodium sulphide with the formation of a soluble complex in accordance with the reaction:

$$HgS + Na_2S \rightarrow Na_2IIgS_2 \tag{12.59}$$

which is susceptible to hydrolysis. Its solubility depends on the presence of alkali, usually NaOH.

The overall leaching reaction of tetrahedrite in an acid medium using ferric ions is very slow and the leaching kinetics is relatively complicated. Elemental sulphur is the main reaction product, if leaching takes place at temperatures of up to 100 °C. The amount of elemental sulphur varies in the range 75–95%, but in the majority of cases it is approximately 90% [182]. A small amount of sulphur is transferred into sulphate form by a parallel leaching reaction:

$$S^{2-} + 2Fe^{3+} \rightarrow S^0 + 2Fe^{2+} \tag{12.60}$$

$$S^{2-} + 8Fe^{3+} + 4H_2O \rightarrow SO_4^{2-} + 8Fe^{2+} + 8H^+ \tag{12.61}$$

The formation of SO_4^{2-} increases the concentration of ferrous ions. Elemental sulphur forms and remains on the leached surface of tetrahedrite in the form of a network of connected globules. Although the globules cover the entire surface, it appears that the sulphur layer is relatively porous and enables the diffusion of reagents manifested in the linear leaching kinetics.

At a higher concentration of the ferric ions, copper is transformed to the bivalent form in accordance with the reaction:

$$Cu^+ + Fe^{3+} \rightarrow Cu^{2+} + Fe^{2+} \tag{12.62}$$

and leaching of the tetrahedrite takes place in accordance with:

$$Cu_{12}Sb_4S_{13} + (24 + 4x)Fe^{3+} \rightarrow 12Cu^{2+} + 4Sb^x + 13S^\circ + (24 + 4x)Fe^{2+} \tag{12.63}$$

where x is a formal number describing the valency of the antimony ions.

The overall concentration of antimony in the solutions of ferric chloride is always less than 20 ppm and this indicates that the majority of antimony precipitates from the solution. Examination of morphology shows [182] that $FeSbO_4$ phase appears and the composition of this phase is similar to the mineral tripuhyite, containing Fe^{3+} and Sb^{5+}. The existence of this phase and the absence of soluble antimony indicate that the principal reaction product is antimony in the pentavalent form:

$$Cu_{12}Sb_4S_{13} + 44Fe^{3+} \rightarrow 12Cu^{2+} + 4Sb^{5+} + 13S^0 + 44Fe^{2+} \tag{12.64}$$

$$Sb^{5+} + Fe^{3+} + 4H_2O \rightarrow FeSbO_4 + 8H^+ \tag{12.65}$$

It is assumed that some form of a compound of the pentavalent antimony forms in accordance with the equation (12.64) and this compound is transferred into the solution where it precipitates on a suitable surface. If a compound with a content of trivalent antimony forms, this compound is transferred into the solution where it is oxidised to Sb^{5+} by ferric ions and subsequently precipitates from the solution. Antimony is easily oxidised by the ferric ions [182] by the reaction:

$$Sb^{3+} + 2Fe^{3+} \rightarrow Sb^{5+} + 3Fe^{2+} \qquad (12.66)$$

in relation to the amount of antimony and ferrous ions in the solution.

The overall reaction of leaching of antimony in ferrous sulphate may be described by the equation:

$$Cu_{12}Sb_4S_{13} + 18Fe_2(SO_4)_3 \rightarrow 12CuSO_4 + 2Sb_2(SO_4)_3 + 13S^{\circ} + 36FeSO_4 \qquad (12.67)$$

Similarly, the following reaction holds for leaching in ferric chloride:

$$Cu_{12}Sb_4S_{13} + 36FeCl_3 \rightarrow 12CuCl_2 + 4SbCl_3 + 13S^0 + 36FeCl_2 \qquad (12.68)$$

The leaching reaction and the formation of tripuhyite obviously take place simultaneously. Hydrolysis may be prevented by the application of acid solutions. On the basis of thermodynamic studies of the Sb–Cl_2–H_2O system, Lin claims [207] that the Sb^{3+} ions are stable only in the acid range at pH values lower than 0.83. On the other hand, this fact indicates that it may be possible to separate copper and antimony directly in the solution when the antimony compounds precipitate in the insoluble residue and copper remains in the solution [208].

Figure 12.29 shows the kinetic relationships of the leaching of copper and antimony from tetrahedrite in the temperature range 40–90 °C in acidified ferric chloride with a concentration of 0.5 M. The kinetic curves of leaching of copper are parabolic and can be described with a high accuracy by the model of the decreasing volume.

The leaching curves of antimony are different. At lower temperatures, i.e., at lower concentrations of leached antimony, the curves are also exponential but at a yield of 5–6% antimony starts to precipitate from the solution in the form of a grey amorphous precipitate. This precipitate was identified by microanalysis as tripuhyite, $FeSbO_4$. This phenomenon in general is also repeated in the examination of the effect of the concentration of the leaching agent and in investigating the effect of the grain size on the leaching of tetrahedrite. The absolute amount of antimony is always the same and this means that the leaching of antimony is limited by its solubility in the given conditions.

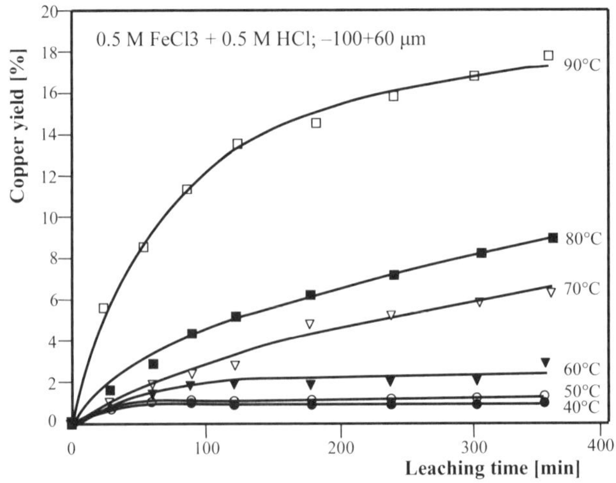

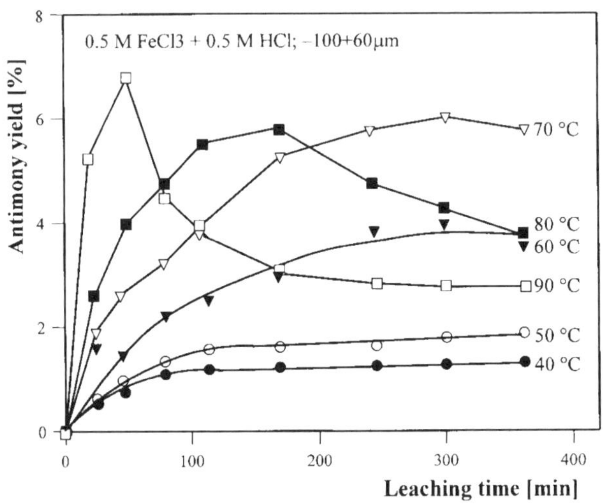

Fig. 12.29. The leaching curves of tetrahedrite. Top – the time dependence of leaching of copper at different temperatures, bottom – time dependence of leaching of antimony at the same temperature.

The kinetic leaching curves showed that both copper and antimony are transferred into the solution at lower temperatures in approximately the same amount and at approximately the same rate. However, whilst copper remains dissolved in the solution, antimony precipitates from the solution, after leaching of a certain amount, in the form of an insoluble compound similar to tripuite. The relationships presented in Fig. 12.30 show that the solubility of antimony in the acid solution of ferric chloride is limited. These values obtained at approximately 5% yield of antimony from the

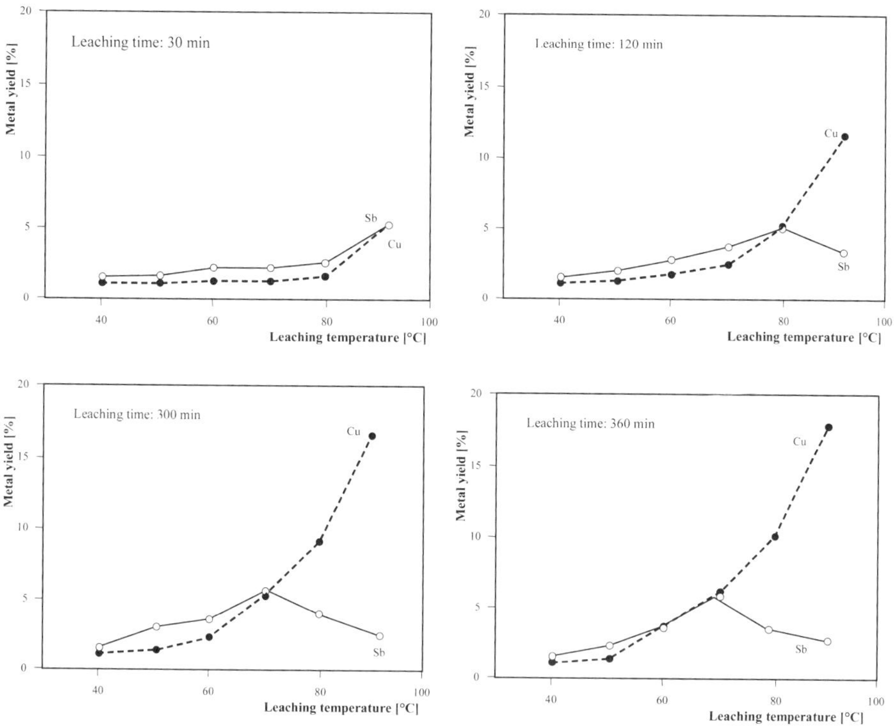

Fig. 12.30. Dependence of the efficiency of leaching metals for different leaching times.

investigated tetrahedrite. After considering the experimental conditions, this amount is approximately 107 mg/l of antimony. This value is reached in relation to the leaching temperature after different periods of time, and decreases with increasing temperature, Fig. 12.30.

These results show quite unambiguously of the possibility of separating copper and antimony in the solution. Whilst the solubility of copper compounds in the given medium is high and forms a basis for the industrial application of ferric chloride in actual hydrometallurgical processes, the solubility of antimony is restricted quite considerably and the small amounts of antimony in comparison with copper may be extracted from the solution using standard methods.

In order to obtain better kinetic parameters of leaching of tetrahedrite, investigations have been carried out with the use of other leaching agents. Mostecky et al. [209] patented a method of processing tetrahedrite concentrates in nitric acid. Antimony was efficiently transferred into the solution, but the method did not provide for the recirculation of HNO_3 which was trapped in the final stage and converted to ammonium nitrate to be used as a fertiliser. However, it was found quite rapidly that this procedure is economically irrational.

The complex procedure of processing tetrahedrite concentrates solvent extraction [210]. The procedure was promising, although further experiments to investigate the case hardening of mercury, extraction, removal of bismuth from the solution, etc.

The application of nitric acid with the addition of potassium nitrate for leaching of tetrahedrite was later studied by Havlik et al. [211]. In comparison with the study [209], in this case, special attention was given to the recirculation of released nitrate gases and the return of these gases into the process in the form of nitric acid.

The chemism of the process may be described as follows:

Pyrite and chalcopyrite are leached with the formation of the iron sulphate or chloride and of elemental sulphur and sulphate sulphur in accordance with the reactions:

$$FeS_2 + \frac{11}{3}HCl + \frac{14}{3}KNO_3 \rightarrow FeSO_4 + KHSO_4 + \frac{11}{3}KCl + \frac{4}{3}H_2O + \frac{14}{3}NO \quad (12.69)$$

$$FeS_2 + \frac{8}{3}HCl + \frac{2}{3}KNO_3 \rightarrow FeCl_2 + \frac{2}{3}KCl + \frac{4}{3}H_2O + \frac{2}{3}NO + 2S^0 \quad (12.70)$$

$$CuFeS_2 + \frac{16}{3}HCl + \frac{16}{3}KNO_3 \rightarrow FeSO_4 + CuSO_4 + \frac{16}{3}KCl + \frac{8}{3}H_2O + \frac{16}{3}NO \quad (12.71)$$

$$CuFeS_2 + \frac{16}{3}HCl + \frac{4}{3}KNO_3 \rightarrow FeCl_2 + CuCl_2 + \frac{4}{3}KCl + \frac{8}{3}H_2O + \frac{4}{3}NO + 2S^0 \quad (12.72)$$

The sulphate form contains 97% of sulphur from pyrite, whereas the elemental sulphur contains only 3% of sulphur. In the case of chalcopyrite, 55% of sulphur is oxidised into the sulphate form and 45% to elemental form.

Tetrahedrite is leached by the reactions:

$$3Cu_{12}Sb_4S_{13} + 12FeCl_3 + 56HCl + 122KNO_3 \rightarrow$$
$$\rightarrow 36CuSO_4 + 3H_2SO_4 + 12FeSbO_4 + 122KCl + 122NO + 40H_2O \quad (12.73)$$

$$3Cu_{12}Sb_4S_{13} + 12FeCl_3 + 80HCl + 44KNO_3 \rightarrow$$
$$\rightarrow 36CuCl_2 + 39S^{\circ} + 12FeSbO_4 + 44KCl + 44NO + 40H_2O \quad (12.74)$$

The formation of ferric compounds has not been considered, only ferrous compounds have been taken into account, since the ferric compounds are permanently reduced.

HNO_3 forms in the regeneration cycle. In this case, the course of the reactions is simpler taking into account the fact that potassium chloride does not form during the reaction. At a low HNO_3 concentration, NO forms in the process, whereas NO_2 forms at higher concentrations. Taking into account the experimental conditions, the formation of NO was considered in this case in the equations.

On the basis of previous considerations regarding the leaching mechanism of tetrahedrite in the given system, it was concluded that both the Na^+ and K^+ ions take active part in the reactions. The sodium ion plays a negative role, potassium forms soluble $KSbO_3$. The intermediate stage of formation in subsequent breakdown of $KSbO_3$ was not considered in the case of tetrahedrite leaching, since it does not cause any problems in the process.

The following reaction takes place in the presence of KNO_3

$$KNO_3 + H_2SO_4 \rightarrow KHSO_4 + HNO_3 \quad (12.75)$$

The behaviour of iron in the next phase of the process is described by the equations:

$$3FeCl_2 + 2KHSO_4 + 6H_2O \rightarrow KFe_3(SO_4)_2(OH)_6 + KCl + 5HCl + \frac{3}{2}H_2 \quad (12.76)$$

$$3FeSO_4 + 2KHSO_4 + 6H_2O \rightarrow KFe_3(SO_4)_2(OH)_6 + 2H_2SO_4 + \frac{3}{2}H_2 \quad (12.77)$$

This leads to the precipitation of iron as potassium jarosite with simultaneous regeneration of the acids.

Ukasik and Havlik [212] leached tetrahedrite in the solution of HCl through which a gas atmosphere containing ozone was bubbled. In comparison with other results obtained in leaching of copper from tetrahedrite in the normal conditions, in this case the yield of copper in the solution was relatively high, whereas leaching of antimony stopped after a certain period of time, as shown in Fig. 12.31. This was explained by the gradual formation of insoluble antimony oxides on the leached surface.

The leaching of tetrahedrite in HCl using ozone is described by the equations:

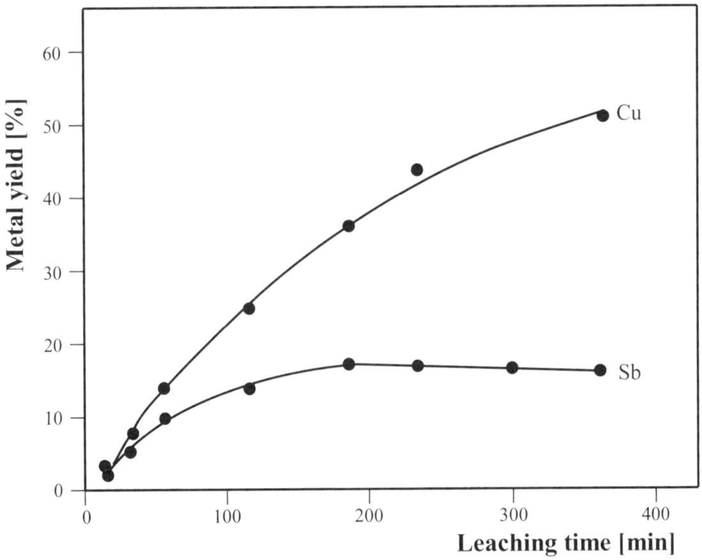

Fig. 12.31. Kinetic curves of leaching of tetrahedrite using ozone.

$$Cu_{12}Sb_4S_{13} + 36HCl + 24O_3 \rightarrow 4SbCl_3 + 12CuCl_2 + 18H_2SO_4 \quad (12.78)$$

$$3Cu_{12}Sb_4S_{13} + 132HCl + 88O_3 \rightarrow 12SbCl_5 + 36CuCl_2 + 66H_2SO_4 \quad (12.79)$$

$$Cu_{12}Sb_4S_{13} + 28HCl + 20O_3 \rightarrow 4SbOCl + 12CuCl_2 + 14H_2SO_4 \quad (12.80)$$

Because of a shortage of thermodynamic data it was possible to evaluate only some of these equations, but the tendency indicates the oxidation of antimony species from the solution to higher oxides:

$$3SbCl_3 + 3H_2O = Sb_2O_3 + 6HCl_{(a)} \qquad \Delta G^0_{25} = -65.85\,kJ \quad (12.81)$$

$$4SbCl_3 + 6H_2O = Sb_4O_6 + 12HCl_{(a)} \qquad \Delta G^0_{25} = -126.125\,kJ \quad (12.82)$$

$$6SbCl_3 + O_{3(a)} + 9H_2O = 3Sb_2O_4 + 18HCl_{(a)} \qquad \Delta G^0_{25} = -856.32\,kJ \quad (12.83)$$

$$6SbCl_3 + O_{3(a)} + 9H_2O = 3Sb_2O_5 + 18HCl_{(a)} \qquad \Delta G^0_{25} = -1230.17\,kJ \quad (12.84)$$

$$2SbOCl + H_2O = Sb_2O_3 + 2HCl_{(a)} \qquad \Delta G^0_{25} = -2.42\,kJ \quad (12.85)$$

The produced oxide precipitates in all likelihood from the solution, since its solubility is almost zero.

The results of acid oxidation leaching of the tetrahedrite concentrate may be summarised as follows:

Ferric chloride is more efficient than ferric sulphate for leaching of tetrahedrite but despite this, the leaching of natural tetrahedrite concentrates in the acid medium of ferric chloride in the temperature range 40–90 °C is slow and characterised by distinctive parabolic kinetics. The increase of temperature improves the kinetic parameters of leaching. The determined apparent activation energy of approximately 38 kJ/mole indicates that the process takes place in the mixed regime.

Copper is leached in the solution with increasing intensity but, despite this trend, the resultant yield is only approximately 35%. Higher yields may be obtained using stronger oxidation agents.

Antimony is leached in the solution only on a limited scale with the yield of approximately up to 5%. On reaching this value, antimony precipitates from the solution in the form of a compound similar to tripuhyite.

Acid oxidation leaching by ferric chloride makes it possible to separate copper and antimony in the solution followed by extraction of copper for production. A weak point is of course the low yield preventing application of this procedure in practice.

For these reasons, investigations have been carried at into the possibilities of increasing the efficiency of selective transfer of copper from tetrahedrite into the solution. One of the possibilities is the thermal treatment of the tetrahedrite concentrate prior to leaching [213–215]. The concept of the application of roasting of tetrahedrite concentrate prior to leaching was not accidental but resulted from the fact that in Slovakia tetrahedrite has been processed for a long period of time by roasting to remove mercury and the residue, after roasting, was stored in heaps in the form of the so-called tetrahedrite calcine in the amount of approximately 6000 t/year. This production was stopped in 1993 for ecological reasons. Although there were attempts to process these materials by adding them to the charge for pyrometallurgical production of copper, the method caused considerable ecological and technological problems.

Figure 12.32 shows the kinetic curves of leaching of copper and antimony from the tetrahedrite calcine. The dependence shows the positive effect of temperature on the leaching of copper and also a high yield, especially at high temperatures, in comparison with the yield from the tetrahedrite concentrate, Fig. 12.29. At the highest temperature (80 °C) the yield of copper from the roasted product was 74% and that of antimony 14%. Com-parison of the ratio of the masses of copper and antimony in the roasted product (1.1:1)

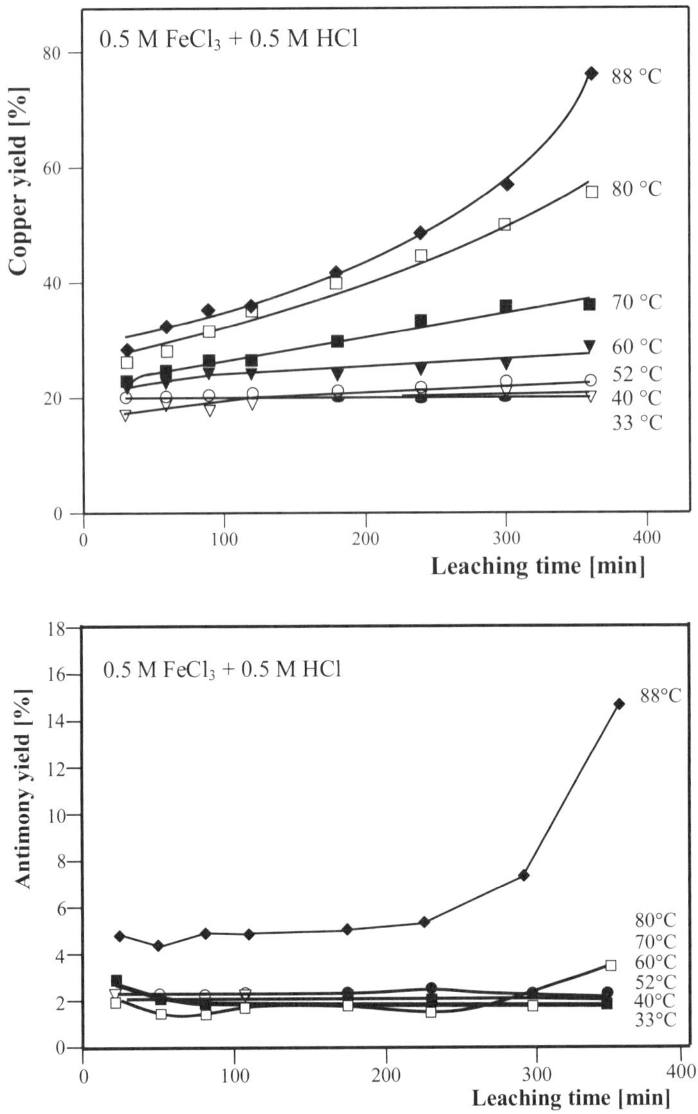

Fig. 12.32. The leaching curves of tetrahedrite calcine. a) time dependence of leaching of copper at different temperatures, b) time dependence of leaching of antimony at the same temperature.

shows that the ratio of the yields is as high as 5.2:1. This means that copper is also leached from other components of the roasted product. In all likelihood, it is leached from cuprous sulphate. The form of the curves fully confirms this. As indicated by Fig. 12.32*a*, more than 20% of copper is leached more or less instantaneously in the initial stages. At low leaching temperatures, this amount increases only slightly and corresponds to the expected course of leaching. This consideration is in agreement with auxiliary experiments in which the roasted product was rinsed with distilled water for 15 min at room temperature and after this time 18.3% of copper was leached in the water. These easily soluble compounds were present in the roasted product as a consequence of high-temperature reactions in roasting of the tetrahedrite starting material in order to vapourise mercury.

The situation in leaching of antimony is slightly different. Figure 12.32*b* shows the kinetic curves of the temperature dependence of leaching of antimony from the tetrahedrite calcine. In comparison with the leaching curves of copper, Fig. 12.30*a*, there are no large amounts of antimony leached from the calcine into the solution in the initial stage and no antimony was leached in water in the already mentioned auxiliary experiment.

In addition to the highest leaching temperature investigated, 80 °C, antimony is leached in the solution to the minimum extent, with the yield of approximately 2%. The yield is higher only of the highest temperature, although the curve does not have the standard form of the curves of leaching of metals in experiments with simple sulphides. This means that in roasting, antimony is transformed to some phase resistant to leaching which reacts only at high leaching temperatures and after a long process time.

The leaching curves of the tetrahedrite concentrate indicate that after leaching a yield of approximately 5%, antimony precipitates from the solution in the form of an insoluble compound. This is almost the same amount as in other cases of leaching of tetrahedrite and confirms the assumption on the limited solubility of antimony.

These results also show the possibility of not only separating copper and antimony in the solution but also the positive effect of thermal processing, i.e. roasting of the tetrahedrite concentrate prior to leaching. Antimony also precipitates from the solution in leaching of the concentrate but, in particular, in leaching of the calcine the amount of antimony in the solution is minimal. At leaching temperature of approximately 80 °C, the solubility of copper compounds on the given medium is high and provides a suitable basis for the industrial application of ferric chloride in actual hydrometallurgical processes. The solubility of the antimony-containing compounds is restricted quite dramatically. It is highly likely that the small amounts of antimony in comparison with copper can be removed from the solution by the standard methods. This would provide a basis for hydrometallurgical processing of tetrahedrite raw materials which so far could not be processed by any other method in order to obtain copper because of serious problems with the separation of antimony from copper.

References

1. Vukotic S.: Solubility of galena, sphalerite, and chalcopyrite in water and in the presence of H_2S, Bull. Bur. Rech. Geol. Minieres, 3, 1961, 12–27.
2. Stanczyk M.H., Rampacek C.: Oxidation leaching of copper sulfides in acidic pulps at elevated temperatures and pressures, U.S. Bur. Mines. Rept. Invest., 6193, 1963.
3. Ichikuni M.: *Bull. Chem. Soc. Japan*, 33, 3, 1960, 1052–1057.
4. Habashi F., Toor T.: *Metallurgical Transactions* B, 10B, March 1979, 49–56.
5. Meyers R.A., Hamersma J.W., Kraft M.L.: *Environm. Sci. & Techn.*, 9, 1, 1975, 70–71.
6. Evans D.J.I., Romanchuk S., Mackiw V.N.: *Bull. Can. Inst. Min. & Met.*, 65, 1974, 105–120.
7. Kuhn M.C., Arbiter N., King H.: *Bull. Can. Inst. Min. & Met.*, 67, 1974, 62–73
8. Haskett P.R. et al.: Hydrometallurgical process for copper recovery from sulfide ores, U.S. Patent 559,942, 1975.
9. Trail R.J., McClelland W.R.: Investigations in ore dressing and metallurgy, Mines Branch Rep. 688, 1928, 101–109.
10. Pike R.D. et al.: *Trans. Am. Inst. Min. Metall*, 90, 1930, 312–345.
11. Brown S.L., Sullivan J.D.: Dissolution of various copper minerals, Bureau of Mines RI 3228, 1934.
12. Sullivan J.D.: *Trans. Am. Inst. Min. Metall*, 106, 1933, 515–546.
13. Klec V.E., Liopo V.A.: *Trudy Irkutsk. Politekh. Instituta*, 27, 1966, 123–130.
14. Yermilov V.V., Tkachenko O.B., *Tseft A.L. Trudy Inst. Metal., Obogashch*, 30, 1969, 3–14.
15. Ichikuni M.: *Bull. Chem. Soc. Japan*, 33, 3, 1960, 1259–1262.
16. Ichikuni M.: *Bull. Chem. Soc. Japan*, 35, 1962, 1765–1768.
17. Hiskey J.B., Wadsworth M.E.: *Met. Trans. B*, 6B, March 1975, 183–190.
18. Dutrizac J.E., MacDonald R.J.C., Ingraham T.R.: *Trans. Met. Soc. AIME*, 245, May 1969, 955–959.
19. Dutrizac J.E., MacDonald R.J.C.: *Met. Trans.*, 2, 1971, 2310–2312.
20. Dutrizac J.E.: *Met. Trans. B*, 9B, September 1978, 431–439.
21. Jones D.L., Peters E.: The leaching of chalcopyrite with ferric sulfate and ferric chloride, Extract. Metall. of Copper, II, Yannopoulos & Agarwal eds., TMS-AIME, New York, 1976, 633–665.
22. Beckstead L.W., Munoz P.B., Sepulveda J.L., Herbst J.A., Miller J.D.: Acid ferric sulfate leaching of attritor-ground chalcopyrite concentrates, Extractive Metallurgy of Copper, vol.II., Yannopoulos & Agarwal eds., 1976, AIME, New York, 612–632.
23. Haver F.P., Wong M.M.: *Journal of Metals*, February 1971, 25–29.
24. Baur J.P., Gibbs H.L., Wadsworth M.E.: *Met. Soc. AIME*, 72-B-96, 1972.

25. Dutrizac J.E.: *Met. Trans. B*, 12B, June 1981, 371–378.
26. Munoz P.B., Miller J.D., Wadsworth M.E.: *Met. Trans. B*, 10B, June 1979, 149–158.
27. Warren G.W., Wadsworth M.E., El-Raghy S.M.: *Met. Trans. B*, 13B, December 1982, 571–579.
28. Warren G.W., Wadsworth M.E., El-Raghy S.M.: Anodic behavior of chalcopyrite in sulfuric acid, Hydrometallurgy, Research, Development and Plant Practice, Osseo-Asare and Miller eds., AIME, 1983, 261–275.
29. Warren G.W., Sohn H.-J., Wadsworth M.E., Wang T.-G.: *Hydrometallurgy*, 14, 1985, 133–149.
30. Wadsworth M.E.: Electrochemical reactions in hydrometallurgy, Metallurgical treatises, Tien and Elliot eds., AIME 1981, 1–22.
31. Ammou-Chokroum M., Sen P.K., Fouques F.: Electrooxidation of chalcopyrite in acid chlorine medium; kinetics, stoichiometry and reaction mechanism, 13th. Int. Min. Proc. Congress, Laskowski and Fuerstenau eds, Elsevier 1981, part 2a, Warszaw 1981, 759–809.
32. Hirato T., Kinoshite M., Awakura Y., Majima H.: *Met. Trans. B*, 17B, March 1986, 19–28.
33. Hirato T., Majima H., Awakura Y.: *Met. Trans. B*, 18B, September 1987, 489-496.
34. Hirato T., Majima H., Awakura Y.: *Met. Trans. B*, 18B, March 1987, 31–39.
35. Braithwaite J.W., Wadsworth M.E.: Oxidation of chalcopyrite under simulated conditions of deep solution mining, Extract. Metall. of Copper, vol. II, Yannopoulos & Agarwal eds., TMS-AIME, 1976, 752–775.
36. Yu P.H., Hansen Ch.K, Wadsworth M.E.: *Met. Trans.*, 4, September 1973, 2137–2144.
37. Dutrizac J.E.: *Met. Trans. B*, 13B, September 1982, 303–309.
38. Linge H.G.: Hydrometallurgy, 2, 1976/1977, 219–233.
39. Dutrizac J.E., MacDonald R.J.C.: *Can. Met. Quart.*, 12, 4, 1973, 409–420.
40. Burkin A.R.: Composition and phase changes during oxidative acid leaching reactions, MINTEK 50, 1984, 519–525.
41. Ferreira R.C.H., Burkin A.R.: Acid leaching of chalcopyrite, In: Leaching and reduction in hydrometallurgy, The IMM London, 1976, 54–66
42. Palmer B.R., Nebo C.O., Rau M.F., Fuerstenau M.C.: *Met. Trans. B*, 12B, September 1981, 595–601.
43. O'Malley M.L., Liddell K.C.: *Met. Trans. B*, 18 B, September 1987, 505–510.
44. O'Malley M.L., Liddell K.C.: A rate equation for the initial stage of the leaching of $CuFeS_2$ by aqueous $FeCl_3$, HCl and NaCl, Hydrometallurgical Reactor Design and Kinetics, Bautista, Wesely and Warren eds., SME-AIME 1986, 67–75.
45. Linge H.G.: *Hydrometallurgy*, 2, 1976, 51–64.
46. Bardt H.: Recovering metals contained in metalliferous ore, Waste Residues, and Alloys, German Patent 353,795 (1919–1922), Canadian Patent 233,566 (1923), Chemical Abstracts 17, 3154.
47. Björling G., Kolta G.A.: Oxidizing leach of sulfide concentrates and other materials catalyzed by nitric acid, Proc. 7th. Int. Mineral Processing Congress, Arbiter N., ed., Gordon & Breach, New York, 1966, 127–138.

48. Björling G., Lesidrenski P.: Hydrometallurgical production of copper from activated chalcopyrite, AIME Annual Meeting, New York, February 1968.
49. Björling G., Kolta G.A.: Wet oxidation as a method of utilization of chalcopyrite, sphalerite, and molybdenite, J. Chemistry U.A.R, 12, 3, 1969, 423–435.
50. Björling G.: Leaching of mineral sulfides by selective oxidation at normal pressure, Int. Symp. on Hydrometallurgy, Evans D.J.I., Shoemaker R.S. eds., AIME. New York,1973, 701–707.
51. Björling G., Faldt I., Lindgren E., Toromanov I.: A nitric acid route in combination with solvent extraction for hydrometallurgical treatment of chalcopyrite, Extractive Metallurgy of Copper, vol. II., Yannopoulos J.C., Agarwal J.C. eds., AIME, New York, 1976, 725–737.
52. Habashi F.: Nitric acid in hydrometallurgy, Int. Conf. Quo Vadis Hydrometallurgy, Havlik et al. eds., Herlany, May 1995, 60–68.
53. Habashi F.: *Trans. Soc. Min. Eng. AIME*, September 1973, 224–227.
54. Sen P.K., Ganguly A., Roy S.: *Trans. Indian Inst. of Metals*, 3, 6, 1980, 473–479.
55. Prater J.D., Queneau P.B., Hudson T.J.: *Trans. Soc. Min., Eng. AIME*, 254, 2, 1973, 127–132.
56. Brennecke H.M.: Recovery of metal values from ore concentrates, US Patent 3,888,748, 1975.
57. Brennecke H.M.: Nitric-sulfuric leach process for recovery of copper from concentrates, Min. Engineering, August 1981, 1259–1266.
58. Pawlek F.E.: The influence of grain size and mineralogical composition on the leachibility of copper concentrates, Extractive Metallurgy of Copper, vol.II,Yannopoulos and Agarwal eds., AIME, New York, 1976, 690–705.
59. McElroy R.O., Duncan D.W.: The Silver Institute Letter, British Columbia Research Council, Vancouver, Canada, 5, 8, September 1975.
60. Snell G.J., Sze M.C.: New oxidative leaching process uses silver to enhance copper recovery, EMJ, 178, 10, 1977, 100–105.
61. Miller J.D., Portillo H.Q.: Silver catalysis in ferric sulphate leaching of chalcopyrite, 13th. Int. Mineral Processing Congress, Warsaw, Laskowski J. ed, Elsevier, 1979, 852–901.
62. Price D.W., Warren G.W.: Hydrometallurgy, 15, 1986, 303–324.
63. Murr L.E., Hiskey J.B.: *Met. Trans. B*, 12B, June 1981, 255–267.
64. Antonijevič M.M., Jankovič Z., Dimitrijevič M.: Hydrometallurgy, 35, 1994, 187–201.
65. Gerlach J.K., Gock E.D., Ghosh S.K.: Activation and leaching of chalcopyrite concentrates with dilute sulfuric acid, Int. Symposium on Hydrometallurgy, AIME, New York, 1973, 403-416
66. Tkáčová K., Baláž P.: *Hydrometallurgy*, 21, 1988, 103–122.
67. Baláž P.: Relationships and technological aspects of mechanical activation of sulphide minerals, PhD Thesis, Institute of Mining, SAV Košice, December 1992.
68. Cobble J.R., Jordan C.E., Rice D.A.: Hydrometallurgical production of copper from flotation concentrates, Bureau of Mines RI 9472, 1993, 1–14.

69. Havlík T., Kammel R.: Behavior of sulfur during leaching of sulfidic mineral, EPD Congres 1996, Warren G.W. ed Anaheim, TMS, 1996, 843–857.
70. Glaum G.V., Walli E.: Process for Oxidizing Metal Sulfides in Aqueous Suspensions, Canadian Patent No. 965964, May 1972.
71. Buttinelli D., Lavecchia R., Lupi C., Pochetti F., Geveci A., Topkaya Y.: *La Chimica & L'Industria*, 72, 1990, 707–713.
72. Buttinelli D., Geveci A., Lupi C., Pochetti F., Stoppa L., Topkaya Y.: Ferric chloride versus cupric chloride leaching of copper–zinc complex sulphide ores, Hydrometallurgy, Fundamentals, Technology and Innovations, Hiskey, Warren eds., AIME, 1993, 971–985.
73. Ozolins L., Rušikina L.P.: Sovershenst. Techn. i Technol., Razrab. Mestorozhd. Polez., Iskop, 1968, 103–120.
74. Khavskij N.N.: Use of ultrasonics in processes of metal extraction from ores, Symp. The Use of Ultrasonics in Extr. Met. Technology Ltd., 1973, 1–12.
75. Khavskij N.N., Zhilin Yu.S., Kanevskij Ye.A.: Kinetics of copper diffusion in sulphuric acid solutions of Fe^{3+}, Symp. The Use of Ultrasonics in Extr. Met. Technology Ltd., 1973, 94-96
76. Khavskij N.N., Zhilin Yu.S., Kanevskij Ye.A., Yakubovich I.A.: Study of the effect of ultrasonic field cavitation on the diffusion of copper in sulphuric acid solutions of Fe^{3+}, Symp. The Use of Ultrasonics in Extr. Met. Technology Ltd., 1973, 86-93
77. Pesic B., Zhou T.: *Met. Trans. B*, 23B, February 1992, 13–22.
78. Subramanian K.N., Jennings P.H.: *Can. Met. Quart*, 12, 2, 1972, 387–400.
79. Warren I.H., Vizsolyi A., Forwrad F.A.: The pretreatment and leaching of chalcopyrite, CIM Bulletin, May 1968, 637–640.
80. Subramanian K.N., Kanduth H.: Activation and leaching of chalcopyrite concentrate, CIM Bulletin, June 1973, 88–91.
81. Habashi F.: Recent advances in hydrometallurgy, Hydrometallurgy and Electrowinning, II, Yannopoulos and Agarwal eds., Part City Press, Baltimore, 1976, 609–635.
82. Bartlett R.W.: *Met. Trans. B*, 23B, June 1992, 241–248.
83. Sohn H.-J., Wadsworth M.E.: *Journal of Metals*, November 1980, 18–22.
84. Baláž P., Špaldon F., Luptáková A., Bastl Z., Škrobian M., Havlík T., Briančin J.: *Rudy*, 6, 38, 1990, 159–163.
85. Baláž P., Špaldon F., Luptáková A., Bastl H., Havlík T., Škrobian M.: *Int. Journal of Mineral Processing*, 32, 1/2, 1991, 133–147.
86. Sutton J.A., Corrick J.D.: *Min. Engng.*, 15, 6, 1963, 37–40.
87. Malouf E.E., Prater J.D.: *Journal of Metals*, 13, 1961, 353–356.
88. Watanabe A., Uchida T., Furuya S.: Identification of sulphur oxidizing bacteria and iron bacteria and properties of iron-oxidizing bacteria in metal mine water, Hakko Kyokaishi, 25, 1967, 431–440.
89. Duncan D.W., Trussell P.C.: *Can. Met. Quart.*, 3, 1964, 43–55.
90. Bryner L.C., Anderson R.: *Ind. Engng. Chem.*, 49, 1957, 1721–1724.
91. Wyckoff R.W.G.: *Bull. Soc. Fr. Miner. Cristallogr.*, 93, 1970, 120–122.
92. Havlík T.: *Neželezné kovy*, 4, 1987, 121–126.
93. Havlík T.: Mechanism of acid oxidation dissolution of chalcocite, Thesis, Faculty

of Metallurgy, Technical University, Košice, 1995.

94. Kammel R.: Private communication.
95. Winand R.: *Hydrometallurgy*, 27, 1991, 285–316.
96. Avraamides J., Muir D.M., Parker A.J.: *Hydrometallurgy*, 5, 1980, 325–330.
97. Havlík T., Škrobian M., Petričko F.: *Acta Metallurgica Slovaca*, 2, 2, 1996, 133–141.
98. Havlík T., Škrobian M., Baláž P., Kammel R.: *International Journal of Mineral Processing*, 43, 1995, 61–72.
99. Havlík T.: *Acta Met. Slovaca*, 4, 1, 1998, 42–48.
100. Havlík T., Kammel R.: *Minerals Engineering*, 8, 10, 1995, 1225–1234.
101. Lowe D.F.: The kinetics of the dissolution reaction of copper and copper-iron sulphide minerals using ferric sulphate solutions, Ph.D Thesis, University of Arizona, 1970
102. Antonijevič M.M., Jankovič Z., Dimitrijevič M.: *Hydrometallurgy*, 71, 2004, 329–334.
103. Dalton R.F., Diaz G., Price R.& Zunkel A.D. *Journal of Metals*, August, 51, 1991.
104. Dalton R.F., Diaz G., Hermana E., Price R. & Zunkel A.D. The Cuprex metal extraction process: Pilot plant experience and economics of a chloride-based process for the recovery of copper from sulphide ores, Proc. Copper 91 - Cobre 91, vol II, 61, 1991.
105. Tuller W.N. : The Sulphur Data Book, p.129. McGraw Hill, 1954.
106. Havlík T., Kmetová D.: Method of acid oxidative sulfide leaching containing of iron, Slovak patent, AO 238 913-19.8.1985
107. Havlík T., Škrobian M.: Rudy, 37, 10, 1989, 295–302.
108. Havlík T., Škrobian M.: *Can. Met. Quart.*, 29, 2, 1990, 133–139.
109. Havlík T., Dvoršíková J., Ivanová Z., Kammel R.: *Metall*, 53, 1–2, 1999, 57–60.
110. Havlík T.: Study of dissolution of Cu sulphides in an acid medium, Thesis, Technical University, Košice, 1982.
111. Havlík T.: Acid oxidation leaching of of chalcopyrite and behaviour of sulphur in this process, PhD Thesis, Technical University, Košice, 1996.
112. Rapoport F., Iljinskaja A.A.: Laboratory methods of roducing clean gases, Gosud. Nauchno-Tekh. Izd. Khim. Literatury, Moscow, 1963, 107–123.
113. Havlík T., Šulek K., Briančin J., Kammel R.: *Metall*, 52, 7–8, 1998, 1–4.
114. Havlík T., Miškufová A., Tatarka P.: *Acta Metallurgica Slovaca*, 4, Special Issue 4/2001, 62–68.
115. Havlík T., Popovičová M.: *Acta Metallurgica Slovaca*, 6, 2, 2000, 171–177.
116. Havlík T., Popovičová M., Ukašík M: *Metall*, 55, 10, 2001, 332–335.
117. Popovičová M., Havlík T.: *Acta Metallurgica Slovaca*, 7, 1, 2001, 27–33.
118. Laubertová M., Havlík T.: *Acta Metallurgica Slovaca*, 8, 2/2002, 124–129.
119. Laubertová M., Havlík T., Galová P.: *Acta Metallurgica Slovaca*, 9, 4/2003, 270–276.
120. Antonucci V., Correa C.: Sulfuric acid leching of chalcopyrite concentrate assited by application of microwave energy, Copper95 – Cobre95, Int. Conf., vol.

III, Santiago de Chile, Nov. 1995, 549–557.
121. Baláž P.: *Chemické listy*, 88, 1994, 508–513.
122. Madhuchhanda M. et al.: *Mineral Metallurgical Processing*, 18, 2, 2001, 106-109.
123. Devi N.B. et al.: *Met. and Mat. Trans. B*, vol. 32B, October 2001, 777–784.
124. Havlik T., Laubertová M., Miškufová A., Kondas J., Vranka F.: *Acta Metallurgica Slovaca*, 10, Special Issue 2/2004, 80–87.
125. Havlik T., Laubertová M., Miškufová A., Kondas J., Vranka F.: *Acta Metallurgica Slovaca*, 10, Special Issue 2/2004, 80–95.
126. Hyvärinen O., Hämäläinen M.: Method for producing copper in hydrometallurgical process, US Patent 6,007,600 (1999) .
127. Hyvärinen O., Hämäläinen M.: *Acta Metallurgica Slovaca*, 10, Special Issue 2, 2004, 107–123.
128. Shirts, M.B., Winter, J.K., Bloom, P.A., Potter, G.M., 1974. Aqueous reduction of chalcopyrite concentrate with metals, USBM RI 7953, US Department of the Interior, Washington, DC.
129. Dreisinger D., Abed N.: *Hydrometallurgy*, 66, 2002, 37–57.
130. Hackl R.P., Dreisinger D.B., Peters E.: Reverse leaching of chalcopyrite, In: Copper 1987—Vol.3: Hydrometallurgy and Electrometallurgy of Copper, Copper, W.C, 1987, 181–200.
131. Baláž P.: Extractive Metallurgy of Activated Minerals, Esevier, Amsterdam, 2000.
132. Baláž P.: *Int. J. Miner. Process.*, 72, 2003, 341–354.
133. Langer E.: Theory of induction and dielectric heat [in Czech], Academia Praha, 1979.
134. Florek I., Lovás M., Múrová I.: Intensification of leaching of minerals by microwave radiation, Proc. New Trends in Processing, Ostrava, 1997, 48–56.
135. Sullivan J.D.: *Trans. Am. Inst. Min. Metall.*, 106, 1933, 515–547.
136. Habashi F.: Principles of Extractive Metallurgy, vol. 2, Hydrometallurgy, Gordon and Breach, New York, 1969.
137. Warren I.H.: *Australian Journal of Applied Science*, 5, 1, 1958, 36–51.
138. Stanczyk M.H., Rampacek C.: Oxidation leaching of copper sulfides in acidic pulps at elevated temperatures and pressures, Bureau of Mines RI, 6193, 1963.
139. Fisher W.W., Roman R.J.: The Dissolution of Chalcocite in Oxygenated Sulfuric Acid Solution, Circ. New Mexico. State Bur. Mines, Miner. Resour., 122, 1971, 28.
140. Dahms J., Gerlach J., Pawlek F.: *Erzmetall*, Vol. XX., 1967, No.5, 203–208.
141. Roman R.J., Brenner B.R.: *Miner. Sci. Eng.*, 5, 1973, 3–24
142. Thomas G., Ingraham T.R., MacDonald R.J.C.: *Can. Met. Quart.*, 6, 3, 1967, 281–285.
143. Moh G.H.: Proc. 7th. General Meeting Int. Mineral Association, Tokyo, Mineral. Soc. of Japan, 1971, 226–232.
144. Tkachenko O.B., Tseft A.L.: Tr. Inst. Metal. Ogogashch., Alma-Ata, 30, 1969, 15–23.
145. Mulak W.: *Roczn.Chem.*, 43, 1969, 1387–1394.
146. Kopylov G.A., Orlov A.I.: *Zh. Irkutsk. Polit. Inst.*, 46, 1969, 127–132.

147. King J.A., Burkin A.R., Ferreira R.C.H.: Leaching of chalcocite by acidic ferric chloride solutions, Leaching and reduction in hydrometallurgy, IMM London, 1975, 36-46
148. Burkin A.R.: *Min. Sci., Engng.*, 1, 1, January 1969, 4–15.
149. Kmetová D., Kuffa T., Havlík T.: Study of fundamentals processes in hydrometallurgy [in Slovak], Final report of the State research project III-6-4/7, Košice, 1980, part I, II.
150. Peters E., Loewen F.: Met. Trans., 4, January 1973, 5–14.
151. Brennet P., Jafferali S., Vanseveren J.-M., Vereecken J., Winand R.: Met. Trans., 5, January 1974, 127–134.
152. Mackinnon D.J.: *Hydrometallurgy*, 1, 1976, 241–257.
153. Lal R., McNicol J.H.: The BHAS copper leach plant, Techn. Paper A87-1, Metall. Soc. AIME, Warrendale, Pa, 1987
154. Meadows N.E., Ricketts N.J., Smith G.D.J.: Oxygen leaching of copper-lead matte in acidic chloride/sulphate solutions. In: Research and Development in Extractive Metallurgy, Aus. Inst. Min. Metall., Adelaide Branch, 1987, 125-120
155. Havlík T.: Study of dissolution of Cu sulphides in an acid medium [in Slovak], Thesis, Faculty of Metallurgy, Technical University Košice, 1982, 144.
156. Havlík T.: *Trans. Tech. Univ. Kosice*, 2, 2, 1992, 231–237.
157. Fiala J., Havlík T., Škrobian M.: *Chemický prumysl*, 9, 40/65, 1990, 488–494.
158. Bartlett R.W.: *Met. Trans. B*, 23B, June 1992, 241–248.
159. Bartlett R.W, Wilson D.B., Savage B.J., Wesely R.J.: A process for enriching chalcopyrite concentrates, Hydrometallurgy Reactor Design and Kinetics, Bautista et al. eds, TMS, Warrendale, PA, 1986, 227–247.
160. Sullivan J.D.: Chemistry of leaching chalcocite, U.S. Bureau of Mines TP-473, 487, 1930.
161. King J.A.: Solid state changes in the leaching of copper sulphide, Ph.D. Thesis, University of London, 1966.
162. Thomas G., Ingraham T.R.: *Can. Met. Quart.*, 6, 2, 1967, 153–165.
163. Baur J.P., Gibbs H.L., Wadsworth M.E.: Met. Soc. AIME, Pamphlet 72-B-96, 1972.
164. Dutrizac J.E., MacDonald R.J.C.: Can. Met. Quart., 13, 3, 1974, 423–433.
165. Mulak W.: *Roczn. Chem.*, 45, 1971, 1417–1424.
166. Lowe D.F.: The kinetics of the dissolution reaction of copper and copper-iron sulphide minerals using ferric sulphate solutions, Ph.D. Thesis, University of Arizona, 1970.
167. Walsh C.A., Rimstidt J.D.: *Canadian Mineralogist*, 24, 1986, 35–44.
168. Cheng Ch.Y., Lawson F.: *Hydrometallurgy*, 27, 1991, 269–284.
169. Kato T., Oki T.: *Denki Kagaku*, 40, 1972, 670–674.
170. Kametani H.: Some aspects of anodic oxidation of sulphide suspensions for direct electrowinning of metals, Joint Meeting on Cell Design in Electrowinning and Electrorefining, Southampton University, July 1974.
171. Mackinnon D.J.: *Hydrometallurgy*, 2, 1976, 65–76.
172. Hirato T., Hiai H., Awakura Y., Majima H.: *Met. Trans. B*, 20B, August, 1989, 485–491.

173. Duda L.L., Bartecki A.: *Hydrometallurgy*, 8, 1982, 341–354.
174. Cho E.H.: *Journal of Metals*, January 1987, 18–20.
175. Mao M.H., Peters E.: Acid pressure leaching of chalcocite, Proc. Hydrom. Res. Devel. Plant. Pract., Osseo-Assare ed, SME-AIME 1983, 243-26
176. Beyke C.J., Wethington A.M., Lee A.Y.: Oxidative leaching and electrowinning of copper with fluosilicic acid, Proc. Copper 91-Cobre 91, III., Hydrometallurgy and Electrometallurgy of Copper, Cooper W.C., Kemp D.J, Lagos G.E., Tan K.G. eds., Pergamon Press, 1991
177. Woodcock J.T.: Some aspects of the oxidation of sulphide minerals in aqueous suspension, The Australasian Institute of Mineral and Metallurgical Processing, No. 197, March 1961, 47-85
178. Walsh C.A., Rimstidt J.D.: *Canadian Mineralogist*, 24, 1986, 35–44.
179. Pattrick R. A. D., Hall A. J.: *Mineralogical Magazine*, 47, 1983, 441.
180. Nishihara G.S.: *Economic Geology*, 9, 1914, 743–757.
181. Brown S. L., Sullivan J. D.: US Bureau of Mines RI 3228, 1934.
182. Dutrizac J.E., Morrison R.M.: The leaching of some arsenide and antimonide minerals in ferric chloride media, Proc. Hydrometallurgical Process Fundamentals, Bautista R.C. (Ed.), Plenum Publish. Corp, 1984, 77–122.
183. Neiva Correia N.M.J., Carvalho J.R., Monhemius A.J.: *Hydrometallurgy*, 2000, 167–179.
184. Havlík T., Škrobian M.: *Trans. Tech. Univ. Kosice*, 2, 1992, 139–144.
185. Havlík T., Škrobian M., Dudáš D.: *Hutnícke listy*, XLVI, 12-12, 1991, 76–80.
186. Havlík T., Škrobian M., Baláž P.: Erz*metall*, 47, 2, 1994, 122–129.
187. Gerlach J., Pawlek F., Rodel R., Schade G., Weddige H.-P.: *Erzmetall*, 35, 9, 1972, 448–453.
188. Caldon F.: Treatment of metal bearing mineral material, US Patent Patent No. 4,084,961, 1978.
189. Scheiner B.J., Smyres G.A., Haskett P.R., Lindstrom R.E.: Copper and silver recovery from a sulfide concentrate by ferrous chloride-oxygen leaching, U.S. Bureau of Mines Rep. Inv. No. 8290, 1978.
190. Bahr A., Priesemann T.: Recovery of silver from refractory ores, Proc. XVI. Int. Min. Processing Cong., Stockholm, Elsevier, Amsterdam, 1988, 1221-1234
191. Neiva Correia M.J., Carvalho J.R., Monhemius A.J.: *Minerals Engineering*, 6, 12, 1993, 1217–1225.
192. Brynner L.C. et al.: *Ind. Engng. Chem.*, 36, 1954, 2587–2592.
193. Frenay J.: Recovery of copper antimony and silver by bacterial leaching of tetrahedrite concentrate, Proc. Copper '91 Int. Symp. 3 Pergamon, New York, 1991, 99–105.
194. Brunner V.H., Strasser H., Pumpel T.: Microbiological leaching of copper and zinc from barite containing tetrahedrite (Kleinkogel/Tirol area) through thiobacillus ferrooxidants, Arch. Lagerstaettenforsch. Geol. Bundesanst. 16, 1993 5–12.
195. Kušnierová M.: *Miner. Slovaca*, 27, 6, 1995 407–412.
196. Kušnierová M.: Biological leaching of tetrahedrite, I. Miedzynarodowa konferencja przerobki kopalin, Zakopane, 1995, 147–152.
197. Kaloč J., Podbora Z., Sýkora V.: *Hutnícke listy*, 22, 9, 1967, 626–629.

198. Dayton S.: *Eng. Min. Journal*, January 1982, 79–83.
199. Anderson C.G., Krys L.E.: Leaching of antimony from a refractory precious metals concentrate, Hydrometall. Fundam. Tech. Innov., Utah 1993, 342–363
200. Ackerman J.B., Anderson C.G., Nordwick S.M., Krys L.E.: Hydrometallurgy at the Sunshine mine metallurgical complex, Hydrometall. Fundam. Tech. Innov., Utah 1993, 477–498.
201. Baláž P., Kammel R., Achimovičová M.: *Metall*, 48, 1994, 217–220.
202. Baláž P., Kammel R., Kušnierová, Achimovičová M.: Mechanochemical treatment of tetrahedrite as a new non-polluting method of metals recovery, Hydrometallurgy'94, Proc. Int. Symp., Cambridge 1994, 209–218.
203. Baláž P., Sekula F., Jakabský S., Kammel R.: *Miner. Eng.*, 8, 12, 1995 1299–1308.
204. Baláž P., Achimovičová M., Ficeriová J., Kammel R., Šepelák V.: *Hydrometallurgy*, 47, 2–3, 1998, 297–307.
205. Baláž P., Ficeriová J., Šepelák V., Kammel R.: *Hydrometallurgy*, 43, 1–3, 1996, 367–377.
206. Baláž P., Kammel R., Labuda L.: *Uhlí-Rudy-Geologický pruzkum*, 2, 1995, 48-52.
207. Lin H. K.: *Hydrometallurgy*, 73, 2004, 283–291.
208. Ivanová Z., Dvoršíková J.: *Acta Metallurgica Slovaca*, 4, Special Issue 1/1998, 90–96.
209. Mostecký J., Kolihová D., Sychra V. et al.: A method of processing tetrahedrite concentrate, AO 177947, 15.03.1979.
210. Bumbálek V., Černák K., Horák V.: Complex processing of Czechoslovak tetrahedrite raw materials [in Czech], Proc. Hornícka Příbram, 1978, 183–188.
211. Havlík T., Škrobian M., Kammel R.: *Metall,* 52, 4, 1998, 210–213.
212. Ukasik M., Havlik T.: *Hydrometallurgy*, 77, 2005, 139–145.
213. Havlík T., Kammel R.: *Acta Metallurgica Slovaca*, 2, 4, 1996, 321–327.
214. Havlík T.: *Acta Met. Slovaca*, Special Issue, 1/1998, 86–90.
215. Havlík T., Ivanová Z., Dvoršíková J., Kammel R.: *Metall,* 53, 7–8, 1999, 390–394.

CHAPTER 13

MORPHOLOGY AND BEHAVIOUR OF SULPHUR IN THE LEACHING OF SULPHIDES

One of the main advantages of the leaching processes is the complete elimination of gaseous emissions of sulphur oxides. In these processes, sulphur is oxidised in most cases to elemental sulphur or, depending on the process, to higher valencies. In the US Bureau of Mines process [1] only 'a small amount' of sulphate is formed and the pyrite in the copper concentrate does not change during leaching. Consequently, it was proposed to produce elemental sulphur by leaching using ammonium sulphide followed by oxidation in order to precipitate sulphur. In the MINTEK process [2], less than 5% of oxidised sulphide is oxidised to the sulphate form. Milner et al. [3] note that in the COMINCO process elemental sulphur is the main reaction product and it is recommended to produce elemental sulphur by flotation. The yield of elemental sulphur was 75–90%. In the CUPREX [4–6] and HENKEL processes [7], the main product is elemental sulphur although the amount of formed SO_4^{2-} has not been accurately defined. The real amount of SO_4^{2-} may depend on the mineralogical composition of the concentrate. In the GCM process [6], elemental sulphur is separated by pressure filtration at 150 °C, but the yield is only 60–70% of the total amount of sulphur. In the CUPREX process [4], sulphur is recovered by flotation followed by filtration of molten flotation sulphur. However, this method is characterised by a large number of difficulties. In all studies, it has been stressed that sulphur is contaminated by many impurities, mainly selenium, substituting sulphur.

In addition to the previously mentioned processes, the CYMET process [8] in which the chalcopyrite concentrate is leached at 98 °C, produces up to 25% SO_4^{2-}. This behaviour has not as yet been explained. Similarly, the CLEAR process [9], in which leaching takes place at 105 °C and subsequently at 140 °C, produces 15–25% of the sulphate.

As already mentioned, the problem of the formation and behaviour of sulphur in the leaching processes has been investigated in several fundamental research studies but no complete explanation has as yet been proposed. In the majority of the studies it is stated that more than 90% and, in some cases, more than 95% of elemental sulphur is formed. Ravi et al [10] reported the formation of up to 99.5% elemental sulphur during leaching of chalcopyrite at 97 °C. Dutrizac [11] investigated ratio of the amount of elemental sulphur and sulphate, formed in leaching of chalcopyrite in ferric chloride. More than 95% of elemental sulphur and up to 5% of the sulphate were formed at temperatures below 100 °C. The relative amounts of SO and SO_4^{2-} are not affected by the leaching time of up to 90 hours, the concentration of $FeCl_3$ of up to 2 M or HCl concentration up to 3 M. Similarly, the presence of air of oxygen in the leaching medium did not have any effect on the amount of formed sulphur and sulphate. Elemental sulphur is not attacked by ferric chloride below the melting point of sulphur [12].

The morphology of formed elemental sulphur is greatly affected by the particle size, the holding time in the medium, and the leaching conditions [11, 13, 14]. Elemental sulphur agglomerates and coats the individual chalcopyrite particles. Although the sulphur layer inhibits the leaching process, the agglomeration of sulphur must have a beneficial effect on the removal of sulphur from leaching residue by filtration. The process of efficient and fast filtration is mentioned in all important studies, despite the fact that the leaching residue contained fine particles. The reasons for the large differences in the relative amount of formed elemental sulphur and sulphate during leaching of chalcopyrite, published in the literature, are not known. Of course, they may reflect large differences in the purity of the chalcopyrite concentrate and the value of the oxidation potential (i.e., the type of oxidation agent) in the experimental solution. The mechanism of the formation of elemental sulphur during leaching has not as yet been completely described, although a number of theoretically possible reactions have been proposed. The morphological studies of elemental sulphur, formed during leaching of galenite using cuprous chloride [15], indicate that H_2S may form as an intermediate product during the reaction. This was taken into account later in the investigations of leaching of galenite and sphalerite by means of $FeCl_3$ [16]. Although a large number of morphological formations of elemental sulphur have been identified in the investigations of leaching of chalcopyrite in $FeCl_3$ [13, 14], it is necessary to investigate whether these formations change during leaching and whether they are influenced by the leaching conditions and, consequently, whether they can influence the process kinetics.

Table 13.1 gives the amount of elemental sulphur, formed in different hydrometallurgical processes or in experiments carried out by different authors.

It may be seen that the amount of elemental sulphur, formed in the processes in the pilot plant or production conditions, is considerably smaller than in laboratory research. The residual sulphur is oxidised into soluble forms, in most cases sulphuric acid. It is the same product as in pyrometallurgy and the efficiency of marketing of the product determines the feasibility of the economic parameters of the process. Unfortunately, at present, sulphuric acid is almost impossible to sell anywhere profitably in the world and processing of the acid requires higher production costs.

The majority of the primary raw materials of nonferrous metals

Table 13.1. Amount of elemental sulphur formed in different hydrometallurgical processes and by different authors [17]

Process or author	Year	Amount of S^0 formed [%]
SHERRITT GORDON	1967	> 85
CYMET	1972	~ 75
COMINCO	1974	75–90
U.S. BOM	1975	62
CLEAR	1980	75-85
Paytner	1974	> 95
Jones, Peters	1976	90
Kunda et al.	1976	> 95
Dutrizac	1978	94–97
Dutrizac	1982	> 95
Ravi	1987	82
Dutrizac	1989	~ 94

include sulphide ores and concentrates with a relatively high sulphur content. In leaching of, for example, chalcopyrite, sulphur in sulphide form is usually transferred into the elemental form according to the reaction:

$$CuFeS_2 + 4FeCl_3 \rightarrow CuCl_2 + 5FeCl_2 + S^\circ \quad (13.11)$$

and coats partially or completely the leached surface. The reagents must diffuse through this layer and this usually greatly hinders the overall kinetics of the process. The leaching processes may be relatively long (expressed in days) at relatively high temperatures (~95 °C) in large volumes of the diluted solutions. These are the main reasons for the high energy costs and the relatively unfavourable economic parameters of hydrometallurgical methods. In addition, the separation of elemental sulphur from the leaching residue requires further expenditure. In addition, recovered sulphur with a content of 90–98% of sulphur, and the residue consists of the oxides of Fe, Pb, Zn, Cu, Hg, Sb, Se, etc. This type of high;y contaminated sulphur is difficult to process further.

However, in reality, sulphide sulphur does not transfer completely to the elemental form in accordance with the equation (13.1). Part of the sulphur is oxidised to the soluble form according to the reaction:

$$CuFeS_2 + 16FeCl_3 + 8H_2O \rightarrow CuCl_2 + H_2SO_4 + 12HCl + 17FeCl_2 \quad (13.2)$$

If the system has a sufficiently high redox potential, for example, using ozone as the oxidation agent, or by high-pressure leaching, sulphur is not precipitated in elemental form, but in the form of a soluble sulphate:

$$3CuFeS_2 + 8O_3 \rightarrow 3CuSO_4 + 3FeSO_4 \quad (13.3)$$

The system is characterised by a relatively high rate of the process because it is not hindered by the diffusion of reactants through the layer of elemental sulphur which is oxidised to sulphuric acid.

The amount of sulphur in the soluble form varies in the range 3–40%, Table 13.1, and the reasons for the scatter have not as yet been completely explained. On the other hand, it is evident that it is strongly affected by various factors, such as purity of chalcopyrite, the oxidation potential of the solution, leaching time, reagent concentration, aeration, mixing rate, erosion, etc. Basically, it may be concluded that specific conditions result in the transfer of a specific amount of sulphur to the soluble form.

Leaching of chalcopyrite in ferric chloride

Figure 13.1 shows unleached chalcopyrite containing, in addition to the main grains, also a number of submicroscopic particles, with the ragged surface of the majority of the grains. In leaching, the reaction rate of the small particles is greater because of the surface area and a larger amount of copper is transferred into the solution. After leaching the small grains, the total reaction rate decreases because only large particles react. This is in accordance with the shape of the leaching curves, as already mentioned. Gradually, with time, the particles are covered by the formed sulphur with the temperature having a significant effect. Figure 13.2*a–d* show chalcopyrite after leaching for six hours at temperatures of 22, 50, 70 and 95 °C indicating clearly not only the gradual leaching of the individual grains but also the fact that the reaction is fast only at temperatures above 90 °C.

The photographs also show that leaching is not uniform and does not take place gradually in all grains but in certain grains it is extensive or complete whereas in others it is not. With increasing temperature the number of reacting grains increases and all the grains react at temperatures above 90 °C, Fig. 13.3*a*, *b*.

The unleached particles show that the crystallographic orientation of the chalcopyrite grains may differ and the rate of leaching through these grains is slower than through the remainder of the grains. On the other hand, it appears that preferential leaching takes place on the particles with sharp edges and ledges on which the charge accumulates and is the driving force of the

Fig. 13.1. Unleached chalcopyrite –100 +60 μm.

Fig. 13.2. Chalcopyrite leached in 0.5 M $FeCl_3$+0.5 M HCl, six hours; a) 22 °C; b) 50 °C; c) 70 °C; d) 95 °C.

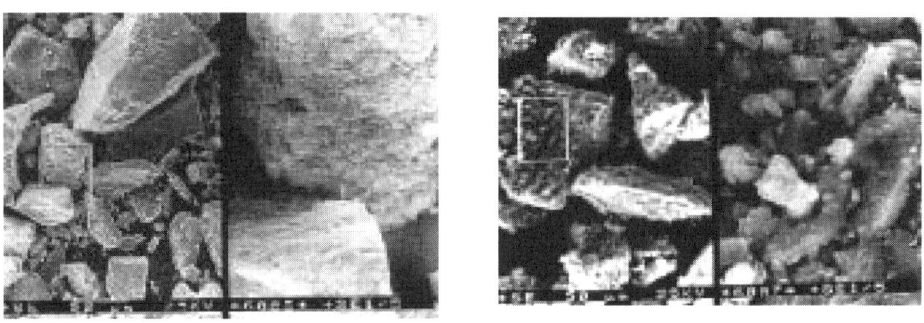

Fig. 13.3. Chalcopyrite leached in 0.5 M $FeCl_3$+0.5 M HCl, six hours; a) 22 °C; b) 95 °C.

electrochemical reaction of leaching of chalcopyrite. Auxiliary experiments show that accurate polished surfaces of the massive chalcopyrite sample were leached only to the minimum extent in the same medium as the one used for the same chalcopyrite, processed by milling. A special role is also played by the accumulation of defects as a result of milling and the support of leaching by the

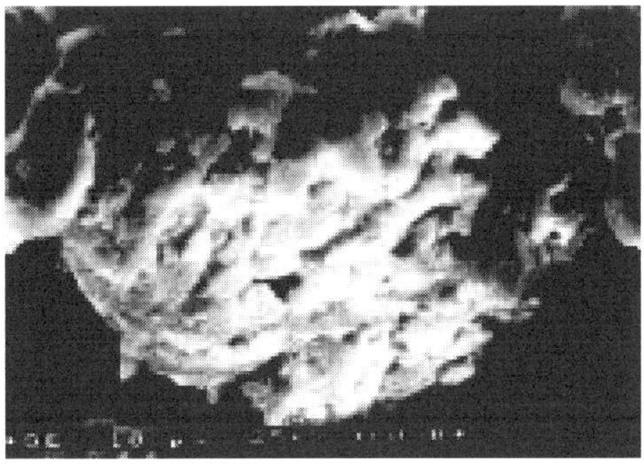

Fig. 13.4. Detail of sulphur on the leached surface of chalcopyrite in 0.5 M $FeCl_3$+0.5 M HCl, 6 h, 95 °C.

defective structure [18].

The occurrence of sulphur in the form of individual globules, Fig. 13.4, shows that elemental sulphur forms by precipitation from the aqueous solution [19, 20], as expressed later.

This mechanism better explains the observed morphology of sulphur than the surface migration of sulphur. Although the leaching reaction could be explained by a different mechanism, the morphological studies show that sulphur is at least partially dissolved and subsequently precipitates. In addition, as mentioned later, another reaction product was identified, CuS. Reaction (13.1) includes the formation of the solid phase and, therefore, preferential nucleation and growth of the grains takes place in the active areas of chalcopyrite. This assumption is also confirmed by examination of the morphology of partially leached specimens at lower temperatures which show the attacked surface, but only in certain grains, Fig. 13.2.

The effect of the concentration of the leaching agent on morphology is shown in Fig. 13.5*a–f*. Experiments were carried out at 80 °C and the concentration of the ferric ion in the solution was 0 M, 0.1 M, 0.25 M, 0.5 M, 0.75 M and 1 M $FeCl_3$ in a 1 M solution of HCl.

As assumed, the chalcopyrite grains are covered by elemental sulphur whose amount increased slowly with increasing concentration of the ferric ion, but it may be concluded unambiguously that the effect of temperature is much stronger than that of the concentration of the ferric ions. However, morphology is similar and the results indicate the formation of a discontinuous layer of sulphur in the globular form. The reaction rate and, therefore, nucleation of sulphur depend on the concentration of the

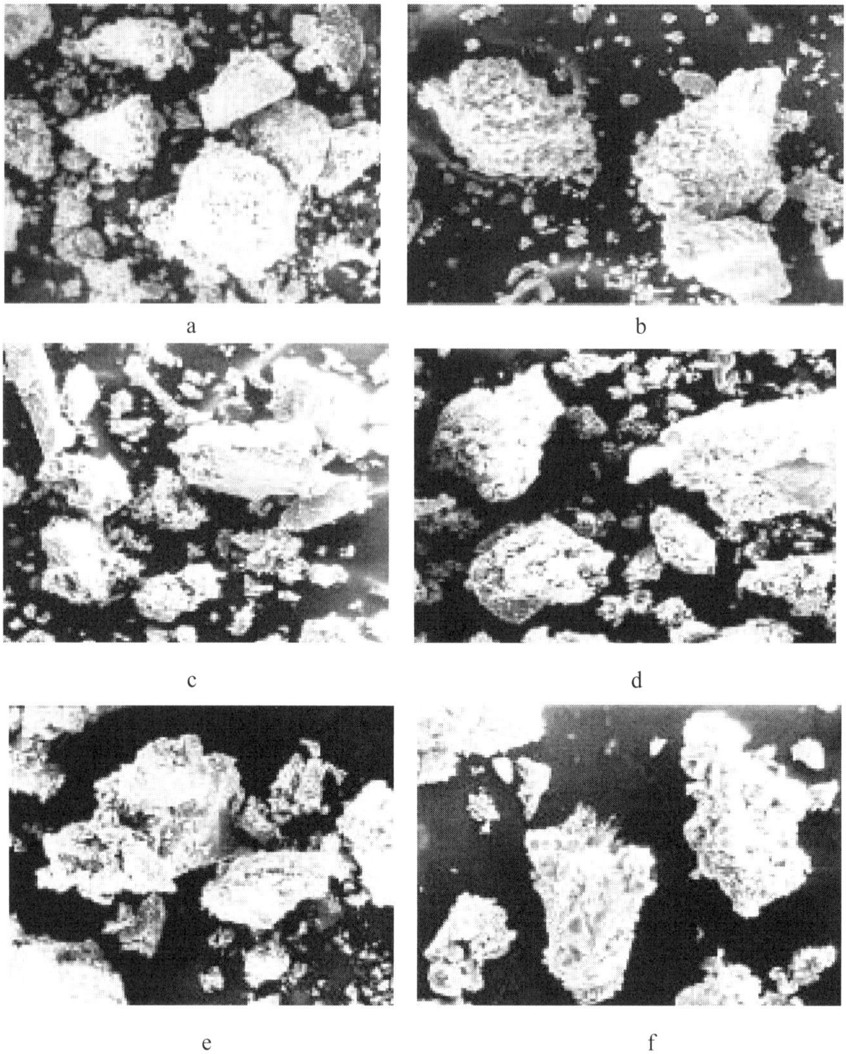

Fig. 13.5. Chalcopyrite leached at 80 °C, 6 hours. a) 1 M HCl without $FeCl_3$; b) 0.1 M $FeCl_3$ +1 M HCl; c) 0.25 M $FeCl_3$ +1 M HCl; d) 0.5 M $FeCl_3$ +1 M HCl; e) 0.75 M $FeCl_3$ +1 M HCl; f) 1 M $FeCl_3$ +1 M HCl.

ferric ions in the solution; Fig. 13.6 shows that the amount of sulphur formed at a lower concentration is lower but the sulphur is in the form of larger globules. On the other hand, Fig. 13.7 shows the formation of a more continuous layer of sulphur in the form of small globules indicating the larger number of nuclei and rapid grain growth.

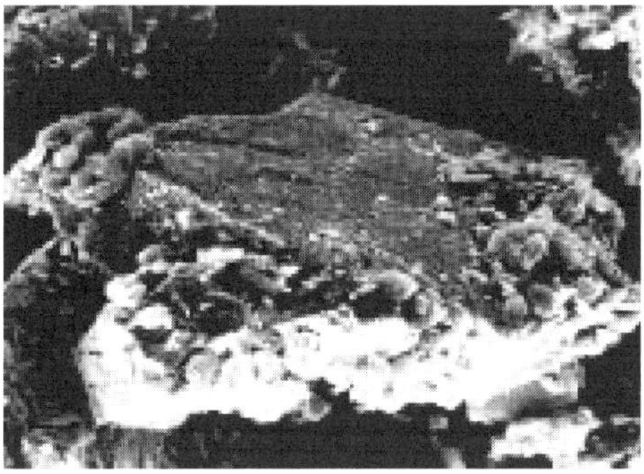

Fig. 13.6. Detailed view of chalcopyrite leached in 0.25 M $FeCl_3$ +1 M HCl, 6 h, 80 °C.

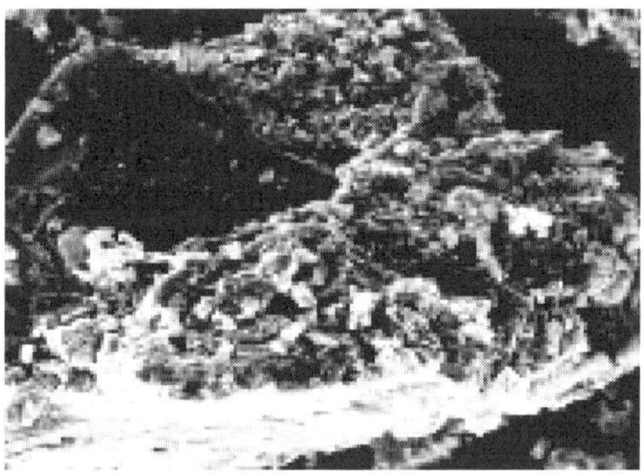

Fig. 13.7. Detailed view of the chalcopyrite leached in 1 M $FeCl_3$ +1 M HCl, 6 h, 80 °C.

Leaching of chalcopyrite in ferrous sulphate solution

Figure 13.8*a–d* shows partially leached chalcopyrite in a sulphate medium at different temperatures, as in the case of chloride.

Generally, the surface of the particles is less attacked in comparison with leaching using the chloride. The presence of a large amount of sulphur is evident only at 99 °C. The morphology of the sulphur particles is different, as in the case of the chloride, detailed view in Fig. 13.9.

In the sulphate medium, the linear kinetics of the process was detected, probably as a result of slower leaching in comparison with the chloride, representing the starting parts of the kinetic curves. The smaller amount of elemental sulphur, formed by the leaching reaction, is at least partly responsible for this course because sulphur does not cover completely the leached surface and, at the same time, since the degree of conversion is low, the interfacial area changes only slightly.

a

b

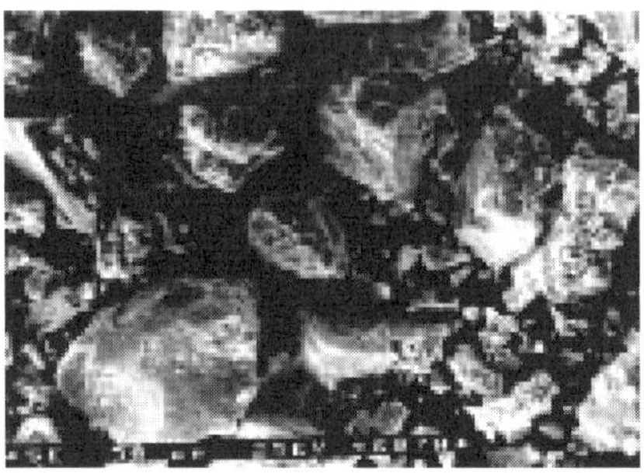

c

d

Fig. 13.8. Chalcopyrite leached in 0.5 M $Fe_2(SO_4)_3$+ 0.5 M H_2SO_4, 6 h: a) 22 °C; b) 50 °C; c) 70 °C; d) 99 °C.

Fig. 13.9. Detail view of chalcopyrite, leached in 0.5 M $Fe_2(SO_4)_3$+ 0.5 M H_2SO_4, 6 h, 99 °C.

Since the resultant sulphur does not completely cover the leached surface, photographs show directly chalcopyrite, Fig. 13.9. Chalcopyrite is characterised by a relatively high degree of defects; roughness of the surface is evident in certain crystallographic directions and, in addition to this, there are also large cracks resulting from the presence of internal stresses in leaching. The morphology of sulphur on the leached surface is not clearly globular, as in the previous case, either as a result of the slower reaction and, consequently, a small amount of precipitated sulphur, or as a result of a different formation mechanism. Taking into account more angular form of sulphur islands, it is likely that the contribution of precipitation of sulphur to this process is less marked. In any case, the different morphology indicates at least the partial formation of the different sulphur allotrope S7, as in the case of the chloride medium in which mostly α-S8 forms.

The amount of SO_4^{2-} in the solution resulting from the oxidation of sulphide sulphur cannot be determined because of the sulphate medium in which leaching takes place. Taking into account the amount of elemental sulphur, estimated from the results of semi-quantitative X-ray diffraction phase analysis, the amount of the formed elemental sulphur was estimated at 90–95%.

Leaching of chalcopyrite in sulphuric acid in the presence of ozone

Figure 13.10*a–d* shows the particles of partially leached

Fig. 13.10. Chalcopyrite leached in 0.5 M H_2SO_4 + 2.5% O_3, 30 hours. a) 22 °C; b) 30 °C; c) 60 °C; d) 75 °C.

chalcopyrite at different temperatures. The overall view of the individual samples partially leached using ozone shows a gradual decrease of the particle size and also rounding of the sharp edges. However, the most important result is the one which shows that none of the particles showed, at the investigated temperature, the presence of elemental sulphur on the leached surface or in the clusters. Consequently, the affected surface of the leached particles can again be clearly observed without any treatment of the samples, and it is clear that the highest degree of 'attack' was recorded for the sample leached at 22 °C. Leached chalcopyrite contains pores and cracks, especially after longer leaching times, and this induces stresses. Semi-quantitative analysis by the EDS method of selected particles shows the presence of only chalcopyrite within the error range of the method.

Since elemental sulphur does not cover the leached surface and the size of the surface rapidly decreases as a result of the high rate of the process, the observed parabolic kinetics indeed copies the model of the decreasing volume although it is obvious that the model cannot be completely exact because of the irregular form of the grains and/or formation of cracks and cavities which change the actual interfacial area.

Despite this, examination by optical microscopy confirmed the formation of particles with different colours, as the main matrix of chalcopyrite, Fig. 13.11*a, b*. Although these products are exceptionally subtile and, consequently, could not be identified exactly by X-ray diffraction phase analysis, the colour of the products indicates that it is covellite or blue covellite. This was confirmed by thermodynamic considerations of the formation of covellite as an interfacial product in leaching of chalcopyrite. This product does not coat the entire leached surface exposed to the leaching reagents and assuming a sufficient amount of the reagents, the reaction takes place to the very end. However, a different situation exists in the resultant pores and cavities in which a shortage of the leaching agent and/or oxidation agent becomes evident. Consequently, the reaction cannot be completed and this results in the formation of an interfacial product–the true or non-stoichiometric covellite. However, this result greatly changes the previously made considerations regarding the leaching mechanism of chalcopyrite and should explain some of the problems in this area.

Careful handling of leached chalcopyrite in a solution containing ferric ions and additional experiments, including measurements of

a

b

Fig. 13.11. The surface of partially leached chalcopyrite, magnification. a) × 100, b) × 240.

the redox potential of leaching, showed that CuS also forms in these cases, i.e., the formation of CuS in leaching is a regular phenomenon.

Leaching of chalcopyrite in ferric chloride solution in the presence of carbon tetrachloride

The role of the organic solvent in the leaching processes is the permanent removal of the resultant elemental sulphur from the particle surfaces.

Figure 13.12*a–d* shows the morphology of partially leached chalcopyrite in a chloride medium in the presence of carbon tetrachloride.

Fig. 13.12. Chalcopyrite leached in 0.5 M $FeCl_3$ +0.5 M HCl + CCl_4, 6 hours; a) 22 °C; b) 50 °C; c) 70 °C; d) 99 °C.

None of the photographs indicates the presence of elemental sulphur. The leached surface shows a relatively high degree of disruption. Taking into account the presence of a relatively large amount of small particles, it may be concluded that leaching is accompanied by the formation of relatively high stresses leading to gradual destruction of the leached particles. A detailed view of a particle of leached chalcopyrite at 90 °C, Fig. 13.13, shows a wrinkled surface, whereas the detailed view of another grain, Fig. 13.14, formed in the same conditions, shows a highly destructed grain.

Figure 13.15 shows the surface of chalcopyrite leached in a ferric sulphate solution in the presence of carbon tetrachloride at 90 °C. It may be seen that elemental sulphur again does not form in this case but, in accordance with the previous considerations, the leached surface is gradually attacked. Taking into account the lower overall leaching rate, the degree of destruction of the surface is lower and resembles more attack on the surface than volume destruction.

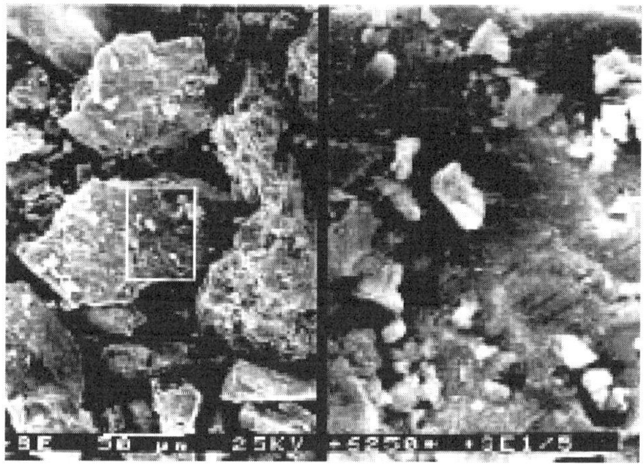

Fig. 13.13. Detail of a chalcopyrite grain leached in 0.5 M $FeCl_3$ +0.5 M HCl + CCl_4, 6 h, 99 °C.

Fig. 13.14. Detail of a chalcopyrite grain leached in 0.5 M $FeCl_3$ + 0.5 M HCl + CCl_4, 6 h, 90 °C.

Leaching of chalcopyrite in a high-frequency field

In order to intensify the process of leaching of chalcopyrite, experiments were carried out with leaching in a high-frequency field, as described previously in [21, 22]. The leaching medium was an aqueous solution of ferric chloride with the concentration in the range 0–1 M $FeCl_3$, acidified with the addition of HCl with different concentrations in the range 0–1 M. However, in this case it is not possible to examine the effect of temperature because the

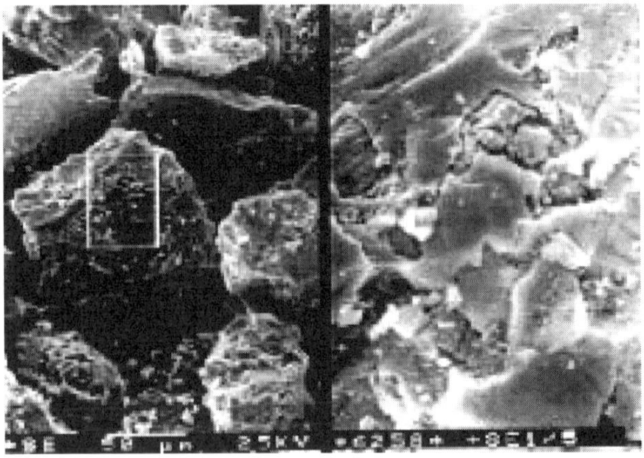

Fig. 13.15. Chalcopyrite leached in 0.5 M $Fe_2(SO_4)_3$ +0.5 M H_2SO_4 + CCl_4, 6 h, 90 °C.

experiments were carried out at 104 °C, as indicated by the principle of high-frequency heating.

Figure 13.16*a–f* shows the morphology of partially leached chalcopyrite with a change of the concentration of the ferric ion in the leaching solution, with the acid concentration being constant, 0.5 M HCl. Figure 13.16*a* shows the morphology of chalcopyrite leached only in the acid without ferric ions.

The effect of the amount of the ferric ions, i.e., the leaching agent, is strong, as indicated by morphological examination. The use of more concentrated solutions results in the gradual covering of the leached surface with elemental sulphur. This sulphur is globular and highly porous, Fig. 13.17.

Figure 13.18*a–d* shows the morphology of the partially leached chalcopyrite in relation to the amount of the acid in the leaching solution. It may be seen that on reaching the required acidity of the solution, represented by the value 0.5 M HCl, the amount of precipitated sulphur is no longer influenced by the acidity of the solution.

Leaching of tetrahedrite concentrate and tetrahedrite calcine in ferric chloride

As discussed previously, investigations were also carried out into the possibilities of acid leaching of the tetrahedrite concentrate in the solution of ferric chloride. X-ray diffraction phase analysis showed, in addition to the tetrahedrite $Cu_{12}Sb_4S_{13}$, that the concentrate contains small quantities of chalcopyrite $CuFeS_2$ and

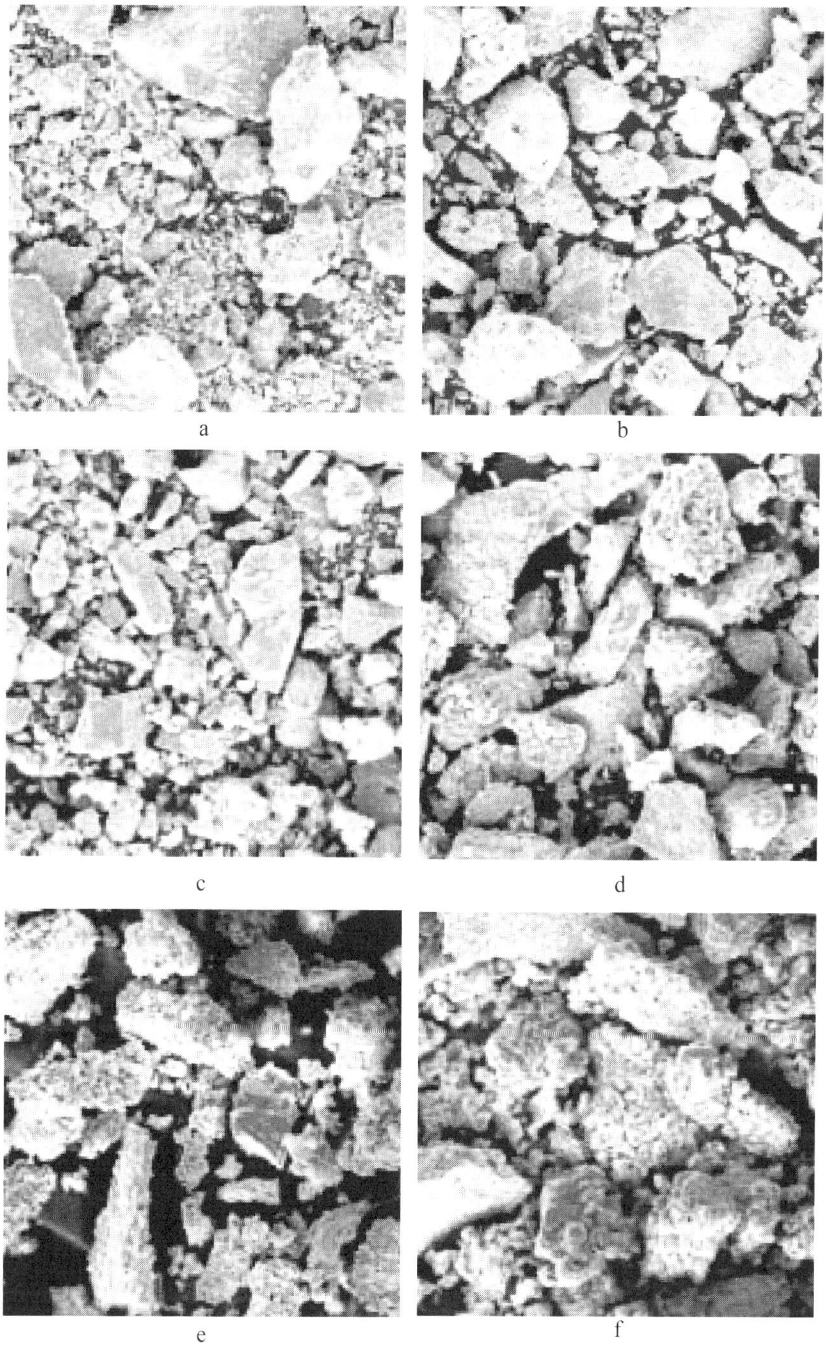

Fig. 13.16. Chalcopyrite leached in a high-frequency field at 104 °C, 0.5 M HCl, 3 h. a) 0.5 M HCl without $FeCl_3$; b) 0.05 M $FeCl_3$ + 0.5 M HCl; c) 0.1 M $FeCl_3$ + 0.5 M HCl; d) 0.5 M $FeCl_3$ + 0.5 M HCl; e) 0.75 M $FeCl_3$ + 0.5 M HCl; f) 1 M $FeCl_3$ + 0.5 M HCl.

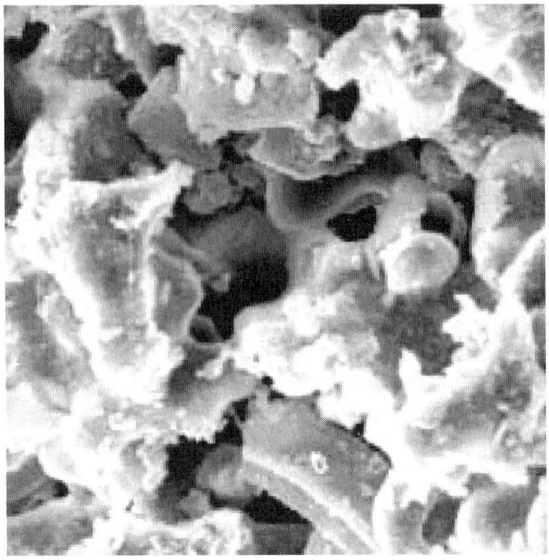

Fig. 13.17. Detailed view of the layer of elemental sulphur, magnification × 6000.

chalcostibite $CuSbS_2$ and also quartz SiO_2. Leaching was carried out using 0.5 M $FeCl_3$ +0.5 M HCl in the temperature range 40–90 °C [23].

Figure 13.19*a–f* (left) shows the change of the morphology of the particles of the leached tetrahedrite concentrate in relation to leaching temperature. The particles breakdown with increasing temperature as a result of extraction of metals into the solution canbe seen. At the same time, the particles are gradually covered by formed sulphur and they are almost completely covered at a leaching temperature of 90 °C, Fig. 13.20*a*.

Microscopic studies of solid residues after leaching of the tetrahedrite calcine in the ferric chloride acid solution, Fig. 13.19*g–l*, show that increasing temperature results in a decrease of the particle size and fragmentation of the particles as a result of destruction by leaching of metals from the tetrahedrite. Comparison with similar micrographs of the partially leached tetrahedrite concentrate, Fig. 13.19*a–f* shows that the degree of destruction of the calcine is considerably higher.

The leached particles are gradually coated with elemental sulphur, formed as a result of leaching. At the highest temperature, 88 °C, the entire surface is coated with elemental sulphur, Fig. 13.20b. This confirms the measured results in the sense that the efficiency of the process is the highest at high temperatures. In

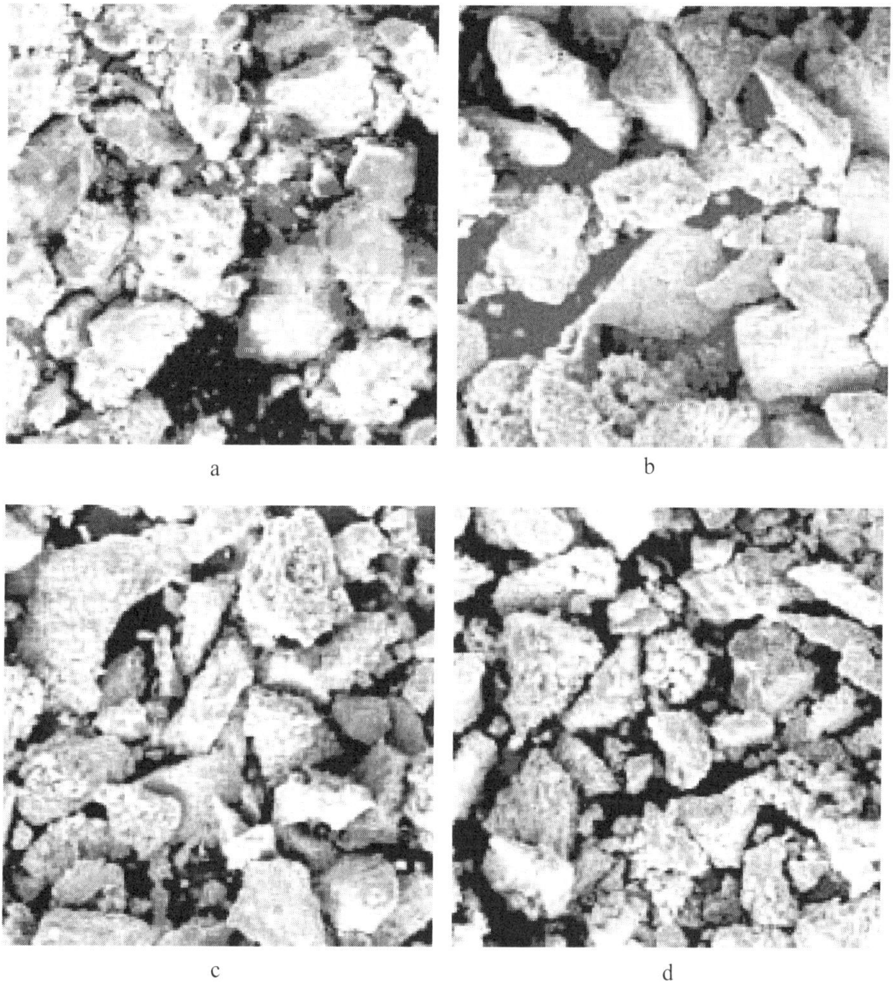

Fig. 13.18. Chalcopyrite leached in a high-frequency field at 104 °C, 0.5 M Fe_3+, 3 h.

a) 0.01 M HCl; b) 0.25 M HCl; c) 0.5 M HCl; d) 1 M HCl.

contrast to the globular form of sulphur, formed in the leaching of the tetrahedrite concentrate, Fig. 13.20a, the sulphur formed in leaching of the tetrahedrite calcine is dendritic and forms probably directly on the leached surface and not by precipitation from the solution [24]. The non-globular morphology, Fig. 13.20b, representing possible precipitates of antimony compounds, is also clearly visible.

a

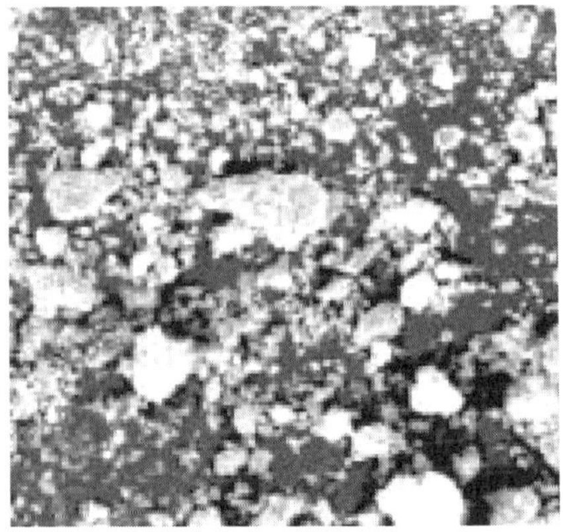

g

b

h

c

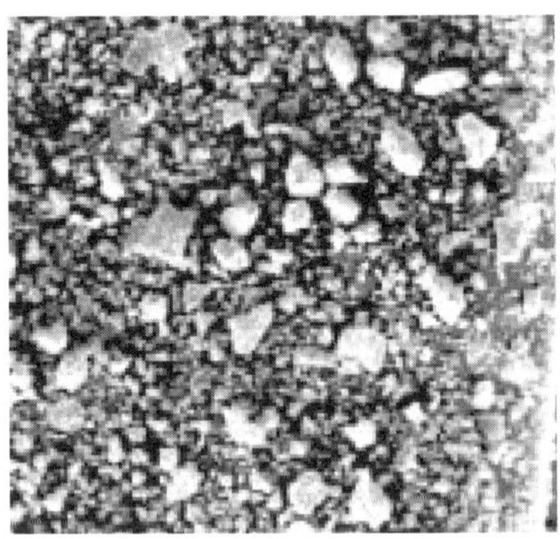

i

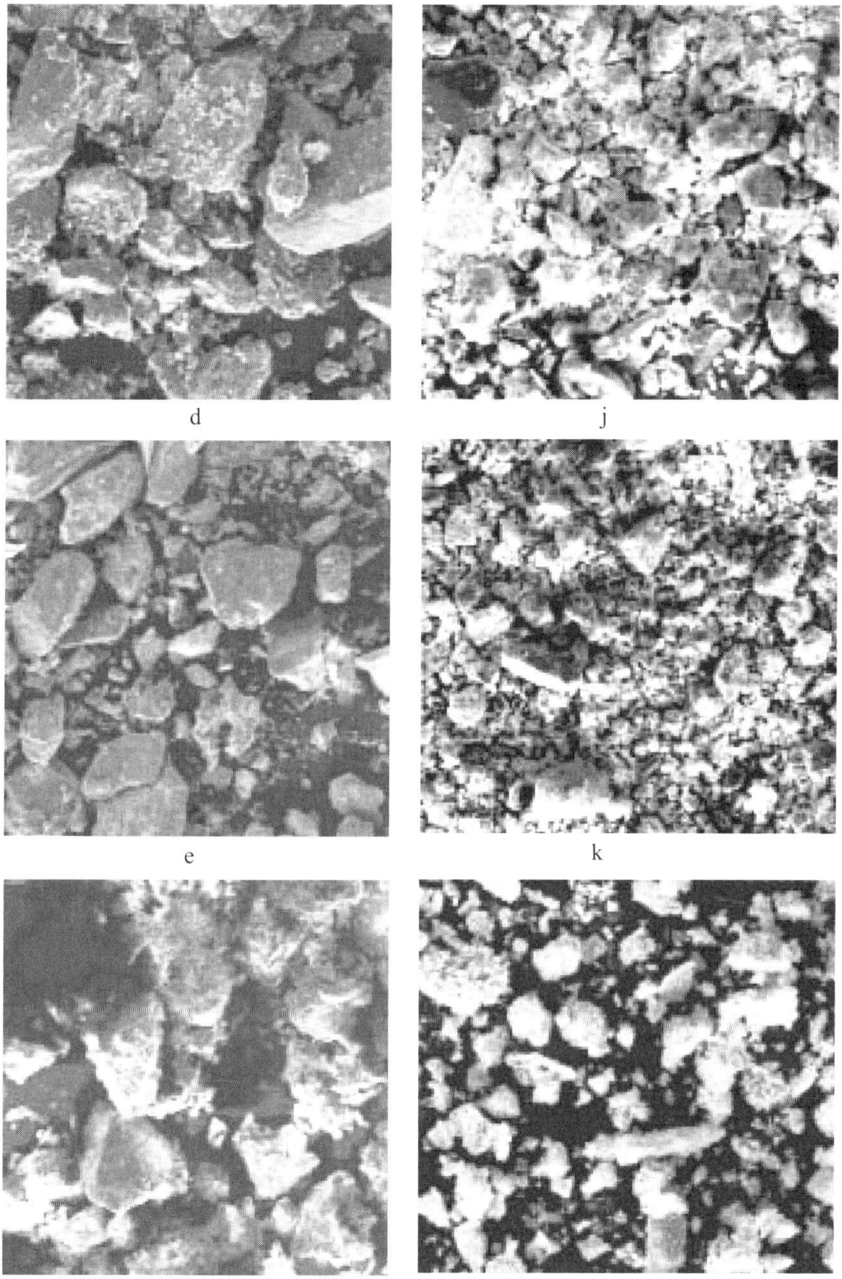

d j e k f l

Fig. 13.19. Micrographs of the partially leached tetrahedrite concentrate (left) and tetrahedrite calcine (right).

Left: a–f) tetrahedrite leached in 0.5 M $FeCl_3$ + 0.5 M HCl, 6 h, a) no leaching; b) 30 °C; c) 50 °C; d) 70 °C; e) 80 °C; f) 90 °C.

Right: g–l: tetrahedrite calcine leached in 0.5 M $FeCl_3$ + 0.5 M HCl, 6 h. g) no leaching; h) 30 °C; i) 50 °C; j) 70 °C; k) 80 °C; l) 88 °C.

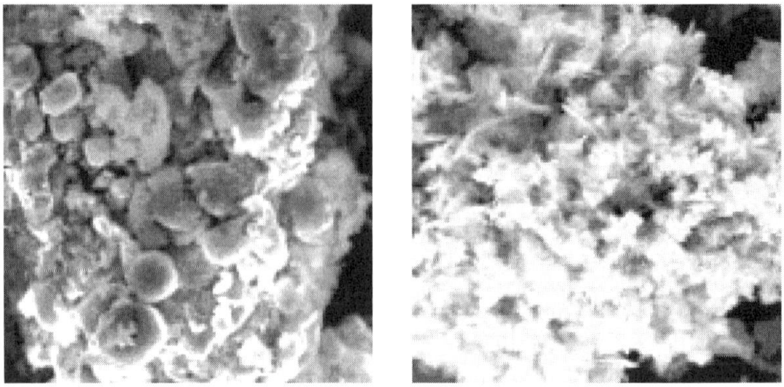

a b

Fig. 13.20. Detailed view of the surface of a leached tetrahedrite particle at 90 °C, coated with elemental sulphur; a) tetrahedrite concentrate; b) tetrahedrite calcine.

References

1. Haver F.P., Baker R.D., Wong M.M.: Improvements in ferric chloride leaching of chalcopyrite concentrates, US Bur. Mines, RI 8007, 1975.
2. Paynter J.C.: *Journal of South Afr. Inst. Min. Metall.*, 74, 4, 1974, 158–170.
3. Milner E.F.G., Peters E., Swinkels G.M, Vizsolyi A.I.: Copper hydrometallurgy, US Patent 3.798.026, 1974.
4. Dalton R.F., Diaz G., Price R., Zunkel A.D.: The Cuprex metal extraction process: Pilot plant experience and economics of a chloride-based process for the recovery of copper from sulphide ores, Proc. Copper 91-Cobre 91, vol II., 1991, 61–69.
5. Dalton R.F., Diaz G., Price R., Zunkel A.D.: *Journal of Metals*, August 1991, 51–56.
6. Craigen W.J.S, Krysa B.D., Brown H.H., Barlin B.: Evaluation of the Great Central Mines (GCM) hydrometallurgical copper process for application to Hudson Bay Mining and Smelting (HBMS) copper concentrates, In: Projects '88, Metall. Soc. Can. Inst. Min. Metall., Montreal, 1988, P7.
7. Fletcher A.W., Sudderth R.B., Olafson S.M.: *Journal of Metals*, 49, 9, 1991, 57–59.
8. Kruesi P.R.: Process for the recovery of metals from sulfide ores through electrolytic dissociation of the sulfides, US Patent 3,673,061, 27.6.1972.
9. Atwood G.E., Curtis C.H.: Hydrometallurgical process for the production of copper, US Patent 3.785.944, 1974.
10. Ravi B.P., Rajasekha A.M., Somasekar B.: $FeCl_3$ leaching of chalcopyrite concentrate of Chitradurga, Karnataka State, Bull. Electrochemistry, 3, 3, 1987, 241–244.
11. Dutrizac J.E.: *Hydrometallurgy*, 23, 1990, 153–176 .
12. Corriou J.P., Kikindai T.: *Journal of Inorganic and Nuclear Chemistry*, 43, 1981, 9–15.
13. Hirato T., Kinoshite M., Awakura Y., Majima H.: *Metallurgical Transactions*

B., 17B, March 1986, 19–28.
14. Majima H., Awakura Y., Hirato T., Tanaka T.: *Can. Metall. Quarterly*, 24, 4, 1985, 283–291.
15. Dutrizac J.E.: *Met. Transactions*, 20B, 1989, 475–483.
16. Lotens J.P., Wesker E.: *Hydrometallurgy*, 18, 1987, 39–54.
17. Havlík T., Kammel R.: Behavior of sulfur during leaching of sulfidic mineral, EPD Congres 1996, Warren G.W. ed, Anaheim, TMS, 1996, 843-857.
18. Balá• P.: Extractive Metallurgy of Activated Minerals, Elsevier, Amsterdam, 2000.
19. Dutrizac J.E.: *Hydrometallurgy,* 23, 1990, 153–176.
20. Dutrizac J.E.: *Met. Trans.*, 20B, 1989, 475–483.
21. Šulek K., Havlík T.: *Acta Met. Slovaca*, 4, 1, 1998, 49-54.
22. Havlík T., Šulek K., Brianin J., Kammel R.: *Metall*, 52, 10-11, 1998, 624–627.
23. Havlík T., Ivanová Z., Dvoršíková J., Kammel R.: *Metall,* 53, 7–8, 1999, 390–394 .
24. Dvoršíková J., Šulek K., Gubric P., Laco T.: *Acta Metallurgica Slovaca*, 2, 2, 1996, 103–114.

Chapter 14

STUDY OF THE FINE STRUCTURE

To ensure the high efficiency of the hydrometallurgical process from the viewpoint of the optimum yield, the purity of the product and also minimal energy consumption, it is necessary to control continuously the key areas of the process. All the areas of existence of solid semi-products and products of the process can be studied efficiently by X-ray diffraction (XRD) phase analysis because of the extremely high information content of the method. In addition to the basic information on the qualitative and/or quantitative content of the individual phases in the examined mixture, X-ray diffraction phase analysis provides information on the actual structure of the investigated materials, the possible existence of metastable components, kinetics of mechanisms of processes taking place, the texture of the products or the grain size and anisotropy of precipitated metals, etc. This type of information forms a suitable basis for the unambiguous explanation of these processes and offers direct control and optimisation of these processes, especially using advanced diffractometers with automatic control and evaluation and efficient collection of information, for example, by means of a position-sensitive detector.

Of course, no method is self-sufficient and, consequently, also examination of the fine structure in metallurgy has its limitations. Generally, the hydrometallurgical process is characterised by the formation of a large number of semi-products with greatly differing properties. This is given by the fact that the energy profile of the kinetic trajectory in pyrometallurgical processes is far more steeper than in the hydrometallurgical processes so that the relatively small differences between the potentials of the individual phases have no significant effect. The small energy requirement and, consequently, higher ecological efficiency of the hydro-metallurgical processes are determined by the flat profile of the kinetic trajectory and the differences in the potentials of the individual phases are controlling in this case.

The most important problem of examination of the fine structure in hydrometallurgical processes is the formation of a large number of compounds often characterised by a high degree of non-stoichiometry. The resultant semi-products and products are often in the hydrated state with an unknown content of crystalline water. Another problem is the frequent formation of metastable and nonequilibrium phases which may in real-time transform in the solid-state with the formation of further compounds with similar properties. The hydrometallurgical processes are very often characterised by the formation of amorphous or semi-amorphous phases (precipitates) with the non-defined fraction of the amorphous component which can, however, recrystallise in real-time in some cases. One of the important problems is that these components form a multicomponent mixture and the apparently 'simple' qualitative or quantitative diffraction phase analysis acquires a qualitatively new dimension.

As already mentioned, leaching does not take place in jumps but in real time. Consequently, the leaching process is defined by the leaching kinetics and mechanism. The kinetics defines the amount of the reacted compound per unit time whereas the mechanism indicates the manner in which this takes place. A very frequent phenomenon in leaching is the formation of a solid reaction product at the interface and/or permanent depletion of the surface layers of the leached material in one of the basic components. Since the reaction semi-product or product forms the 'thick' layer, X-ray radiation cannot penetrate through this layer and the resultant information describes only the surface condition. On the other hand, the reaction product seldom covers the leached surface by an absolutely compact layer so that we can obtain useful information from various areas. The morphology and grain size of these intermediate layers are also interesting.

It may be necessary to say that the leaching of minerals, especially more complex materials, takes place in fact in the black box system and in reality we only guess the leaching mechanism. Information is obtained at entry and exit but, generally, no information is obtained from the process itself characterised, for example, by the already mentioned formation of reaction products and coating of the interface and, consequently, by diffusion and/or local depletion of the concentration of the leaching agent resulting in a different local reaction mechanism. The process may be accompanied by internal diffusion of certain components of the leached minerals in the direction to the leached surface thus increasing the stresses and causing the subsequent formation of cracks and pores through which the reaction products diffuse, but the effect of capillary forces also results in penetration of the leaching agent into the pores, their local exhaustion and subsequent inhibition of the reaction. In addition, as a result of previous preparation of the charge (milling), the structure contains more or less cumulated defects resulting from mechanical or mechanochemical effects. The effect of the formation of microgalvanic cells in the leaching process and the role of these cells in electron transfer are also often discussed.

All these difficulties reflect the real structure of the components of the process. Elemental analysis provides only a relatively small amount of information and is basically restricted to the quantitative determination of the yield of metals transferred into the solution. Despite these difficulties, only the examination of the fine structure by sophisticated methods may provide information on the mechanism of the hydrometallurgical process. The main method used at the present time are the X-ray diffraction methods, with neutron and electron radiation used less frequently. However, taking into account the complicated situation, complementary methods, such as electron microscopy, thermogravimetry, differential thermal analysis, infrared spectroscopy, etc, are also used.

14.1. Examples of the application of X-ray diffractometry in hydrometallurgy

Chalcopyrite, $CuFe_2$, is the most common copper mineral in sulphidic copper ores. As already mentioned several times, the principal products of leaching chalcopyrite are the soluble ions of copper and iron and elemental sulphur. The resultant sulphur covers the leached surface and hinders the diffusion of the individual leaching reagents to and from the leached surface. This results in the often contradicting interpretation of the kinetics of mechanisms of leaching of chalcopyrite.

Study of the fine structure of chalcopyrite in the process of leaching provides, at first sight, a relatively small amount of information. Chalcopyrite is highly resistant to processing as a result of its tetragonal, close-packed structure and does not undergo any obvious phase transformation in the leaching process and, consequently, the previously published results of X-ray diffraction phase analysis for chalcopyrite in leaching have been restricted to commenting on the presence or absence of elemental sulphur on the leached surface. Generally, the results of leaching of chalcopyrite were described as follows: copper and iron are leached proportionally to their content from chalcopyrite, and no kinetic relationships of iron leaching have been published. This is understandable, because in the great majority of published studies leaching was carried out using the ferric ion as the oxidant. Since the amount of iron the solution is several times greater than the absolute amount of iron leached from the sample, it is obviously not possible to investigate these and no attempts have been made. However, different situation exists in the case of using ozone as the oxidant in sulphuric acid solutions. The total amount of iron in the leaching medium originates from the leached specimens so that it is possible to investigate the leaching kinetics. The results show that the amount of iron leached from chalcopyrite is greater than that of copper, as indicated by Table 14.1 and Fig. 14.1.

This would mean that there must be certain changes in the leached chalcopyrite, but no special attention has been paid to this. The first exact confirmation of this behaviour must be logically the stress. Indeed, microscopic examination has confirmed the presence of large cracks on the leached surface as a result of stresses, Fig. 14.2 [2, 3].

Table 14.1. The amount of leached copper and iron from chalcopyrite [1]

Leaching time [min]	Leaching temperature 22 °C		
	Amount of Cu leached [%]	Amount of Fe leached [%]	Δ[%]
0.5	3,52	1.32	2.2
1	6.87	2.87	–4
3	15.1	11.52	–3.58
5	22.59	21.37	–1.22
10	35.4	39.48	+4.08
15	46.43	52.18	+5.75
20	55.23	62.29	+7.06
25	62.7	70.83	+8.13
30	62.9	78.3	+15.4

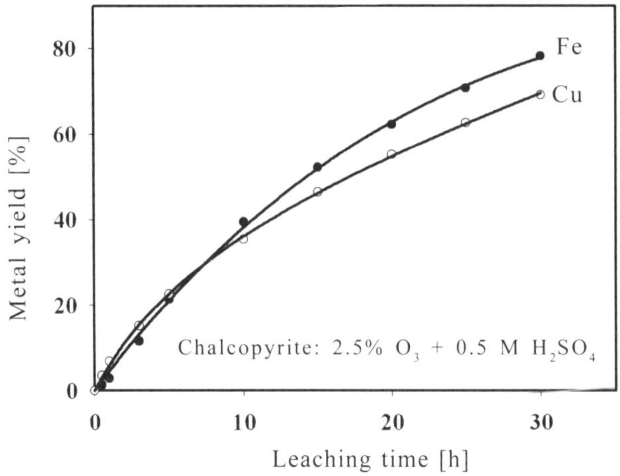

Fig. 14.1. The kinetic curves of leaching of copper and iron from chalcopyrite according to Table 14.1.

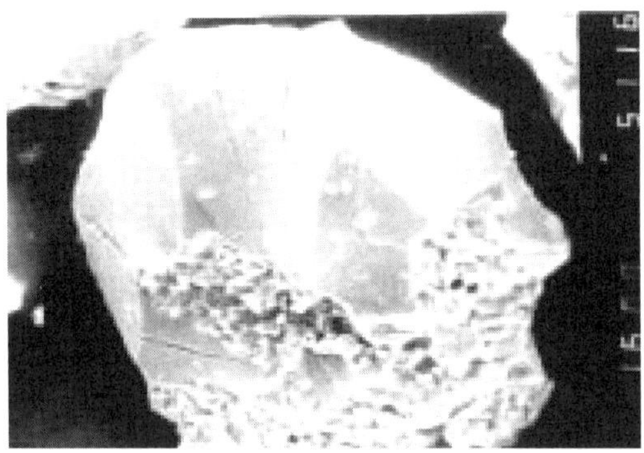

Fig. 14.2. The micrographs of partially leached chalcopyrite using ozone. The grain contains a large number of cracks.

On the basis of these observations it may be concluded that the problem is quite the reverse: how is it possible that phenomenological measurements indicate a shortage of copper or iron in the structure of chalcopyrite and its diffraction pattern does not change, as indicated by the conclusions obtained in the studies published so far. Examination of the fine structure of partially leached chalcopyrite may confirm whether this is the case.

Samples of partially leached chalcopyrite in the chloride or sulphate medium were subjected to X-ray diffraction phase analysis, and the results indicate that leaching is accompanied by the formation of elemental sulphur, with the highest intensity of the process observed at leaching temperatures above 60 °C. The leaching of chalcopyrite is not accompanied by the cxplicit formation of some intermediate product. However, the precision determination of the structure parameters of partially leached chalcopyrite shows that the leaching process is accompanied by changes in the structural parameters of chalcopyrite depending on leaching temperature. This is clearly indicated by the dependence of the ratio of the structure parameters *c*/*a* to conversion of chalcopyrite at different temperatures, Tables 14.2 and Fig. 14.3. The dependence of the parameters *c* and *a* in relation to the conversion of chalcopyrite indicates that as the parameter *a* increases, parameter c decreases, in the manner identical with the behaviour of the ratio *c*/*a*. A peculiar type of distortion takes place in this case.

The situation in leaching using ozone and in microwave leaching is somewhat different. In this case, the structure is also distorted, as indicated by Table 14.3 and Fig. 14.4, but in the reverse direction relating to temperature. The explanation is simple: the degree of deformation follows the variation of the amount of leached copper and iron. In this case, the increase of temperature reduces the efficiency of leaching as a result of a decrease of the solubility of ozone in the solution.

It should also be mentioned that despite the fact that the changes of the structural parameters are very small, and could be regarded as the scatter of measurements, it nevertheless appears that these relationships are real. This is confirmed by both relationships,

Table 14.2. The changes in the ratio of structural parameters *c*/*a* in relation to temperature for chloride and sulphate leaching media [1]

$FeCl_3$ + HCl		$Fe_2(SO_4)_3 + H_2SO_4$	
t [°C]	*c*/*a*	t [°C]	*c*/*a*
22	1.9680895	3.5	1.970518
40	1.9692186	22	1.969864
50	1.9690750	40	1.969854
60	1.9681049	50	1.970094
70	1.9640318	55	1.968762
80	1.9595708	60	1.967725
90	1.9510115	70	1.967383
		80	1.965819

Table 14.3. The variation of the structural parameters of leached chalcopyrite using ozone [1]

Leaching medium: O_3+ H_2SO_4		Leaching medium: Microwave radiation + $FeCl_3$ + HCl	
t [°C]	*c/a*	Microwave radiation [min]	*c/a*
3.5	1.9685795	1	1.970875
15	1.9689806	3	1.970626
20	1.9693550	5	1.970398
30	1.9697305	10	1.970369
50	1.9698416	15	1.970005
60	1.9699571	30	1.970101
75	1.97	45	1.971006
		60	1.970129

following the trend of the temperature dependence of leaching independently measured on different sets of the specimens of different nature.

The changes of the structural parameters of partially leached chalcopyrite depend directly on the explicit manifestations of the leaching time which also corresponds to the change of the structure of leached chalcopyrite. The amount of sulphur was estimated from the values of the intensity of diffraction of sulphur and chalcopyrite from the diffraction patterns of the partially leached samples. Although these changes are not very dramatic because they are expressed by the values of the change of the parameters of the structure in thousandths of a nanometre, the trend is evident and is in agreement with the data obtained in previous studies [3, 4]. In addition to this, this relationship correlates directly with the indicated change of the reaction mechanism of leaching of chalcopyrite [1, 4].

From the viewpoint of X-ray diffraction phase analysis, measurements of the partially leached specimens of chalcopyrite are made more difficult by the presence of elemental sulphur. Although the sulphur can be partially removed to some extent by rinsing, for example, in carbon disulphide, processing of the

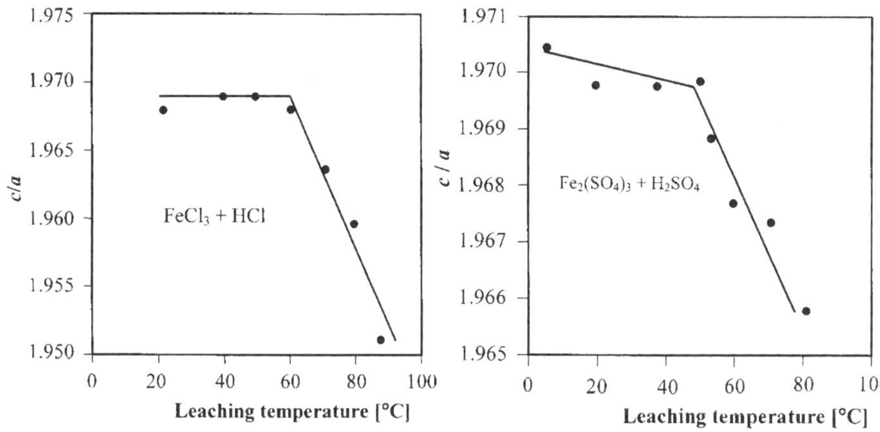

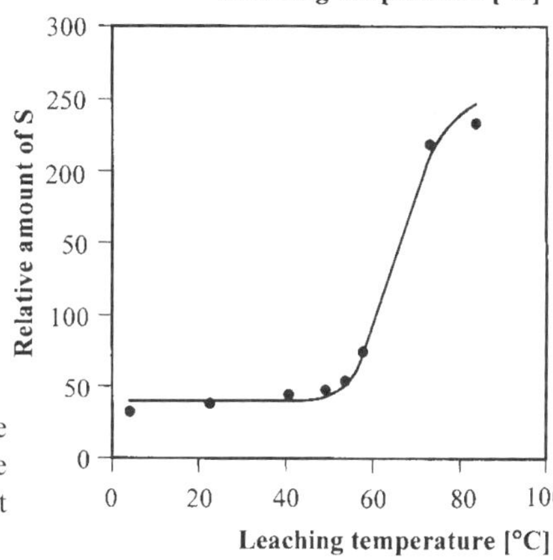

Fig. 14.3. Temperature dependence of the change of the structural parameter in relation to the conversion of chalcopyrite. the relative amount of sulphur (below).

specimen of this type results in suppression of these changes. However, X-ray diffraction phase analysis can be carried out efficiently on specimens of partially leached chalcopyrite using ozone not only because of the complete absence of elemental sulphur but also because of the very long leaching time and the high degree of conversion of copper and iron.

Therefore, it is necessary to answer the following fundamental question: why the deformation of the diffraction pattern of partially leached chalcopyrite is not more 'dramatic' when the assumed deformation of the structure is expected and measured in accordance with Table 14.1 and Fig. 14.1. These results show that after a sufficiently long leaching time (after 30 hours) at 22 °C, the conversion of copper is 69% and that of of iron 78%, i.e., the difference is approximately 9% in favour of iron. This is already a relatively large difference and results unavoidably in the enrichment of the chalcopyrite matrix with copper.

However, it is also necessary to answer the question whether the

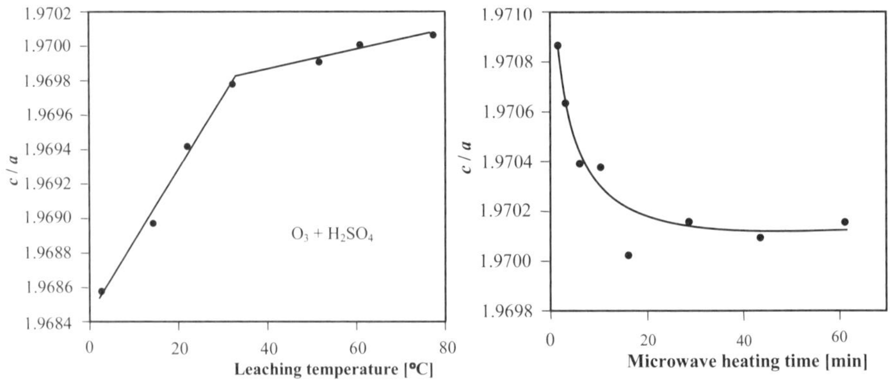

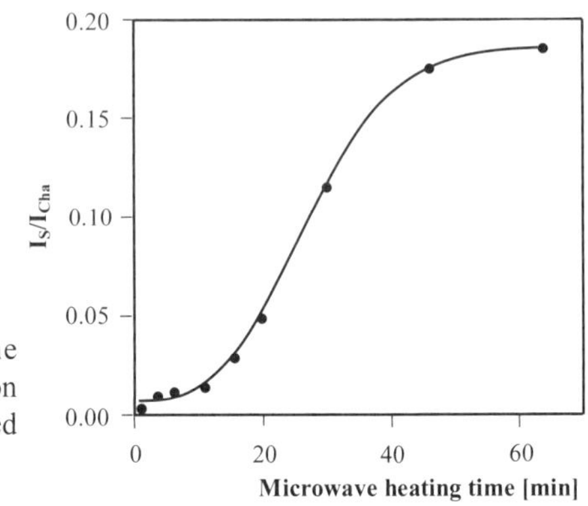

Fig. 14.4. Dependence of the change of the structural parameter of the degree of conversion of chalcopyrite; the relative amount of produced sulphur (below).

measured values are a manifestation of a measurement error or, after all, the presence of the interface product covellite (without any iron) on the leached surface. If chalcopyrite contains structural defects, this should be reflected in its diffraction patterns if the measurements accurate.

If some of the constructional particles of copper, iron and sulphur leave the structure disproportionately, this should have some effect, although a small one, on the measured diffraction pattern. At first sight, the diffraction pattern of partly leached chalcopyrite show the retreat of chalcopyrite peaks and also peaks of other substances due to its quantitative volume [5, 6]. The number of the effects, already mentioned, almost completely prevents the exact analysis of the obtained diffraction pattern. However, satisfactory possibilities could be obtained from the theoretical considerations assuming knowledge of the structure of the investigated substances. This could be achieved by the following considerations [7]:

The integral intensity of diffraction of any plane (*hkl*) may be expressed as follows:

$$I_{\mathrm{obs}(hkl)} = s\, m_{(hkl)} LPG_{(hkl)} \left| F_{(hkl)} \right|^2 \qquad (14.1)$$

where s is the universal scaling factor for all diffracting planes, $m_{(hkl)}$ is the frequency factor for the given plane (hkl), $LPG(hkl)$ is the Lorentz polarisation factor for the plane (hkl) and $|F(hkl)|^2$ is the square of the absolute value of the structural factor of the plane (hkl).

Since the values of $m(hkl)$ and $LPG(hkl)$ are *a priori* known in the equation (14.1), if the Miller indexes h, k, l and planar spacing $d(hkl)$ are available, the previous equation can be transformed to the form:

$$\left| F_{\mathrm{obs}(hkl)} \right|^2 = s \left| F_{(hkl)} \right|^2 \qquad (14.2)$$

and $|F\ (hkl)|^2$ may be expressed as follows:

$$\left| F_{(hkl)} \right|^2 = A^2_{(hkl)} + B^2_{(hkl)} \qquad (14.3)$$

Here $A_{(hkl)}$ is the real and $B_{(hkl)}$ is the imaginary part of the structural factor, and:

$$\mathrm{A}_{(\mathrm{hkl})} = \sum \mathrm{e_i f}_{i(\mathrm{hkl})} - 2\mathrm{B_i} \frac{\sin^2\theta}{\lambda^2} \cos\left(2\pi(\mathrm{hx_i} + \mathrm{ky_i} + \mathrm{lz_i})\right) \qquad (14.4)$$

$$B_{(\mathrm{hkl})} = \sum e_{\mathrm{i}} \mathrm{f}_{i(\mathrm{hkl})} - 2\mathrm{B_i} \frac{\sin^2\theta}{\lambda^2} \sin\left(2\pi(\mathrm{hx_i} + \mathrm{ky_i} + \mathrm{lz_i})\right) \qquad (14.5)$$

where e_1 is the occupation factor of the i-th atom in the given crystallographic position; $f_{i(hkl)}$ is the scatter factor of the i-th atom, B_i is the isotropic temperature factor of the i-th atom, λ is the wavelength of the X-ray radiation used, and x_i, y_i, z_i are the fraction coordinates of the position of the i-th atom in the elementary cell.

On the basis of these mathematical considerations, it is possible to predict the value of the non-scaled integral intensity of any plane (hkl), providing the position of the atoms in the elementary cell, and the temperature and occupation factors of the atoms (the scatter factors of the atoms are tabulated values) are known. At the same time, these mathematical facilities make it possible to analyse the reasons to explain if there is a difference between the detected and predicted values of the integral intensities. However, it should be stressed that in addition to structural changes (the change of the position of the atoms in the elementary cell, the change of the occupation factors for non-stoichiometric compounds, etc), this agreement between the observed and predicted intensities is caused mainly by the measurement error and the effect of the real structure and of preferential orientation. In the majority of cases, the possible errors of measurement and the effect of preferential orientation can be eliminated by careful preparation of the samples for measurement, or they can be corrected to the extent at which they can be ignored. Therefore, the effects of the real structure, including non-stoichiometry, become the main reason.

In this case, the possible changes in the form of the diffraction patterns under the effect of the changes in the structure of chalcopyrite were determined using the method of modelling and predicting the form of the diffraction diagram with the simulation of the shortage of the individual constructional particles in the structure by computer simulation [8] enabling the calculation of theoretical diffraction patterns on the basis of structural characteristics of the individual simulated substances.

Figure 14.5*a–d* show simulation diffraction patterns of chalcopyrite from the lattice of which individual atoms of copper or iron were taken away resulting in large changes of the form of the diffraction patterns of these non-stoichiometric compounds which would enable these changes to be also recorded accurately in reality. However, the situation is more complicated; it must be remembered that both iron and copper atoms are transferred simultaneously into the solution and after long-term leaching there were differences in the amount of leached iron and copper of approximately 10% in favour of iron. This condition corresponds to the modelled diffraction pattern, Fig. 14.5*d*. The potential changes should be evident mainly on the diffractions of (101), (004), (220) and (312) crystallographic planes.

Comparison of the simulated and actual diffraction patterns of partially leached chalcopyrite in a chloride medium and 95°C, Fig. 14.6*a*, indicates the presence of manifestations predicted by simulation. In addition, these manifestations are not observed in the specimens of initial chalcopyrite, Fig. 14.6*b*.

It appears that X-ray diffraction phase analysis may be used to prove disproportionate leaching of chalcopyrite. From the theoretical viewpoint, this is obviously a positive result; previous studies

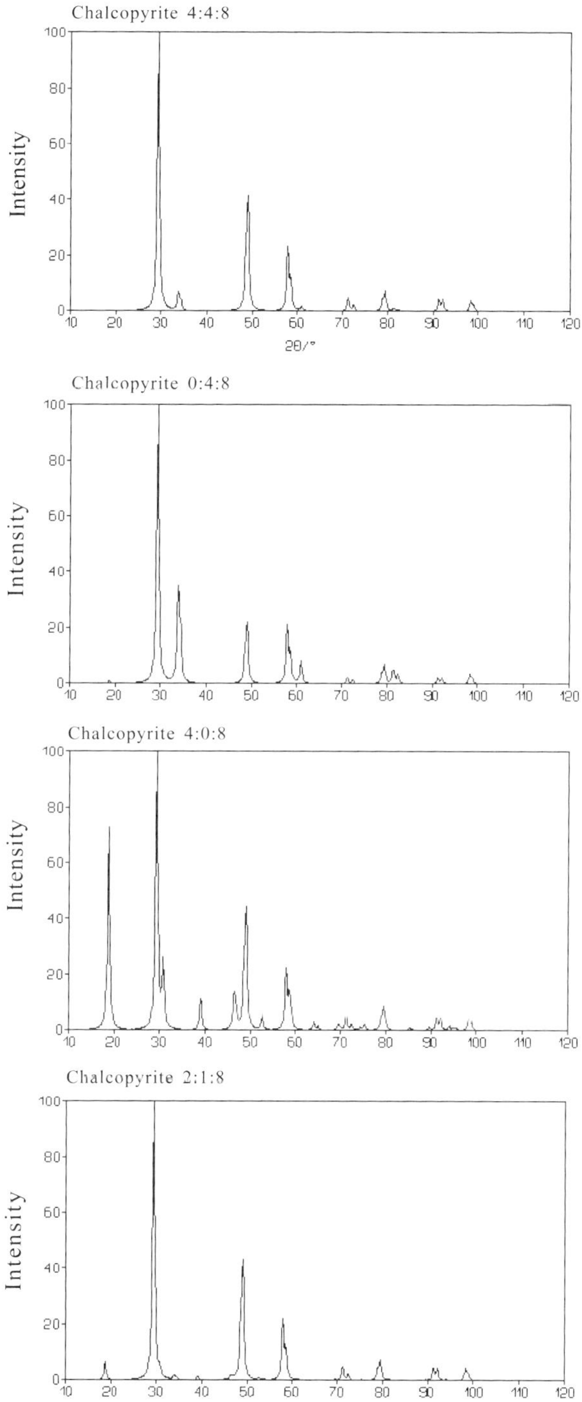

Fig. 14.5. Simulated diffraction patterns of partially leached chalcopyrite.

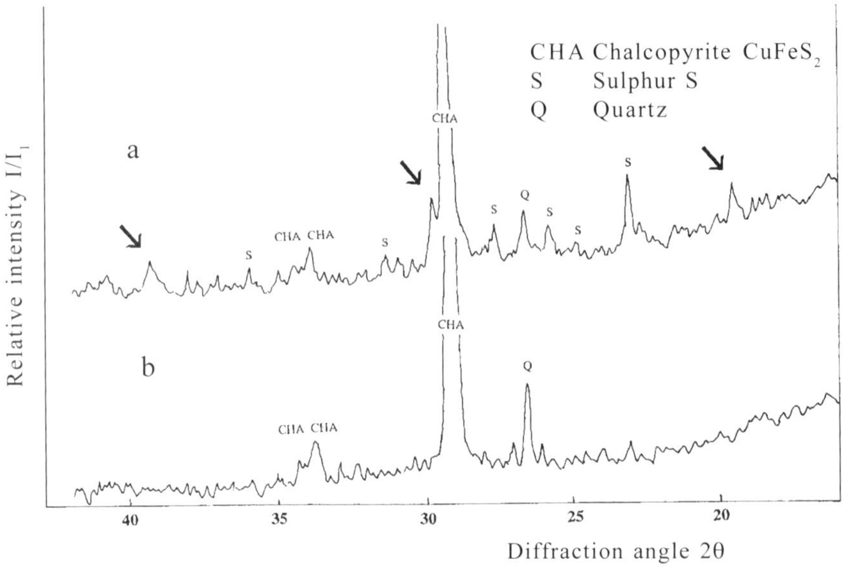

Fig. 14.6. Part of the actual diffraction pattern: a) partially leached chalcopyrite; b) initial specimen.

bypassed this problem and did not pay any attention to it. This was caused by the complicated nature of the problem and the insufficient number of suitable tools. The method described here may be used to model in advance the diffraction patterns of predicted non-stoichiometric substances and this may be used to compare the course of the process and support it by the facts regarding the mechanism. Until now, these processes have been considered in the hypothetical form or on the basis of phenomenological measurements, enabling only the evaluation of the explicit behaviour of the substances.

Here it is useful to mention other practical importance of this procedure. The sulphides and, therefore, also chalcopyrite, act as carriers of precious metals in polycomponent concentrates. At present, ore mines are being closed and only high profitability may keep them open. The content of precious metals in these materials is not as low as it appears, although extraction of these metals is relatively demanding and, consequently, costly. Every decrease of the cost of production of initial materials has a strong effect up to the level of national economy. If it is possible to predict the position of a specific admixture in the structure and estimate the moment at which the admixture leaves the structure during processing, this

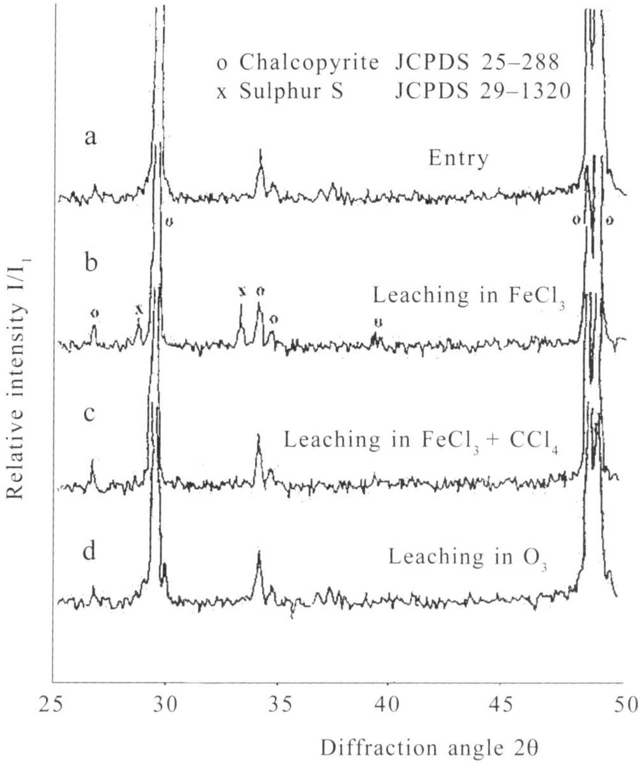

Fig. 14.7. Diffraction patterns. a) initial chalcopyrite; b) partially leached chalcopyrite in $FeCl_3$; c) partially leached chalcopyrite in $FeCl_3$ + CCl_4; d) leaching using ozone.

would help to considerably modify technology and direct and secondary expenditure on extraction of precious metals. The considerations and experiments presented here indicate that it is possible.

Elemental sulphur, as an important product of leaching, may be identified by X-ray diffraction phase analysis. If the specimens of chalcopyrite, leached in a chloride medium without CCl_4, show, on the leached surface, elemental sulphur as a leaching product (Fig. 12.5), then after leaching in the presence of CCl_4 (Fig. 12.12) no elemental sulphur was detected, as confirmed by X-ray diffraction phase analysis. The diffraction patterns of chalcopyrite in Fig. 14.7 indicate that after leaching in the solution of ferric chloride without carbon tetrachloride, the partially leached sample contains elemental sulphur, Fig. 14.7*b*, whereas Fig. 14.7*c*, showing the diffraction pattern of partially leached chalcopyrite in the presence of CCl_4, and Fig. 14.7*d*, showing the diffraction pattern of partially leached chalcopyrite by means of ozone, and Fig. 14.7a (the initial sample of chalcopyrite) do not show the presence of any elemental sulphur [9, 10].

The comparison of the diffraction patterns of the chalcopyrite leached in the chloride (Fig. 14.8) and sulphate (Fig. 14.9) media using the RIFRAN system [11] and taking into account the calculated data for all available allotropes of elemental sulphur, gives interesting results. Whilst in the case of the chloride medium the sulphur in the leaching residue consists mainly of orthorhombic sulphur S_8 in a mixture with monoclinic sulphur S_8, in leaching in the sulphate medium S_7 is found in a mixture with S_8. This is a completely new result; until now, the published studies mentioned only the presence of S_8, which is understandable because of the absence of the standard X-ray diffraction data.

Orthorhombic sulphur α-S is the conventional, most widely encountered form stable at room temperature and atmospheric pressure, whereas monoclinic sulphur, β-S is a high-temperature form, although undercooling and the presence of certain impurities may cause it to remain in the monoclinic form for a number of weeks. Cycloheptasulphur γ-S7 transforms gradually to α-S8, although at low temperatures it is stable for several weeks. However, it is important to note that this allotrope polymerises at 45°C. In any case, the properties of the identified sulphur allotropes differ and this evidently influences the leaching process and its kinetics. These results are also confirmed by morphological studies, Fig. 12.4 and 12.9, which show clearly the differences in the morphology of elemental sulphur as the leaching product in the chloride and sulphate media.

The leaching of other copper sulphides, such as bornite Cu_5FeS_4 [12, 13] or chalcocite Cu_2S [14, 15] is, in contrast to chalcopyrite, accompanied by a distinctive solid state transformation. Leaching takes place in two stages, for example, according to the reactions:

$$Cu_2S + Fe^{3+} \rightarrow CuS + Fe^{2+} + Cu^{2+} \tag{14.6}$$

$$CuS + Fe^{3+} \rightarrow Cu^{2+} + Fe^{2+} + S^{\circ} \tag{14.7}$$

However, in this case, covellite CuS does not form suddenly and forms by means of a continuous change through a number of non-stoichiometric sulphides of the type Cu_xS, represented by compounds known as minerals: chalcocite (Cu_2S), djurleite ($Cu_{1.97}S$), digenite (Cu_9S_5), anilite (Cu_7S_4), geerite ($Cu_{1.6}S$), spionokopite ($Cu_{39}S_{28}$),

R I F R A N version 2.01 (c) 1992 by FPW	OUTPUT OF ANALYSIS

Date : 04-17-1996	Time : 11:22

S A M P L E

Comment : Sira na luh. zvysku TA18 - CHA eliminovany
Lambda : 1.54059 [A']
Number of bigest peaks : 5
Agreement in peak(s) : 3
Error-Window : 3.0
Matching range : < 9 , 45 >
Stop after : 200
Active PDF : C:\R-T-G\RIFRAN\RIFDATA\DRSC
Total standards : 359

Number of found : 3

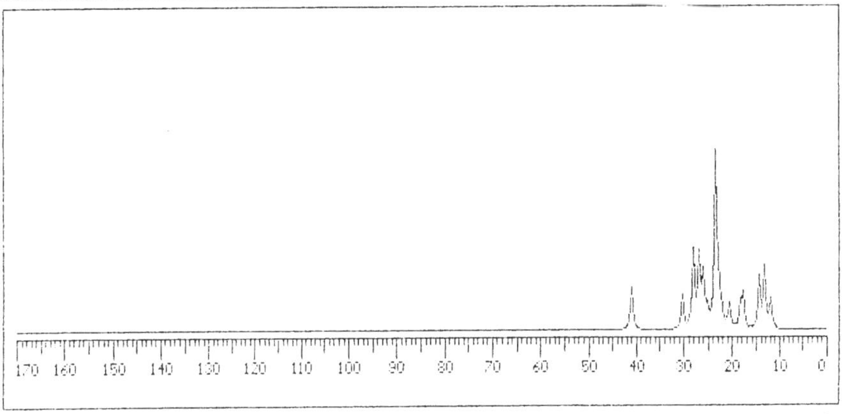

n	d	2theta	Irel	n	d	2theta	Irel
1	7.655	11.550	18	9	3.847	23.100	100
2	6.805	13.000	35	10	3.552	25.050	10
3	6.189	14.300	30	11	3.444	25.850	30
4	5.021	17.650	20	12	3.336	26.700	40
5	4.884	18.150	15	13	3.207	27.800	45
6	4.350	20.400	15	14	2.967	30.100	20
7	4.037	22.000	10	15	2.212	40.750	25
8	3.917	22.680	20				

Positive standards

@ 24-733	Sulfur S	Sulfur, Syn 0.1652
@ 8-247	Sulfur S	Sulfur, Syn 0.1580
@ 96-15	Sulfur S8 (monoclinic)	- 0.0836

Fig. 14.8. The results of XRD phase analysis of partially leached chalcopyrite in a chloride medium at 90°C after 6 h using the RIFRAN system; the presence of chalcopyrite is ignored.

R I F R A N version 2.01 (c) 1992 by FPW OUTPUT OF ANALYSIS

Date : 04-17-1996 Time : 11:26

S A M P L E

Comment : Sira na luh. zvysku TB12 (marec 1996)
Lambda : 1.54059 [A']
Number of bigest peaks : 6
Agreement in peak(s) : 4
Error-Window : 2.0
Matching range : < 9 , 45 >
Stop after : 200
Active PDF : C:\R-T-G\RIFRAN\RIFDATA\DRSC
Total standards : 359

Number of found : 6

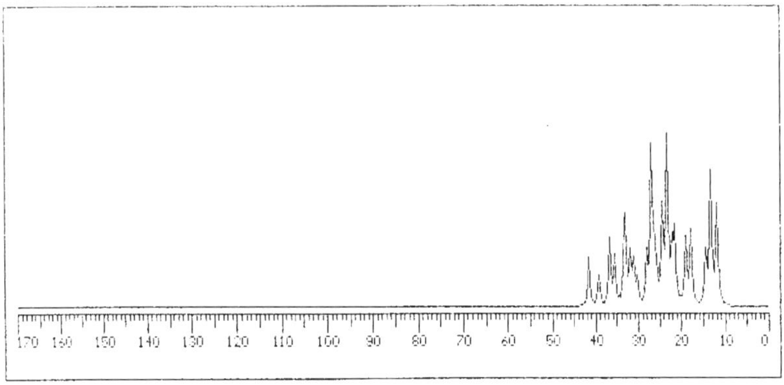

n	d	2theta	Irel		n	d	2theta	Irel
1	7.525	11.750	60		12	3.348	26.600	90
2	6.831	12.950	80		13	3.224	27.650	30
3	6.232	14.200	30		14	2.975	30.010	15
4	5.007	17.700	45		15	2.905	30.750	25
5	4.754	18.650	40		16	2.823	31.672	30
6	4.178	21.250	40		17	2.732	32.750	40
7	4.055	21.900	30		18	2.704	33.100	30
8	3.864	23.000	100		19	2.541	35.300	30
9	3.697	24.050	55		20	2.463	36.450	40
10	3.497	25.450	15		21	2.305	39.050	20
11	3.424	26.000	29		22	2.187	41.250	30

Positive standards		
@ 24-1251	Sulfur S	0.3759
@ 23-562	Sulfur (S)	0.2585
@ 96-12	Sulfur S7 (gamma)	- 0.1500
@ 24-733	Sulfur S	Sulfur, Syn 0.1133
@ 8-247	Sulfur S	Sulfur, Syn 0.0950
@ 96-5	Sulfur S13	- 0.0783

Fig. 14.9. The results of XRD analysis of partially leased chalcopyrite in a sulphate medium at 90°C after 6 h using the RIFRAN system, the presence of chalcopyrite is ignored.

yarrowite (Cu_9S_8), blue covellite ($Cu_{1.05}S$), and covellite (CuS).

During hydrometallurgical processing, leaching is accompanied by a continuous change of the structure of the leached matrix of chalcocite. Understanding the relationships governing these changes when using different types of reagent is an important step in explaining the entire process of leaching of chalcocite. It is therefore necessary not only to know and define which non-stoichiometric compounds of copper sulphide form during leaching, but also determine the kinetic parameters of the process.

Figure 14.10 shows the formation of sulphides, depleted in copper on the leached surface of chalcocite. The leaching agent was hydrogen peroxide whose high oxidation potential ensured the oxidation of resultant elemental sulphur to the sulphate (soluble) form so that the leached surface was not coated with sulphur.

Introducing the degree of conversion of leached chalcocite, which is the amount of the substance reacted with time in the manner shown in Fig. 14.8, gives the time dependence of conversion shown in Fig. 14.11.

The results of analysis of modelling kinetic equations of leaching and also the results of microscopic examination were used in determining the leaching mechanism which may be described by the model of the shrinking core using the relationship:

$$[1-(1-R)^{\frac{1}{3}} = k\ \tau \qquad (14.8)$$

as described in the chapter dealing with the kinetics of the leaching processes. The results were used to determine the temperature dependence of leaching and the apparent activation energy, which was 23 kJ/mole. This value is in good agreement with the published results, but the method makes it also possible to determine the partial values of activation energy of the formation of the individual intermediate products of leaching. The methods of chemical analysis of the amount of leached copper in the solution do not enable this to be carried out.

X-ray diffraction analysis may also be used to examine the processes of extraction of metals from the solution. From the metallurgical viewpoint, electrowinning of metals from a solution may be characterised as a reduction of the metal cation to its metallic form followed by precipitation on the cathode. This process is relatively complicated and takes place through several partial reactions. The metal precipitates at high current densities, of the

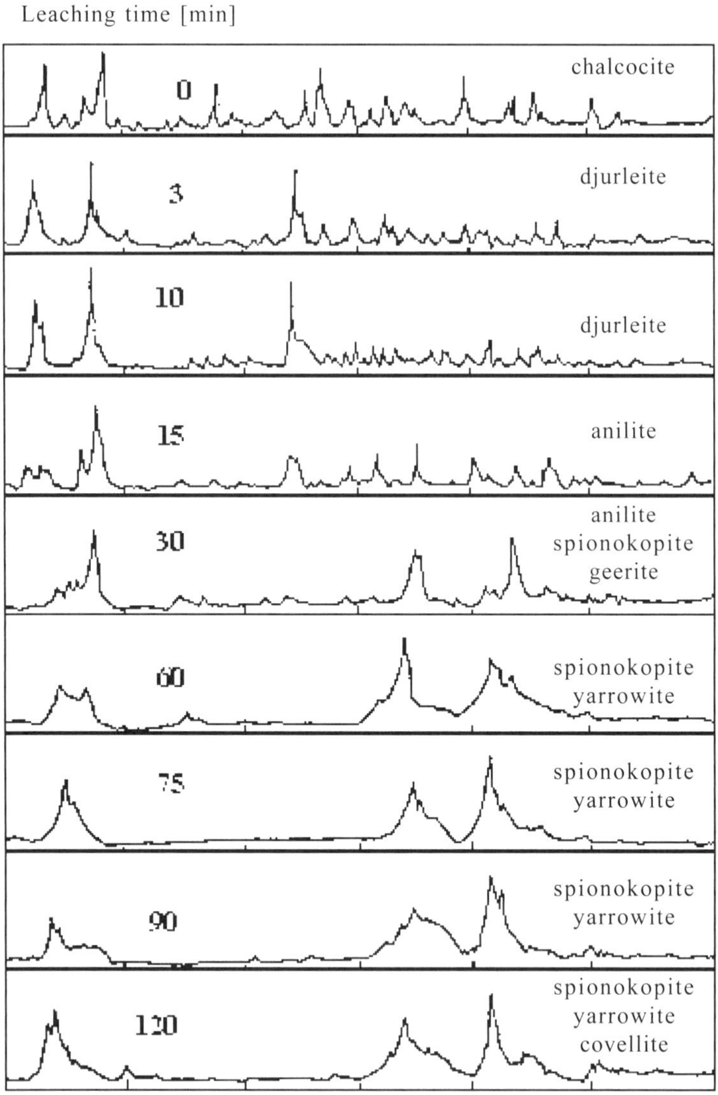

Fig. 14.10. Diffraction patterns of leaching residue of chalcocite at different stages of the process.

order of hundreds of A/m^2, in the electrode. Consequently, the entire system is far away from equilibrium. The local current and potential conditions may stimulate the formation of metastable and nonequilibrium phases whose formation would be thermodynamically unlikely under the normal conditions.

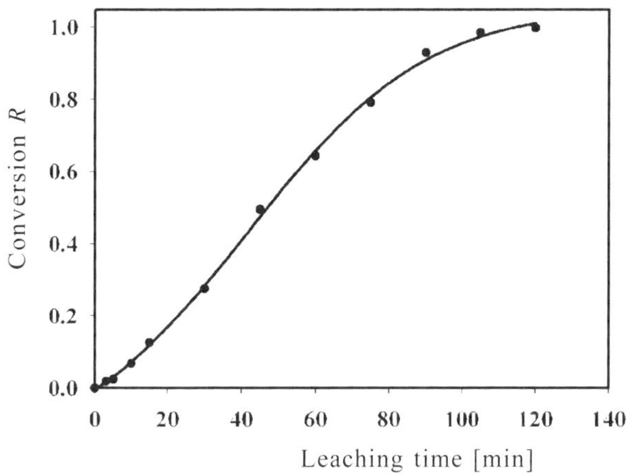

Fig. 14.11. Time dependence of the conversion of chalcocite in leaching at 60°C.

X-ray diffraction phase analysis was also used for the characterisation of resultant products in the monitoring the orientation of extracting metal and in identification of produced a binary alloys and also for the identification of metallic and the metallic products on electrodes. In examination of the effect of impurities [16, 17], present in the electrolyte, on the process of precipitation of zinc from acid sulphate electrolytes the results were used to characterise the morphology of the product in four categories, and the current efficiency and the preferential orientation of the deposit were also investigated (Table 14.4).

The degree of preferential orientation was determined using the empirical relationships:

$$I_{rel(hkl)} = \left(\frac{I/I_{0(hkl)}}{I/I_{0(101)}} \right)_{sample} \times \left(\frac{I/I_{0(101)}}{I/I_{0(hkl)}} \right)_{ASTM} \tag{14.9}$$

The angle of inclination of the deposit was determined from the empirical relationship:

$$u = \frac{\sum I/I_{0(hkl)} \cdot \alpha_{(hkl)}}{\sum I/I_{0(hkl)}} \tag{14.10}$$

where $\alpha(hkl)$ is the angle of the (hkl) plane in relation to the substrate. The increase of parameter u (the dominance of the planes with a high angle of inclination in relation to the substrate)

Table 14.4. Morphological types of the zinc deposit with the corresponding orientation and current efficiency (preferential orientation in relation to the ASTM data for zinc dust)

Morphology	Angle in relation to electrode	Preferential orientation	Current efficiency
Basal	0-30°	[002], [103], [105]	usually low
Transition	30–70°	[114], [112], [102]	usually high
Triangular	~70°	[101]	high
Vertical	70–90°	[100], [110]	high

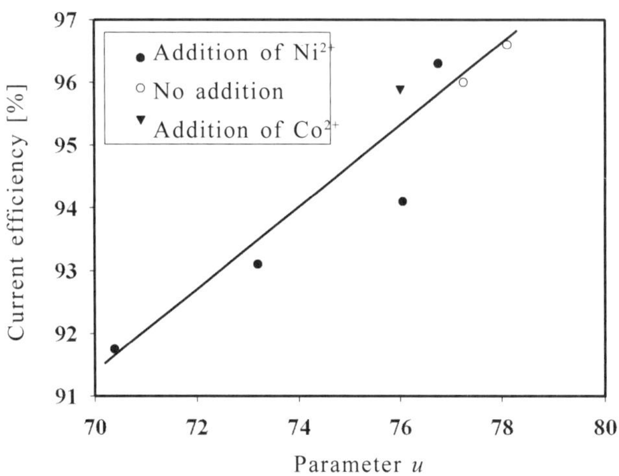

Fig. 14.12. The effect of orientation of electrolytically precipitated zinc on current efficiency.

also increases the current efficiency of the pure electrolyte as well as of the electrolytes with the additions of Ni and Co, as shown in Fig. 14.12 [18, 19].

There are only a few data on the texture of electrolytically precipitated metal in the literature, but there is a larger amount of information on the orientation of the precipitated metal than the data obtained from empirical description of relationships. For example, examination of the texture of electrolytically precipitated copper shows that the copper, precipitated at lower temperatures, is characterised by a lower sharpness of the main component of the growth texture <110>, higher stresses, the smaller size of crystals and smaller thermal scatter and a higher solidification rate in comparison with copper, precipitated and higher temperatures.

X-ray diffraction phase analysis provides a large amount of information also in the investigations of electrolytically produced alloys. It has been shown that the areas of stability of the individual phases, produced by electrowinning, greatly differ in comparison with the alloys prepared by melting. If an equilibrium phase diagram indicates the existence of several phases in the Zn–Ni system, some of the phases cannot be prepared by electrolytic procedures, for example β'. Similarly, the Cu–Cd system produced from an ethylenediamine electrolyte at room temperature shows the presence of a solid solution with up to 22 at.% of cadmium, although the equilibrium composition of the phase is supposed to be only 0.3 at.%. The resultant non-equilibrium electrolytically precipitated phase broke down after annealing into a mixture of $Cu_2Cr+\alpha$-phase in accordance with the binary Cu–Cd phase diagram. Therefore, it appears that the method of X-ray diffraction analysis is a suitable tool for the nondestructive analysis of the stability of the phase region of non-equilibrium states of the materials.

In addition to these applications of X-ray diffraction analysis in hydrometallurgical procedures of producing metals and alloys, the method can also be used efficiently in the determination of the size of crystals of the metals obtained by electrowinning, in the characterisation of the stress state in these metals, in monitoring of recrystallisation processes, in identification of nonmetallic fractions in the produced metal, and so on [20–24].

These examples show that in general the frequent results of hydro- and electrometallurgical processes are the products in a non-equilibrium state. In some cases, they show a tendency for phase transformations already after small changes in the conditions or after a relatively short period of time. This greatly reduces the possibility of determining the mechanism of a specific process. Although the methods of *in situ* X-ray diffraction, such as high-temperature, low temperature, and high-pressure methods have been yielded quite reliable results in the area of research using commercially available systems, *in situ* investigations of hydrometallurgical processes are rare, especially because of the demanding experimental procedure. They are based on the existence of an electrolytic cell directly on the goniometer of the X-ray diffractometer, and the electrode on which the metal precipitates is periodically rotated in the electrolyte or in the beam of diffracted X-ray radiation thus ensuring instantaneous recording of the actual state on the electrode. Another method is the application of an electrolytic cell below a foil through which diffracted radiation penetrates to the electrode on which the electrolytically precipitated metal settles [25–27]. The process of recording the diffraction pattern is not interrupted during electrolysis. This has been used to explain changes in the phase composition of PbO_2 in H_2SO_4 [28].

References

1. Havlík T.: Acid oxidation leaching of chalcopyrite and behaviour of sulphur in this process, PhD dissertation, Košice, April 1996.
2. Havlík T., Škrobian M.: *Rudy*, 37, 10, 1989, 295–302.
3. Havlík T., Škrobian M.: *Canadian Metallurgical Quarterly*, 29, 2, 1990, 133–139.
4. Havlík T., Škrobian M., Balá• P., Kammel R.: *Int. Journal of Mineral Processing*, 43, 1995, 61–72.
5. Havlík T.: X-ray diffractometric examination of the mechanism of leaching of sulphide polycomponent concentrate, Kolokvium '92, Real structure of substances, Vojtichov, May 1992, 28–35.
6. Havlík T.: Behaviour of chalcopyrite in leaching of complex tetrahedrite concentrate, Proc. Biohydrometalurgia II, BaU SAV Košice, October 1992, 84–90.
7. Havlík T., Škrobian M.: Hydrometallurgy, In: Industrial Applications of X-Ray Diffraction, Chung F.H., Smith D.K. eds., Marcel Dekker. Inc., New York.
8. Smrcok L., Weiss Z.: *Journal of Applied Crystallography*, 26, 1993, 140–141.
9. Havlik T., Kammel R.: *Minerals Engineering*, 8, 10, 1995, 1125–1134.
10. Havlik T., Kammel R.: Behavior of sulfur during leaching of sulfidic mineral, Proc. EPD Congres 1996, Warren G.W. ed., TMS 1995, 843–849.
11. Havlík T., Škrobian M., Petricko F.: *Ceramics–Silikáty*, 3, 37, 1993, 127–135.
12. Pesic B., Olson F.A.: *Hydrometallurgy*, 12, 1984, 195–215.
13. Pesic B., Olson F.A.: *Met.Trans. B*, 14B, December 1983, 577–588.
14. Fiala J., Havlik T., Skrobian M.: *Chemický prùmysl,* 9, 40/65,1990, 488–494.
15. Whiteside L.S., Goble R.J.: *Canadian Mineralogist*, 24, 1986, 247–258.
16. Ault A.R., Frazer E.J.: *Journal of Applied Electrochemistry*, 18, 1988, 583–589.
17. Ault A.R. et al: *Journal of Applied Electrochemistry*, 18, 1988, 32–37.

18. Baker R.G., Holden C.A.: *Plating and Surface Finishing*, 72, 3, 1985, 54–57.
19. Felloni L. et al: *Journal of Applied Electrochemistry*, 17, 1987, 574–582.
20. Swathirajan S.: *Journal of Electroanalytical Chemistry*, 221, 1987, 211–228.
21. MacKinnon D.J. et al: *Journal of Applied Electrochemistry*, 17, 1987, 1129–143
22. Murr L.E., Hiskey J.B.: *Met. Trans. B*, 12B, June 1981, 255–267.
23. Hall D.E.: *Plating and Surface Finishing*, 70, 11, 1983, 59–65.
24. Hsu G.F.: *Plating and Surface Finishing*, 71, 4, 1984, 52–55.
25. Lagiewka E., Budniok A.: *Surface and Coatings Technology*, 27, 1986, 57–66.
26. Lagiewka E., Pajak L.: Structure of electrodeposited Cu–Cd alloys after annealing, Materials Science Forum, 79–82, 1991, 581–588.
27. Lagiewka E., Pajak L.: Structure of electrolytic Cu–Cd alloys obtained in galvanostatic conditions, in: Diffraction Methods in Material Science, Hašek J. ed., Nova Science Publishers, Inc., New York, 1992, 201–209.
28. Herron M. E. et al: In-situ X-ray diffraction studies of lead dioxide in sulphuric acid during potential cycling, Phase Transitions, 39, 1992, 135–144.

Chapter 15

MECHANISM OF LEACHING OF COPPER SULPHIDES IN AN ACID MEDIUM

Mechanism of chalcopyrite leaching

Generally, the mechanism of chalcopyrite leaching in an acid oxidation medium may be characterised as follows:

Placing chalcopyrite in an aqueous solution with a high oxidation potential results in a reaction leading to partial formation of a sulphate:

$$CuFeS_2 + 8H_2O \rightarrow Cu^{2+} + Fe^{2+} + 2SO_4^{2-} + 16H^+ + 16e^- \qquad (15.1)$$

and, in particular, elemental sulphur

$$CuFeS_2 \rightarrow Cu^{2+} + Fe^{2+} + 2S^o + 4e^- \qquad (15.2)$$

At a lower value of the redox potential, CuS or Cu_xS can form in the solution according to

$$CuFeS_2 + 2e^- \rightarrow CuS + Fe^{2+} + S^o + 2e^- \qquad (15.3)$$

If the redox pair is formed by Fe^{3+}/Fe^{2+} the anodic reaction is as follows:

$$CuFeS_2 \rightarrow Cu^{2+} + Fe^{2+} + 2S^0 + 4e^- \qquad (15.4)$$

whilst the following reactions take place on the cathodic half-cell (in chloride media):

$$Fe^{3+} + e^- \rightarrow Fe^{2+} \qquad (15.5)$$

$$FeCl^{2+} + e^- \rightarrow Fe^{2+} + Cl^- \qquad (15.6)$$

$$FeCl_2^+ + e^- \rightarrow Fe^{2+} + 2Cl^- \qquad (15.7)$$

$$FeCl_{3(aq)} + e^- \rightarrow Fe^{2+} + 3Cl^- \qquad (15.8)$$

The rate of the process is controlled by the reaction from the leached surface and partially depends also on the concentration of the ferric ion in the solution. Above a certain level of the concentration of hydrochloric acid, ferrous chloride or the content of other chlorides, such as magnesium chlorides [1], the rate is no longer dependent. However, on the other hand, the presence of cuprous chloride increases the leaching rate, and small amounts of sulphate ions reduce this rate. At a high content of sulphate ions the leaching solution behaves as a 'slow' leaching medium of ferric sulphate. Figure 15.1 shows schematically the mechanism of leaching of chalcopyrite in ferric chloride.

If the redox pair is formed by Cu^{2+}/Cu^{+}, the main reaction is:

$$CuFeS_2 + 3\,Cu^{2+} \xrightarrow[H_2O]{NaCl} 4\,Cu^{+} + Fe^{2+} + 2\,S^0 \qquad (15.9)$$

However, reaction may not take place to the end because the previously mentioned reversed reaction does not take place but the copper sulphide forms in accordance with reaction:

$$2\,Cu^{+} + S^0 \rightarrow CuS + Cu^{2+} \qquad (15.10)$$

The high value of the Cu^{2+}/Cu^{+} ratio is supported by the high

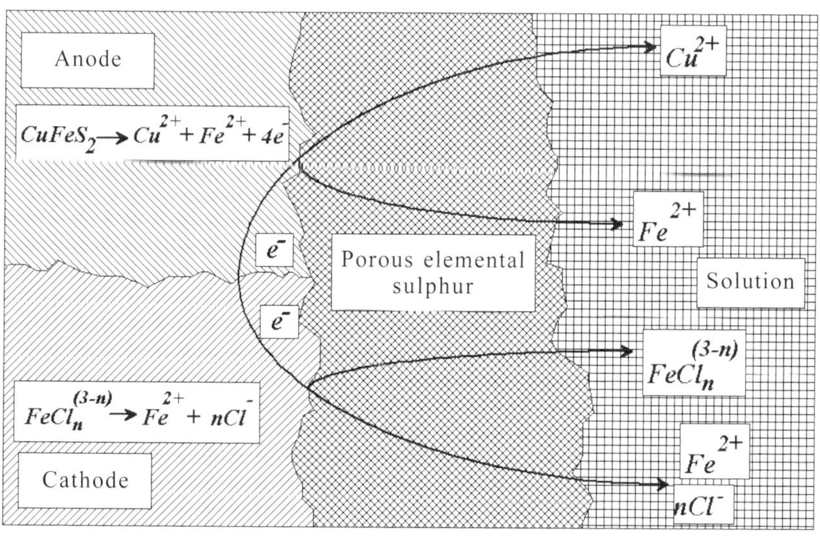

Fig.15.1 Leaching mechanism of chalcopyrite by a direct reaction with ferric chloride.

concentration of the chloride, high temperature, low pH and short leaching time. Because of the very low content of Cu^{2+} in the solution it is necessary to carry out two-stage leaching or cementing with copper scrap.

The reduction of chalcopyrite by Cu^+ to the sulphide Cu_xS or bornite Cu_5FeS_4 may take place for example as follows:

$$CuFeS_2 + 4Cu^+ \rightarrow 2Cu_2S + Fe^{2+} + Cu^{2+} \quad (15.11)$$

The reaction is fast on fresh surfaces but greatly slows down with the formation of a thin layer of Cu_xS or bornite on the leached surface. If the reaction takes place in the presence of metallic copper [2], the reaction is fast and results in the formation of a product which can be leached much more efficiently by conventional oxidants than chalcopyrite. The reaction takes place in accordance with the equation:

$$CuFeS_2 + Cu^0 + 2Cu^+ \rightarrow 2Cu_2S + Fe^{2+} \quad (15.12)$$

However, the shape of the kinetic curves of leaching of chalcopyrite in the acid solution of ferric chloride, Fig. 12.3, shows that the process takes place in at least two stages which are distinctive especially at higher temperatures.

Initially, the rate of the process is very high and a large amount of copper is leached into the solution. Gradually, the rate of the process decreases and the process becomes more or less linear. The presence of soluble copper (and iron) ions and also of elemental sulphur, as a reaction product in the solution was confirmed. Examination of the surface of partially leached chalcopyrite in the chloride medium by scanning electron microscopy (SEM) shows the presence of sulphur in the form of individual globules, Fig. 13.4.

This shows that elemental sulphur probably forms by precipitation from the solution by the following mechanism [4]: in the initial stages, the main process is the extremely fast reaction of chalcopyrite including partial steps:

$$Cu^+Fe^{3+}S_2 + 4HCl \rightarrow CuCl + FeCl_3 + 2H_2S \quad (15.13)$$

$$CuCl + FeCl_3 \rightarrow CuCl_2 + FeCl_2 \quad (15.14)$$

The hydrogen sulphide formed by the reaction (15.13) dissolves in the medium at the interface and is gradually oxidised by ferric chloride:

$$H_2S_{(aq)} + 2FeCl_3 \rightarrow 2FeCl_2 + 2HCl + S^o \qquad (15.15)$$

It is not known whether dissolution takes place by the completely indirect mechanism (reactions 15.13, 15.14) or by some combination of the indirect mechanism and direct effect of the ferric ion on chalcopyrite. Part of hydrogen sulphide oxidises up to sulphuric acid

$$H_2S_{(aq)} + 8\,FeCl_3 + 4\,H_2O \rightarrow H_2SO_4 + 8\,HCl + 8\,FeCl_2 \qquad (15.16)$$

for which there are perfect thermodynamic conditions.

At the same time, the bivalent copper from the solution is precipitated by the produced H_2S on the surface of chalcopyrite with the formation of CuS according to the equation

$$CuCl_2 + H_2S \rightarrow CuS + 2\,HCl \qquad (15.17)$$

CuS was sucessfully identified on the leached surface, Fig.13.11.

Reaction (15.16) is thermodynamically more advantageous than the formation of FeS. If FeS does form because of the large surplus of $FeCl_3$ and possible formation of $FeCl_2$ by reaction (15.15), cementing of CuS takes place in accordance with the equation:

$$FeS + CuCl_2 \rightarrow CuS + FeCl_2 \qquad (15.18)$$

As a result of this stage, the surface of chalcopyrite is gradually coated with elemental sulphur and CuS which then form a solid solution at the interface.

During the next leaching stage the leaching rate decreases. The copper leaching kinetics gradually changes but the form of the kinetic curves of leaching of copper and iron becomes similar during this phase. The amount of produced CuS is smaller because the amount of resultant H_2S decreases.

Comparison of the sudden decrease of the reaction rate after the first phase with the formation of sulphur has confirmed that the increase of the thickness of the sulphur layer at the interface does not cause any decrease of the reaction rate. It appears that the porous layer of elemental sulphur has no direct effect on the kinetics. Especially at lower temperatures, sulphur does not completely cover the leached surface not mentioning the fact that it is strongly 'attacked' as a result of constant mixing and rubbing. Diffusion of the dissolved oxidant in the solution within the sulphur pores cannot be the rate-controlling step because the activation

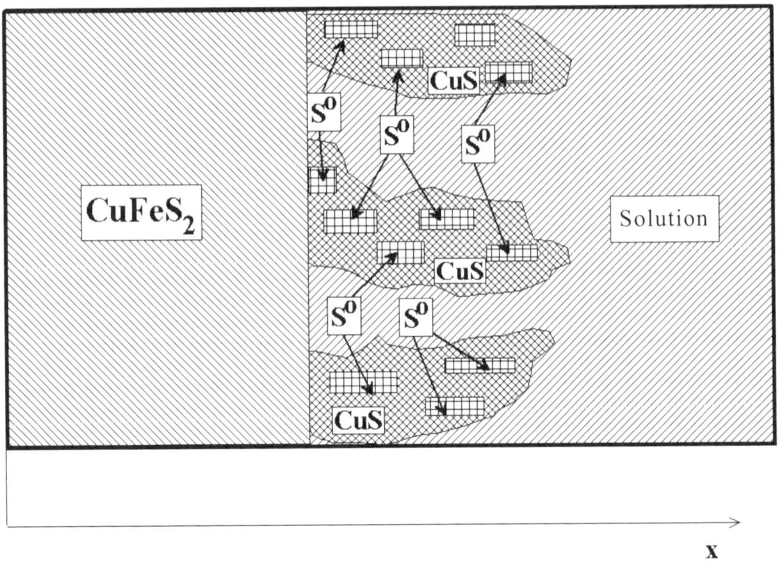

Fig.15.2 Model of leaching of chalcopyrite in the presence of CuS.

energy of the reaction when using sulphate is too high and in the case of chloride is also closer to the kinetic than diffusion region.

The precipitation of CuS, formed as an intermediate product, blocks the pores in the layer of elemental sulphur at the interface and the sulphur layer thus efficiently isolating the surface of chalcopyrite against the supply of the oxidant to the leached surface. Of course, this means that in a later phase chalcopyrite is leached only by the acid reaction. In addition, this reaction may continue only with the elimination of CuS. Since CuS remains at least partially at the interface, since CuS was identified, and the rate of the acid reaction of chalcopyrite must be unavoidably higher than that of oxidation leaching of CuS. Therefore, this reaction controls the overall leaching rate. Also, on reaching the steady state, the amount of CuS decreases in comparison with the initial amount in the starting phase. This is limited by the volume of the pores of the produced elemental sulphur.

In the stabilised conditions, two difference reactions take place at the interface, Fig. 15.2, i.e. the acid reaction at the chalcopyrite-solution interface and chemical oxidation of CuS at the solid–solution interface. Consequently, elemental sulphur forms exclusively by the oxidation CuS which plays the role of an intermediate compound between H_2O and S^0. In this stage, sulphide ions are oxidised not only from chalcopyrite but also from CuS. The measured activation energy values (~ 50 kJ/mol [3]), are in agreement with the chemical reaction of oxidation of CuS.

This mechanism explains better the observed morphology of sulphur than surface migration of sulphur. It can be used to explain the behaviour of chalcopyrite despite the fact that the result sulphur layer is porous and relatively pervious and the form of sulphur precipitated on the surface and also the incomplete extraction of sulphur in elemental form may be explained. To explain stoichiometric anomalies of the initial stage of oxidation and the differences in the leaching of copper and iron it is not necessary to consider the hypothetical mechanism of diffusion in the solid state. The precipitation of CuS which has been confirmed describes these phenomena quite accurately.

The electrochemical mechanism of leaching of chalcopyrite is based on the transfer of electrons and holes in the region of the conductive, forbidden and valency bands of the semiconductors. Since the dissolution current of the contribution of holes from the oxidant Fe^{3+} is greater than that of the contribution of the electrons on Cu^{2+} to the solution, it is assumed that the primary stage of dissolution of chalcopyrite is controlled by the contribution of the holes. The chalcopyrite ion Fe^{3+} is linked with the conductivity band and the chalcopyrite ion Cu^{2+} with the valency band, and therefore Fe^{3+} is released electropositively into the solution and Cu^{+} remains after the initiating step. After initial transfer of holes *h*, electron transfer continues according to the reaction

$$CuFeS_2 + 3h^+ \rightarrow Fe^{3+} + \bullet CuS_2 \tag{15.19}$$

followed by the transfer of holes

$$\bullet CuS_2 + 2h^+ \rightarrow Cu^{2+} + 2S \tag{15.20}$$

or electron transfer

$$\bullet CuS_2 \rightarrow Cu^{2+} + 2S + 2e^- \tag{15.21}$$

The first step illustrates breaking of the Fe–S bond and the second one breaking of the Cu–S bond. $\bullet CuS_2$ is an unstable transition radical. This mechanism shows that copper remains in the lattice because it is bonded with the valency band. This is in agreement with the observations that the Fe/Cu ratio in the solution is higher than 1 in the initial leaching stage.

The transition radical $\bullet CuS_2$ may break down into stable transition CuS according to:

$$\bullet CuS_2 \rightarrow CuS + S \quad (15.22)$$

Biegler and Swift published [5] cyclic voltamograms of pure natural chalcopyrite showing a high anodic effect potentials higher than 1.0 V. Prior to these effects there were less marked effects at 0.8 V interpreted as the formation of a layer of copper sulphide.

Leaching mechanism of chalcocite

The leaching method of chalcocite in an acid oxidation medium may be described as follows:

Placing Cu_2S in an aqueous solution with a high oxidation potential results in the anodic reaction:

$$Cu_2S \rightarrow CuS + Cu^{2+} + 2\,e^- \quad (15.23)$$

with the formation of CuS on the Cu_2S surface.

If oxygen is the oxidant, the cathodic reaction consumes hydrogen ions and oxygen at the surface according to:

$$2\,H^+ + 1/2\,O_2 + 2\,e^- \rightarrow H_2O \quad (15.24)$$

If the cathodic pair includes ferric ions, then the following phenomenon takes place without any consumption of hydrogen ions:

$$2\,Fe^{3+} + 2\,e^- \rightarrow 2\,Fe^{2+} \quad (15.25)$$

If the kinetics is sufficiently fast and there are sufficiently strong oxidation conditions and sufficiently high temperatures, polarisation of the surface is sufficient for the joint formation of CuS and S^0. Sulphur forms in accordance with the reaction:

$$CuS \rightarrow Cu^{2+} + S^o + 2\,e^- \quad (15.26)$$

leading to the position of a sulphur layer on the surface of CuS.

Figure 15.3 shows schematically these reactions: reaction (15.23) takes place on interface I with the formation of a porous layer of CuS which enables the diffusion of cuprous ions and electrons to the interface II with the oxidant (O_2 or Fe^{3+}) reduced by the reactions (15.24) and (15.25). The potential is stabilised by the reaction (15.23). After the entire amount of Cu_2S has reacted, the

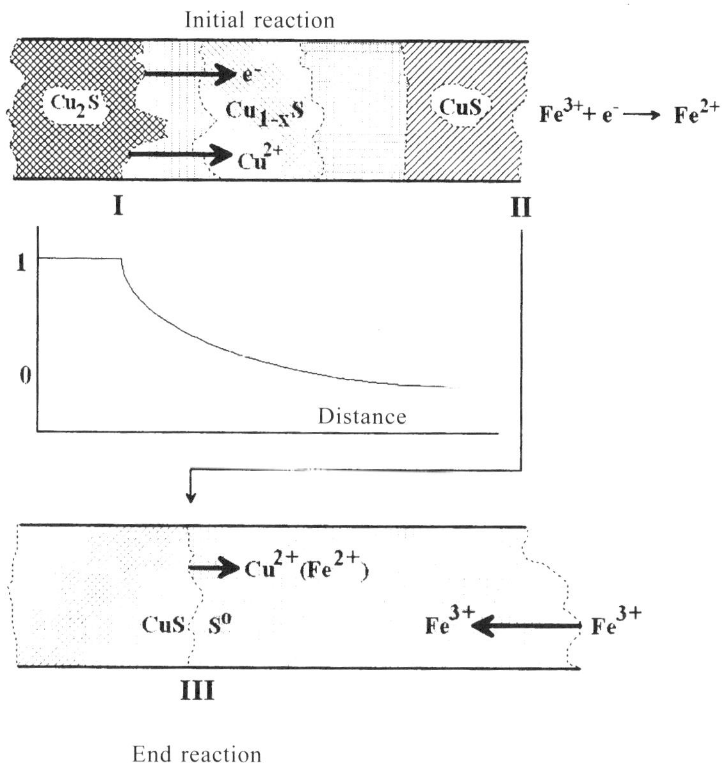

Fig.15.3 Schematic representation of leaching of chalcocite.

potential will decrease and a non-conducting sulphur layer will form. The oxidant must diffuse through the sulphur layer and this results in a large decrease of the reaction rate. Equation (15.26) represents the reaction at interface III.

The process takes place in two stages, the first fast one through a number of non-stoichiometric compounds of the type Cu_xS with a low value of the activation energy indicated that the process is controlled by diffusion process, mostly by diffusion of reactants through the layer of the reaction product – elemental sulphur. The rate of the second stage is considerably lower and is controlled by the chemical or electrochemical reaction at the reaction surface. The processes differ in the rates of the individual stages in relation to temperature and pressure and also the oxidation potential in the solution. The oxidation potential is secured using some suitable agent, mostly the ferric ion. The use of other oxidation agents with high oxidation potentials may accelerate the dissolution kinetics or, if the oxidation potential in the solution exceeds +0.44 V, the sulphur may be oxidised to a higher (other than zero valent) stage. This will have a strong effect on accelerating the leaching process because the inhibiting effect of diffusion of reactants through the elemental sulphur layer, coating the leached surface, is eliminated.

Mechanism of bacterial leaching [6]

The view that the bio-oxidation of sulphide minerals also includes oxidation of the mineral by trivalent iron is receiving more and more support amongst the investigators. This oxidation takes place in accordance with the relationship:

$$MeS_x + Fe^{3+} \rightarrow Me^{x+} + Fe^{2+} + S(?) \qquad (15.27)$$

In addition to the transfer of metal ions into the solution, bivalent iron and part of elemental sulphur also form. They form a substrate for the growth of bacteria in accordance with the reactions:

$$4Fe^{2+} + O_2 + 4H^+ \rightarrow 4Fe^{3+} + 2H_2O \qquad (15.28)$$

$$S(?) + O_2 \rightarrow SO_4^{2-} \qquad (15.29)$$

The transfer of bivalent iron to the trivalent form probably takes place in the extra-cell layer formed by polysaccharides. This outer layer is formed by a micro-organism bonded with the sulphide surface.

Thiosulphate mechanism (FeS_2, MoS_2, WS_2)

$$FeS_2 + 6Fe^{3+} + H_2O \rightarrow S_2O_3^{2-} + 7Fe^{2+} + 6H^+ \qquad (15.30)$$

$$S_2O_3^{2-} + 8Fe^{3+} + 5H_2O \rightarrow 2SO_4^{2-} + 8Fe^{2+} + 10H^+ \qquad (15.31)$$

This mechanism, alternative oxidation of thiosulphate according to:

$$2Fe^{3+} + 2S_2O_3^{2-} \leftrightarrow 2FeS_2O_3^{+} \rightarrow 2Fe^{2+} + S_4O_6^{2-} \qquad (15.32)$$

is also possible. The ions containing sulphur may transform, after a series of complex reactions, to elemental sulphur.

Polysulphide mechanism (ZnS, $CuFeS_2$, $FeAsS$, MnS, etc.)

$$MeS + Fe^{3+} + H^{+} \rightarrow Me^{2+} + 0.5H_2S_n + Fe^{2+} \quad (15.33)$$

$$0.5H_2S_n + Fe^{3+} \rightarrow 0.125S_8^o + Fe^{2+} + H^{+} \quad (15.34)$$

$$0.125S_8^o + 1.5O_2 + H_2O \rightarrow SO_4^{2-} + 2H^{+} \quad (15.35)$$

The schematic representation of both bio-leaching mechanisms is shown in Fig. 15.4.

The sulphide minerals are attacked either by ferric ions of ferric ions and protons. The role of the micro-organisms is to regenerate the ferric ion and maintain a sufficiently high redox potential for the reaction to take place in oxidised sulphur and, consequently, form suitable conditions for reducing pH. In other words, the role of bacteria is to produce ferric ions and protons consumed in leaching reactions. This takes place probably in the extra-cell polysaccharide layer surrounding the micro-organism bonded with the surface of the mineral.

In bioleaching of sulphides, an important role is played by the part of the bacteria bonded to the solid substrate containing sulphur. These bacteria are in a special situation because in addition to being able to oxidise iron in the solution, they may use the solid sulphide

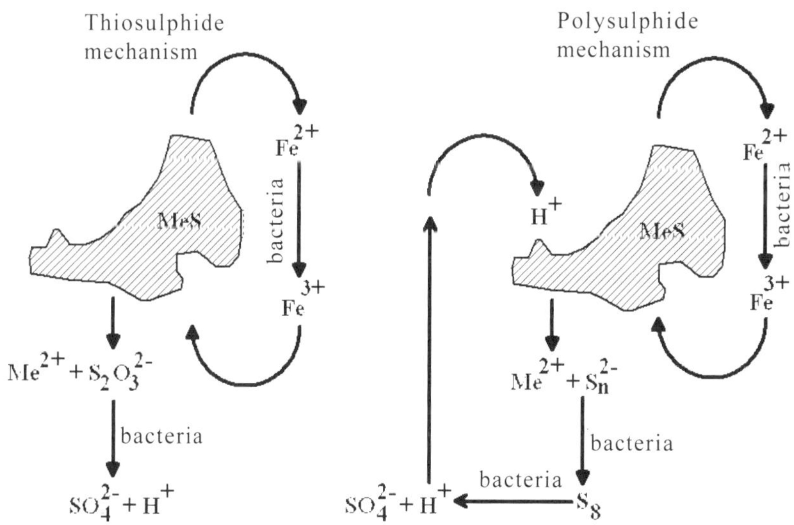

Fig.15.4 Schematic representation of bioleaching mechanisms.

as an alternative energy source.

In normal aerobic conditions, the bacteria bonded on the surface layer may use the bivalent iron ion in the solution as an energy substrate (reaction 15.28). In this case, the bacteria receives energy for its growth from the transfer of electrons from Fe^{2+} to O_2. The bonded bacteria may also receive energy by direct oxidation of the solid sulphide substrate. It may be concluded that the bonded bacteria in a privileged position for oxidation of primary sulphur compounds or any other secondary product containing sulphur which forms in accordance with the previously mentioned equations.

However, the bacteria may also grow in anaerobic conditions using the trivalent iron as an electron acceptor. The well known bacteria, used in hydrometallurgy, *Thiobacillus ferrooxidans* may grow using the redox of the Fe^{3+}/Fe^{2+} either as an electron donor or electron acceptor.

References

1. Winand R.: Hydrometallurgy, 27, 1991, 285–316.
2. Avraamides J., Muir D.M., Parker A.J.: *Hydrometallurgy*, 5, 1980, 325–336.
3. Havlík T.: Acid oxidation leaching of chalcopyrite and behaviour of sulphur in this process, PhD dissertation, Košice, April 1996, 291.
4. Havlík T.: *Acta Metallurgica Slovaca*, Special Issue 3/1998, 62–68.
5. Biegler T., Swift D.A.: *Journal of Applied Electrochemistry*, 9, 1979, 545–551.
6. Štofko M., Štofková M.: Nonferrous metals, Emilena Košice, June 2000, s.293.

Chapter 16

THE CURRENT STATE AND PROSPECTS OF HYDROMETALLURGICAL PROCESSES

The selection of methods for the leaching of raw materials depends mainly on the value of the metal value in the raw material, the richness of the ore, the cost of extraction, the cost of processing the initial materials, such as milling, roasting, enrichment, and the leaching capacity of the material. It is likely that preference will be given to the method with the lowest capital and production costs, together with the high yield and favourable environmental aspects. In the case of copper minerals, malachite $CuCO_3{\cdot}Cu(OH)_2$ and azurite $2CuCO_3{\cdot}Cu(OH)_2$, are easily leached in diluted sulphuric acid, whereas the sulphides covellite, CuS, chalcocite Cu_2S and chalcopyrite $CuFeS_2$ are more difficult to leach and leaching requires the presence of an oxidation reagent. The most frequent mineral in sulphidic copper ore, chalcopyrite, is a highly refractory material and efficient leaching is very demanding. This also dictates the development of technological processes ensuring the optimum leaching processes in their entirety.

Table 16.1 shows the examples of general characteristics of leaching methods.

Agitation leaching

Agitation leaching is carried out using a finely ground charge, mostly because of the high reaction rate due to the large reaction interfacial area. Agitation (stirring) minimises the thickness of the diffusion layer and if the gas phase takes place in leaching, it maximises the gas–liquid phase interfacial area. When using higher pressures, the leaching temperature is higher than 100 °C resulting in the acceleration of the leaching reactions. High-temperature pressure leaching is performed in autoclaves and is the most expensive leaching method from the viewpoint of capital expenditure and production costs [2].

There are several types of autoclaves, but horizontal pressure

Table 16.1 Leaching methods [1]

Leaching method	Grain size	Characteristic rate	Cost
Agitation leaching	< 0.5 mm	90–95 % in 24 h	High capital and production costs
Pressure leaching	1–45 μm	90–95 % in 24 h	
Percolation leaching	< 10 mm	~ 80 % per week	High capital costs
Heap leaching	Debris	~ 50 % in several months	Low capital and production costs
Dump leaching	As mined	~ 50 % in 1–2 years	
In-situ leaching	Coarse crushed material	Various	

vessels are used in most cases. This type contains usually three or four sections, with each section containing its own stirrer for agitation. Figure 16.1 shows the scheme and Fig. 16.2 the overall view of an autoclave.

The pulp is introduced under pressure into one end of the autoclave and flows through the vessel from section to section. Heating is usually realised by supplying superheated steam and if the required temperature is reached, the supply of steam is interrupted during exothermic reactions. In reality, it is also necessary to control temperature by cooling and this is generally achieved by the internal circulation of water in cooling coils.

In some cases, the horizontal autoclaves are rotated in order to prevent the formation of sediments, and this method is also used for spherical pressure vessels. Other modifications include the addition of inert spheres for the removal of insoluble intermediate products on the leached surface to facilitate leaching.

Long tubular horizontal autoclaves are not divided into sections so that the pulp flows through the autoclave at a relatively high rate. These autoclaves are used mainly for the leaching of bauxite.

Continuous pressure leaching processes utilise several autoclaves in a series. This capacity to carry out pressure leaching in a continuous regime in comparison with the batch regime is a significant advantage because it eliminates inter-operational procedures and reduces the time and also the energy demand. In

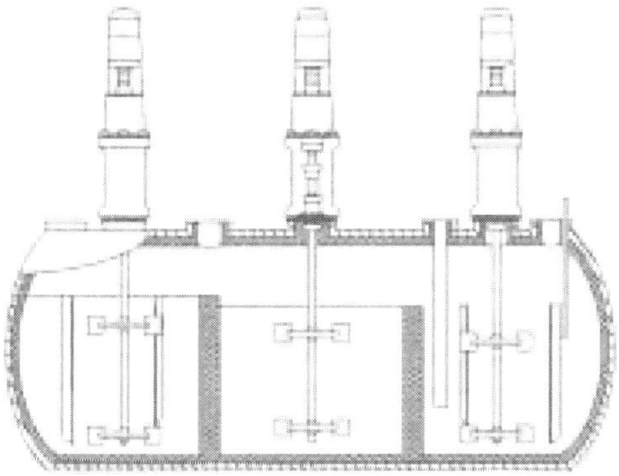

Fig. 16.1. A horizontal autoclave with three sections.

Fig. 16.2. Industrial horizontal autoclave, Adelaide, Australia.

addition, the process can be automatically controlled.

Efficient leaching of metal from ore requires chemically aggressive leaching conditions which in turn require extremely high quality materials for constructing autoclaves. These materials must be capable of withstanding long-term loading, corrosion and stress erosion. Usually, they are produced from special carbon alloyed steels and titanium structural elements may also be included. The steels are highly resistant to alkaline media but in the acid medium, especially when using chlorides, they are attacked by corrosion. Since the acid-resistant materials, such as tantalum, niobium or the already mentioned titanium, are very expensive, these autoclaves use acid-resistant lining with a lead coating [3].

Agitation leaching is used often in the normal atmospheric conditions. The reaction vessels are fitted with mechanical stirrers

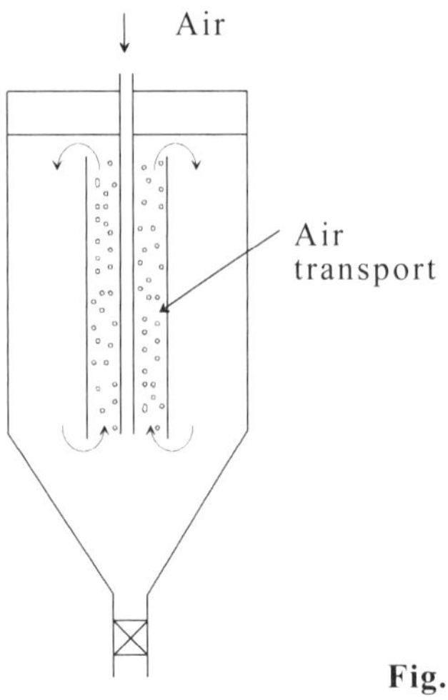

Fig. 16.3. Diagram of the Pachuca tank.

or air blowing may also be used for stirring. This method is highly efficient in the so-called Pachuca tank, Fig. 16.3.

In this type of reactor, the pulp is slowly heated with the flow of superheated steam brought together with air so that the required temperature of 70–80°C is easily obtained. Agitation leaching may be carried out in the steady metered regime or as a continuous process. The initial trends were directed to high-value small-volume processes but large capacity operations, which can easily be also automated, were preferred at a later stage. The intensity of stirring in this case is sufficient to obtain the conditions of the reactor ensuring the continuous flow and stirring of the pulp (continuous-flow stirred tank reactor). To satisfy all the required conditions of the process, it is necessary to operate several reactors in a series. After completing leaching, the solid should be separated from the liquid phase, mostly by filtration in drum rotation filters, disk or horizontal band filters. If sludge forms in the process, the process is very slow and expensive, despite adding various flocculants. An alternative of filtration may be the counterflow settling, which also includes rinsing and thickening [4].

Percolation leaching

Percolation leaching is used for relatively lean ores with a large grain size (9–12 mm), and this is also the origin of another name, sand leaching. The ore is processed in large vats or containers in which the bottom is produced from a filter from the ore through which the solution penetrates.

The vats are usually lined with lead or mastic. The capacity of the vat is approximately 10000 m^3, the depth up to 6 m. Larger containers may take up to 12 000 tonnes of ore and are filled and emptied mechanically. Small production plants use the dosing regime, but in large plants, only the semi-continuous or continuous regime is efficient. In the continuous regime, there is at least one vat outside the circuit and it is emptied to remove the leached ore and a new charge is placed in. Leaching takes place in a different container or containers. Charging and emptying last approximately 15 h and leaching approximately a week.

This method is advantageous because it does not require expensive filtration systems and is capable of producing a relatively highly concentrated solution using counterflow leaching.

Heap leaching

Heap and dump leaching processes are approximately identical although they differ in the type and amount of leached material. In heap leaching, the ore is crashed and piled up into a relatively small heaps (up to approximately 100 000 t) on an impervious base or concrete or asphalt surface with drainage channels and blow holes for the removal of the saturated solution into the collection container. The dumps are larger and consist of the mined ore and ballast, usually from surface mines. This mining waste usually contains a sufficiently large number of valuable minerals for efficient processing. The dump is periodically sprayed with a leaching solution on the surface or through perforated pipes into the dump, or with drainage channels. The leaching cycle in the case of a heap may last several months, but a dump may be leached even several years before the sufficient concentration of the solution is reached.

Both types of leaching are used extensively for mining waste and lean copper ore. The process has a long history, with origins at least in the 17th century in Rio Tinto in Spain (Table 1.2). Heap leaching was realised considerably later for efficient extraction of uranium and gold. However, since the leaching of gold requires a highly toxic cyanide solution, exceptional safety measures have been applied to these productions.

Figure 16.4 shows schematically the situation in heap and dump leaching.

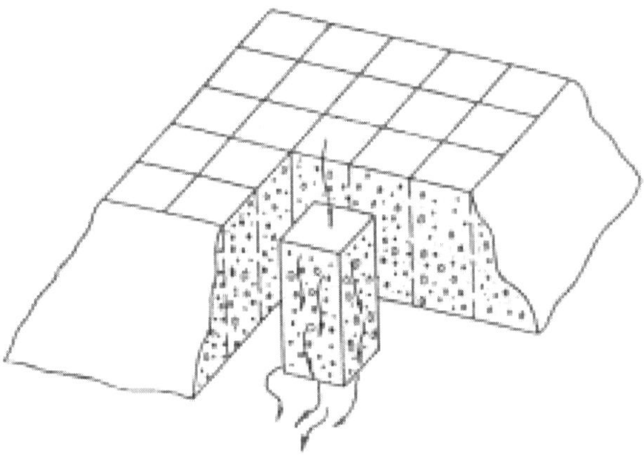

Fig. 16.4. Schematic of a leach heap.

Leaching in situ

This method is used for leaching directly in the area of the deposit prior to mining the ore. The mine should have a sufficiently impervious base to prevent the penetration of the leaching solution into underground water.

In the Miami Mine in Arizona, copper was extracted from residual lean ores after completing mining in 1954. The leaching solution penetrated 3–4 days into the ore and was collected in a central sump from where it was pumped to the surface. Copper was produced from the solution by cementation, using iron scrap [5].

In Slovakia, this method was used in the Austro-Hungarian empire in a copper mine in Smolnik in eastern Slovakia.

In general, the question of the grain size of the leached material is very important because milling is one of the most expensive operations. The empirical dependence of the grain size on the richness of the ore and the leaching method, shown in Fig. 16.5, was determined for the individual leaching methods [6].

16.1. History of hydrometallurgical processes

Although previous chapters included extensive discussions of the possibilities of increasing the efficiency of leaching processes using unconventional reagents, or by the application of new promising methods and equipment, it must be said that only several principal leaching methods are used at the present time and this is determined by the type of processed starting material. In most cases, lean oxidic and/or complex ores, or ore with a content of noble metals,

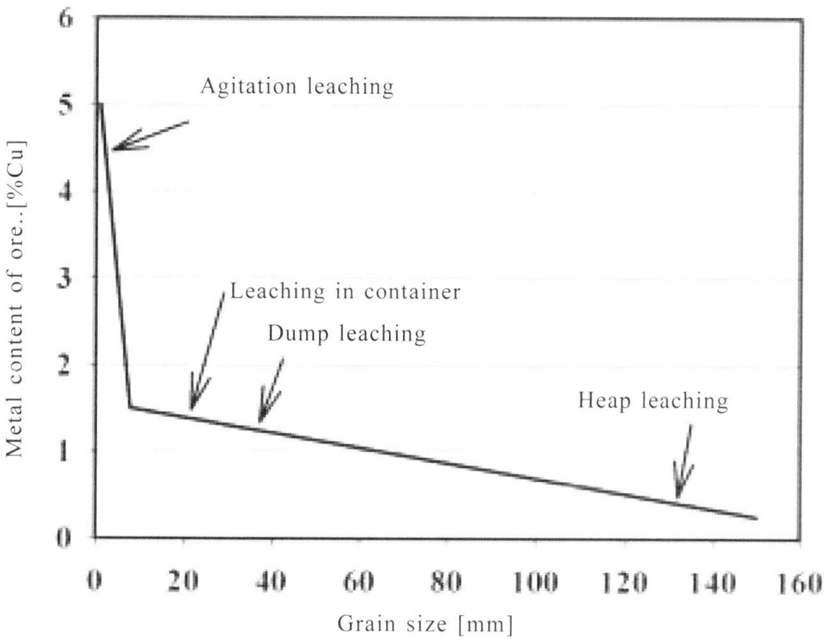

Fig. 16.5. Dependence of the grain size and the method of hydrometallurgical processing on richness of the ore.

are processed, whereas the extent of processing of sulphide ore or concentrate is smaller. Of course, an important factor is also the type of metal present in the ore; efficient technologies have been developed for zinc raw materials, but in the case of copper the situation is more complicated. Generally, oxide ores are processed in most cases. As regards the sulphides, covellite and chalcocite ores are utilised here. There are no sufficiently efficient technologies for the concentrates, and the largest challenge to hydrometallurgists are the chalcopyrite raw materials because of their extremely high refractoriness and also problems with the removal of iron.

Chalcopyrite is the main sulphide mineral of copper. The majority of the sulphide copper minerals also contain iron sulphides, usually pyrite, and many of them also contain economically significant amounts of gold and silver. The richness of copper in a typical sulphide ore is relatively low, usually lower than 1.0%. Prior to pyrometallurgical processing, the copper sulphides are enriched into concentrates with 20–40% of copper, depending on the mineralogy of the individual sulphide. The typical chalcopyrite–pyrite concentrate may contain approximately 26% of copper, 31% of iron, and 36% of sulphur. Pyrometallurgical practice includes the typical processes such as melting, converter treatment, casting of anodes and electrorefining to produce high-purity copper. The melting and refining processes use efficient technologies, have high energy efficiency (although this was not always the case), and high efficiency in extracting the metal, including gold and silver. Iron and the majority of impurities are efficiently transferred into stable slags. However, the melting processes are highly demanding on investment and require large capacity operations. As mentioned previously, in the processing of a typical concentrate, the production of 1 t of copper is accompanied by the formation of almost 2.8 t of SO_2 or approximately 4.2 t of sulphuric acid. In the past, the majority of SO_2 was discharged into the atmosphere, together with other pollutants. Therefore, with increasing environmental protection pressure and requirement for strict control of the emission of sulphur oxides, many companies have started to consider hydrometallurgy as an alternative to the pyrometallurgical production of copper from sulphides.

Leaching of copper from sulphide ore deposits by extraction of copper from natural waters of these deposits, was already realised in 1086 by Chinese metallurgists [7]. Similar large-scale operations enabled extraction of copper prior to 1737 from acid waters in Rio Tinto in Spain. In the 19th century, increasing interest in pure metals resulted in the application of electrolytic refining, and the method of electrowinning of metals was introduced at the beginning of the 20th century. The rapid increase of the volume of production of copper in the previous century, especially in the USA and Chile, stimulated the development of new leaching technologies and of methods of extraction of copper from aqueous solutions. In the last quarter of the 20th century, it became possible to extract metals from aqueous solutions by solvent extraction. In addition to this, work was also carried out on other types of leaching solutions, such as sulphuric acid. This development shifted the hydrometallurgical production of copper from the research to industrial level.

The earliest mention of hydrometallurgy is probably in Agricola's book De Re Metallica [8] and describes the separation of silver from gold by selective dissolution of silver and decantation of the solution into a copper vessel resulting in cementation of silver. This was used as a basis for the cementation of copper which played a significant role in the production of copper in the following centuries (Table 1.2).

The last decade of the 19th century and, in particular, the first two decades in the 20th century, mark the beginning of the era of the majority of leaching operations using oxide ore. The first method was applied in Clifton, Arizona, and included leaching in a vessel and cementation with iron. Another early production plant was Anaconda where roasting, leaching and precipitation of iron into tailings was used. The first application of electrolytic precipitation was in 1915 from a leaching solution used for oxide ores in Ajo, Arizona, and also in Chuquicamate, Chile.

Ammonia leaching was used for the first time in 1916 in Kennecott, Alaska, for tailings from gravitational separation of carbonate ores and in Calumet and Hecle, Michigan, also for tailings from gravitational separation using natural copper ores [9]. Both technologies produced cuprous oxide by boiling of ammonia from the solution using superheated steam. Ammonia leaching was also used for carbonate–silicate copper ores in Bwana M'Kubwa (Zambia) [10], with pre-reduction in order to remove silicates.

The first application of ferric sulphate in leaching in a vat with mixed sulphide–oxide ores with subsequent electrowinning of copper started in 1926 in Inspiration, Arizona [11]. Further improvement, tested in 1929 and applied commercially in 1934 [12], was the combination of leaching with flotation of mixed oxide-sulphide ores or tailings (LPF), applied for the first time to mixed ores in Miami, Arizona, and followed by many other similar production plants using this example. The first production by agitation leaching of the oxide ores followed by electrolysis was applied in Panda and in Belgian Congo (now Zaire). This method was tested in 1921 [13]. Table 1.2 in Chapter 1 shows the historical review of leaching production plants.

The Sherritt-Gordon company used commercial ammonia leaching for sulphide concentrates of nickel at elevated temperatures and pressure [14], and the procedures were also tested on zinc and copper concentrates. The Anaconda Arbiter process was introduced in the 70s of the 20th century [15–18] and its development was stimulated by several environmental restrictions regarding the emission of copper oxides.

Oxidation conditions are essential for the rapid and complete transfer of copper sulphides and, in particular, chalcopyrite. These conditions are efficiently ensured in production using ferric chloride or cuprous chloride, or a combination of both compounds. This results in a high leaching rate at moderate temperatures and pressures with a high metal yield. Pyrite is not attacked and remains in the undissolved residue together with elemental sulphur. In addition, the leaching solution may again be oxidised for recycling during electrolysis, either by anodic oxidation or by chloride, generated on the anode. Taking these potential advantages into account, special attention was paid to the development of these technologies in the last 20 years of the previous century in order to find a 'clean' alternative for smelting.

Technologies used at normal conditions of temperature and pressure

US Bureau of Mines [19] and Mintek [20] have developed a relatively simple procedure for processing chalcopyrite concentrate. Leaching is carried out in a single step in the vicinity of the boiling point in the solution of ferric chloride (at approximately 105°C). The high yield of copper is assured by milling 95% concentrate to the grain size below 1 mm and by extending the leaching time to 8 hours. The parameters of the yield of metal and the removal of iron were investigated. In the BoM process, copper is produced directly by electrolysis from the chloride solution, the Mintek method uses solvent extraction with the LIX-64N agent for transforming the cuprous chloride to sulphate for conventional electrolysis. Both processes remove iron by oxidation and hydrolytic precipitation. Fine milling is the essential first stage of both processes.

The flowsheet of the BoM process is shown in Fig. 16.6.

Cominco company [21] carried out low-capacity tests and developed a process of leaching copper concentrates in ferric chloride; 97% of the charge is milled to the grain size smaller than 1 mm. Leaching takes place in the counterflow system, in two stages, with the temperature in each stage being 95°C. Leaching for 9–12 hours transferred 99% of copper into the solution, containing 50–200 gl of $FeCl_3$. The long reaction time was probably the result of using the temperature of 95°C. As shown later, the leaching of chalcopyrite depends greatly on temperature and the increase of temperature to the boiling point (i.e.> 105°C) greatly increased the efficiency of leaching. The spent leaching solution of iron chlorides was regenerated under pressure at a temperature of 135–165°C and the surplus iron was precipitated as Fe_2O_3 in the iron oxidation state. In the Cominco process, the solution filtered in the hot state was processed using metallic copper to reduce the entire amount of copper to the single valency form. After cooling

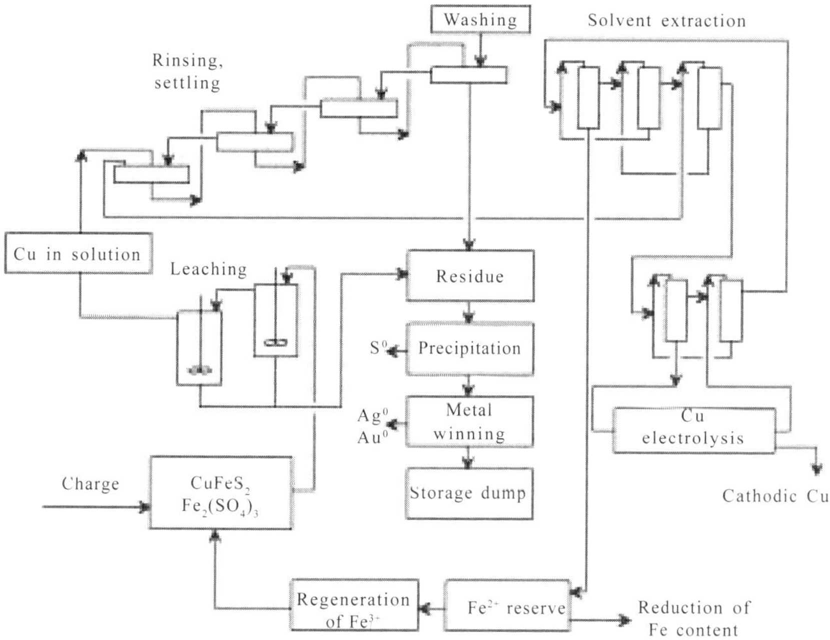

Fig. 16.6. The flowsheet of the BoM process.

of the solution, pure crystals of copper chlorides were produced. The reduction of the cuprous chloride by hydrogen during the generation of HCl in the autoclave resulted in the formation of copper powder. The Cominco process shows the possibility of direct production of high purity copper and also of reducing the cost of electrolysis.

The Sherritt-Cominco Copper process was developed by the companies Sherritt Gordon Mines Ltd and Cominco Ltd in order to compete with smelting processes [22, 23]. The special features of the process is the 98% yield of copper of the quality comparable with electrolytic copper, a comparable gain of gold and silver with smelting operations, and the yield of 90% of sulphur in elemental form, with the remainder transferred into deponable tailings. Another feature is the extraction of related metals molybdenum on zinc, and acceptable environmental parameters. It was also shown possible to proces materials with a large richness range and composition of the concentrate.

The process efficiently removes iron from the concentrate prior to leaching copper. This is achieved by thermal activation of the concentrate in which iron is transferred into the form soluble in sulphuric acid. The calcine is leached in sulphuric acid solution to remove iron. Two-stage counterflow leaching must be used for this operation. After separating the liquid and solid phases, iron is precipitated from the solution in the form of ammonium jarosite. The solid leaching residues were pressure-leached in an acid solution with the overpressure of oxygen so that copper was transferred from the sulphide to sulphate form. The yield of copper was 99% and the maximum yield of sulphur to elemental form was achieved. The residual solution contains only 10–15 g/l of H_2SO_4 thus avoiding hydrolytic precipitation of ferric ions and maintaining the iron in the form of ferrous ions for further processing of the solution. The impurities Te, As, Bi, Sn, Pb and Se were removed by co-precipitation with Fe_2O_3. Copper was produced electrolytically from purified solutions at a high current density.

This process may be used for processing a wide range of copper concentrates. The yield of copper from medium-rich and rich concentrates may be comparable with that in the melting processes and is better than in the processes of melting lean and pyrite charges.

The flowsheet of the Sherritt–Cominco processes shown in Fig. 16.7.

The Cyprus Metallurgical Corporation (CYMET) developed a process based on two-stage counterflow leaching in the acid solution $FeCl_3$–$CuCl_2$–NaCl [24, 25]. The concentrate is leached for three hours in both stages at 98°C, the yield of copper is 99%. The supersaturated solution contains 100–150 g/l Cu and 110–160 g/l of Fe. After thickening and filtration, the solution is cooled in vacuum to approximately 40°C and half of copper crystallises as CuCl. The crystalline CuCl is centrifuged, rinsed, dried and reduced with hydrogen in a fluid reactor at 500°C. Sand is added to the charge in the fluid layer in order to prevent sintering of the charge, and the resultant copper–sand mixture is subsequently remelted into copper pigs. The spent electrolyte with the content of approximately 50 g/l Cu and 110–160 g/l Fe is oxidised at 95°C with oxygen and HCl from the fluid reactor. Repeated oxidation of the solution and processing of iron by precipitation to sodium jarosite and β-FeOOH lasts three hours. The CYMET process is realised in a small continuous plant processing approximately 20 t of chalcopyrite concentrate per day. It was shown in the pilot plant that the copper from the chloride solution may be also efficiently extracted by a non-electrolysis processes. The CYMET plant was closed in 1982 and the process was not developed any further in this form.

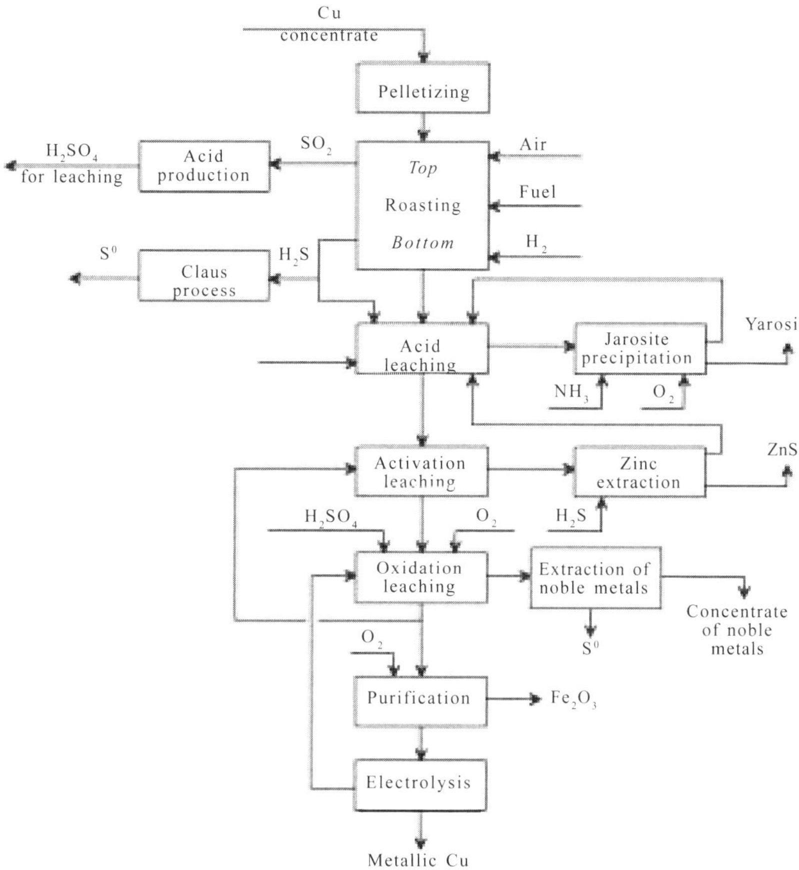

Fig. 16.7. The flowsheet of the Sherritt-Cominco process.

The flowsheet of the CYMET process is shown in Fig. 16.8.

A chloride hydrometallurgical process of processing chalcopyrite concentrates was developed by the company Duval Corporation and is known as the CLEAR process (Copper Leaching Electrowinning and Recycle), was used for 6 years and finally reached the productivity of 100 t copper per day [26–28]. The process is patented by US patent 3879272 [29]. It is based on two-stage counterflow leaching to obtain a high copper yield. The first stage operates at 105 °C with a solution with the composition 20 g/l $CuCl_2$, 4 g/$FeCl_2$, 80 g/l NaCl and 44 g/l KCl. In the development of the process it was found that the high overall concentration of the chlorides permits the use of high leaching temperatures resulting in a high yield. In the first leaching stage, metallic copper is added to the solution to ensure that the entire amount of copper is present as Cu^+ and the purified solution is sent for electrolysis. The residue from the first leaching stage is subjected to repeated leaching at 105°C at the oxygen overpressure of 330 kPa. The high-temperature and overpressure of oxygen lead to the situation in

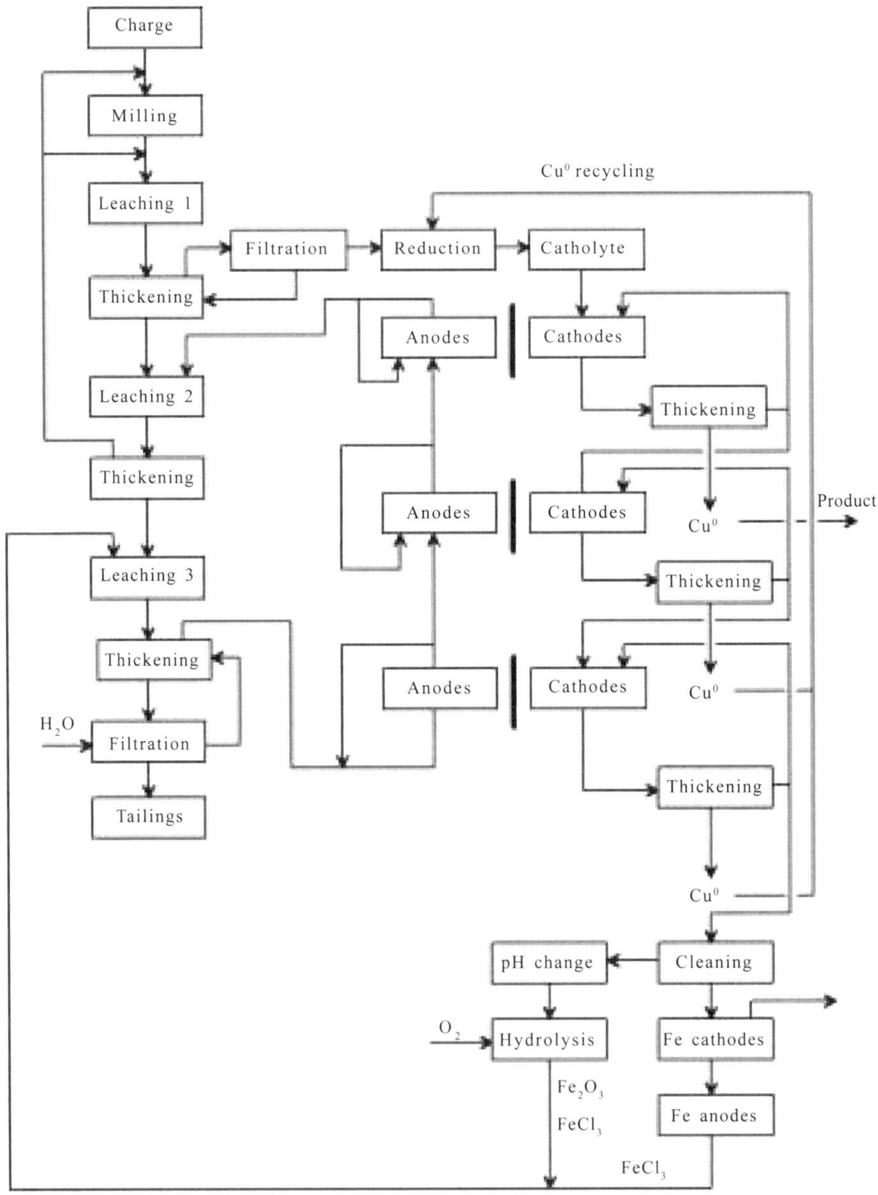

Fig. 16.8. The flowsheet of the CYMET process.

which a large portion of sulphur is oxidised to the sulphate form. The precipitation of jarosite and β-FeOOH in the second leaching stage influence the control and regulation of the presence of sulphate and iron. Metallic copper was produced from the saturated solution in the first leaching stage by diaphragm electrolysis using high current density and mechanical mixing in order to prevent the formation of copper crystals on the cathode. On the anode, half of the copper ions oxidise to cuprous chloride which is recycled in the second leaching stage.

Figure 16.9 shows the flowsheet of the CLEAR process.

The results obtained in the CLEAR pilot plant show that the process of chloride leaching makes it possible to obtain a high yield of copper from the chalcopyrite concentrates without the emission of SO_2. Although the electrolysis reactions should theoretically lead to the formation of only Cu and $CuCl_2$, part of gaseous Cl_2 was also released. This resulted in hygiene problems in the plant and also in

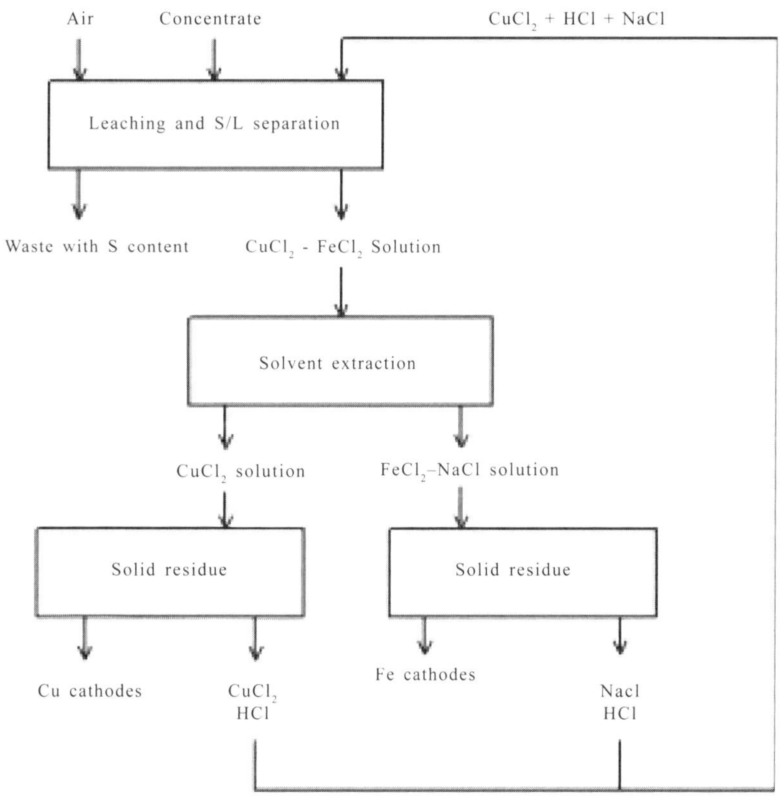

Fig. 16.9 The flowsheet of the CLEAR process.

extensive corrosion of equipment. It was shown that it is absolutely essential to ensure efficient control of the concentration of CuCl and the electrolysis parameters in order to prevent the formation of chlorine. It was also shown that the electrolytically produced copper contains the majority of silver released from the concentrate and the efforts for the selective extraction of silver were not successful. It was therefore necessary to refine copper further. The leaching residue with thc content of jarosite, FeOOH and sulphur were processed using a small amount of H_2SO_4.

The CLEAR pilot plant was the largest milestone in the development of chloride leaching technologies. Later, commercial continuous production systems demonstrated the general feasibility of chloride procedures on refractory sulphides such as chalcopyrite. The CLEAR process also showed the need for further work in solving the problem of the purification of the solution and extraction of secondary products and also better methods of controlling cathodic and anodic reactions in the course of electrolysis. The CLEAR pilot plant was closed in 1982, mainly due to a drop in the price of copper.

Minemet Recherche operated the leaching process with $CuCl_2$ to solve several problems, identified in the CYMET and CLEAR processes [30]. Leaching was realised in a counterflow system in two stages at approximately 100°C for 3 h in each stage. The solution contains 50 g/l of Cu^{2+} and 250 g/l NaCl. The yield of copper was higher than 98% and elemental sulphur was preferentially formed in the process. The purified saturated solution was processed by solvent extraction with the LIX-65N reagent. Oxygen was added to the solution to ensure that all the copper is transferred into the Cu^{2+} bivalent form. The acid, released by the extraction of Cu^{2+} ions, was used for the oxidation of Cu^{+} ions. The added organic agent was rinsed in order to remove the chloride and was subsequently stripped by means of H_2SO_4, producing the $CuSO_2$-H_2SO_4 solution for conventional electrolysis. Silver was not removed together with the Cu^{2+} ions and was collected in the rafinnate.

The main advantage of the Minemet process is that high-purity copper is obtained in compact form by sulphate electrolysis using efficient technology. Although the theoretical requirements on energy in electrowinning of copper from the solutions of cupric sulphate are considerably higher than in the case of electrolysis from cuprous chloride, the purity of the produced metal is higher. Therefore, if the larger disadvantage of the process is the purification of the chloride leaching solutions and the control of the quality of the electrolysis product, the transfer of copper to cuprous sulphate may be a suitable guide of how to solve the problems with the quality of copper and extraction of silver in chloride technology.

A slightly different approach was used in the CUPREX process [31,32] based on the leaching–solvent extraction–electrolysis procedure. The chalcopyrite concentrate is leached in $FeCl_3$ to produce elemental sulphur and dissolve $CuCl_2$. The important aspect is that the process is characterised by the excess of $FeCl_3$ to ensure that the entire amount of copper is present in the bivalent form. The leaching reactions favour $CuCl_2$ and $FeCl_2$, although the system contains several compounds in equilibrium [33]. After separation of the liquid and solid phases, the solution is cooled down, processed by $CaCl_2$ to precipitate gypsum and this was followed by adding the Acorga CLX50 agent. The neutral cuprous chloride is selectively separated and it is not necessary to keep the required value of pH of the solution during extraction. After removing minor compounds, the organic agent was stripped with water at 65°C to produce an aqueous solution of $CuCl_2$ with > 100 g/l of Cu^{2+}, used for the electrolytic extraction of copper.

Although the principal electrode reactions are the precipitation of powder copper and the release of gaseous chlorine, a portion of the cuprous ions is reduced in the form of cupric ions. Subsequently, it is necessary to oxidise the catholyte and extract the rest of copper in a separate operation of solvent extraction ('reforming') using a high organic/aqueous phase ratio. The organic phase from reforming is subsequently mixed with the organic agent for main solvent extraction.

The CUPREX process has several advantages. It is used for the production of high-purity copper powder with a high current yield (94%) and a low energy consumption. The silver content of copper is < 1 ppm and the CUPREX process is suitable for producing a rich copper secondary product from the rafinnate, depleted in copper. The solvent extraction circuit is relatively straight and does not require pH control. However, the overall scheme is complicated and the ion-selective membranes may represent a weak area in satisfying the environmental requirements on production. A weak point of this production system may be two shortcomings: electrolysis is carried out in a solution of cupric ions and coarse-grained copper is produced.

The flowsheet of the CUPREX process is shown in Fig. 16.10.

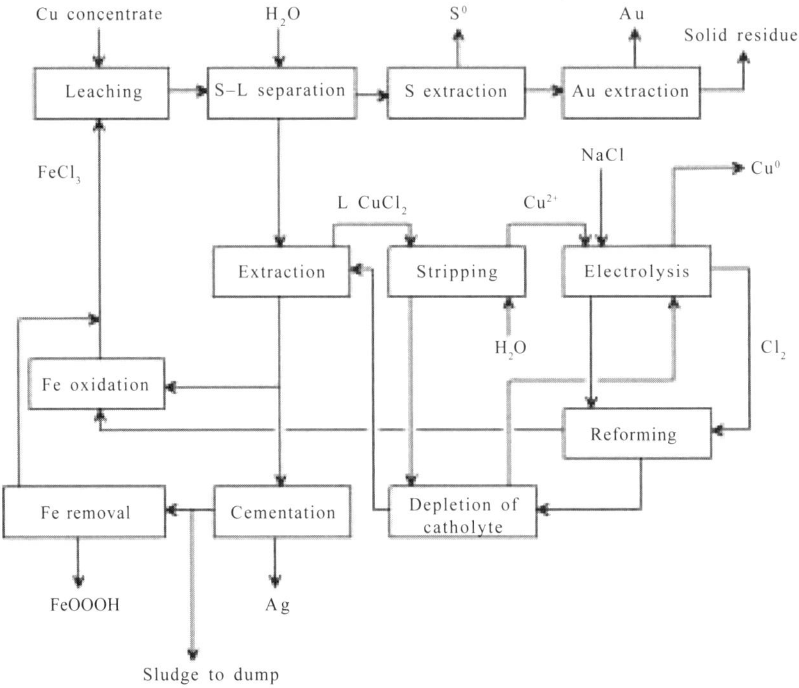

Fig. 16.10. The flowheet of the CUPREX process.

The Norwegian process Elkem [34] includes five stages in which a complex copper concentrate with a zinc content is leached by the counterflow method in brine in two stages resulting in the formation of a chloride solution, containing zinc and copper. Copper is produced as powder by electrolysis. A later variant of the process used a solution of cuprous chloride from which copper is extracted by solvent extraction prior to electrolysis. Silver is transferred into the rafinnate, and high-purity copper is also produced. Electrolysis is carried out in a fluid reactor. Zinc is extracted from the chloride solution using tributylphosphate. Iron is removed in the oxide form by oxidation in air. Lead is produced by crystallisation in the form of lead tetrachloride.

The Great Central Mines (GCM) process also produces coarse-grained copper but it does not use purification by solvent extraction [35]. As in the majority of chloride leaching processes, the first stage is fine milling of at least 80% of the concentrate smaller than 20 μm. The milled concentrate is subsequently leached in the $FeCl_3$–NaCl solution, where 99% of Cu and 93% of Ag is leached out.

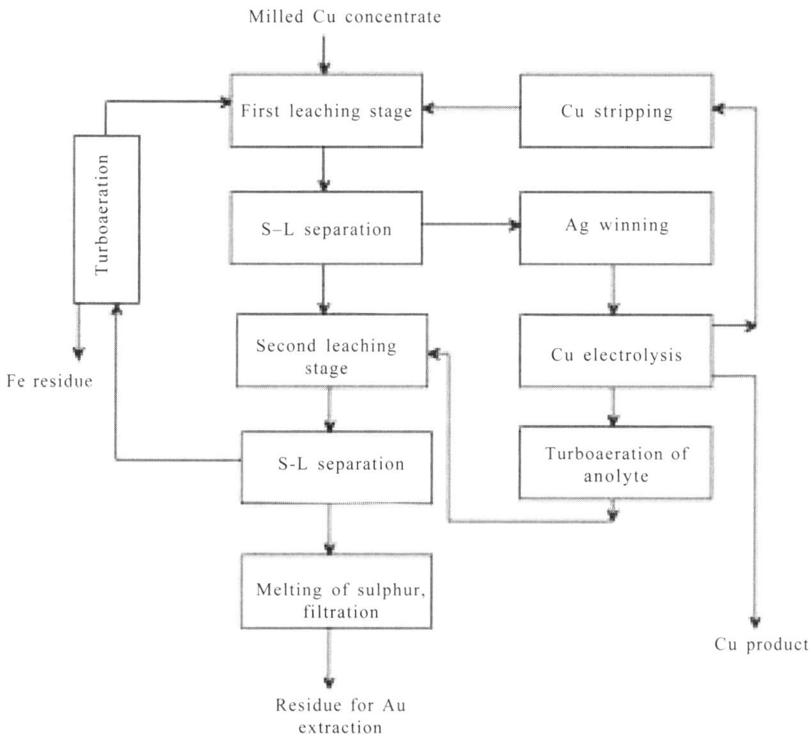

Fig. 16.11. Flowsheet of the GSMprocess.

Leaching is realised by the counterflow method in two stages so that the saturated solution contains only cuprous chloride and ferrous chloride. The content of impurities is controlled by recycling part of the precipitate containing iron into the second leaching stage. The chloride solution is electrolysed by diaphragm electrolysis producing granulated copper and also regenerating part of the $FeCl_3$ solution. Electrolysis produces approximately 85% of copper and the rest of copper is obtained in stripping halls where the spent catholyte from the main electrolysis stage is processed. The solution from the anolyte section is transferred to turbo-aeration unit in order to regenerate the solution. Iron does not precipitate in this operation. In this process, the total iron content from chalcopyrite is dissolved, together with 0–12% of iron from pyrite. Excess iron was removed by oxyhydrolysis to Fe_2O_3.

The flowsheet of the GSM process is shown in Fig. 16.11

The GCM process was developed on the laboratory scale. Silver is removed from the saturated solution with solid iodine resulting in a high yield of coarse-grained copper. The process is suitable for specific conditions with a shortage of sulphuric acid.

Henkel Corporation [36] developed a mixed organic extrahent for the transfer of copper from a chloride medium to the conventional sulphate solution $CuSO_4$–H_2SO_4 for electrolysis. Generally, the solvent extraction circuit may be included in any leaching process using ferric chloride with the excess of Fe^{3+} in the saturated solution to ensure the presence of the entire amount of copper in the form of $CuCl_2$. These conditions can be easily achieved in any previously described process by setting the ratio of the liquid and solid phases. The first component (di-isodecyl ester of pyridine dicarboxyl acid) of the mixed organic agent extracts neutral species of $CuCl_2$ from the concentrated acid leaching medium. The reagent is selective in respect of copper from the majority of impurities and also silver remaining in the solution. The silver-enriched product may be prepared from such a solution by cementation, precipitation of sulphides, etc. After extracting silver, the raffinate is subjected to oxidation hydrolysis to regenerate the leaching agent $FeCl_3$ and to precipitate excess iron. The principal oxidation agent is gaseous chlorine, which forms in a different stage of the process. Copper from the organic phase is transferred into the solution of NaOH with pH = 2.2 to wash away the chlorides. Since it is necessary to maintain the accurate value of pH in this operation, copper is transferred from the first reagent to the chelate extrahent (LIX 860) and the chloride ions released in the process are also removed by rinsing. The aqueous phase from rinsing is directed into conventional electrolysers for chlorine–alkaline electrolysis for regeneration of NaOH and for obtaining chlorine for oxidation of iron. Finally, the extracted product is transferred to the spent electrolyte H_2SO_4–$CuSO_4$ to remove copper in the form of cuprous sulphate. Copper is extracted in the form of a high purity cathode by the process of efficiently developed electrolysis of copper [37, 38].

General disadvantages of such an approach are well-known. It is necessary to employ extensive operations of solvent extraction and copper must be obtained by electrolysis from the solution of cupric ions. Subsequently, it is necessary to carry out separate electrolysis to obtain Cl_2 and NaOH. These general disadvantages must be compared with the advantage of obtaining high-density, high-purity cathodes and the secondary silver product.

Broken Hill Associated Smelters Pty Ltd (BHAS) developed a process for extracting copper from copper matte produced in a lead-making shaft furnace [39]. The fine ground matte is leached by sulphuric acid containing 20 g/l of chloride at the boiling point and atmospheric pressure in the presence of oxygen. When the chloride concentration is below 1 M, it is possible to use solvent extraction for direct extraction of copper using Acorga M5640 agent and subsequently strip the solution by the conventional procedure using the recirculating electrolyte. Complex Cu/Pb/Zn sulphide concentrates with chalcopyrite may also be leached for 6 hours using a similar procedure. Copper is obtained by solvent extraction using Oxim with the formation of a solution suitable for conventional electrolyte, whereas zinc can be obtained by solvent extraction using 2-diethylhexyl phosphoric acid or by precipitation as a basic zinc salt. Lead remains in the leaching residue together with a small amount of non-reacted sphalerite and chalcopyrite, elemental sulphur, gold and silver. The residue maybe processed in the copper smelting plant, or by flotation or gravitational methods.

The FCL leaching process was developed by CANMET for processing of Cu–Zn concentrates by the solution of ferric chloride [40]. This technology may be generally applied to ores and concentrates with a content of pyrite or pyrrhotite containing copper and silver, or nickel and cobalt. The process is based on two-stage leaching in the $FeCl_3$ solution with the removal of the residue into the waste or for extraction of gold. Copper and silver are cemented from the solution using iron and zinc is extracted by tributyl phosphate. The supplied organic compound is stripped by the spent electrolyte from the zinc chloride electrolysis and the chlorine, generated on the anode, is recycled. The solution for electrolysis is depleted in iron by solvent extraction using ethyl–hexyl phoshate and, finally, by precipitation using zinc powder. The refined product from liquid zinc extraction is again oxidised by recycled chlorine and returned for leaching, with iron removed from it by oxyhydrolysis.

The flowsheet of the FCL process is shown in Fig. 16.12.

Oxidation leaching of the ores or concentrates in fluorosilicate acid in the Extramet process [41] enables highly efficient extraction of metals (100% of lead and silver, 90–95% of copper and zinc). They may be separated by a simple procedure and extracted with a high yield (at least 99%) by cementation or electrolysis. The cementation of copper or silver takes place without any problems in the given medium by the ACTIMAG process. The precipitation of goethite from this medium is also quite easy. The electrolysis of lead is possible in the same conditions as classic electrowinning with the efficiency of approximately 99% and, at the same time, the fluosilicate acid is regenerated by anodic oxidation. Advantages include the high solubility of metals, such as Pb, Cu, Zn, Sn and Ag, the possibility of obtaining various metals in the pure condition, high selectivity of lead in the electrolysis stage, small volumes taking into account the high solubility of metals in this medium, the yield of sulphur in useful form (at least 20%), the possibility of extracting

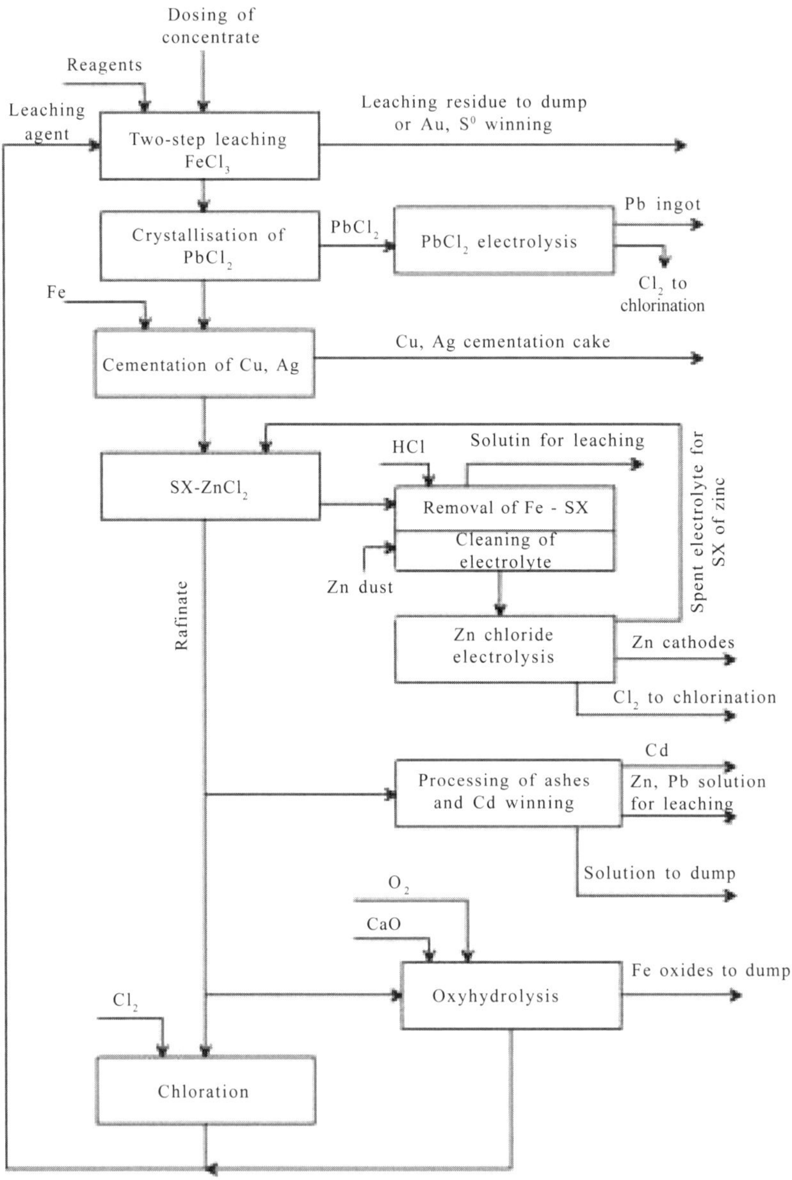

Fig. 16.12. Flowsheet of the FCL process.

noble metals, and the flexibility of processed sulphide ores.

The CENIM-LNETI process [42, 43] is based on oxidation leaching in concentrated solutions of ammonium chloride, using oxygen. The process includes leaching of metals (Cu, Zn, Pb and Ag) together with the production of ammonia and sulphate which are returned to the solution, elemental sulphur, which remain in the leaching residue. One of the important characteristics of the process this is that it operates in almost a neutral medium, with pH between 6 and 7.

Leaching is carried out in two counterflow stages. The most important working conditions are the temperature of 105°C and the partial pressure of oxygen of 150 kPa. After completing neutral leaching, the process continues without oxygen so that a part of the residual sulphides reduces the majority of leached Cu^{2+} to Cu^{+} in order to prevent the precipitation of cuprous diamine during cooling. This is followed by solvent extraction, and silver and mercury are obtained by cementation of copper. Lead is separated as chloride and crystallises during cooling of the solution to 50°C by vacuum evaporation. The cuprous chloride is again dissolved and lead is cemented by granulated zinc or zinc powder. The concentrated solution of the zinc chloride from the cementation stage is used for rinsing powder zinc from the organic phase during extraction of this metal.

The sulphates, formed during leaching, are removed by precipitation with lime in the presence of citric acid. The operation should be carried out prior to the removal of zinc and at a stoichiometric shortage of lime. This results in the required level of the content of sulphates in the solution.

Zinc is obtained by the DEHPA process at 50°C. This operation also removes Ca and Cu^{2+} in the relatively large quantitics. The joint precipitation of these elements is minimised by using the organic phase in an amount similar to that required for saturation with zinc. Another important factor in this stage is that the majority of copper present is in the non-extractable cuprous form and, also, there is a minimum amount of the sulphate for maintaining the low concentration of Ca. In any case, the organic phase present is washed away in the solution of the zinc chloride, which forms during the cementation of copper, for elimination of both precipitated copper ad lime. Subsequently, the organic phase is rinsed with water to eliminate chlorides and finally stripped using a recirculating zinc electrolyte. After removing zinc, C^{+} is oxidised with oxygen or air to Cu^{2+} which is subsequently extracted using LIX 65N. The

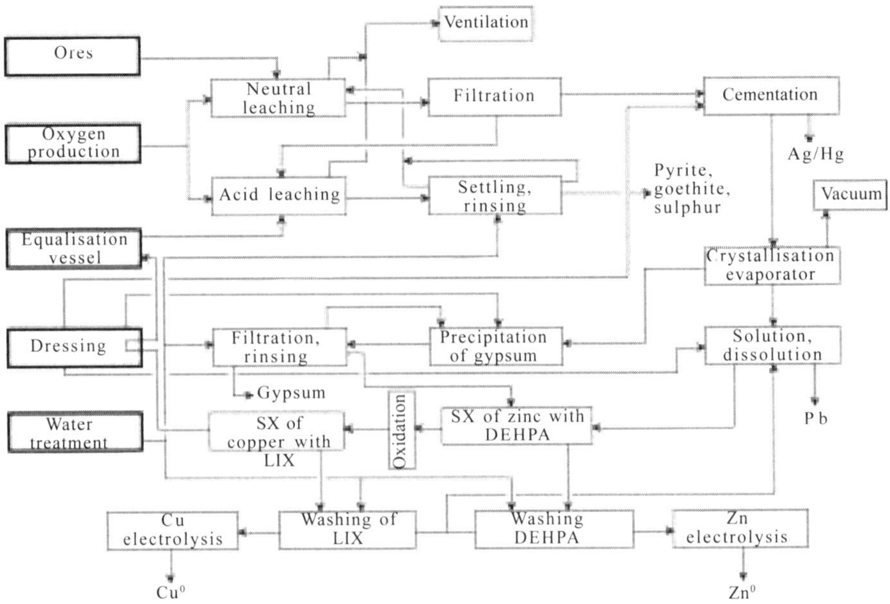

Fig. 16.13. Flowsheet of the CENIM-LNETI process.

resultant acid solution after extraction of copper is recycled in the acid leaching stage.

The process was tested on the laboratory scale with satisfactory results. The leaching and the processes of elimination of the sulphate and lead (II) chloride have been demonstrated in the batch regime and solvent extraction in continuous regime.

The flowsheet of the process is shown in Fig. 16.13.

The INTEC process (Sydney, Australia) is the result of co-operation of a consortium including 11 companies from Australia, Thailand, Japan, USA, Canada and Great Britain. The process is patented by the International patent number WO 94/00606 [44]. The process has a number of improvements, especially in the area of elimination of impurities and obtaining valuable components (gold, silver).

In the majority of electrolysis applications, the anodic reaction is loss-making, and the produced oxygen is discharged into the atmosphere. The INTEC technology uses anodic energy for the conversion of sulphides to the soluble form and also for the formation of elemental sulphur. Leaching is carried out using a concentrated solution of sodium chloride with the anodically produced oxidation agent, producing high-purity copper with side products containing gold and silver. The primary oxidation agents

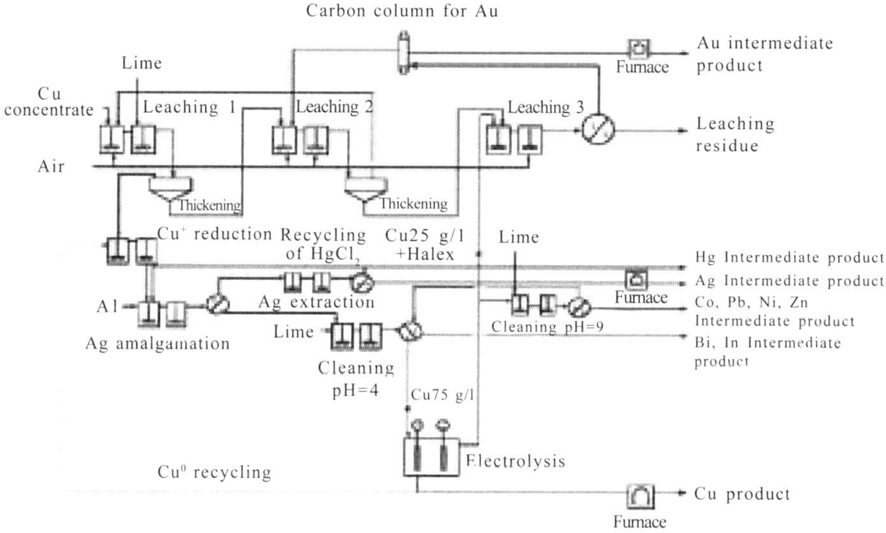

Fig. 16.14. The flowsheet of the INTEC process.

are cupric ions and halogen complexes (halexes) which form on the anode in electrolysis.

The three-stage process is based on counterflow leaching at 80-85 °C, and iron is precipitated as a goethite product using air at atmospheric pressure. The conversion of sulphide sulphur to the elemental form reaches approximately 95% without formation of any sulphates in the leaching agent or residue after purification. Gold is leached and precipitated in the metallic form with negligible contamination by copper, or silver on active carbon. The purification stage produces metallic silver and various mixtures of contaminants. The electrolysis stage produces high-purity copper on the cathode and cuprous ions on the anode. If the concentrates contain lead or zinc, technology may be modified in order to produce both zinc and lead in metallic form.

The flowsheet of the process is shown in Fig. 16.14.

Leaching of chalcopyrite takes place in three stages in $CuCl_2$–NaCl–NaBr brine, with air used as the oxidation agent. In the third leaching stage, the potential is modified by adding $BrCl^{2-}$ complex. The higher potential is used to promote leaching of the gold into the solution in the form of a tetrabromide complex.

In the first leaching stage, chalcopyrite is leached by oxygen with the formation of covellite and cupric ions. In the second stage, in addition to oxygen, the bivalent copper ion also takes part in the oxidation of chalcopyrite. In the third stage, the $BrCl^{2-}$ complex is added to accelerate leaching. Copper is produced by chloride electrolysis from a pre-purified electrolyte. Since copper is in the bivalent form in the electrolyte, prior to electrolysis it is necessary to reduce copper. This takes place either by means of metallic copper or iron. Since the electrolyte contains the bromine anion, the potential precipitation of chlorine on the anode is suppressed as a result of the formation of a bromine–chloride complex.

The chemistry of this process and the leaching conditions in the individual stages are favourable for the direct removal of some impurities together with the precipitated goethite. They include mainly: arsenic (in the form of ferric arsenate), antimony, molybdenum, tin, titanium, selenium, and tellurium. Other impurities are removed in the precipitation stage where changing pH to approximately 7 results in precipitation from the solution of alkaline oxides of bismuth, cadmium, cobalt, indium, tin, nickel and zinc. The present sulphide (approx. 5% of sulphur is oxidised to the sulphate form) precipitates in the form of gypsum. Gold is extracted by absorption on active charcoal prior to the leaching agent reaching the second stage of leaching, and silver is produced in the final stage of purification of the electrolyte in the form of amalgam by cementation on aluminium foil.

Since copper is in the bivalent form in the solution prior to electrolysis, modification of pH makes it possible to ensure the precipitation of impurities, either directly in the purification stage of the electrolyte or in the leaching process.

The main advantages of the INTEC process include:

- the entire process is designed to ensure the removal of certain impurities present in the leaching solution directly in the leaching process;
- although the current density of chloride electrolysis is still higher, the energy balance is much more favourable than in sulphate electrolysis;
- the process is designed to produce precious metals from the solution (gold and silver) in technological stages which are the integral part of the leaching process;
- the favourable economic parameters of the process.

The main disadvantages are:

- regarding the form of gold in the raw material, the particles containing gold may be coated by produced sulphur. In the case of gold-bearing pyrite, which is more resistant to the effect of the leaching solution, part of gold is conserved in the volume of the non-reacted particles. In both cases, the gold-bearing fraction must be subjected to additional processing and it is desirable to separate the fraction from the other solid components of the leaching residue;

- both pyrite and sulphur require additional processing, because they cannot be stored for long periods of time due to ecological restrictions;
- the application of amalgamation requires strict control of mercury in the process (environmental and hygiene requirements) and also in produced copper (contamination of the product).

At present, the results of successful pilot plant and demonstration trials are used by INTEC to construct commercial production systems, including the Ivanhoe Turquoise Hill project in Mongolia.

The final and successful hydrometallurgical process is the HydroCopper process [45] developed by Outokumpu, Finland. This process provides a new approach, in particular to the method of extracting copper from the solution. The process is protected by the US patent number 6007600 [46]. The concept of the process is shown in Fig. 16.15.

The main difference in comparison with other processes is that in this process, copper is extracted by the precipitation of Cu_2O instead of relatively complicated chloride electrolysis of the solution containing CuCl. The first stage of the process is leaching of the chalcopyrite concentrate. This consists of the three-stage counterflow leaching at normal pressure and a temperature of 80–100 °C; in order to increase the efficiency of leaching copper above 96%, relatively long leaching times are essential [47]. The leaching solution is represented by the acid solution $CuCl_2$ + NaCl. In the first stage, the chalcopyrite is leached by Cu^{2+} with the formation of Cu^+, Fe^{2+} and elemental sulphur. Iron is removed from the solution by oxidising it to the trivalent form by air and precipitation in the form of goethite or haematite. The accompanying elements present in the concentrate (for example, zinc sulphide) are also leached and transferred into the solution. Sulphur is transferred during leaching into the leaching residue in the elemental form, and a small part is oxidised to sulphate. The main part of Cu^{2+} is reduced to Cu^+ with pH control.

In the next stage, after refining the solution, copper present in the solution in the form of CuCl is precipitated as Cu_2O using NaOH. NaOH is produced by chlorine–alkaline electrolysis in accordance with the reaction:

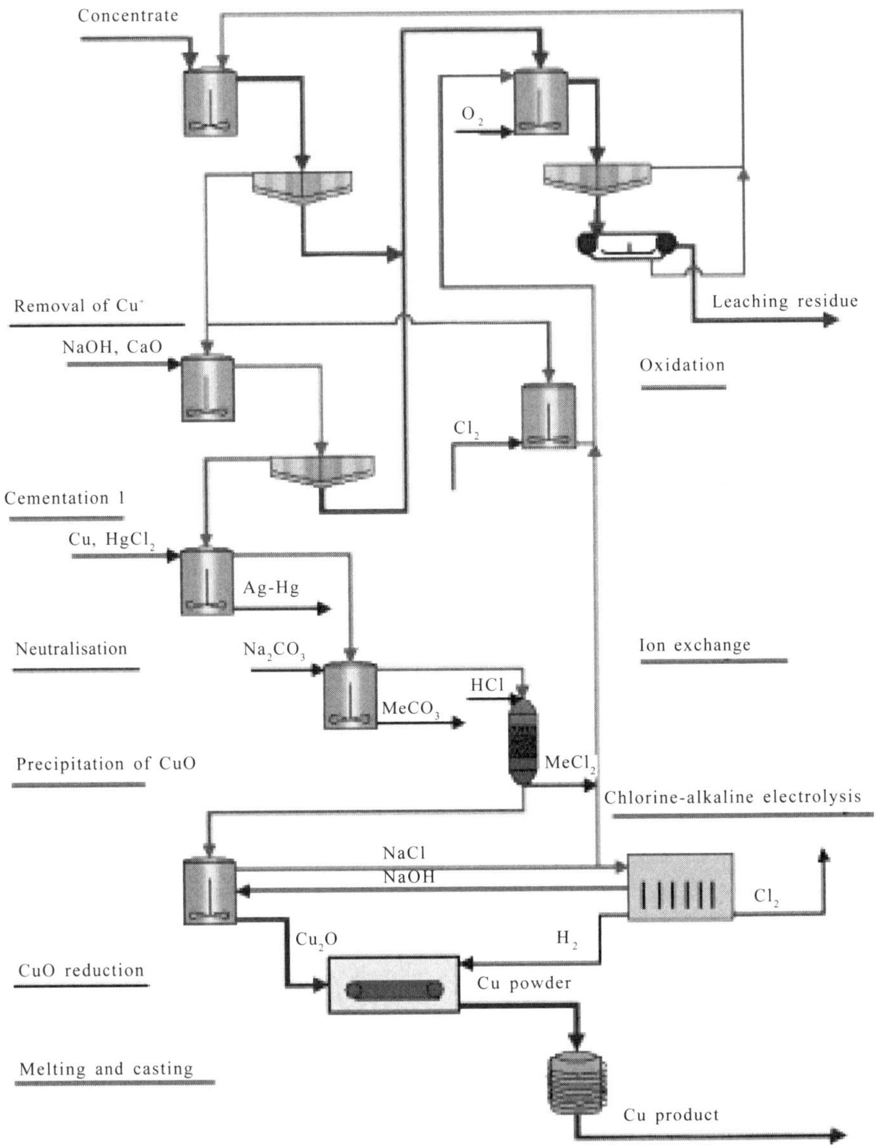

Fig. 16.15. The flowsheet of the HydroCopper process.

$$2NaCl + 2H_2O \rightarrow 2NaOH + Cl_2 + H_2 \tag{16.1}$$

and other products of electrolysis are also utilised in the technological process: hydrogen is used for the reduction of Cu_2O and chlorine is used for the regeneration of the leaching medium.

This method does not require any acids or alkali and, from this viewpoint, it is a closed technological process.

The main advantages of the HydroCopper process include:

- replacement of chloride electrolysis by precipitation of copper oxide and regeneration of the solution by chlorine–alkaline electrolysis which is easier to control;
- due to the fact that chlorine is the expected product of electrolysis, equipment is modified to avoid excessive corrosion and also to fulfil all the safety and hygiene requirements;
- electrolysis is realised at high current density but as there is no copper electrolysis, common problems with dendritic growth of copper and its eventual reoxidation are avoided;
- the overall consumption of electrical energy in electrolysis is lower in comparison with chloride electrolysis;
- the process produces high-quality copper powder which can be directly remelted and cast to produce high purity copper;
- the reduction process may be followed directly by remelting or continuous casting of copper.

The disadvantages of the process include:

- investment for the sections of equipment that are in contact with gaseous chlorine;
- the requirement for the strict control of all operations with gaseous chlorine (hygiene and ecological requirements);
- the presence of the univalent copper ion in the leaching solution requires control of the acidity of the medium (this applies to all technologies in which Cu^+ is present);
- relatively high sensitivity of chloride technology to the quality of the starting materials and the presence of accompanying elements;
- the need for efficient purification of the solution to remove impurities prior to precipitation of Cu_2O.

Although the HydroCopper technology has a number of problems associated with the chloride processes, some of the disadvantages mentioned in the previous sections have been eliminated. They include mainly:

- since precipitation results in the formation of Cu_2O, there are no problems with the reoxidation of copper which forms in chloride electrolysis;
- it is possible to use high current density in chlorine–alkaline electrolysis without the problems of the formation of dendritic copper which are encountered in chloride electrolysis;
- in the final analysis, the energy balance of electrolysis is better than that of chloride electrolysis which is not capable

at present of saving more than 50% in comparison with sulphate electrolysis;
- the construction of electrolysers for chlorine–alkaline electrolysis is easier to control (closed and compact units), and the electrolysis process does not require any special attention;
- despite the effort to conduct chloride electrolysis without any need for purification of the electrolyte, the presence of certain important elements in the solution (silver, selenium, tellurium) requires purification of the solution in the final analysis – in the HydroCopper process, this operation is an integrated part of the entire process.

After successful pilot plant trials, a demonstration production section was constructed in Pori, Finland, which produces 1 t of copper every day and commercial production is planned.

Special attention is being paid to the development of technical and economic feasibility of the process based on the bacterial leaching to produce copper, antimony and silver from a concentrate containing in particular chalcopyrite, pyrite and tetrahedrite. It is well-known that *Thiobacillus ferroxidans* oxidises sulphides or elemental sulphur directly on the surface of the mineral [48] and it also oxidises ferrous and soluble sulphur species as tetraionates and thiosulphates, provided they are in the solution [49]. Although these organisms are most active in acid solutions (at pH = 1–3), their intracell pH must be close to the neutral values of pH. pH decreases in anaerobic conditions and also with a decrease of the concentration of the ferrous ions and pyrite, i.e., clearly said, for obtaining higher pH values, both oxygen and suitable reductant are necessary.

This may be used as a basis for proposing a process of bacterial leaching, as shown in Fig. 16.16 [50]. The essential biochemical reactions have been studied extensively [51, 52], but some of the aspects of these reactions are not yet completely clear. The bioleaching mechanism may be described as follows: sulphides or sulphur and iron in bivalent form are oxidised on the surface of the cell or in the outer membranes of the cell [51, 52] thus releasing the electrons which are transported through the wall of the cell by the biochemical mechanism [53]. The hydrogen ions and oxygen diffuse through the cell wall, depending on the activity gradient, and oxygen inside the cell is reduced by electrons thus removing the excess of hydrogen ions and obtaining a relatively high value of pH. If the course of the chemical reaction is taken into account, it may

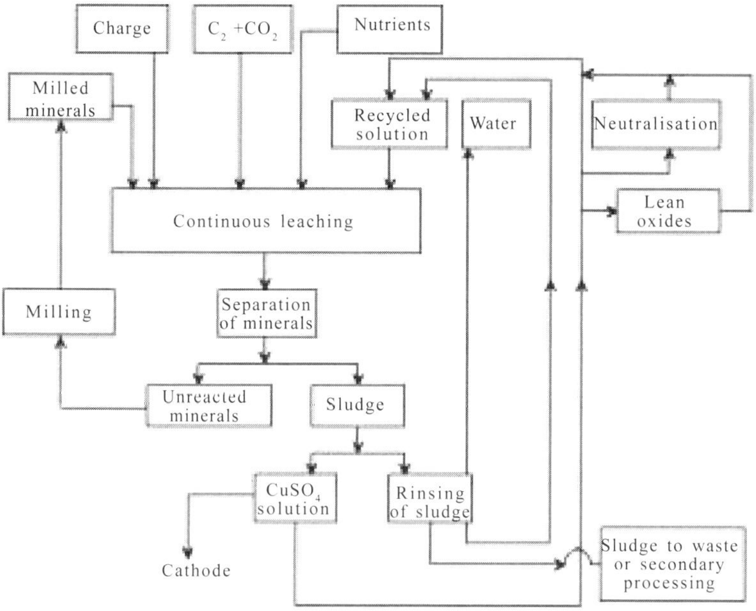

Fig. 16.16. Principal diagram of bacterial leaching.

be concluded that although part of the oxidation reactions takes place inside the cell and part of the reduction reactions takes place away from the cell, the metabolic process is dominant. If sulphur is also oxidised inside the cell, this means that pH would decrease during oxidation of iron and the resultant ferric ions would be hydrolysed and precipitated and this would be accompanied by the release of hydrogen ions which would interfere with the activity of the cell.

If the high pH value is efficiently maintained inside the cell, the reaction of the half-cell of formation of water should prevail in the cell over the reactions of oxidation and hydrolysis of iron. Similarly, although some oxygen reduction takes place outside the cell, in most cases, reduction must take place inside the cell. The separation of the oxidation and reduction reactions characterises the effect of bacteria as basically electrochemical phenomena [53].

Like *Thiobacillus ferrooxidans*, other bacteria also induce alternative reaction processes replacing the release of free energy by the reduction processes. These bacteria have not been studied sufficiently by metallurgists, but they may prove to be useful for the processing of effluents. For example, they may be used to reverse

Fig. 16. 17. The view of the BacTech/Mintek process.

oxidation processes of acid mine drainage waters (AMD waters) in the precipitation of sulphides.

The companies BacTech and Mintek have developed jointly the process BacTech/Mintek of bioleaching in a tank in Penoles, Monterey, Mexico. The Penoles demonstration plant bacterial leaching followed by conventional solvent extraction and electrowinning (SX/EW) produces 500 kg of cathode copper per day. The commercial production of 25 000 t of copper and zinc per annum is being prepared at Penoles.

The view of the BacTech Mingtek process is shown in Fig. 16.17.

The company GeoBiotics in Lakewood, Colorado, has developed an alternative approach to bioleaching. The GEOCOAT process, developed by company, includes the coating of concentrates with a suitable substrate, usually ballast rock, and placing the material in a conventional heap. The heap is moistened with an acid solution containing iron and nutrient materials, and air is also supplied under a low pressure to the bottom of the heap. This technology was initially developed for the extraction of gold and extensive studies of the possibilities of leaching of copper sulphides started at a later date.

The flowsheet of the GEOCOAT process is shown in Fig. 16.18.

The leaching of oxide copper ores followed by the production of cathodic copper by solvent extraction and electrolysis – SX/EW technology, has been sufficiently developed as a cheap technology of extraction of copper. Later, the technology was used efficiently also in the leaching of mixed oxide and chalcocite ores, in particular, in Chile at Cerro Colorado, in Quebraba Blanca and in Zaldivar. At

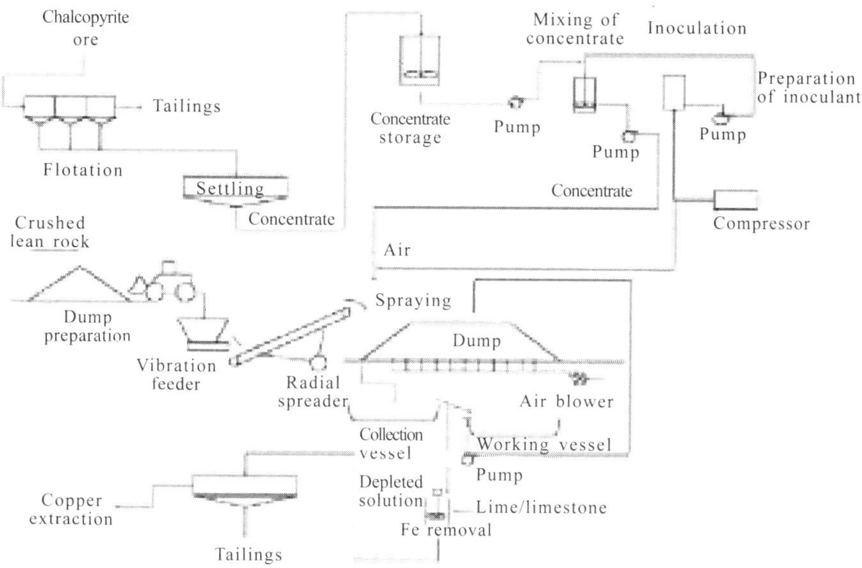

Fig. 16.18. The flowsheet of the Geocoat process.

present, work is continued to apply heap leaching also to chalcopyrite ores. The success of heap leaching together with SX/EW method of extraction of copper has been used as an impetus in the development of the second-generation of hydrometallurgical processes for the processing of chalcopyrite concentrates.

The volume of production of primary copper from oxide ores and tailings and, later, from chalcocite raw materials, by the processes of leaching and solvent extraction and electrolysis has increased greatly in the last three decades and at present equals approximately 20% of the overall world production of copper. The majority of these plants are in Chile, USA and Australia. The typical cost of these operations is 1–1.5 US$/kg Cu, depending on the richness of the ore, extraction costs, the yield of copper, the price of the reagents and energy.

Commercial utilisation of pressure hydrometallurgy for sulphides

The first commercial application of pressure hydrometallurgy in the processing of sulphides appeared in 1950. The nickel refinery Sherritt at Fort Saskatchewan, Canada, started using in 1954 the method of ammonia pressure oxidation leaching for the extraction of copper, nickel, and cobalt from pentlandite–chalcopyrite concentrates. At present, the output of the refinery is considerably greater using relatively rich nickel–cobalt sulphides. Nickel and cobalt are extracted as high-purity metals, copper as a sulphide semi-product material, and sulphur is converted to ammonium sulphate used as a fertiliser.

In 1950, the first two plants for pressure leaching in sulphuric acid solution were built, both in the USA, for the processing of the sulphide concentrates of cobalt. Garfield Cobalt Refinery (Utah) of the Calera Mining Co, started in 1953 processing of cobaltite (CoAsS)–chalcopyrite–pyrite concentrate and The National Lead Company (Missouri) started processing in 1954 the siegenite ($[Co, Ni]_3Si_4$)–chalcopyrite–pyrite concentrate. Both technologies used high temperatures (190–245 °C) and high pressure, generated by compressed air acting as an oxidation agent. This procedure was developed by the Chemical Construction Company (Chemico), initially as a method of extraction and production of cobalt. In these 'aggressive' leaching conditions, the entire amount of sulphide sulphur is completely oxidised to sulphate. The metals are extracted from the solution by the technology of reduction with hydrogen, developed by Sherritt and Chemico. These plants depended partially on various economical factors and stopped production mainly because of the economical reasons and the insufficient amount of raw material for processing.

Dynatec (formerly Sherritt) solved many research projects in laboratory, miniplant and pilot plant scale, in order to find economically efficient and competitive and environmentally acceptable technology of acid oxidation leaching of sulphide ores and concentrates containing copper and iron. The majority of these process included easy recovery of copper and precious metals and conversion of sulphur into the elemental form, or its fixing to gypsum.

One of the initial studies was performed at the beginning of the 60s of the previous century and was concerned with the processing of a chalcocite–pyrite concentrate containing 22% of copper, 28% of iron and 39% of sulphur. Mineralogically, the concentrate contained approximately 29% of chalcocite and 60% of pyrite, with the pyrite containing approximately 90% of the total amount of sulphur. Pyrometallurgical processes resulted in the formation of approximately 3.4 t of SO_2 for every 1 t of copper. The hydrometallurgical process included pressure leaching under slightly oxidising conditions in diluted sulphuric acid (at 80–105°C and 140–

350 kPa oxygen pressure). This enabled selective extraction of up to 97% of copper, with less than 10% of pyrite taking part in the reaction. After partial neutralisation of the leaching solution to remove iron, it was believed that the solution is suitable for extracting copper by direct electrolysis with recycling of the spent electrolyte for pressure leaching. Although this process was not commercially viable, it was used as a basis for the flowsheet of the process developed for ores from Las Cruces.

The original process of Dynatec Corporation, Fort Saskatchewan, Canada, was developed in 1960 for the direct leaching of chalcopyrite-containing concentrates. The process was developed for processing of fine milling material of a charge to 98% below 44 μm with leaching at approximately 110°C with a relative surplus of the concentrate in relation to the amount of acid. This resulted in the formation of a copper-rich solution with a low iron content, directly suitable for electrolysis. Prior to separating the solid from the liquid phase, the pulp heated to temperatures higher than the melting point of sulphur was oxidised (~117 °C), cooled down and the non-reacted sulphides, which were agglomerated with elementary sulphur, were again extracted and separated from the iron oxides and ballast by screening and flotation. After removing elemental sulphur, the non-reacted sulphides were recycled for pressure leaching. As with chalcocite, the spent electrolyte was recycled for producing the acid for pressure leaching. The flotation tailings were subjected to cyaniding to extract gold and silver. The process, although it was efficient mainly in the case of the chalcopyrite concentrates, encountered problems in the processing of the charge containing pyrite, mainly because of the extensive oxidation of pyrite to ferric sulphate and sulphuric acid. In addition, the extraction of noble metals was not very efficient.

At present, the Dynatec process is based on pressure oxidation at 150°C. In this case, low-quality coal is added as an agent for the dispersion of liquid sulphur. The charge is milled to a particle size of 30–40 μm and the non-reacted sulphides are flotated and recycled to obtain the maximum yield of copper. Copper is produced by the SX/EW procedure, although direct solvent extraction is being studied extensively. Miniplant tests have been carried out. Dynatec has also suggested a flowsheet for the MK Gold's Las Cruces project in Spain. The process uses chalcocite–bornite ore in combined atmospheric and low-pressure autoclave leaching after milling to a particle size of 105 μm.

In the 70s, the increase of the cost of energy and increasing stringent restrictions on the emission of SO_2, mentioned previously, resulted in special interest and activity in the development of hydrometallurgical processes by many other companies. The most active companies were, in addition to others, Anaconda Copper, Cominco, Cyprus, duPont, Duval, Freeport, Inco, Kennecott, Lurgi, Noranda, Sherritt/Dynatec, Sunshine, and also a large number of government, research institutes and universities. In the majority of processes they tested chalcopyrite concentrates with the aim to to precipitate the maimn part of sulphur in elemental form and recover copper by electrowinning. Iron was removed in the form of the hydrated iron oxide (goethite, haematite) or as jarosite. The leaching system included atmospheric and pressure leaching in acid chlorides of sulphates and also in ammonium sulphate, and in this case, sulphur was oxidised to the soluble sulphate. Because of the refractory nature of chalcopyrite, many of these processes included preliminary processing of chalcopyrite into a form with higher reactivity. Some of the processes contained preliminary heat treatment with metallic copper, iron or elemental sulphur because of the transformation of chalcopyrite to chalcocite, covellite, or idaite, whilst others included hydrometallurgical conversion using metallic copper or iron in the atmospheric conditions, or of cuprous sulphate at elevated temperatures. The 70s of the 20th century also relate to the discovery of ultrafine milling (grain size of the product smaller than 10 μm) in order to 'activate' sulphide minerals. Other development directions included the additions of various reagents to the leaching systems, such as sulphur, chloride ions or nitrates, to ensure accelerated leaching of chalcopyrite. Some of them have been developed to the pilot-plant or demonstration level.

Although these processes are technically very similar and eliminated problems with the contamination of environment by the sulphur oxides, the majority of the processes are based on a complex scheme, which often contains highly corrosive conditions in certain operation stages. Some of the companies prefer the purity of the copper product and also secondary sulphur products whereas some other companies prefer the environmental stability or acceptability of certain solid waste. The extraction of noble metals is complicated and less efficient than in the pyrometallurgical operations.

However, at the same time, it should be mentioned that the development of pyrometallurgical procedures of production of copper at that time was highly dynamic. The development of technologies such as flash smelting and continuous converting enabled a far more extensive and efficient extraction of sulphur in comparison with the older processes, so that the large number of the environmental restrictions and emission limits have been fulfilled and, therefore, the pyrometallurgical methods have again become important.

In early stages, only two hydrometallurgical plants reached the commercial or semi-commercial level of direct processing of copper sulphide concentrates of copper in the 70s of the 20th century, both in the USA. One of them was Anaconda's Arbiter Plant [15–18] which started production in 1974 with a capacity of 36 000 t of cathodes per annum. The process was based on the ammonia oxidation leaching of the mixed concentrate, consisting of chalcocite, bornite and chalcopyrite with a large amount of pyrite. Copper was extracted from the ammonia leaching solution by solvent extraction and electrolysis (SX/EW). Ammonia leaching was based on the extraction of easy to leach copper from chalcocite and bornite (70-80% of copper), extraction of chalcopyrite (and the majority of gold and silver) from the leaching residue by flotation for processing in the smelting plant, with the removal of pyrite, containing almost 70% of the sulphur of the original concentrate. The plant was closed in 1977 as a result of high production costs, changes in the mineralogical composition and also because of complications associated with handling of sulphates.

Another plant, already described previously, was based on chloride technology under the name CLEAR by Duval Company [27–29], developed for the processing of chalcopyrite concentrates, which started in 1978. The annual capacity of the plant was 32 000 t of copper and has been regarded as technically successful, although modifications have been made to ensure economic efficiency. The plant, regarded by Duval Company as a demonstration prototype, was also closed after several years.

Although many of the hydrometallurgical processes developed for the processing of copper and iron containing sulphides did not meet the expectations, several processes of pressure leaching have been developed for simple sulphides and have been commercially successful. One of these processes is the plant constructed by company Inco CRED (Copper Refinery Electrowinning Department) [54] in Sudbury, Ontario, processing the noble metal concentrate, containing copper sulphide similar to chalcocite as regards the composition. The plant started production in 1973 and the process was based on the leaching of solid constituents in a spent electrolyte at a temperature of 105°C using oxygen in a cuprous sulphate solution to produce a solution for electrolysis. Another product were the residue containing PM and elemental sulphur for further enrichment by organic extraction of sulphur. Problems were encountered in the removal of sulphur when the leaching conditions were changed because of the shortage of sulphuric acid in the system, in the oxidation of sulphidic sulphur and the conversion of copper to the basic copper sulphate salts. The basic salts were subsequently dissolved in the spent electrolyte forming a rich concentrate of noble metals from the solid residue.

Dynatec/Sherritt developed acid pressure leaching processes for the processing of metals of the platinum group present in the nickel-copper matte. The plant was operated by five producers PGM Impala (1968), Rustenburg (1981), Lonmin (1985) and Northam (1992), all in South Africa, and Stillwater (1996) in the USA [54]. The second stage consisted of processing of the chalcocite residue by acid leaching at temperatures between 175–160°C in order to dissolve copper and oxidise copper to the sulphate leaving consequently the solid residue rich in the metals of the platinum group for final refining.

The flowsheet of the process is shown in Fig. 16.19.

In 1984, The Sunshine Mining Co. started a new plant for acid oxidation pressure leaching of a silver-carrying chalcocite intermediate product from preliminary alkaline leaching of the tetrahedrite concentrate for primary production of antimony. This was a dosing leaching operation using sulphuric acid with the addition of nitric acid starting at approximately 90°C and completed at around 150°C, used for the dissolution of silver and also copper and iron with conversion of part of copper to elemental form. Silver was produced by precipitation and copper was extracted by solvent extraction and electrolysis. Another innovation, introduced several years later, was the fine grinding of the concentrate with 80% below 10 μm and the replacement of nitric acid by sodium nitrate which is a more efficient catalyst. This plant operated for approximately 16 years prior to closure because of several decreases in the price of silver and antimony on the market.

As already mentioned, these examples of commercial operations are specialised cases of processing of the chalcocite type charge with complete oxidation of sulphur (in the majority of cases) to the sulphate. As such, they have not been used for the direct processing of the chalcopyrite charge.

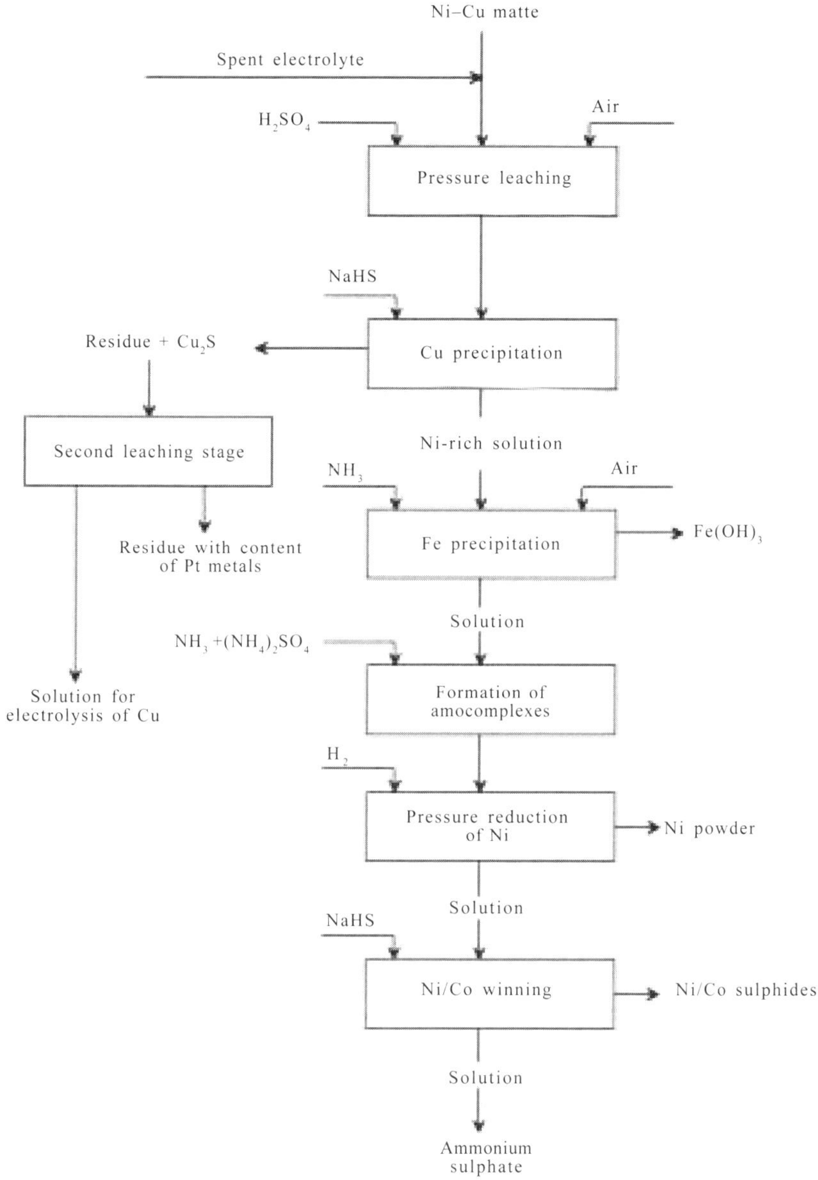

Fig. 16.19. Flowsheet of the Dynatec/Sherrit process.

The encouraging results obtained in the development of the leaching methods for various copper raw materials and/or concentrates have expanded the possibilities of these methods to the production of other sulphides. The 80s and 90s of the 20th century witnessed several successful commercial transformations of the autoclave processes of other natural sulphides, mostly zinc concentrates and refractory gold ores and concentrates. Four companies adapted the process of pressure leaching of zinc Dynatec (Cominco 1981, Kidd Creek 1983, Ruhr Zink 1991 and Hudson Bay Mining and Smelting 1993) for acid sulphate leaching of zinc sulphide concentrates using a recirculating electrolyte at a temperature of 150°C with oxygen in which approximately 98% of zinc was leached and the majority of sulphur was transformed to the elemental form.

In the case of refractory gold ores and concentrates, in which a significant portion of gold is closed in pyrite and/or arsenopyrite, more 'aggressive' conditions (usually between 190 and 225 °C) were used. The sulphides and arsenides were completely oxidised, thus releasing gold for production by conventional cyaniding. The first plant, using pressure oxidation for gold, started production in 1983 and this was followed by introduction of further 12 plants. It is interesting to note that the conditions used in acid pressure oxidation processing of refractory gold-bearing raw materials are very similar to those used more than 30–40 years ago for the leaching of cobalt in Calera and Fredericktown, with considerably more sophisticated and effective operations.

In the 90s of the 20th century, special interest was paid again to pressure acid leaching of primary copper-ferrous sulphides and also chalcopyrite. This was partially the result of the efficient implementation of autoclave technologies for leaching of zinc and refractory gold ores and also the development of constructional materials, more efficient mills for fine and ultrafine milling of sulphides, mainly because of the availability of reagents for solvent extraction, capable of extracting copper from concentrated solutions.

This period is characterised by the development of several 'new' acid sulphate pressure leaching processes for the chalcopyrite concentrates based on the development results in the 70s. The main features of these processes were pressure leaching in three general temperature ranges: 100–110 °C (i.e., below the melting point of elemental sulphur) with the formation of elemental sulphur; 150 °C (i.e., above the melting point of sulphur, as in the case of zinc concentrates), also producing elemental sulphur and utilising suitable surface agents for the prevention of agglomeration of non-reacted sulphides by molten sulphur; high-temperature oxidation (220 °C) used for refractory gold-bearing ores and concentrates with complete oxidation of sulphur to the sulphate.

The low temperature leaching processes, newly installed, such as the ACTIVOX process [55, 65], are based on the fine grinding of chalcopyrite to 80% below 10 μm or finer and leaching at 100°C in sulphuric acid. Potential acceleration by chloride ions to copper extraction and conversion of sulphides to elemental sulphur is also possible. Iron is removed as goethite.

Two main protagonists work in the area of the processes operating at medium temperatures, i.e. approximately 150 °C. The CESL process [56, 65] is based on chloride-accelerated oxidation pressure leaching with the control of the amount of the acid to ensure that copper is transferred to the basic cupric sulphate salts, iron to haematite and sulphur to the elemental form, which is similar to the leaching process developed by Noranda in the 70s. After separating the liquid from solid phase, the basic cupric sulphate is dissolved by atmospheric leaching in the acid raffinate from solvent extraction for obtaining copper by the SX/EW process. The residues, containing elemental sulphur, small amounts of the non-reacted sulphides and haematite are flotated to produce sulphide–sulphur fractions for the removal of sulphur in the organic solvent. The desulphurised fraction is mixed with the haematite flotation tailings for further leaching at 220 °C to complete the extraction of copper and produce the residue with a content of noble metals for cyaniding. This process was used for several years as a high-intensity miniplant and also as a demonstration plant with various concentrates.

The flowsheet of the CESL process is shown in Fig. 16.20.

The Dynatec process [57, 65] applied for the chalcopyrite concentrates developed originally for pressure leaching of zinc from sulphide concentrates, includes oxidation leaching in the spent electrolyte or, preferably, the acid raffinate from solvent extraction at 150 °C using small amounts of coal as an efficient anti-agglomeration agent. Fine milling results easily in a high yield of copper (more than 90%) in a single stage. Alternatively, as tested in the continuous circuit of the miniplant, the first stage was limited to approximately 85% yield and the non-reacted sulphide was obtained by flotation. Elemental sulphur was removed by melting and filtration for the recycling of sulphides to leaching, finally with more than 98% efficiency of producing copper. The noble metals remained in the haematite flotation tailings from which gold and, to a smaller extent, silver were efficiently produced by cyaniding with a low consumption of reagents despite the small amount of residual

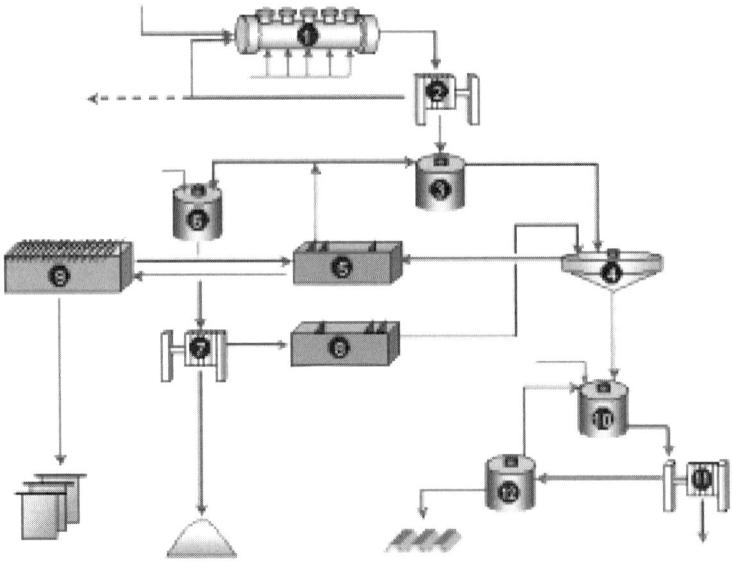

Fig. 16.20. The flowsheet of the CESL process.

elemental sulphur.

The Placer Dome Co. presented the process of complete oxidation resulting in a high degree of extraction of copper with conversion of the majority of iron to haematite residue with a relatively low sulphur content used to produce noble metals by cyaniding. Copper is also produced by the SX/EW process. Complete oxidation using water converts the majority of the amount of sulphide sulphur to cupric sulphate and sulphuric acid which increases the attractiveness of the process mainly in the applications in which the diluted acid may be used for the leaching of copper oxide deposits.

A common characteristic of these processes is that the impurities, such as antimony, bismuth, mercury and several other elements, are extensively precipitated during leaching together with the iron oxides which makes these processes interesting for the processing of 'non-pure' concentrates. Another feature is that the process is capable of direct processing of leaner charges in comparison with those required by hydrometallurgical processes. This makes it possible to reach a higher yield of copper and, in particular, noble metals from the ores which would have to be pretreated later to ensure that the content of the noble metals is available for cyaniding. Although Dynatec Company does not expect the pressure leaching of chalcopyrite to completely replace smelting, the process could be useful for small charges, poor ores, especially with a high impurity content.

Of all renewed activities in processing chalcopyrite in the 90s of the previous century, only one pressure process reached the commercial level, although with a chalcocite charge. It is the chalcocite project Mt Gordon [58] of the company Western Metals, Ltd in Australia. The charge consists of a chalcocite–pyrite ore, with approximately 8.5% of copper, mostly in the chalcocite form. In commercial operation, the fresh ore is mixed in the overheated raffinate product from solvent extraction, and a large portion of copper is leached in this stage by the ferric ion in the raffinate. Subsequently, the pulp is leached in the autoclave at a temperature of 90–95 ssw°C in the presence of oxygen at a total pressure of approximately 700 kPa, with the extraction of the majority of residual copper. In addition, part of copper is leached from the pulp when it flows through the autoclave. Copper is obtained from a solution containing 25–30 g/l of copper by the SX/EW process. In many aspects, i.e., the mineralogical composition of the charge, the high efficiency of extraction of copper with the minimum 'attack' on pyrite and in the pressure leaching conditions, the process is similar to the process developed by the Sherritt company in the 60s. The Mt. Gordon plant reached and exceeded the annual level of production of 45 000 t of copper cathodes in May 1999.

At the 90s of the previous century, the Dynatec company developed a hydrometallurgical process for the processing of large amounts of sulphide ores of the Las Cruces deposit in the vicinity of Seville in Spain. The deposit with 16 million t, with approximately 6% of copper, was originally operated by the British company Rio Tinto PLC who sold the deposit later to the American company MK Gold Company [59]. The main mineral of the copper in the ores is the chalcocite, and the pyrite is the most significant admixture. The attempts of Rio Tinto's Anamet Services to enrich copper minerals by flotation proved to be economically inefficient because of the low yield of copper and the high content of the mixtures in the concentrate. Despite this, the original studies of Anamet, expanded by Dynatec, indicate that the direct acid oxidation leaching of ore in 'moderate' conditions results in a sufficiently high yield of copper with the minimum degree of decomposition of pyrite. In 1997–1998, Dynatec used the process for the development of a scheme of commercial production by testing the extensive metered, continuous or fully integrated regime in the miniplant. Depth analysis and comparison of two hydrometallurgical processes, i.e., Rio Tinto process of atmospheric leaching using ferric sulphate and the Dynatec process of oxidation pressure leaching for the processing of ores, indicate that higher yields of copper were obtained in pressure leaching and, consequently, the process was selected for the construction of a commercial plant. When the Las Cruces deposit was acquired by MK Gold Company, Dynatec further demonstrated the efficiency of the process by a second campaign in the continuous miniplant in 2000.

The Rio Tinto company [60] developed a process on the pilot plant scale for acid oxidation leaching in the normal conditions of pressure and temperature using ferric sulphate as the oxidation agent for various sulphate minerals of copper present in lean complex ores, especially chalcocite, covellite, enargite and pyrite. The leaching agent is regenerated *in situ* using air. Copper is obtained from the solution by solvent extraction and electrolysis. The spent leaching agent with a surplus of the acid is neutralised with limestone or carbonates with the formation of gypsum.

The flowsheet of the Rio Tinto process is shown in Fig. 16.21.

The Dynatec process consists of two-stage counterflow leaching of the ore, primarily in the conditions of the Sherritt process, developed in the 60s of the previous century, leading to the formation of a copper-containing solution suitable for extraction of copper by the SX/EW process. The milled ore is leached at atmospheric pressure and at approximately 80°C with oxygen in the solution containing ferric ions and the acid, recycled from subsequent pressure leaching operations. In atmospheric leaching ~55% of copper is leached by the reaction of ferric ions with chalcocite and covellite, with the formation of cupric sulphate and ferrous sulphate. Part of the acid is consumed by the components of the tailings and by oxidation of the ferrous ions to ferric ions. After separating the solid and liquid phases, the solution is enriched in copper and depleted in the acid from atmospheric leaching and is suitable for subsequent solvent extraction.

The solid residue of atmospheric leaching is transferred into the autoclave for pressure operations at a pressure of 250 kPa and a temperature of approximately 90 °C for further leaching with oxygen in the raffinate from solvent extraction containing iron and acid. After emptying the autoclave, the pulp is transferred to the stage of atmospheric leaching to produce further quantities of copper. In pressure leaching and dressing steps, the majority of covellite is

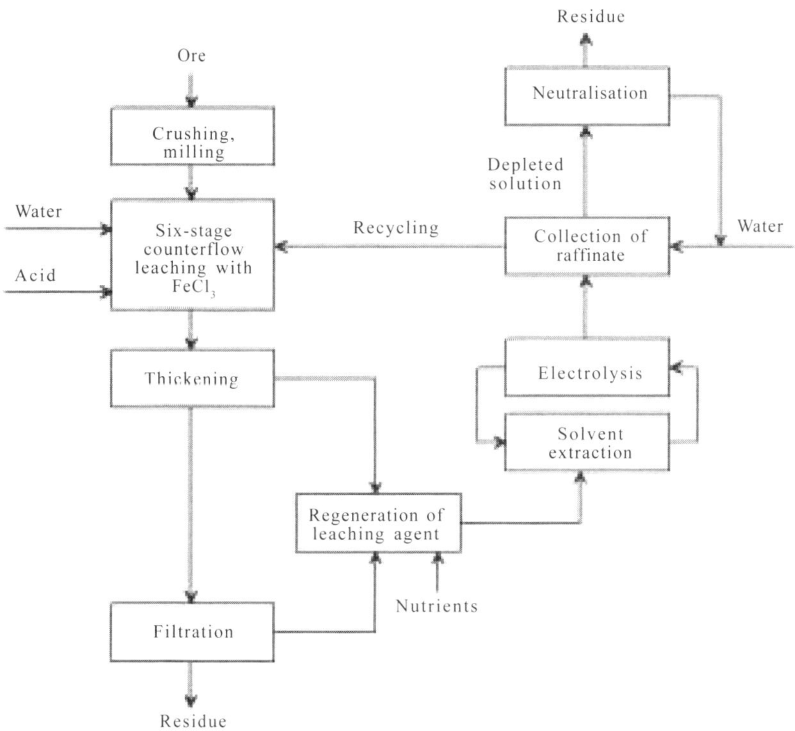

Fig. 16.21. Flowsheet of the Rio Tinto process.

processed, together with part of the chalcopyrite with copper extraction efficiency from the ore over 90%. The end leaching residue is separated and completely rinsed prior to placing on an environmentally acceptable dry heap.

The residual solution from the atmospheric leaching stage is transferred into two stages of solvent extraction for separation and extraction of copper. Initially, the solution is processed in the circuit for primary solvent extraction producing ~94% of copper. To maintain the equilibrium of water as well as the control of the level of iron, sulphate and other impurities on the requested values, a small part of the primary raffinate is processed in the second solvent extraction circuit and the produced copper solution is added to the spent solution and recycled into the pressure leaching circuit.

The operations of neutralisation of the acid and precipitation of the metal from the waste are carried out using air, limestone and lime. The neutralisation residues, containing gypsum and magnetite, are separated and rinsed prior to placing on the environmentally acceptable dry heap. The depleted solution, water in principle, is recycled inside the overall production process and/or subjected to environmentally safe processing. The pure cathode copper is produced from the primary solution of solvent extraction using conventional electrolysis. The spent electrolyte is returned to the stripping section of the primary solvent extraction circuit.

The MK Gold Company is developing a complete commercial production system of this type for the processing of ores from the Las Cruces deposit with the planned annual capacity of 72 000 t of metallic copper.

The process of leaching the iron-rich copper concentrate, developed by Lurgi-Mitterberg company, is based on the slower reaction with oxygen in diluted sulphuric acid in the temperature range below the melting point of elemental sulphur (~119°C) [61]. Even at the oxygen overpressure of 2000 kPa and a leaching time of 2–3 h, only approximately 20% of copper reacts. A relative increase of the degree of leaching of copper becomes evident only at temperatures above 180 °C. However, at the same time, a part of copper is oxidised in the sulphuric acid. In this case, the leaching capacity was increased by milling of the dry concentrate in vibration mills. The application of vibration milling in the process flowsheet enables the L-M process to be used in a single stage at temperatures below the melting point of elemental sulphur. Sulphur is agglomerated and screened, Fe_2O_3 is filtered-off, and copper is obtained by electrolysis with the formation of sulphuric acid essential for leaching.

The increase in the number of the hydrometallurgical processes indicates the effort to control the prices of copper produced from the primary sulphide concentrates besides smelting plants. The entry of the copper giants Phelps Dodge, BHP Billiton and Codelco indicates the successful completion of long-term development in this direction.

The second-largest world producer of copper, Phelps Dodge [62], constructed for 40 million US dollars a demonstration plant of pressure oxidation leaching at Bagdad, Arizona, which processes approximately 136 t of concentrate every day, i.e. 16 000 t of cathode copper annually by the method of conventional solvent extraction and electrolysis, Fig. 16.22. This amount represents approximately 15% of the output of the Bagdad plant. This plant uses the technology developed by Phelps Dodge and Placer Dome.

During this period, the largest world producer of copper Codelco merged with BHP Billiton and formed Alliance Copper [63] for the development and application of a biotechnology for the processing

Fig. 16.22. View of the Phelps Dodge pressure leaching plant.

of copper and molybdenum ores and concentrates. The first project is the prototype of the bioleaching process near Chuquicamata in Chile with the total cost of 60 million US dollars for the processing of 77 200 t of concentrate per annum and the production of 20 000 t of cathode copper per annum with plans for the construction of commercial production plants. Prior to this stage, Billiton already operated a pilot plant at Chuquicamata using its BioCOP technology.

Other large groups are MIM and Anglo American. The MIM/ Highlands Albion Process (Nenatech) [64] is used jointly by MIM and Highlands Pacific and was initially developed for the Frieda River project in PNG. The process uses fine milling for reducing the grain size of the charge to approximately 16–18 μm, followed by a leaching in ferric sulphate at approximately 80°C and atmospheric pressure using oxygen or blown air. Copper is produced by conventional SX/EW technology. The process has been applied efficiently on the pilot plant scale. At present, work is being carried out to determine the economic efficiency of commercial production at Mt Isa and further applications for the zinc concentrate in the McArthur River are being planned.

The Anglo-American Corporation and the University of British Colombia have developed the AAC/UBC Hydrometallurgy Process [65] of medium-pressure oxidation leaching at 160 °C with the addition of a surface agent for the separation of liquid sulphur. The charge is subjected to fine milling to 10–20 μm and copper is produced by conventional SX/EW technology. The pilot plant operates using equipment AARL in South Africa.

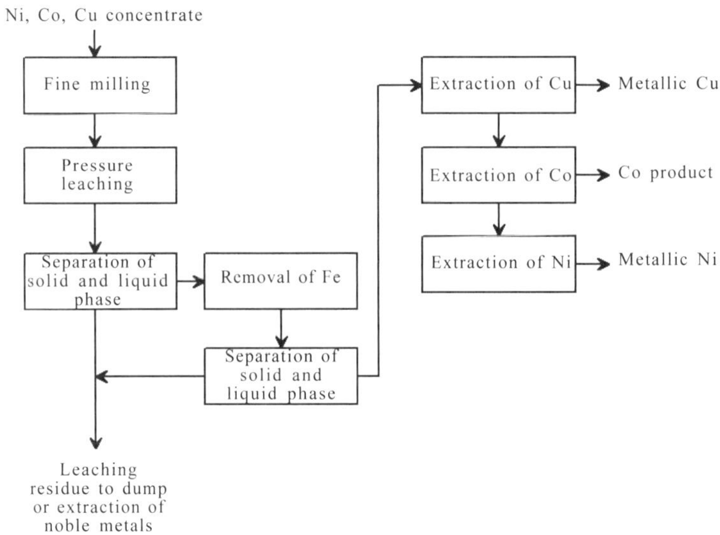

Fig. 16.23. The flowsheet of the Activox process.

The NSC process [65] is based on medium-pressure leaching at 125–155°C, catalysed by nitrous compounds obtained from sodium nitrite milled to 10 μm. Copper is obtained by the conventional SX/EW process. A small commercial plant operated up to the middle of the 90s in Sunshine Mining and Refining in Montana. Further work is planned in the Centre for Advanced Mineral and Metallurgical Processing, Montana, USA, and includes the existing Sunshine plant for the processing of cobalt and chalcopyrite concentrates for the company Capital Corporation, Idaho Cobalt Project.

Western Minerals Technology in Perth, Australia, developed the Activox process [65], i.e. low-pressure oxidation leaching operating at 100 °C, Fig. 16.23. The charge is subjected to ultrafine milling to 5–15 μm. Copper is produced by the conventional SX/EW process. Pilot plant trials have been successfully completed. Western Minerals Technology is paying special attention to nickel–cobalt–copper ores and copper concentrates with the impurities, which are difficult to process by pyrometallurgical processes. In future, copper- and gold-bearing concentrates will be processed.

The process Mt Gordon [58] has been used successfully commercially at Western Metals' Mt Gordon in Queensland and produces annually approximately 50 000 t of high-quality copper cathodes from chalcopyrite ores by bioleaching in ferric chloride by

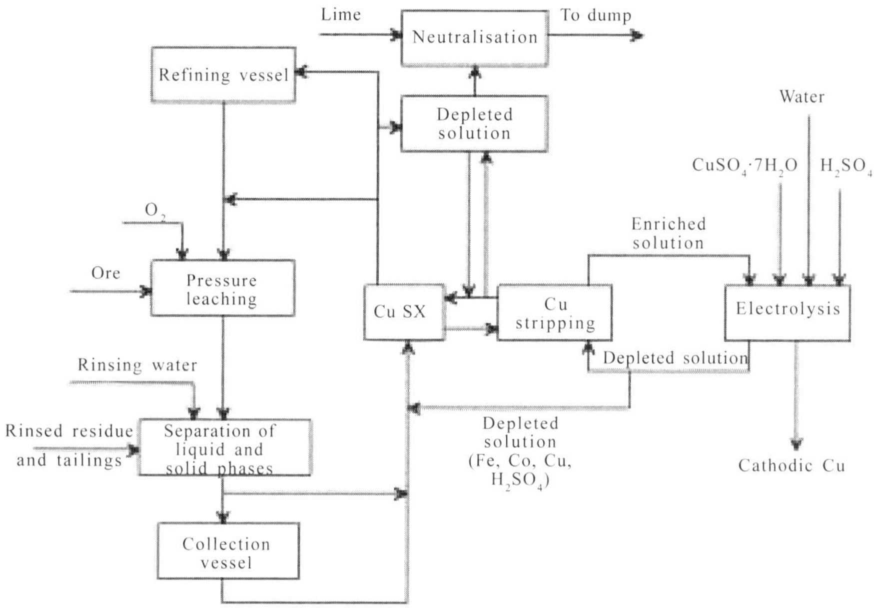

Fig. 16.24. The flowsheet of the Mt Gordon process.

low-pressure oxidation leaching. The charge is milled to 75–106 μm and copper is produced by the conventional SX/EW process. In future, it is planned to process chalcocite concentrates from lean ores and later to process chalcopyrite concentrates.

The flowsheet of the Mt Gordon process is shown in Fig. 16.24.

In any case, it may be concluded that the interest paid by the large companies, such as Phelps Dodge, BHP Billiton, Outokumpu and Codelco, indicates that the copper industry has found its approach to 'cutting the Gordic knot' of the hydrometallurgical processing of sulphide concentrates. This will probably result in a significant shift on the worldwide scene of copper production.

The following tables summarise the new hydrometallurgical processes [66].

Table 16.2 New sulphate processes of hydrometallurgical processing of chalcopyrite concentrates

Process	Pressure conditions	Sulphur products	Mechanism	Development stage	Comment
Activox	Low pressure	S^0, SO^{2-}_4	Direct	Laboratory	Fine milling
Nenatech	Atmospheric	S^0, SO^{2-}_4	Direct	Laboratory	Fine milling
Dynatec	Medium pressure	S^0	Direct	Laboratory	With coal
AAC/UBC	Medium pressure	S^0	Direct	Pilot plant	With surfactant
Placer Dome	High pressure	SO^{2-}_4	Direct	Demonstration	High temperature
Biocop	Atmospheric	SO^{2-}_4	Indirect	Commercial	Bacterial
Bactech/ Mintek	Atmospheric	S^0, SO^{2-}_4	Indirect	Demonstration	Bacterial
Geocoat	Atmospheric	SO^{2-}_4	Indirect	Pilot plant	Bacterial

Table 16.3 New sulphate-chloride processes of hydrometallurgical processing of chalcopyrite concentrates

Process	Leaching system	Reagents	Pressure conditions	Sulphur products	Development stage
Antler Noranda	$HCl–CuCl_2–CuSO_4$	O_2, MnO_2	Pressure	S^0	Laboratory
BHAS	$H_2SO_4–Cl^-$	O_2, NaCl	Atmospheric	S^0	Commercial (for matte)
CESL	$H_2SO_4–Cl^-$	O_2, NaCl	Pressure	S^0	Demonstration

Table 16.4 New chloride processes of hydrometallurgical processing of chalcopyrite concentrate

Process	Leaching system	Reagents	Extraction of noble metals	Copper extraction	Development stage
CLEAR	$CuCl_2$–NaCl	O_2, air	Au to residue, Ag to Cu	Electrolysis	Commercial 1976–1982
CYMET	$FeCl_3$–$CuCl_2$ –NaCl	H_2, sand	Au to residue, Ag precipitate	H_2 reduction	Pilot plant
CUPREX	$FeCl_3$–NaCl	Cl_2	Au to residue, Ag to Cu	Electrolysis	Laboratory
Hydrocopper	$CuCl_2$–NaCl	Cl_2, air	Au to residue, Ag to Cu	Electrolysis	Pilot plant
INTEC	NaCl–BrCl	air, $CaCO_3$	Au and Ag to solution	Reduction, precipitation, electrolysis	Pilot plant

Table 16.5 Commercial plants for bioleaching of copper

Plant	Capacity [t/day]	Company	Ore richness [%Cu]	Duration
Lo Aquirre, Chile	16000	Sociedad Minera Pudahuel	1.5	1980-to now
Mt.Leyshon, Australia	1370	Normandy Poseidon	0.15	1992-1995
Cerro Colorado, Chile	16000	Rio Algom	1.4	1993-to now
Girrilambone, Australia	2000	Straits Resources	3	1993-to now
Ivan, Chile	1500	Glamis Gold	2.1	1994-to now
Quebrada Blanca, Chile	17300	Cominco	1.3	1994-to now
Andacollo, Chile	16000	Dayton Mining	1.0	1996-to now
Dos Amigos, Chile	3000	Cemin	2.5	1996-to now
Cerro Verde, Peru	32000	Cyprus-AMAX	0.7	1996-to now
Zaldivar, Chile	20000	Placer Dome	1.4	1998-to now
S&K Copper, Myanmar	15000	Myanmar Government, Ivanhoe Copper	1.25	1998-to now

Finally, for better vision, the following table gives information on the financial aspects of the Mt Gordon plant [67] as at 2001.

Table 16.6 Financial budget of the Mt Gordon plant (in millions of $)

Operation	Reagents	Output	Energy	Plant	Others	Total	US $ / kg Cu	%
Crushing/milling	0.6	0.3	0.7	0.8	0.3	2.6	0.067	10.8
Leaching	1.8	0.5	1.0	2.5	0.7	6.4	0.1634	26.7
SX / EW	1.4	1.0	6.1	0.9	0.9	10.4	0.2625	42.9
Tailings	2.1	0.1	0.1	0.4		2.8	0.0723	11.6
Others					1.9	1.9	0.0482	8.0
Direct cost	6.0	1.9	7.8	4.6	3.8	24.1	0.6135	100
Price %	25.1	8.0	32.3	19.0	15.6	100		

References

1. Jackson E.: Hydrometallurgical Extraction and Reclamation, Ellis Horwood Ltd., New York, 1986, pp.253.
2. Habashi F.: Recent advances in pressure hydrometallurgy, Int. Conf. on advances in chemical metallurgy, Bombay, January 1979.
3. Evans L.: Selecting engineering materials for chemical and process plant, Business Books, Ltd., 1974.
4. Evans L.: Solid – liquid separation, Eng. Min. Journal, 181, 6, 1980, 120–124
5. Fletcher J.B.: In place leaching – Miami Mine, Symp. Arizona Section Meeting, AIME, April 1962.
6. Dresher W. H.: Producing copper nature's way: Bioleaching, On line magazine from the Copper Development Association, May 2004.
7. Lung T.N.: *Hydrometallurgy*, 17, 1986, 113–129.
8. Agricola G.: De Re Metallica, 1556, Hoover, H.C and Hoover L.H. transl, Dover, New York.
9. Benedict C.H., Kenny H.C.: *Trans. AIME*, 70, 1924, 595–610.
10. van der Poel C.S.: Metallurgical operations at Bwana M'Kuba, Proc. 3rd. Empire Mining and Metallurgical Congress, South Africa, part 3, 1930, 427–449.
11. Aldrich H.W., Scott W.G.: *Trans. AIME*, 106, 1933, 650–677.
12. Bean J.J.: LPF Method of concentration at Miami Copper Co, Quarterly Colorado School of Mines, 56, 1961, 263–282.
13. Wheeler A.E., Eagle H.Y.: *Trans AIME,* 106, 1933, 609–650.
14. Boldt J.R. et al: Extractive metallurgy of sulphide ores, The Winning of Nickel, Van Nostrand, New York, 1967, 299–313.
15. Arbiter N.: ANACONDA's Ammonia leach process, Proc. AMC, Denver, September 1973.
16. Arbiter N.: ANACONDA's ammonia leach process, Proc. AIME Meeting in Dallas, February, 1974.
17. Kuhn M.C., Arbiter N., Kling H.: Anaconda's Arbiter process for copper, CIM Bulletin, february 1974, 62–73.
18. Kuhn M.C., Arbiter N., Kling H.: Physical and chemical separations via the Arbiter Process, Proc. XI. Int. Mineral Processing Congress, Cagliari, 1975, 831–848.
19. Haver F.P., Baker R.D., Wong M.M.: Improvements in ferric chloride leaching chalcopyrite concentrates, US Bureau of Mines RI 8007, 1975.
20. Paytner J.C.: *J. S. Afr. Inst. Min. Metall.*, 74, 1974, 158–170.
21. Milner E.F.G., Peters E., Swinkels G.M., Viszolyi A.I.: Copper hydrometallurgy, US Patent 3,798,026, March 3th., 1974.
22. Swinkels G.M., Berezowsky R.M.G.S.: The Sherrit-Cominco copper process–Part I: The Process CIM Bulletin, February, 1978, 105–121.
23. Maschmeyer D.E.G., Milner E.F.G., Parekh B.M.: The Sherrit-Cominco copper process–Part III: Commercial Implications, CIM Bulletin, February, 1978, 131–

139.
24. Kruesi P.R., Allen E.S., Lake J.L.: CYMET process – Hydrometallurgical conversion of base-metal sulphides to pure metals, CIM Bulletin, june 1973, 81–87.
25. Kruesi P.R., Goens D.N.: Process for the recovery of copper from its sulfide ores, US Patent 3,901,776, August, 26th., 1975.
26. Atwood G.E., Curtis Ch.H.: Hydrometallurgical Process for the Production of Copper, US Patent 3,785,944, 15.1.1974.
27. Atwood G.E., Livingston R.W.: *Erzmetall*, 33, 5, 1980, 251–255.
28. Schweitzer F.W., Livingston R.: Duval's CLEAR hydrometallurgical process, Chloride Electrometallurgy, Parker ed., TMS-AIME, New York, 1982, 221–227.
29. Atwood G.E., Curtis C.H.: Hydrometallurgical process for the production of copper, US Patent 3,879,272, 22.4.1975.
30. Demarthe J.M., Gandon L., Georgeaux A.: A new hydrometallurgical process for copper, In: Extractice Metallurgy of Copper, Yannopoulos and Agarwal eds., TMS-AIME, New York, 1976, 825–848.
31. Dalton R.F., Diaz G., Price R., Zunkel A.D.: *Journal of Metals*, august 1991, 51–56.
32. Dalton R.F., Diaz G., Hermana E., Price R., Zunkel A.D.: The CUPREX metal extraction process: Pilot plant experience and economics of a chloride-based process for the recovery of copper from sulphide ores, In: Hydrometallurgy and Electrometallurgy of Copper, Cooper et al eds., Pergamon, New York, 1991, 61–69.
33. McDonald G.W., Udovic T.J., Dumesic J.A., Langer S.H.: *Hydrometallurgy*, 13, 1984, 125–135.
34. Andersen E., Boe G.H., Danielssen T., Finne P.M.: Production of base metals from complex sulphide concentrates by the ferric chloride route in a small continuous pilot plant, In: Complex Sulphide Ores, Jones ed., Inst. Min. Metal., London, 1980, 186–192.
35. Craigen W.J.S., Krysa B.D., Brown H.H., Barin B.: Evaluation of the Great Central Mines (GCM) hydrometallurgical copper process for application to Hudson Bay Mining and Smelting (HBMS) copper concentrates, In: Projects '88, Metall. Soc. Can. Inst. Min. Metall., Montreal, 1988, 7p.
36. Fletcher A.W., Sudderth R.B., Olafson S.: *Journal of Metals*, 43, 8, 1991, 57–59.
37. Ledeboer B.J.: Factors in the design of copper electrowinning tankhouse, In: Hydrometallurgy and Electrometallurgy of Copper, Cooper et al eds., Pergamon, New York, 1991, 563–574.
38. Paschen P., Langfellner M., Mori G.: Productivity increase and energy conservation in copper electrowinning, In: Hydrometallurgy and Electrometallurgy of Copper, Cooper et al eds., Pergamon, New York, 1991, 575–591.
39. Meadows N.E, Valenti M.: The BHAS copper lead matte treatment plant, Proc. Non-ferrous Smelting Symp., Australian IMM, Victoria, 1989, p153.
40. Craigen W.J.S., Kelly F.J., Bell D.H.: Evaluation of the CANMET ferric chloride leach (FCL) process for treatment of complex base-metal, Sulphide Ores Sulphide Deposits, 1990, 255–271.

41. Noual P., Bienvenu G., Rizet L.: Recovery of metals from complex sulfide concentrates in fluosilicic medium by Extramet proces, Proc. Complex Sulfides, Zunkel et al eds., TMS-AIME, 1985, 925–937.
42. Limpo J.L., Figueiredo J.M., Amer S., Luis A.: *Hydrometallurgy*, 28, 1992, 149–161.
43. Limpo J.L., Luis A., Gomez C.: *Hydrometallurgy*, 28, 1992, 163–178.
44. Everett P.: Production of metals from minerals, International Patent No. WO 94/00606.
45. Hyvärinen O., Hämäläinen M., Leimala R.: Outokumpu HydroCopperTM process – A novel concept in copper production, In: Chloride Metallurgy 2002, Vol. II, 32nd Annual Hydrometallurgy Meeting (Ed. Peek E., Van Weert G.), 2002, 609–613.
46. Hyvärinen O., Hämäläinen M.: Method for producing copper in hydrometallurgical process, US Patent 6,007,600, December 28, 1999.
47. Muir D. M.: Basic principle of chloride hydrometallurgy, In: Chloride Metallurgy 2002, Vol.II, 32nd Annual Hydrometallurgy Meeting (Ed. Peek E., Van Weert G.), 2002, 759–791.
48. Espejo R.T., Romero R.: *Applied and Environmental Microbiology*, 53, 1987, 1907–1912.
49. Eccleston M., Kelly D.P.: *J. Bacteriology*, 134, 1978, 718–727.
50. Thorma A.E.: The role of Thiobacillus ferrooxidans in hydrometallurgical processes, In: Advances in Biochemical Engineering, Ghose, Fiechter, Blakebrough eds., vol. 6, 1977, Springer-Verlag, Berlin.
51. Tuovinen O.H., Kelly D.P.: *Zeitschrift f. Allg. Mikrobiologie,* vol.12, 4, 1972, 311–346.
52. Kelly D.P.: Biochemistry of the chemolithotropic oxidation of inorganic sulfur, Phil. Trans., Royal Society, London, B, 298, 1982, 499–528.
53. Groudev S.N.: Electrochemical oxidation system during the leaching of chalcopyrite by Thiobacillus ferrooxidans, Comtes rendus de l'Academie bulgare des Sciences, vol.33, 6, 1980, 829–831.
54. Berezowsky R.: Acid pressure leaching of copper sulfides, Part 1, Pincock perspectives, 26, January 2002, 1–4.
55. http://www.wmt.com.au/activox-cuau.htm.
56. Defreyne J., Barr G., McCunn G.: CESL Copper process – Moving from pilot plant to production scale operation, http://www.cesl.com/english/copperprocess.html
57. http://www.dynatec.ca/.
58. Shaw D.R., Dreisinger D.B., Lancaster T., Richmond G.D., Tomlinson M.: *Journal of Metals*, July 2004, 38–38.
59. Berezowsky R.: Acid pressure leaching of copper sulfides, Part 2, Pincock perspectives, 27, February 2002, 1–4.
60. Smalley N., Davis G.: *Minerals Engineering*, 13, 6, 2000, 599–608.
61. Türke P., Fischer P.: Hydrometallurgical Treatment of Complex Copper Sulphide Concentrates with Special Reference to the Lurgi - Mitterberger Process, In: Proc. Int. Symp. "XComplex Metallurgy '78, Jones M.J. ed., Bad Harzburg 1978, 101–112.
62. http://www.phelpsdodge.com/AboutUs/OurCompanies/Mining/Technology.htm

63. http://www.codelco.com/hacia_futuro/tecnologia_alianzas.asp
64. Taylor A.: Big guns now in copper concentrate leaching race, Alta Metallurgical Services, September, 2002, 1–4.
65. Milbourne J., Tomlinson M., Gormely L: Use of Hydrometallurgy in direct processing of base metal / PGM concentrates, Hydrometallurgy 2003, 5th. Int. Conference in Honor of Prof. Ian Ritchie, vol. 1., Young C.A., Alfantazi A.M., Anderson C.G., Dreisinger D.B., Harris B., James A. eds., TMS, 2003.
66. Peacey J., Guo X.J., Robles E.: Copper Hydrometallurgy – Current status, preliminary economics, future direction and positioning versus smelting, http://www.hatch.ca/Non_Ferrous/articles/default.html.
67. Dreisinger D., Richmond G., Hess F., Lancaster T.: The competitive position of the Mt.Gordon copper process in the copper industry, Alta Copper Symposium, Perth, Australia, May 2002.

Index